Trends in Mathematics is a series devoted to the publication of volumes arising from conferences and lecture series focusing on a particular topic from any area of mathematics. Its aim is to make current developments available to the community as rapidly as possible without compromise to quality and to archive these for reference.

Proposals for volumes can be sent to the Mathematics Editor at either

Birkhäuser Verlag
P.O. Box 133
CH-4010 Basel
Switzerland

or

Birkhäuser Boston Inc.
675 Massachusetts Avenue
Cambridge, MA 02139
USA

Material submitted for publication must be screened and prepared as follows:

All contributions should undergo a reviewing process similar to that carried out by journals and be checked for correct use of language which, as a rule, is English. Articles without proofs, or which do not contain any significantly new results, should be rejected. High quality survey papers, however, are welcome.

We expect the organizers to deliver manuscripts in a form that is essentially ready for direct reproduction. Any version of TEX is acceptable, but the entire collection of files must be in one particular dialect of TEX and unified according to simple instructions available from Birkhäuser.

Furthermore, in order to guarantee the timely appearance of the proceedings it is essential that the final version of the entire material be submitted no later than one year after the conference. The total number of pages should not exceed 350. The first-mentioned author of each article will receive 25 free offprints. To the participants of the congress the book will be offered at a special rate.

Advances in Analysis and Geometry

New Developments Using Clifford Algebras

Tao Qian
Thomas Hempfling
Alan McIntosh
Frank Sommen
Editors

Springer Basel AG

Editors' addresses:

Tao Qian
Faculty of Science and Technology
University of Macau
P.O. Box 3001
Macau
e-mail: fsttq@umac.mo

Thomas Hempfling
Basel, Switzerland
e-mail: hempfling@birkhauser.ch

Alan M[c]Intosh
Centre for Mathematics and its Applications
Mathematical Sciences Institute
Australian National University
Canberra ACT 0200, Australia
e-mail: Alan.McIntosh@maths.anu.edu.au

Frank Sommen
Department of Mathematical Analysis
University of Ghent
Galglaan 2
9000 Ghent, Belgium
e-mail: fs@cage.rug.ac.be

2000 Mathematical Subject Classification 11Fxx, 11G15, 14L24, 15Axx, 17B35, 30A05, 30F45, 30G35, 31Bxx, 32Axx, 32C10, 32U05, 35G15, 35Jxx, 35Q60, 42Bxx, 42C10, 44A35, 46E15, 46F10, 47Axx, 47Bxx, 51K05, 53B40, 53Cxx, 58G30, 58Jxx

A CIP catalogue record for this book is available from the
Library of Congress, Washington D.C., USA

Bibliographic information published by Die Deutsche Bibliothek
Die Deutsche Bibliothek lists this publication in the Deutsche Nationalbibliografie;
detailed bibliographic data is available in the Internet at <http://dnb.ddb.de>.

ISBN 978-3-0348-9589-7 ISBN 978-3-0348-7838-8 (eBook)
DOI 10.1007/978-3-0348-7838-8

Originally published by Birkhäuser Verlag, Basel, Switzerland in 2004

Printed on acid-free paper produced from chlorine-free pulp. TCF ∞

9 8 7 6 5 4 3 2 1 www.birkhäuser-science.com

Preface

On the 16th of October 1843, Sir William R. Hamilton made the discovery of the quaternion algebra $H = q_0 + q_1 i + q_2 j + q_3 k$ whereby the product is determined by the defining relations

$$i^2 = j^2 = -1, \quad ij = -ji = k.$$

In fact he was inspired by the beautiful geometric model of the complex numbers in which rotations are represented by simple multiplications $z \to az$. His goal was to obtain an algebra structure for three dimensional visual space with in particular the possibility of representing all spatial rotations by algebra multiplications and since 1835 he started looking for generalized complex numbers (hypercomplex numbers) of the form $a + bi + cj$. It hence took him a long time to accept that a fourth dimension was necessary and that commutativity couldn't be kept and he wondered about a possible real life meaning of this fourth dimension which he identified with the scalar part q_0 as opposed to the vector part $q_1 i + q_2 j + q_3 k$ which represents a point in space. In any case the product of two pure vectors $q = q_1 i + q_2 j + q_3 k$, $q' = q'_1 i + q'_2 j + q'_3 k$ gives a quaternion of which the scalar part is minus the inner product $-\langle q, q' \rangle$ and the vector part is the cross product $q \times q'$, which are the two products that are considered in the so called vector algebra of triples (x_1, x_2, x_3) developed by Josiah W. Gibbs since 1881–84 as part of vector calculus. Hamilton also already introduced in 1846 the Nabla operator $\nabla = i\partial x + j\partial y + k\partial z$ from which he also obtained the three most important operators in vector calculus: the gradient of a scalar function, the divergence of a vector valued function and the curl of a vector valued function, operations which play a crucial role in the formulation of electromagnetism.

One could hence argue that quaternion algebra already contained all the elements of vector calculus, yet vector calculus can and was developed independently from quaternions and became part of the standard mathematical education, together with the matrix algebra which was introduced by Arthur Cayley in 1858. Since then, for a long time quaternion algebra was considered as a rather esoteric and somewhat suspect topic. This would probably not have been the case, should computers have been developed at that time. Inspired by the complex multiplication and before the development of matrices, Hamilton was looking for a representation of rotations as products of the form $q \to aq$, q being a vector and a being a unit quaternion, but noticed that this product can be a rotation if and only if q is orthogonal to the vector part of a. Later in 1845 however, Cayley discovered that all rotations in R^3 may be represented by quaternion products of the form $q \to aqa^{-1}$. These are called spin-rotations as opposed to the standard representation of rotations in terms of matrix products $x \to Ax$, which is part of classical vector calculus. For a long time the importance of spinor rotations was not understood. Only if it comes down to problems in which a huge number of rotations has to be computed in real time, the quaternion representation of rotations turns out to be much better than the matrix representation. Examples are problems having to do with aircraft orientation, spacecraft stabilization, robotics, computer vision, virtual reality and so on. It is well known that the use of quaternions for aircraft motion applications was a well hidden secret during the cold war. But to incorporate both translations and rotations into a single multiplication, quaternions are not enough; one needs the geometric algebra of William K. Clifford, also called Clifford algebra for short.

Inspired by the quaternions of Hamilton, Clifford searched for a higher dimensional algebra in which the higher dimensional generalizations of the inner product and the cross product can be obtained from the algebra product.

A higher dimensional generalization of the cross product was already introduced by Hermann G. Grassmann in 1844, namely the exterior algebra (at that time called extensive algebra) in which the basis elements of space satisfy the product rules $e_j e_k = -e_k e_j$ for all $j, k = 1, \ldots, m$.

This algebra enables the definition of higher dimensional volume elements and was later incorporated in the algebra of differential forms introduced by Élie J. Cartan. But the inner product is missing and it is the geometric algebra of Clifford R^m, which contains both the inner and wedge products. The defining relations for R^m are given by

$$e_j e_k + e_k e_j = -2\delta_{jk}, \quad j, k = 1, \ldots, m,$$

and the inner and wedge products of vectors $x = \sum_j e_j x_j$, $y = \sum_j e_j y_j$ are given by

$$xy + yx = -2\langle x, y\rangle, \quad xy - yx = 2x \wedge y.$$

For $m = 1$, one re-obtains the complex numbers, while for $m = 2$ the quaternions by making the identification $i \to e_1$, $j \to e_2$, $k \to e_1 e_2$. But this identification is asymmetrical in the sense that i, j are vectors while k is a bivector. Also the inner product is obtained with a minus sign. Therefore, Clifford suggested the use of the following alternative definition for the Clifford algebra:

$$e_j e_k = -e_k e_j, \quad j \neq k \quad \text{and } e_j^2 = +1.$$

This choice of signature was adopted later on by David O. Hestenes as the basis for what he called geometric calculus. In the three dimensional case one has as basis elements

scalar 1
vectors e_1, e_2, e_3
bivectors $e_2 e_3$, $e_3 e_2$, $e_1 e_2$
trivector (pseudoscalar) $e_1 e_2 e_3$

and already Clifford identified the quaternions with the bivectors

$$i = e_2 e_3, \quad j = e_3 e_1, \quad k = e_1 e_2.$$

In this way the quaternion unit sphere that is used to represent rotations coincides with the spin group Spin(3). In general one may introduce the spin group Spin(m) as a compact subgroup of elements of the form $s = w_1 \cdots w_{2l}$, $l \in \mathbb{N}$, w_j being unit vectors, and rotations in R^m may be represented as Clifford multiplications $x \to sxs^{-1}$. But nowadays for the sake of geometry as well as analysis and in any case for mathematical physics, one needs to consider the Clifford algebra $R^{p,q}$ of mixed signature with defining relations

$$e_j e_k = -e_k e_j,\ j \neq k;\quad e_j^2 = +1,\ j = 1, \ldots, p,\quad e_j^2 = -1,\ j = p+1, \ldots, p+q.$$

Special examples are: the Hestenes' space-time algebra $R^{1,3} = H(2)$ which is isomorphic to the algebra of Dirac matrices (that is itself an extension of the algebra of Pauli matrices representing the quaternions) and the alternative version of the space time algebra $R^{3,1}$ that is isomorphic to the algebra $R(4)$ of Majorana matrices.

For the general development of Clifford algebra and spinor theory we also mention the work by É. Cartan, R. Lifschitz, R. Brauer, H. Weyl, C. Chevalley, M. Riesz, I. Porteous and the late P. Lounesto, who incorporated the Cayley-octonions into Clifford algebra. Of course the list is incomplete.

The consideration of a mixed signature is also essential for the algebraic representation of rotations, translations and other conformal transformations by Clifford multiplications. We hereby mention work by L. Ahlfors, H. Maass and more recent work by D. Hestenes, P. Lounesto, J. Cnops, J. Maks and many others. Recently the Spin(3,1) model for the three dimensional conformal group is used extensively by geometry-oriented people who are interested in applications in robotics, artificial vision cristallographics etc. Also there are the research groups inspired by Hestenes' work. See the results of G. Sobczyk, S. Gull, A. Lasenby, J. Lasenby, C. Doran, G. Sommer, L. Dorst, H. Li and also to the proceedings of conferences edited by these people.

"Clifford analysis" was the title of a paper by John Ryan and of a monograph by Fred Brackx, Richard Delanghe and Frank Sommen presenting the basic function theory for the generalized Cauchy–Riemann operators with values in a Clifford algebra with applications to distributional boundary value theory and Fourier analysis. Since then and up to now the name Clifford analysis refers to an autonomous mathematical discipline centered on the study of the Clifford algebra valued Dirac operator $\partial x = \sum_j e_j \partial x_j$, applications of this operator to problems in physics and engineering and more generally involving the study of any system of special partial differential operators that arises naturally from the use of Clifford algebra as a calculus tool. Also the theory of the Dirac operator on manifolds, in which Clifford bundles are employed, is considered by some authors as part of Clifford analysis.

Before the eighties, the function theory of the Dirac operator was considered as part of hypercomplex analysis which was the term used for any generalization of the Cauchy–Riemann system $(\partial x + i\partial y)f(x + iy) = 0$. It included quaternion analysis in the sense of K. Rudolph Fueter, the Moisil–Theodorescu matrix-system in three dimensions which is equivalent to the quaternion system $(i\partial x + j\partial y + k\partial z)f = 0$ and also Clifford algebra valued analysis which was already developed by V. Iftimie and Fueter and his students, in particular by W. Nef. Apart from that the name hypercomplex analysis is also used for various other generalizations of the analysis of one complex variable such as the theory of Vekua systems and the more general Douglis systems which were investigated by R.P. Gilbert, H. Begehr, G. Hile and others. In this connection we also mention the work by K. Habetha on function theory for general first order elliptic systems.

Hence one could say that Clifford analysis found its origin in quaternion analysis as developed by Fueter, student of Hilbert from 1928. He was mostly motivated by problems in number theory which he tried to generalize from the complex to the quaternion case. Another important generalization of complex analysis was the theory of holomorphic functions of several complex variables, initiated by H. Behnke. For a while there was a kind of competition between quaternion analysis and several complex variables as to which should be the best generalization of complex analysis. Several complex variables quickly evolved into a major mathematical discipline while the theory of regular or "monogenic" functions of a quaternion variable, solving $(\partial q_0 + i\partial q_1 + j\partial q_2 + k\partial q_3)f(q) = 0$ remained somewhat marginal perhaps due to the fact that the product of monogenic functions is no longer monogenic. Also the theory of monogenic functions in Clifford analysis suffers from the same defect, which is a serious barrier in the generalization of complex analysis to higher dimensional Clifford analysis compared with several complex variables. On the other hand, the holomorphic functions of several complex variables may be seen as special monogenic functions. One may go further and consider monogenic functions in several complex variables.

Hence the old competition between complex analysis and quaternion and Clifford analysis faded away.

In fact the real breakthrough of Clifford analysis came during the last three decades of the 20th Century, starting with the development of the basic theory of monogenic functions by R. Delanghe and F. Brackx, followed by important contributions from the "Ghent school" (Delanghe, Brackx, Sommen, Cnops, Van Hamme, Pincket, Constales, Van Acker, Van Lancker) including two books and various papers. Also the complexified version of Clifford analysis as developed by V. Soucek, J. Bures and J. Ryan in combination with the conformal geometry lead to important progress. Also important work on boundary value problems with applications in engineering and physics is one of the cornerstones of the success of Clifford analysis. Hereby we mention several interesting books and papers by the German school with W. Sprößig, K. Gürlebeck, S. Bernstein, H. Malonek, U. Kähler and P. Cerejeiras.

Clifford analysis was introduced to the study of the L^2 boundedness of singular integrals on Lipschitz surfaces by Margaret Murray in her thesis (1983) written under the supervision of Raphy Coifman. She showed how Clifford analysis could be used to prove the L^2 boundedness of the double layer potential operator on surfaces with small Lipschitz constants by using multilinear expansions, a method extended to all Lipschitz constants by Alan M^c^Intosh.

Together with Chun Li, Tao Qian and Stephen Semmes, M^c^Intosh went on to give more direct proofs and related results, including Fourier transformation and multiplier theory of monogenic kernels and bounded holomorphic functional calculi of Dirac operators forming operator algebras of singular integrals on Lipschitz graphs. This study was highlighted at the Clifford analysis conference held in Arkansas in 1993. The analogous theory on spheres and their Lipschitz perturbations was developed by Qian using Fueter's theorem and its generalizations on inducing monogenic functions.

The use of Clifford analysis helped in the proof of Rellich inequalities, which are used to show that operators such as the double layer potential operator are invertible on a Lipschitz surface. This leads to an increased understanding of boundary value problems for the Laplacian as well as for Maxwell's equations.

There is extensive research involving the interplay between Clifford and harmonic analysis by many people including J. Gilbert, G. David, J.-L. Journè, G. Gaudry, R. Long, P. Auscher, P. Tchamitchian, Z. Wu, T. Tao, Yu Y. Qiao, S. Huang and A. Axelsson. Marius Mitrea made many contributions to this theory and gave a good presentation of the field in a book.

There were also books and papers on boundary values and operator theory in Clifford analysis written by M. Shapiro, N. Vasilevski, V. Kisil, V. Kravchenko with important applications to electromagnetism and strong interaction.

There is in addition a long list of people involved in the theory of Dirac operators on manifolds initiated by Sir M.F. Atiyah, Sir R. Penrose, D. Hestenes and T. Friedrich.

Recently there are also important contributions from Clifford analysis into the theory of wavelets, in particular by M. Mitrea, L.Z. Peng, H.P. Liu, the current Ghent school (including N. De Schepper), and, in relation to Paley–Wiener and Shannon interpolation theorems, by T. Qian and K.I. Kou.

Also the theory of Dirac operators on the hyperbolic unit ball is an important topic of current investigation, mentioning hereby H. Leutwiler, T. Hempfling, S.-L. Eriksson-Bique J. Cnops, P. Van Lancker, F. Sommen and D. Eelbode, P. Cerejeiras and U. Kähler and J. Ryan and S. Liu and others. In connection with Rarita–Schwinger systems there is important progress by the group Bures–Soucek–Sommen–Van Lancker–Bernardes, while in connection with hyperfunctions and algebraic analysis we mention the Italian school D. Struppa, I. Sabadini, F. Colombo including F. Sommen.

Some aspects of octonion analysis have been studied by P. Dentoni, M. Sce, C.A. Manogue, T. Dray, L.Z. Peng and X.M. Li.

Last but not least there are important application of Clifford analysis in number theory with Eisenstein series and higher dimensional modular forms by S. Kraußhar.

Needless to say this list of developments is highly incomplete, but we think it does give evidence of the fact that nowadays Clifford analysis has become a major discipline of mathematics with a vast field of applications.

This volume is devoted to analytic and geometric aspects of Clifford analysis and its applications. There are two sources of the collected papers. One is the satellite conference to the International Congress for Mathematicians, August 2002, Beijing, held August 15–18 at the University of Macau; the other articles are invited contributions not presented at the satellite conference. All the selected papers were strictly refereed, and all contain unpublished new results. Some of them include comprehensive surveys in the particular research areas of the authors.

The three parts of the book reflect the diverse aspects and applications of Clifford analysis all of which use Clifford algebras as a main ingredient.

Part A on *Differential Equations and Operator Theory* starts with a survey by A. Axelsson and A. M^c^Intosh on the L_2-theory of boundary value problems for the exterior and interior derivative operators on a bounded, weakly Lipschitz domain in $\mathbb{R}^n$. Particularly, the Hodge decompositions of the derivative operators play an important role for the development of a theory of nilpotent operators in these L_2-spaces.

S. Bernstein shows that monogenic functions of bounded mean oscillation in the unit ball of $\mathbb{R}^n$ can be properly defined. A duality theorem is proved which connects to the Hardy space of right monogenic functions. In a further joint article with K. Gürlebeck, L.F. Reséndis and L.M. Tovar S., monogenic $B^{p,q}$-functions are characterized in terms of harmonic majorants. This is mainly done in the unit disc of $\mathbb{R}^3$ using quaternions. The same setting is used for the study of series expansions of monogenic B^q-functions in the contribution by K. Gürlebeck and A. El Sayed Ahmed.

The survey by F. Brackx, R. Delanghe and F. Sommen summarizes what is known on distributions in the Clifford analysis setting and introduces in a unifying way new kinds of distributions generalizing the known ones and crucially making use of spherical monogenics. Further, particularly the action of Dirac operators on these distributions is studied.

Hypermonogenic functions, introduced by H. Leutwiler in the beginning of the 1990s, have a lot of nice properties — particularly, powers of vectors are hypermonogenic functions — but so far missed a Cauchy-type theorem. Two papers by S.-L. Eriksson-Bique, the first of these jointly written with H. Leutwiler and included in this volume, lead to integral formulas for hypermonogenic functions.

S. Liu and T. Qian examine a new approach to the pointwise convergence of Fourier series on the unit sphere in $\mathbb{R}^4$. The usage of Clifford analysis leads to analogs of Riemann–Lebesgue and Dini type theorems.

Conformally flat manifolds are examples of objects where explicit calculations can be performed, such as for Cauchy kernels, integral formulas and Dirac operators in the Clifford analysis setting. J. Ryan considers in his contribution conformally flat manifolds that are conformally equivalent to open subsets of the n-dimensional sphere in $\mathbb{R}^{n+1}$.

Several extensions of Clifford analysis are proposed in the survey of I. Sabadini and F. Sommen. The authors collect known facts under the important view of monogenic forms and then introduce to the theory of several vector variables and to Clifford algebra valued systems of differential equations defined on bivectors and ℓ-vectors.

In Part B on *Global Analysis and Differential Geometry* Y. Homma is concerned with the construction of Bochner–Weitzenböck formulas for hyper-Kählerian gradients, using Lie theory and the Clifford algebra representation of symplectic groups as main tools.

The sheaf of harmonic spinors on complex two-dimensional space lies in the heart of T. Kori's article. A Runge type approximation theorem, cohomology groups and a Riemann–Roch type theorem are proved for conformally flat manifolds.

M. Martin surveys the diverse connections between Clifford analysis, spin geometry and seminormal systems of Hilbert space operators. Particularly, a new approach to joint seminormality is given, leading to a Putnam type inequality for Riesz transform models.

The contribution by Zhong Chunping and Zhong Tongde defines a mean value Laplacian for strongly Kähler–Finsler manifolds which generalizes the Kählerian Laplacian. Coordinate representations and several properties are proved, and the definition is extended to Hodge–Laplace operators on differential forms.

Applications to artificial intelligence, number theory, numerical analysis, and physics are given in Part C of the book.

Non-commutative determinants of Moore and Dieudonné are applied in S. Alesker's work to characterize plurisubharmonic functions and Monge–Ampére equations in quaternionic variables.

K. Gürlebeck and W. Sprößig study Galpern–Sobolev equations and their discretization, leading to an implicit approximation scheme and to representation formulas for the solutions.

Surprising links to number theory are provided in R.S. Kraußhar's survey. Clifford algebra valued automorphic forms are intensively discussed and a wide range of applications is given. Eisenstein and Poincaré type series play a fundamental role in this new theory of modular forms.

Clifford algebras also serve as framework for automated geometric theorem proving in the contribution of H. Li. One reason is that these algebras are perfectly suitable for invariant computations in metric geometry. Li introduces to Clifford bracket algebras and Clifford expansions and gives several examples.

An introduction to the theory of (quaternion-valued) wavelets is offered by L. Peng and J. Zhao. Multiresolution analysis, scaling functions and filters are discussed for smooth orthogonal wavelets with short support and symmetry.

The main sponsors of the satellite conference are the University of Macau and the Chinese Mathematical Society represented by Zhiming Ma. The co-sponsors include the Center for Mathematics and Its Applications, ANU, Canberra, the Fund for Scientific Research, Belgium, the State University Ghent, and Guangdong New Oasis Life Science and Technology Co. Ltd., joint with Gold Arch International (Group) Co. Ltd. represented by Wai Hung Au. Special thanks for the hospitality and the funding, to both the conference and the publication of this volume, are addressed to the University of Macau, represented by Rector and Chair of the Organizing Committee Vai Pan Iu, to the committee members Yiping Li, Che Man Cheng and Ieng Tak Leong, and to the local helpers at the Department of Mathematics of the University of Macau. The editors wish to specially dedicate their thanks to Prof Rui Paulo Da Silva Martins and Mr Iat Long Lai, who conveyed the first supporting message of the university, and gave an impetus to initializing the local organization work. Last but not least, the editors are grateful to the referees for their extremely valuable assistance in creating this volume.

Contents

Dedicated to Pertti Lounesto

Part A
Differential Equations and Operator Theory

Trends in Mathematics: *Advances in Analysis and Geometry*, 3–29

Hodge Decompositions on Weakly Lipschitz Domains

Andreas Axelsson, Alan McIntosh

Abstract. We survey the L_2 theory of boundary value problems for exterior and interior derivative operators $d_{k_1} = d + k_1 e_0 \wedge$ and $\delta_{k_2} = \delta + k_2 e_0 \lrcorner$ on a bounded, weakly Lipschitz domain $\Omega \subset \mathbf{R}^n$, for $k_1, k_2 \in \mathbf{C}$. The boundary conditions are that the field be either normal or tangential at the boundary. The well-posedness of these problems is related to a Hodge decomposition of the space $L_2(\Omega)$ corresponding to the operators d and δ. In developing this relationship, we derive a theory of nilpotent operators in Hilbert space.

Mathematics Subject Classification (2000). 35J55, 35Q60, 47B99.

Keywords. Maxwell's equations, boundary value problem, Hodge decomposition, Lipschitz domain, first order system, nilpotent operator, exterior derivative.

1. Introduction

The aim of this paper is to survey and further develop the Hilbert space theory of boundary value problems (BVP's) for the exterior (d) and interior derivative (δ) operators in a bounded domain $\Omega \subset \mathbf{R}^n$ with a boundary $\Sigma = \partial\Omega$ of minimal regularity. The BVP we have in mind is the following. Given a $(j-1)$-vector field $G \in L_2(\Omega; \wedge^{j-1})$ which is k_2-divergence free, i.e. $\delta_{k_2} G = 0$, find a j-vector field $F \in L_2(\Omega; \wedge^j)$ such that

$$\begin{cases} d_{k_1} F = 0 & \text{in } \Omega, \\ \delta_{k_2} F = G & \text{in } \Omega, \\ \nu \wedge f = 0 & \text{on } \Sigma. \end{cases} \qquad (1)$$

Here $f := F|_\Sigma$ and ν denotes the outward pointing unit vector field on Σ. Thus the boundary condition $\nu \wedge f = 0$ means that F is normal on the boundary.

The authors were supported by the Australian Government through the Australian Research Council. The research was conducted at the Centre for Mathematics and its Applications at the Australian National University, Canberra.

The differential operators are the zero order perturbations $d_{k_1} = d + k_1 e_0 \wedge$ and $\delta_{k_2} = \delta + k_2 e_0 \lrcorner$ of the operators d and δ defined in Section 2, where $k_i \in \mathbf{C}$ are the wave numbers and $e_0 \in \wedge^1$ is the time-like vector. An important property of these operators is that they are nilpotent, i.e. $d_{k_1}^2 = \delta_{k_2}^2 = 0$.

Example 1.1. When $j = 1$ and $k_1 = k_2 = k$, the BVP (1) is essentially the Dirichlet BVP for the Helmholtz equation $(\Delta + k^2)U = G$, where $U : \Omega \to \wedge^0 = \mathbf{C}$. To see this, let $F = d_k U = \nabla U + k e_0 U$ so that

$$\delta_k F = \delta_k d_k U = (\Delta + k^2)U = G.$$

Since $u = U|_\Sigma = 0$ implies that f is normal, i.e. $\nu \wedge f = 0$ on Σ, we see that the two BVP's are equivalent. Note that since G in this case is a scalar function, the condition $\delta_{k_2} G = 0$ is automatically satisfied.

Similarly, the Neumann problem corresponds to (1) with tangential boundary conditions, i.e. $\nu \lrcorner f = 0$.

Example 1.2. When $n = 3$ and $j = 2$, the BVP (1) coincides with the electromagnetic BVP for time-harmonic ($\frac{\partial}{\partial t} = -i\omega$) Maxwell's equations with frequency $\omega \in \mathbf{C}$ and $\Omega^- = \mathbf{R}^3 \setminus \overline{\Omega}$ being a perfect conductor. Assuming that Ω is composed of a linear, homogeneous, isotropic, possibly conducting material with permittivity $\epsilon > 0$, permeability $\mu > 0$ and conductivity $\sigma \geq 0$, we consider the electromagnetic field

$$F = \sqrt{\epsilon_*}(-ie_0) \wedge (E_1 e_1 + E_2 e_2 + E_3 e_3) \\ + \tfrac{1}{\sqrt{\mu}}(B_1 e_2 \wedge e_3 + B_2 e_3 \wedge e_1 + B_3 e_1 \wedge e_2) : \Omega \longrightarrow \wedge^2 \mathbf{R}^4,$$

where E and B are the electric and magnetic fields and $\epsilon_* := \epsilon + i\sigma/\omega$. If we let $k_1 = k_2 = k = \omega\sqrt{\epsilon_* \mu}$, then Faraday's induction law and the magnetic Gauss' law combine to $d_k F = 0$ whereas Maxwell's–Ampère's law and Gauss' law combine to $\delta_k F = G$, where the four-current $G = \frac{i}{\sqrt{\epsilon_*}}\rho e_0 - \sqrt{\mu} J$ satisfies the continuity equation $\delta_k G = 0$.

In this paper we investigate BVP's from the point of view of *splittings of function spaces* following our earlier work Axelsson–Grognard–Hogan–McIntosh [4], Axelsson [2] and [1]. The splittings relevant to this paper are Hodge type decompositions of the Hilbert space $L_2(\Omega; \wedge)$. For simplicity, assume $k_2 = -k_1^{\mathrm{c}}$. Then the operators $d_{k_1,\overline{\Omega}}$ and $\delta_{k_2,\Omega}$ are adjoint, where $d_{k_1,\overline{\Omega}}$ denotes d_{k_1} with normal boundary conditions and $\delta_{k_2,\Omega}$ denotes δ_{k_2} without boundary conditions in Ω, as in Definition 4.1. Consider the following diagram.

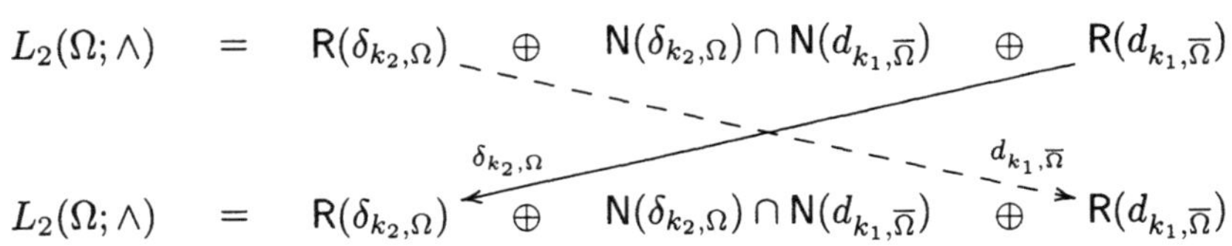

What is needed here is to prove a Hodge decomposition, i.e. that the space $\mathsf{N}(\delta_{k_2,\Omega}) \cap \mathsf{N}(d_{k_1,\overline{\Omega}})$ of "harmonic forms" is finite dimensional and that the ranges $\mathsf{R}(\delta_{k_2,\Omega})$ and $\mathsf{R}(d_{k_1,\overline{\Omega}})$ are closed, or equivalently that $\mathsf{R}(\Gamma)$ is a closed subset of finite codimension in the null space $\mathsf{N}(\Gamma)$ for both choices $\delta_{k_2,\Omega}$ and $d_{k_1,\overline{\Omega}}$ for Γ.

Note that Fredholm well-posedness of the BVP (1), put into operator theoretic language means that

$$\delta_{k_2,\Omega} : \mathsf{N}(d_{k_1,\overline{\Omega}}) \longrightarrow \mathsf{N}(\delta_{k_2,\Omega})$$

is a Fredholm map. Clearly this holds if we have a Hodge decomposition as above. The Hodge decomposition in the case $k_1 = k_2 = 0$ for a general weakly Lipschitz domain is due to Picard [21], and the extension to the case $k_2 = -{k_1}^{\mathrm{c}}$ is straightforward. Although the Hodge decomposition is not valid for general $k_1,\ k_2 \in \mathbf{C}$, nevertheless the BVP (1) is well-posed in the Fredholm sense. Indeed the following result will be proved in Section 4.

Theorem 1.3. *Let $\Omega \subset \mathbf{R}^n$ be a bounded weakly Lipschitz domain, as in Definition 2.1, and let $k_1,\ k_2 \in \mathbf{C}$. Then $\mathsf{R}(\delta_{k_2,\Omega})$ and $\mathsf{R}(d_{k_1,\overline{\Omega}})$ are closed subspaces of finite codimension in $\mathsf{N}(\delta_{k_2,\Omega})$ and $\mathsf{N}(d_{k_1,\overline{\Omega}})$ respectively. The maps*

$$\delta_{k_2,\Omega} : \mathsf{N}(d_{k_1,\overline{\Omega}}) \longrightarrow \mathsf{N}(\delta_{k_2,\Omega}) \tag{2}$$

$$d_{k_1,\overline{\Omega}} : \mathsf{N}(\delta_{k_2,\Omega}) \longrightarrow \mathsf{N}(d_{k_1,\overline{\Omega}}) \tag{3}$$

are Fredholm maps with compact Fredholm inverses.

For the Hodge decomposition with tangential boundary conditions, i.e. with $\delta_{k_2,\Omega}$ and $d_{k_1,\overline{\Omega}}$ replaced by $\delta_{k_2,\overline{\Omega}}$ and $d_{k_1,\Omega}$ as in Definition 4.1, the corresponding result holds.

W.V.D. Hodge's pioneering work on harmonic integrals on Riemannian manifolds during the 1930's was published in his book [12]. The splitting of a differential form into its exact, coexact and harmonic parts, now referred to as the Hodge decomposition, was in this book proved using Fredholm's theory of linear integral equations. The connection between splittings of function spaces such as the Hodge decomposition and boundary value problems in potential theory was early recognised by Weyl [27]. Here it was shown how the classical Dirichlet minimum principle could be replaced by the construction of orthogonal projections in Hilbert space.

In the present paper, we treat Hodge decompositions from a purely first order, operator theoretic point of view. By first order we mean that the focus is on nilpotent operators (see Definition 3.1 below) such as the exterior derivative d and not on the Hodge–Laplace operator $\Delta = d\delta + \delta d$. An early investigation along these lines is Friedrichs [8], where the operators d and δ were introduced as closed unbounded operators. Other references we would like to mention are Kodaira [14], where the weak Hodge decomposition (14) appears, and Gaffney [9] and [10] which introduced the a priori estimate

$$\|F\|_{W_2^1} \lesssim \|dF\|_{L_2} + \|\delta F\|_{L_2} + \|F\|_{L_2}. \tag{4}$$

For a domain with boundary, we discuss this inequality in Theorem 4.10. For further early literature on the Hodge decomposition, we refer to Chapter 7 in Morrey [20].

On a domain with non-smooth boundary, the Gaffney–Friedrich inequality (4) is in general not valid. The first proof of the Hodge decomposition on domains with non-smooth boundaries without using (4) is Weck [26]. The extension to general weakly Lipschitz domains is due to Picard [21].

Returning the Example 1.2, we remark that the standard approach to the Maxwell BVP uses the Maxwell operator M acting on a pair of divergence-free vector fields. An investigation of M on domains with non-smooth boundary, from the point of view of the Weyl decomposition (essentially the Hodge decomposition of vector fields) can be found in Birman–Solomyak [5], [6]. They show (in the language of the present paper) how M constitute part of the elliptic Dirac operator $\mathbf{D}_{\Omega^\perp}$ from Example 4.6.

For further literature on the connection between Hodge decompositions and BVP's, we refer to Schwarz [23] in the case of smooth domains and to Mitrea–Mitrea [17] in the case of strongly Lipschitz domains.

The key idea in this paper is that not only do we treat Hodge decompositions from a pure first order point of view, but we show that by investigating the "half-elliptic" operators d and δ separately, one can easily prove the Hodge decomposition on a domain with weakly Lipschitz boundary. Indeed, it is not necessary to use the given adjoint δ operator in proving that $\mathsf{R}(d_\Omega)$ is closed and of finite codimension in $\mathsf{N}(d_\Omega)$. As in Remark 3.12, we may equally well choose to work with the adjoint given by a metric in which Ω has a smooth boundary.

The first step in the proof of Theorem 1.3 uses the duality theorem 3.3 from general operator theory. As we show in Proposition 3.11, this duality result proves that the maps (2) and (3) have the same properties concerning closed range and compact inverse. The second step in the proof of Theorem 1.3 is Lemma 3.13, where we use the basic differential geometric fact that the exterior derivative is independent of the Riemannian metric, here given in the form of Proposition 2.6. These two steps show that the general case of a weakly Lipschitz domain in Theorem 1.3 can be reduced to the case of a smooth domain Ω. This reduction technique has been used by Picard [21]. We also provide some basic density results for the d and δ operators in Proposition 4.3 and construct extension maps in Proposition 4.8. Finally, we survey three different ways to prove Theorem 1.3 under certain additional regularity and topological assumptions on Σ.

- Theorem 4.10: The classical Gaffney–Friedrichs a priori estimate technique, which gives optimal $W_2^1(\Omega;\wedge)$ regularity for fields in $\mathsf{D}(d_{\overline{\Omega}}) \cap \mathsf{D}(\delta_\Omega)$ if the domain has a smooth boundary.
- Theorem 4.13: The boundary integral equation method, which gives optimal regularity $W_2^{1/2}(\Omega;\wedge)$ in the class of strongly Lipschitz domains by using Rellich estimates.

- Theorem 4.17: A path integral method for a star shaped domain. This method, which is based on the classical Poincaré lemma, seems new. Although it does not give optimal regularity, it has the advantage of being entirely explicit.

2. Preliminaries

Throughout this paper $\Omega = \Omega^+ \subset \mathbf{R}^n$ denotes a *bounded* open set, separated from the exterior domain $\Omega^- = \mathbf{R}^n \setminus \overline{\Omega}^+$ by a weakly Lipschitz interface $\Sigma = \partial\Omega^+ = \partial\Omega^-$, defined as follows.

Definition 2.1. The interface Σ is *weakly Lipschitz* if, for all $y \in \Sigma$, there exists a neighbourhood $V_y \ni y$ and a global bilipschitz map $\rho_y : \mathbf{R}^n \to \mathbf{R}^n$ such that

$$\begin{aligned} \Omega^+ \cap V_y &= \rho_y(\mathbf{R}^n_+) \cap V_y, \\ \Sigma \cap V_y &= \rho_y(\mathbf{R}^{n-1}) \cap V_y, \\ \Omega^- \cap V_y &= \rho_y(\mathbf{R}^n_-) \cap V_y. \end{aligned}$$

In this case Ω is called a bounded weakly Lipschitz domain.

If $\rho : \mathbf{R}^n \to \mathbf{R}^n$ is a global bilipschitz map, then $\rho(\mathbf{R}^{n-1})$ is called a *special weakly Lipschitz* surface/interface and $\rho(\mathbf{R}^n_\pm)$ are called special weakly Lipschitz domains.

By Rademacher's theorem, a weakly Lipschitz surface Σ has a tangent plane and an outward (into Ω^-) pointing unit normal $\nu(y)$ at almost every $y \in \Sigma$.

Example 2.2. We now give two examples of weakly Lipschitz surfaces which are not strongly Lipschitz, i.e. not locally the graph of a Lipschitz function.

(i) Let $\rho_0 : S^{n-1} \to S^{n-1}$ be a bilipschitz homeomorphism of the unit sphere. Consider the conical surface

$$\Sigma := \{x \in \mathbf{R}^n \setminus \{0\} \,;\, x/|x| \in \rho_0(S^{n-1} \cap \mathbf{R}^{n-1})\} \cup \{0\}.$$

The natural parametrisation here is $\rho(r\omega) := r\rho_0(\omega)$, $r \geq 0$, $\omega \in S^{n-1}$. Using the identity $|r\omega - r'\omega'|^2 = |r-r'|^2 + rr'|\omega - \omega'|^2$, it is straightforward to show that $\rho : \mathbf{R}^n \to \mathbf{R}^n$ is a bilipschitz map. Thus Σ is a weakly Lipschitz surface.

An important special case is the "two brick" domain, defined as the interior of

$$\{(x,y,z) \in \mathbf{R}^3 \,;\, y \leq 0,\, z \leq 0\} \cup \{(x,y,z) \in \mathbf{R}^3 \,;\, x \leq 0,\, z \geq 0\}.$$

Indeed, the intersection with the unit sphere S^2 is a two dimensional strongly Lipschitz domain. Nevertheless, the boundary of the two brick domain is not locally a graph of a Lipschitz function around 0.

(ii) Let $a_i > 0$ and $e^{-2\pi}a_2 < a_1 < a_2$ and consider the *logarithmic spiral*

$$\Omega := \{re^{i\theta} \,;\, r > 0,\, \theta \in \mathbf{R},\, a_1 e^{-\theta} < r < a_2 e^{-\theta}\} \subset \mathbf{R}^2.$$

To see that Ω is a special weakly Lipschitz domain, define the maps

$$\rho_s(x,y) := (x\cos(s\ln r) - y\sin(s\ln r), x\sin(s\ln r) + y\cos(s\ln r)),$$

where $r^2 = x^2 + y^2$, or in complex notation $\rho_s : z \mapsto ze^{is\ln|z|}$. We see that $\rho_s \circ \rho_t = \rho_{s+t}$, $s, t \in \mathbf{R}$ and that $|\nabla \otimes \rho_s| \le C$. In particular $\rho_{-1} : \mathbf{R}^2 \to \mathbf{R}^2$ is a bilipschitz map. Since

$$\Omega = \rho_{-1}(\{z \in \mathbf{C} \,;\, \ln a_1 < \arg(z) < \ln a_2\}),$$

this shows that Ω is a special weakly Lipschitz domain. But clearly $\partial\Omega$ is not locally a graph of a Lipschitz function around 0.

In this paper we make use of the three differential operators d, δ and $\mathbf{D}$ as described below. These operators act on functions $F : \Omega \to \wedge$ which take values in an exterior algebra $\wedge$, sometimes referred to as (multivector-)fields. Boundary traces and fields on Σ will be written with small letters, for example f. We here use the complexified exterior algebra

$$\wedge = \wedge_{\mathbf{C}}\mathbf{R}^{n+1} = \wedge^0 \oplus \wedge^1 \oplus \ldots \oplus \wedge^{n+1}$$

for $\mathbf{R}^n$ spacetime. Let $\{e_s \,;\, s \subset \{0,1,\ldots,n\}\}$ be the standard basis for $\wedge_{\mathbf{C}}\mathbf{R}^{n+1}$. Here $e_0 \in \wedge^1$ is interpreted as a (imaginary time-like) vector and the space of j-vectors $\wedge^j$ is the span of $\{e_s \,;\, |s| = j\}$. Furthermore, let $\langle\cdot,\cdot\rangle$ denote the standard complex bilinear pairing, u^c denote component-wise complex conjugation and $u^\neg$ denote *involution*. Concretely, if we expand the multivectors u, $v \in \wedge_{\mathbf{C}}\mathbf{R}^{n+1}$ as $u = \sum_s u_s e_s$ and $v = \sum v_s e_s$, then

$$\begin{aligned} \langle u, v\rangle &= \sum u_s v_s, \\ u^c &= \sum u_s^c e_s, \\ u^\neg &= \sum (-1)^{|s|} u_s e_s. \end{aligned}$$

Definition 2.3. Introduce the counting function $\sigma(s,t) := \#\{(s_i,t_j) \,;\, s_i > t_j\}$, where $s = \{s_i\}$, $t = \{t_j\} \subset \{0,1,\ldots,n\}$. Basic complex bilinear products on the algebra $\wedge$ are the following.

(i) The *exterior product* of two basis multivectors e_s and e_t is

$$e_s \wedge e_t = (-1)^{\sigma(s,t)}\, e_{s\cup t} \quad \text{if } s \cap t = \emptyset \text{ and otherwise zero.}$$

(ii) The left (right) *interior product* $u \lrcorner v$ ($u \llcorner v$) is the unique bilinear (non-associative) product for which $\langle u \lrcorner x, y\rangle = \langle x, u \wedge y\rangle$ and $\langle x \llcorner u, y\rangle = \langle x, y \wedge u\rangle$ respectively for all u, x, $y \in \wedge$. The action on two basis vectors e_s and e_t is

$$e_s \lrcorner e_t = (-1)^{\sigma(s,t\setminus s)}\, e_{t\setminus s}, \qquad e_t \llcorner e_s = (-1)^{\sigma(t\setminus s,s)}\, e_{t\setminus s},$$

if $s \subset t$ and otherwise zero.

(iii) The *Clifford product* of two basis multivectors e_s and e_t is

$$e_s \vartriangle e_t = (-1)^{\sigma(s,t)}\, e_{s\vartriangle t},$$

where $\triangle$ denotes the symmetric difference when acting on index sets. When there is no risk of confusion we will use the standard short-hand notation $uv := u \triangle v$ for the Clifford product.

Proposition 2.4. *For a vector $a \in \wedge^1$ and for general multivectors u, v and $w \in \wedge$ the following hold.*

$$u \lrcorner (v \lrcorner w) = (v \wedge u) \lrcorner w \tag{5}$$

$$a \triangle u = a \lrcorner u + a \wedge u \tag{6}$$

$$a \lrcorner u = -u^{\neg} \llcorner a = \tfrac{1}{2}(a \triangle u - u^{\neg} \triangle a) \tag{7}$$

$$a \wedge u = u^{\neg} \wedge a = \tfrac{1}{2}(a \triangle u + u^{\neg} \triangle a) \tag{8}$$

$$a \lrcorner (u \wedge v) = (a \lrcorner u) \wedge v + u^{\neg} \wedge (a \lrcorner v) \tag{9}$$

These basic geometric algebra identities are essentially well known and we omit the proof. Here (5) is the associativity property of the interior product. The formulae (7) and (8), which are inverse to (6), are sometimes referred to as *Riesz' formulae.* The formula (9) is the derivation property for the interior product. A classical example of (9) is when a, $b = u$ and $c = v$ are vectors in a three-dimensional space. Using the Hodge complement $u^{\perp} := u \lrcorner e_{123}$ (usually called the Hodge star $*u$), and the vector product $b \times c = (b \wedge c)^{\perp}$, we get the well known identity

$$-a \times (b \times c) = a \lrcorner (b \wedge c) = \langle a, b\rangle c - \langle a, c\rangle b.$$

Throughout this paper we make use of the nabla symbol $\nabla = \sum_{j=1}^{n} e_j \partial_j$. We recall that the products $\wedge$, $\lrcorner$ and $\triangle$ induce differential operators

$$dF(x) := \nabla \wedge F(x) = \sum_{j=1}^{n} e_j \wedge (\partial_j F)(x),$$

$$\delta F(x) := \nabla \lrcorner F(x) = \sum_{j=1}^{n} e_j \lrcorner (\partial_j F)(x),$$

$$\mathbf{D}F(x) := \nabla \triangle F(x) = \sum_{j=1}^{n} e_j \triangle (\partial_j F)(x) = dF(x) + \delta F(x).$$

In the same spirit we also denote the full differential of F by $\nabla \otimes F(x) = \sum e_j \otimes (\partial_j F)(x) \in \mathbf{R}^n \otimes \wedge$. Here the formal adjoint of the *exterior derivative* operator d is the negative of the *interior derivative* δ; this differs from the standard convention. Sometimes we refer to d as (generalised) curl and to δ as (generalised) divergence. The (elliptic) *Dirac operator* $\mathbf{D} = d + \delta$ is formally skew-adjoint. Here $\mathbf{D}$ is a square root of the Hodge–Laplace operator $\Delta = d\delta + \delta d$.

The most important property of the differential operators d and δ is that they commute with a change of variables if we change the direction of the field in an appropriate way.

Definition 2.5. Let $\rho : U \to V$ be a diffeomorphism between two open sets U and $V \subset \mathbf{R}^n$. Denote by $\underline{\rho}_x$ the Jacobian matrix of ρ at $x \in U$ and extend this linear map $T_x\mathbf{R}^n \to T_{\rho(x)}\mathbf{R}^n$ to a $\wedge$-isomorphism $\underline{\rho}_x : \wedge \to \wedge$ such that $\underline{\rho}_x(e_0) = e_0$ and

$$\underline{\rho}_x(e_{i_1} \wedge \ldots \wedge e_{i_k}) = (\underline{\rho}_x e_{i_1}) \wedge \ldots \wedge (\underline{\rho}_x e_{i_k}),$$

if $\{i_1, \ldots, i_k\} \subset \{0, 1, \ldots, n\}$. To a field $F : V \to \wedge$ we associate the *pullback* $\rho^* F : U \to \wedge$ and *push forward* $\rho_*^{-1} F : U \to \wedge$ of F as follows.

$$(\rho^* F)(x) := (\underline{\rho}_x)^*(F(\rho(x))), \qquad (\rho_*^{-1} F)(x) := (\underline{\rho}_x)^{-1}(F(\rho(x))).$$

For convenience, we also define the *reduced push forward*

$$\tilde{\rho}_*^{-1} F := J(\rho)\rho_*^{-1} F : U \longrightarrow \wedge,$$

where $J(\rho)(x)(e_1 \wedge \ldots \wedge e_n) = \underline{\rho}_x(e_1 \wedge \ldots \wedge e_n)$ denotes the Jacobian determinant.

Proposition 2.6. *If ρ and F are as in Definition 2.5, then we have commutation properties*

$$d(\rho^* F) = \rho^*(dF), \qquad \delta(\tilde{\rho}_*^{-1} F) = \tilde{\rho}_*^{-1}(\delta F), \tag{10}$$

and homomorphism properties

$$\rho^*(F \wedge G) = \rho^* F \wedge \rho^* G, \qquad \rho_*^{-1}(F \wedge G) = \rho_*^{-1} F \wedge \rho_*^{-1} G, \tag{11}$$

$$\rho^*(F \lrcorner G) = \rho_*^{-1} F \lrcorner \rho^* G, \qquad \rho_*^{-1}(F \lrcorner G) = \rho^* F \lrcorner \rho_*^{-1} G. \tag{12}$$

In particular, if $F^\perp := F \lrcorner e_{01\ldots n}$ denotes the complement of F, then $\rho^*(F^\perp) = (\tilde{\rho}_*^{-1} F)^\perp$.

Proof. Note that we have the two pairs of adjoint operators

$$\underline{\rho}_x : T_x\mathbf{R}^n \longrightarrow T_{\rho(x)}\mathbf{R}^n, \qquad (\underline{\rho}_x)^* : T_{\rho(x)}\mathbf{R}^n \longrightarrow T_x\mathbf{R}^n,$$

and

$$\tilde{\rho}_* : L_2(U; \wedge) \longrightarrow L_2(V; \wedge), \qquad \rho^* : L_2(V; \wedge) \longrightarrow L_2(U; \wedge),$$

if $\nabla \otimes \rho$, $\nabla \otimes \rho^{-1} \in L_\infty$. The identities $d(\rho^* F) = \rho^*(dF)$, $\rho^*(F \wedge G) = \rho^* F \wedge \rho^* G$ and $\rho_*^{-1}(F \wedge G) = \rho_*^{-1} F \wedge \rho_*^{-1} G$ are well known facts from the theory of differential forms. The remaining identities follow by duality. □

In order to treat Stokes' type theorems in a unified way, we record the following theorem, here referred to as the *boundary theorem*.

Theorem 2.7. *Let V be a finite dimensional linear space and let $F : \overline{\Omega} \to V$ be a function in Ω smooth up to $\Sigma = \partial\Omega$ with boundary trace $f := F|_\Sigma$. Then we have*

$$\int_\Sigma \nu(y) \otimes f(y)\, d\sigma(y) = \int_\Omega \nabla \otimes F(x)\, dx,$$

where the integrand is $\mathbf{R}^n \otimes V$ valued, ν is the outward pointing normal and $d\sigma$ is the scalar surface measure.

Remark 2.8. (i) Note that, via a limiting argument, the boundary theorem can be extended to less regular functions.

(ii) Recall that this theorem is universal in the sense that for any given finite dimensional linear space W and bilinear form $L : \mathbf{R}^n \times V \to W$, L can be lifted to a linear map $L : \mathbf{R}^n \otimes V \to W$. Applying this to the formula in the boundary theorem gives the special case $\int_\Sigma L(\nu(y), f(y))\, d\sigma(y) = \int_\Omega L(\nabla, F(x))\, dx$.

We end this section with a discussion, preliminary to Section 4, about natural boundary conditions for d and δ. Let $F : \Omega \to \wedge$ be a multivector field in Ω, smooth up to Σ, and extend it by zero to a field F_z on $\mathbf{R}^n$. If σ denotes the surface measure on Σ, it follows that in distribution sense we have

$$d(F_z) = dF|_\Omega - (\nu \wedge f)\sigma \quad \text{and} \quad \delta(F_z) = \delta F|_\Omega - (\nu \lrcorner f)\sigma.$$

For example, the first identity follows from the boundary theorem, using $V = \wedge \otimes \wedge$ and the linear map $L : \mathbf{R}^n \otimes (\wedge \otimes \wedge) \to \mathbf{C} : a \otimes (F \otimes G) \mapsto (a \wedge F, G)$, since

$$(\nabla \wedge F_z, \Phi) = -\int_\Omega \langle F, \nabla \lrcorner \Phi\rangle = \int_\Omega \langle \nabla \wedge F, \Phi\rangle - \int_\Sigma \langle \nu \wedge f, \phi\rangle,$$

for any $\Phi \in C_0^\infty(\mathbf{R}^n; \wedge)$. Thus, requiring that $d(F_z) \in L_2(\mathbf{R}^n; \wedge)$ means that $dF \in L_2(\Omega; \wedge)$ and that $\nu \wedge f = 0$, i.e. the field F is *normal* to Σ. Similarly, requiring that $\delta(F_z) \in L_2(\mathbf{R}^n; \wedge)$ means that $\delta F \in L_2(\Omega; \wedge)$ and that $\nu \lrcorner f = 0$, i.e. the field F is *tangential* to Σ. We note that each boundary condition refers to half of the components (in the full exterior algebra $\wedge$) of the field vanishing on Σ.

When $F \in L_2(\Omega; \wedge)$ and $d(F_z) \in L_2(\mathbf{R}^n; \wedge)$, although the field F is normal to Σ, it does not necessarily have a well defined normal component $\nu \lrcorner f$ on Σ. To see this, consider the vector field

$$F(x) := \begin{cases} e_n, & 1/(2j+1) < x_n \leq 1/(2j), \\ 0, & 1/(2j) < x_n \leq 1/(2j-1), \end{cases}$$

locally around $x = (x', x_n) = 0$. Then $F_z \in L_{2,\mathrm{loc}}(\mathbf{R}^n; \wedge)$ and $d(F_z) = 0$, but clearly F does not have a well defined trace.

Similarly, control of F and $\delta(F_z)$ is not enough for defining the tangential part of the trace.

3. Nilpotent operators in Hilbert spaces

In this section we develop the operator theory for a nilpotent operator Γ. This is then applied to the d and δ operators in Section 4.

Recall the following basic spaces associated with a linear operator $A : \mathcal{H}_1 \to \mathcal{H}_2$ between Hilbert spaces $\mathcal{H}_i$.

- Domain $\mathsf{D}(A) := \{x \in \mathcal{H}_1 \,;\, Ax \text{ is defined}\}$
- Null space $\mathsf{N}(A) := \{x \in \mathsf{D}(A) \,;\, Ax = 0\}$
- Range $\mathsf{R}(A) := \{Ax \,;\, x \in \mathsf{D}(A)\}$
- Graph $\mathsf{G}(A) := \{(x, Ax)^t \in \mathcal{H}_1 \oplus \mathcal{H}_2 \,;\, x \in \mathsf{D}(A)\}$

If A_1 and A_2 are two linear operators, then we write $A_1 \subset A_2$ if $\mathsf{G}(A_1) \subset \mathsf{G}(A_2)$.

Definition 3.1. An operator $\Gamma : \mathcal{H} \to \mathcal{H}$ in a Hilbert space $\mathcal{H}$ is said to be *nilpotent* if it is closed (i.e. $\mathsf{G}(\Gamma)$ is closed), densely defined (i.e. $\mathsf{D}(\Gamma)$ is dense in $\mathcal{H}$) and if $\mathsf{R}(\Gamma) \subset \mathsf{N}(\Gamma)$. In particular, $\Gamma^2 \subset 0$.

Recall that $\mathsf{N}(A)$ always is closed in $\mathcal{H}_1$ for a closed operator. For a nilpotent operator Γ in $\mathcal{H}$, we have inclusions

$$\mathsf{R}(\Gamma) \subset \overline{\mathsf{R}(\Gamma)} \subset \mathsf{N}(\Gamma) \subset \mathsf{D}(\Gamma) \subset \mathcal{H}. \tag{13}$$

Note carefully that $\mathsf{R}(\Gamma)$ may not be closed. Our main work will be to prove that $\mathsf{R}(\Gamma)$ is closed when Γ is one of the d and δ operators in Ω.

From (13) we also see that Γ acts as a bounded nilpotent operator $\Gamma : \mathsf{D}(\Gamma) \to \mathsf{D}(\Gamma)$, where $\mathsf{D}(\Gamma)$ is a Hilbert space with the graph norm $\|x\|^2_{\mathsf{D}(\Gamma)} := \|x\|^2 + \|\Gamma x\|^2$.

Definition 3.2. Let $A_1 : \mathcal{H}_1 \to \mathcal{H}_2$ and $A_2 : \mathcal{H}_2 \to \mathcal{H}_1$ be two linear operators. We say that A_1 and A_2 are (maximal) *adjoint* operators if

$$\mathsf{G}(A_1) = \{(x, A_1 x)^t \in \mathcal{H}_1 \oplus \mathcal{H}_2 \,;\, x \in \mathsf{D}(A_1)\}$$
$$I\mathsf{G}(A_2) = \{(-A_2 y, y)^t \in \mathcal{H}_1 \oplus \mathcal{H}_2 \,;\, y \in \mathsf{D}(A_2)\}$$

are orthogonal complements in $\mathcal{H}_1 \oplus \mathcal{H}_2$, where $I(x,y)^t := (-y, x)^t$. In particular both A_1 and A_2 are closed, densely defined operators, and if $x \in \mathsf{D}(A_1)$ and $y \in \mathsf{D}(A_2)$, then $(x, A_2 y) = (A_1 x, y)$.

Given a closed, densely defined operator A in $\mathcal{H}$, we define $\mathsf{G}(A^*) := (I\mathsf{G}(A))^\perp$. Since A is densely defined, $\mathsf{G}(A^*)$ is the graph of a closed linear operator A^*, and since A is closed it follows that A^* is densely defined. We say that A^* is the (maximal) adjoint operator of A.

A fundamental result for adjoint operators is the following, which for example can be found in Kato [13].

Theorem 3.3. *Let A and A^* be adjoint closed, densely defined Hilbert space operators. Then $\mathsf{R}(A)^\perp = \mathsf{N}(A^*)$ and $\mathsf{R}(A^*)^\perp = \mathsf{N}(A)$. Moreover, $\mathsf{R}(A)$ is closed if and only if $\mathsf{R}(A^*)$ is closed.*

Corollary 3.4. *If Γ is a nilpotent operator, then so is Γ^*.*

Note that a nilpotent operator acts

$$\Gamma : \mathsf{N}(\Gamma)^\perp = \overline{\mathsf{R}(\Gamma^*)} \longrightarrow \mathsf{R}(\Gamma) \subset \mathsf{N}(\Gamma),$$

where the restriction of Γ is injective. Thus $\mathsf{N}(\Gamma)$ is at least "half" of $\mathcal{H}$.

Proposition 3.5. *Let Γ be a nilpotent operator in a Hilbert space $\mathcal{H}$, with adjoint Γ^*. For each $\alpha \in \mathbf{C}$ with $|\alpha| = 1$, let the corresponding* swapping operator *be*

$\Pi_\alpha := \Gamma + \alpha\Gamma^*$ *with domain* $D(\Pi_\alpha) := D(\Gamma) \cap D(\Gamma^*)$. *Then we have* $\mathcal{H}$*-orthogonal decompositions*

$$\mathcal{H} = \overline{R(\Gamma^*)} \oplus \big(N(\Gamma^*) \cap N(\Gamma)\big) \oplus \overline{R(\Gamma)}, \tag{14}$$

$$D(\Gamma) = \big(D(\Gamma) \cap \overline{R(\Gamma^*)}\big) \oplus \big(N(\Gamma^*) \cap N(\Gamma)\big) \oplus \overline{R(\Gamma)}, \tag{15}$$

$$D(\Gamma^*) = \overline{R(\Gamma^*)} \oplus \big(N(\Gamma^*) \cap N(\Gamma)\big) \oplus \big(D(\Gamma^*) \cap \overline{R(\Gamma)}\big), \tag{16}$$

$$D(\Pi_\alpha) = \big(D(\Gamma) \cap \overline{R(\Gamma^*)}\big) \oplus \big(N(\Gamma^*) \cap N(\Gamma)\big) \oplus \big(D(\Gamma^*) \cap \overline{R(\Gamma)}\big). \tag{17}$$

The swapping operator is a closed, densely defined operator in $\mathcal{H}$ *with null space* $N(\Pi_\alpha) = N(\Gamma) \cap N(\Gamma^*)$ *and range* $R(\Pi_\alpha) = R(\Gamma) \oplus R(\Gamma^*)$. *The adjoint of* Π_α *is* $\alpha^c \Pi_\alpha$. *Thus* Π_1 *is a self adjoint operator and* Π_{-1} *is a skew adjoint operator.*

The swapping operator Π_α *is unitary equivalent to both* $\sqrt{\alpha}\Pi_1$ *and* $-\Pi_\alpha$. *In particular, the spectrum* $\sigma(\Pi_\alpha)$ *is contained in the line* $\sqrt{\alpha}\mathbf{R}$ *and it is symmetric with respect to* 0.

Proof. From Theorem 3.3 we obtain the two orthogonal splittings

$$\mathcal{H} = \overline{\mathsf{R}(\Gamma^*)} \oplus \mathsf{N}(\Gamma) = \mathsf{N}(\Gamma^*) \oplus \overline{\mathsf{R}(\Gamma)}.$$

Using (13), we get inclusions $\overline{\mathsf{R}(\Gamma^*)} \subset \mathsf{N}(\Gamma^*)$ and $\overline{\mathsf{R}(\Gamma)} \subset \mathsf{N}(\Gamma)$. Therefore taking the intersection of the two splittings gives (14). Now write

$$\begin{aligned}
\mathcal{H}_1 &:= \overline{\mathsf{R}(\Gamma^*)} = \mathsf{N}(\Gamma)^\perp \approx \mathcal{H}/\mathsf{N}(\Gamma),\\
\mathcal{H}_0 &:= \mathsf{N}(\Gamma^*) \cap \mathsf{N}(\Gamma) \approx \mathsf{N}(\Gamma)/\overline{\mathsf{R}(\Gamma)} \approx \mathsf{N}(\Gamma^*)/\overline{\mathsf{R}(\Gamma^*)},\\
\mathcal{H}_2 &:= \overline{\mathsf{R}(\Gamma)} = \mathsf{N}(\Gamma^*)^\perp \approx \mathcal{H}/\mathsf{N}(\Gamma^*),
\end{aligned}$$

and let P_i denote the orthogonal projection onto $\mathcal{H}_i$. To prove the decomposition (15), note that the inclusion $\supset$ is trivial. For the opposite inclusion, decompose $x \in \mathsf{D}(\Gamma)$ with (14) as $x = x_1 + x_0 + x_2$, where $x_i \in \mathcal{H}_i$. Since $x_2 \in \overline{\mathsf{R}(\Gamma)} \subset \mathsf{D}(\Gamma)$ and $x_0 \in \mathsf{N}(\Gamma^*) \cap \mathsf{N}(\Gamma) \subset \mathsf{D}(\Gamma)$ we deduce that $x_1 = x - x_0 - x_2 \in \mathsf{D}(\Gamma)$.

The decomposition of $\mathsf{D}(\Gamma^*)$ follows similarly, and taking the intersection of (15) and (16) yields (17).

To determine $\mathsf{N}(\Pi_\alpha)$, note that the inclusion $\supset$ is trivial and $\subset$ follows since $\mathsf{R}(\Gamma)$ and $\mathsf{R}(\Gamma^*)$ are orthogonal. For $\mathsf{R}(\Pi_\alpha)$, the inclusion $\subset$ is trivial. On the other hand if $y = \Gamma x_1 + \alpha\Gamma^* x_2$, then $y = \Pi_\alpha(P_1 x_1 + P_2 x_2)$ where $P_1 x_1 + P_2 x_2 \in \mathsf{D}(\Gamma) \cap \mathsf{D}(\Gamma^*)$.

We now show that Π_α and $\alpha^c\Pi_\alpha$ are maximal adjoint operators. First note that $(\Pi_\alpha x, y) = (x, \alpha^c\Pi_\alpha y)$ if $x, y \in \mathsf{D}(\Pi_\alpha)$, i.e. $\mathsf{G}(\Pi_\alpha)$ and $I\mathsf{G}(\alpha^c\Pi_\alpha)$ are orthogonal. To prove that $\mathsf{G}(\Pi_\alpha)^\perp \subset I\mathsf{G}(\alpha^c\Pi_\alpha)$, let $(-z, y)^t \in \mathsf{G}(\Pi_\alpha)^\perp$ and decompose $y = y_1 + y_0 + y_2$ with (14). We see that $y_1 \in \overline{\mathsf{R}(\Gamma^*)} \subset \mathsf{D}(\Gamma^*)$, $y_0 \in \mathsf{N}(\Gamma^*) \cap \mathsf{N}(\Gamma) \subset \mathsf{D}(\Gamma^*) \cap \mathsf{D}(\Gamma)$ and $y_2 \in \overline{\mathsf{R}(\Gamma)} \subset \mathsf{D}(\Gamma)$. To verify that $y_2 \in \mathsf{D}(\Gamma^*)$, let $x \in \mathsf{D}(\Gamma)$ and calculate

$$(\Gamma x, y_2) = (\Pi_\alpha P_1 x, y_2) = (\Pi_\alpha P_1 x, y) = (x, P_1 z).$$

This proves that $(y_2, P_1 z)^t \in (I\mathsf{G}(\Gamma))^\perp = \mathsf{G}(\Gamma^*)$. Similarly it follows that $y_1 \in \mathsf{D}(\Gamma)$ and thus $y \in \mathsf{D}(\Pi_\alpha)$.

That Π_α is closed and densely defined now follows from the adjointness of Π_α and $\alpha^c\Pi_\alpha$ (or can be verified directly). Furthermore, note that for any $\beta \in \mathbf{C}$, $|\beta| = 1$ we have

$$\Pi_\alpha(P_1 + P_0 + \beta^c P_2) = (P_1 + P_0 + \beta^c P_2)\beta\Pi_{(\beta^c)^2\alpha}.$$

The case $\beta = -1$ shows that Π_α and $-\Pi_\alpha$ are unitary equivalent, and the case $\beta = \sqrt{\alpha}$ shows that Π_α and $\sqrt{\alpha}\Pi_1$ are unitary equivalent. □

Remark 3.6. We have chosen the name "swapping operator" since we have the following mapping diagram

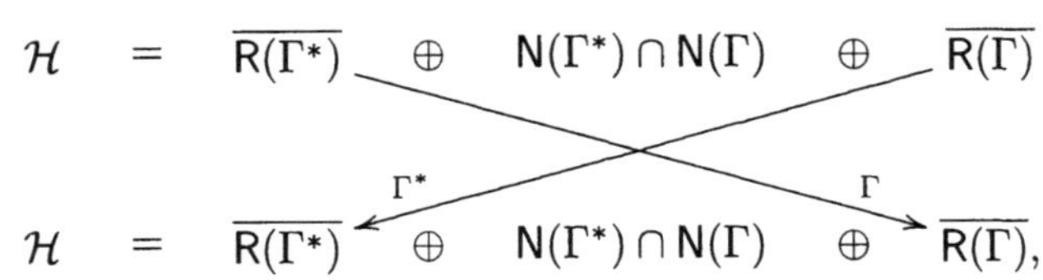

in which Π_α swaps the subspaces $\overline{\mathsf{R}(\Gamma^*)}$ and $\overline{\mathsf{R}(\Gamma)}$.

We now investigate when a nilpotent operator is maximal in the sense that it is "half elliptic". More precisely we make the following definitions.

Definition 3.7. Let $A : \mathcal{H}_1 \to \mathcal{H}_2$ be a closed, densely defined operator between Hilbert spaces. We say that A is a Fredholm operator if the null space $\mathsf{N}(A)$ and the cokernel $\mathcal{H}_2/\mathsf{R}(A)$ are finite dimensional and the range $\mathsf{R}(A)$ is closed (which follows from $\dim(\mathcal{H}_2/\mathsf{R}(A)) < \infty$).

Proposition 3.8. *Let $A : \mathcal{H}_1 \to \mathcal{H}_2$ be a closed, densely defined operator between Hilbert spaces. Then A is a Fredholm operator if and only if there exist bounded operators $T_1, T_2 : \mathcal{H}_2 \to \mathcal{H}_1$ and compact operators $K_1 : \mathcal{H}_1 \to \mathcal{H}_1$ and $K_2 : \mathcal{H}_2 \to \mathcal{H}_2$ such that $\mathsf{R}(T_2) \subset \mathsf{D}(A)$ and*

$$\begin{aligned} T_1 A &= I + K_1 && \text{on } \mathsf{D}(A) \subset \mathcal{H}_1, \\ AT_2 &= I + K_2 && \text{on } \mathcal{H}_2. \end{aligned}$$

In this case, the following are equivalent.

- *The embedding $\mathsf{D}(A) \hookrightarrow \mathcal{H}_1$ is compact.*
- *The left inverse T_1 is compact.*
- *The right inverse T_2 is compact.*

The Fredholm inverses T_1 and T_2 satisfies $T_1 + T_1K_2 = T_2 + K_1T_2$.

Two references on Fredholm operator theory are Schechter [22] and Kato [13].

Definition 3.9. Let $A : \mathcal{H}_1 \to \mathcal{H}_2$ be a Fredholm operator between Hilbert spaces. We say that A is *diffuse* if its domain $\mathsf{D}(A)$ is compact in $\mathcal{H}_1$, or equivalently if it has a compact Fredholm inverse.

Definition 3.10. Let Γ be a nilpotent operator in a Hilbert space $\mathcal{H}$. We say that Γ is a *Fredholm-nilpotent operator* if the reduced operator

$$\widetilde{\Gamma} : \mathcal{H}/\mathsf{N}(\Gamma) \longrightarrow \mathsf{N}(\Gamma)$$

with domain $\mathsf{D}(\widetilde{\Gamma}) := \mathsf{D}(\Gamma)/\mathsf{N}(\Gamma)$ is a Fredholm operator. If $\widetilde{\Gamma}$ is a diffuse Fredholm operator, then Γ is said to be a *diffuse Fredholm-nilpotent operator.*

Proposition 3.11. *Let Γ and Π_α be as in Proposition* 3.5. *Then the following are equivalent.*

(i) Γ *is a* Fredholm-nilpotent operator.
(i′) Γ^* *is a* Fredholm-nilpotent operator.
(ii) Π_α *is a* Fredholm operator.

When this holds, Γ induces a Hodge type decomposition (or splitting) *of $\mathcal{H}$, i.e.*

$$\mathcal{H} = \mathsf{R}(\Gamma^*) \oplus (\mathsf{N}(\Gamma^*) \cap \mathsf{N}(\Gamma)) \oplus \mathsf{R}(\Gamma),$$

where the ranges $\mathsf{R}(\Gamma^*)$ and $\mathsf{R}(\Gamma)$ are closed and $\mathsf{N}(\Gamma^*)\cap\mathsf{N}(\Gamma)$ is finite dimensional. If in addition $\mathsf{N}(\Gamma^*) \cap \mathsf{N}(\Gamma) = \{0\}$, then the splitting is said to be exact.

The equivalence of (i), (i′) *and* (ii) *remains true if "Fredholm(-nilpotent) operator" is replaced by "diffuse Fredholm(-nilpotent) operator". In this case, we also have the following.*

(iii) *The spectrum $\sigma(\Pi_\alpha)$ is a discrete set consisting of eigenvalues only.*
(iv) *If Γ_0 is a bounded, nilpotent operator such that $\Gamma\Gamma_0 + \Gamma_0\Gamma = 0$ on $\mathsf{D}(\Gamma) = \mathsf{D}(\Gamma + \Gamma_0)$, then the perturbed operator $\Gamma + \Gamma_0$ is also a diffuse Fredholm-nilpotent operator.*

Proof. Split $\mathcal{H} = \mathcal{H}_1 \oplus \mathcal{H}_0 \oplus \mathcal{H}_2$ as in the proof of Proposition 3.5.

Theorem 3.3 shows that (i) and (i′) are equivalent since

$$\mathcal{H}_0 = \mathsf{N}(\Gamma) \cap \mathsf{R}(\Gamma)^\perp = \mathsf{N}(\Gamma^*) \cap \mathsf{R}(\Gamma^*)^\perp.$$

Furthermore, since $\Gamma : \mathcal{H}_1 \to \mathcal{H}_2$ and $\Gamma^* : \mathcal{H}_2 \to \mathcal{H}_1$ are adjoint operators, it follows that $(\Gamma^*)^{-1} = (\Gamma^{-1})^* : \mathcal{H}_1 \to \mathcal{H}_2$ is compact if and only if $\Gamma^{-1} : \mathcal{H}_2 \to \mathcal{H}_1$ is. Therefore Γ is a diffuse Fredholm-nilpotent operator if and only if Γ^* is.

Note that since $\mathsf{R}(\Gamma)$ and $\mathsf{R}(\Gamma^*)$ are orthogonal, they are both closed if and only if $\mathsf{R}(\Pi_\alpha) = \mathsf{R}(\Gamma^*) \oplus \mathsf{R}(\Gamma)$ is closed. As $\mathsf{N}(\Pi_\alpha) = \mathcal{H}_0 = \mathcal{H}/\overline{\mathsf{R}(\Pi_\alpha)}$ it follows that (i) and (i′) are equivalent with (ii). Moreover $\mathsf{D}(\Pi_\alpha) = \mathsf{D}(\widetilde{\Gamma}^*) \oplus \mathcal{H}_0 \oplus \mathsf{D}(\widetilde{\Gamma})$, so Π_α is a diffuse Fredholm operator if and only if both Γ and Γ^* are diffuse Fredholm-nilpotent operators.

The discreteness result (iii) follows from the identity

$$(\sqrt{-\alpha} - \lambda)^{-1} - (\sqrt{-\alpha} - \Pi_\alpha)^{-1} = (\sqrt{-\alpha} - \lambda)^{-1}(\lambda - \Pi_\alpha)(\sqrt{-\alpha} - \Pi_\alpha)^{-1}$$

which shows that $(\lambda - \Pi_\alpha)$ fails to be invertible if and only if $(\sqrt{-\alpha} - \lambda)^{-1} \in \sigma(K)$. But since $K := (\sqrt{-\alpha} - \Pi_\alpha)^{-1}$ is compact, its spectrum is discrete.

To prove (iv), let Π_α and Π'_α be swapping operators corresponding to Γ and $\Gamma + \Gamma_0$. Then $\mathsf{D}(\Pi'_\alpha) = \mathsf{D}(\Pi_\alpha)$ is compactly embedded in $\mathcal{H}$. Lemma 3.14 below now shows that $\Gamma + \Gamma_0$ is a diffuse Fredholm-nilpotent operator. □

Remark 3.12. An important observation here is that the statement (i) is independent of which Hilbert norm on $\mathcal{H}$ we are using (as long as it induces the same topology), whereas in (i′) and (ii), the adjoint operator Γ^* and Π_α depends on the scalar product.

We finish this section with two techniques to establish Fredholm-nilpotence of a given nilpotent operator Γ.

Lemma 3.13. *Let $\mathcal{H}$ and $\mathcal{H}_0$ be two Hilbert spaces and consider the diagram*

$$\begin{array}{ccc} \mathcal{H} & \overset{T}{\underset{S}{\leftrightarrows}} & \mathcal{H}_0 \\ \downarrow{\scriptstyle \Gamma} & & \downarrow{\scriptstyle \Gamma_0} \\ \mathcal{H} & \overset{T}{\underset{S}{\leftrightarrows}} & \mathcal{H}_0, \end{array}$$

where Γ and Γ_0 are closed, densely defined operators in $\mathcal{H}$ and $\mathcal{H}_0$ respectively and T and S are bounded maps such that $TS = I_{\mathcal{H}}$. If $T\Gamma_0 \subset \Gamma T$ and $S\Gamma \subset \Gamma_0 S$, then we have the following.

(i) *If Γ_0 is a nilpotent operator, then so is Γ.*
(ii) *If Γ_0 is a Fredholm-nilpotent operator, then so is Γ.*
(iii) *If Γ_0 is a diffuse Fredholm-nilpotent operator, then so is Γ.*

The proof of this intertwining lemma is straightforward and we omit it.

Lemma 3.14. *Let Π_α be a swapping operator as in Proposition* 3.5. *If the embedding $\mathsf{D}(\Pi_\alpha) \hookrightarrow \mathcal{H}$ is compact, then Π_α is a diffuse Fredholm operator with index zero.*

Proof. Consider the operators

$$\lambda I - \Pi_\alpha : \mathsf{D}(\Pi_\alpha) \longrightarrow \mathcal{H}. \tag{18}$$

Since $\sigma(\Pi_\alpha) \subset \sqrt{\alpha}\mathbf{R}$ by Proposition 3.5, (18) is an isomorphism when $\lambda \notin \sqrt{\alpha}\mathbf{R}$. Now observe that $\lambda I : \mathsf{D}(\Pi_\alpha) \to \mathcal{H}$ is a compact operator. Thus $\Pi_\alpha : \mathsf{D}(\Pi_\alpha) \to \mathcal{H}$ is a Fredholm operator with index zero. □

4. Hodge decompositions for d and δ

In this section we apply the general theory for nilpotent operators from Section 3 to the following d and δ operators in a bounded weakly Lipschitz domain Ω.

Definition 4.1. (i) Let d_Ω and δ_Ω be the closed, nilpotent d and δ operators (without boundary conditions) in $L_2(\Omega; \wedge)$ with natural domains, i.e.

$$\mathsf{D}(d_\Omega) := \{F \in L_2(\Omega; \wedge) \,;\, dF \in L_2(\Omega; \wedge)\},$$

and similarly for δ_Ω.

(ii) Let $d_{\overline{\Omega}}$ (d with normal boundary conditions) and $\delta_{\overline{\Omega}}$ (δ with tangential boundary conditions) be the closed, nilpotent d and δ operators in Ω with domains

$$\mathsf{D}(d_{\overline{\Omega}}) := \{F \in L_2(\Omega; \wedge) \; ; \; d(F_z) \in L_2(\mathbf{R}^n; \wedge)\},$$
$$\mathsf{D}(\delta_{\overline{\Omega}}) := \{F \in L_2(\Omega; \wedge) \; ; \; \delta(F_z) \in L_2(\mathbf{R}^n; \wedge)\},$$

where $F_z \in L_2(\mathbf{R}^n; \wedge)$ denotes the zero-extension of F to $\mathbf{R}^n$.

Remark 4.2. If $F \in \mathsf{D}(d_{\overline{\Omega}})$, then F is normal to Σ. The nilpotence of $d_{\overline{\Omega}}$ shows that not only is $d_{\overline{\Omega}}F$ curl free, it is also normal to Σ. Similarly, if $F \in \mathsf{D}(\delta_{\overline{\Omega}})$, then F is tangential to Σ. The nilpotence of $\delta_{\overline{\Omega}}$ shows that not only is $\delta_{\overline{\Omega}}F$ divergence free, it is also tangential to Σ.

Examples of nilpotent operators considered in this paper are, for each wave number $k \in \mathbf{C}$, the four operators

$$d_{k,\Omega} = d_\Omega + ke_0\wedge, \qquad \delta_{k,\Omega} = \delta_\Omega + ke_0 \lrcorner,$$
$$d_{k,\overline{\Omega}} = d_{\overline{\Omega}} + ke_0\wedge, \qquad \delta_{k,\overline{\Omega}} = \delta_{\overline{\Omega}} + ke_0 \lrcorner \, .$$

Obviously we have $d_{k,\overline{\Omega}} \subset d_{k,\Omega}$ and $\delta_{k,\overline{\Omega}} \subset \delta_{k,\Omega}$.

Proposition 4.3. *The operators d_Ω and $d_{\overline{\Omega}}$ have cores (i.e. a subset of the domain which is dense in graph norm)*

$$C_0^\infty(\mathbf{R}^n; \wedge)|_\Omega \subset D(d_\Omega), \qquad C_0^\infty(\Omega; \wedge) \subset D(d_{\overline{\Omega}})$$

respectively. In particular, the inclusions

$$d(C_0^\infty(\mathbf{R}^n; \wedge)|_\Omega) \subset R(d_\Omega), \qquad d(C_0^\infty(\Omega; \wedge)) \subset R(d_{\overline{\Omega}})$$

are dense. We also have dense subspaces

$$\{F|_\Omega \; ; \; F \in C_0^\infty(\mathbf{R}^n; \wedge), \, \operatorname{supp} dF \subset\subset \Omega^-\} \subset N(d_\Omega),$$
$$\{F \in C_0^\infty(\Omega; \wedge) \; ; \; dF = 0\} \subset N(d_{\overline{\Omega}}).$$

The same holds true when d is replaced by δ.

Before giving the proof, we note the following important corollary.

Corollary 4.4. *The two operators $d_{\overline{\Omega}}$ and $-\delta_\Omega$ are adjoint in the sense of Definition 3.2 and so are d_Ω and $-\delta_{\overline{\Omega}}$. For the zero order perturbations we have $d^*_{k,\overline{\Omega}} = -\delta_{-k^c,\Omega}$ and $d^*_{k,\Omega} = -\delta_{-k^c,\overline{\Omega}}$.*

Proof. Consider the first pair. By Definition 3.2 we need to prove that $\mathsf{G}(d^*_{\overline{\Omega}}) = \mathsf{G}(-\delta_\Omega)$, where $\mathsf{G}(d^*_{\overline{\Omega}}) = I\mathsf{G}(d_{\overline{\Omega}})^\perp$.

To show $d^*_{\overline{\Omega}} \subset -\delta_\Omega$, let $(U, F)^t \in \mathsf{G}(d^*_{\overline{\Omega}})$. Then in particular

$$\int_\Omega \langle U, d\Phi^c \rangle = \int_\Omega \langle F, \Phi^c \rangle \quad \text{for all } \Phi \in C_0^\infty(\Omega) \subset \mathsf{D}(d_{\overline{\Omega}}).$$

Thus $-\delta_\Omega U = F \in L_2(\Omega; \wedge)$ in distribution sense, which proves $(U, F)^t \in \mathsf{G}(-\delta_\Omega)$.

Conversely, to show $d^*_{\overline{\Omega}} \supset -\delta_\Omega$, by Proposition 4.3 it suffices to prove that $I\mathsf{G}(d_{\overline{\Omega}})$ and $\{(\Phi, -\delta_\Omega\Phi)^t \; ; \; \Phi \in C_0^\infty(\mathbf{R}^n;\wedge)|_\Omega\}$ are orthogonal. If $U \in \mathsf{D}(d_{\overline{\Omega}})$ and $\Phi \in C_0^\infty(\mathbf{R}^n;\wedge)|_\Omega$, we get

$$\int_\Omega \langle \Phi, d_{\overline{\Omega}} U^c\rangle = \int_{\mathbf{R}^n} \langle \Phi, dU^c_z\rangle = \int_{\mathbf{R}^n} \langle -\delta\Phi, U^c_z\rangle = \int_\Omega \langle -\delta_\Omega \Phi, U^c\rangle.$$

This shows that $d_{\overline{\Omega}}$ and $-\delta_\Omega$ are adjoint. The adjointness of d_Ω and $-\delta_{\overline{\Omega}}$ follows similarly. Moreover, since $(A+T)^* = A^* + T^*$ whenever A is a closed, densely defined operator and T is a bounded operator, this proves the rest of the corollary. □

To prove Proposition 4.3, we use *Lie flows* $t \mapsto \alpha_t^*$ and $t \mapsto \tilde{\alpha}_{t*}^{-1}$ constructed as follows.

Lemma 4.5. *There exists a family* $\alpha_t : \mathbf{R}^n \to \mathbf{R}^n$ *of bilipschitz maps (all being identity outside a compact set),* $|t| < T$, *with the following properties.*

$$\begin{aligned} \overline{\alpha_t(\Omega)} &\subset \Omega, && 0 < t < T, \\ \alpha_t(\Omega) &\supset \overline{\Omega}, && -T < t < 0, \\ \|\alpha_t^* F - F\|_{L_2(\mathbf{R}^n;\wedge)} &\longrightarrow 0, && t \longrightarrow 0, \quad F \in L_2(\mathbf{R}^n;\wedge), \\ \|\tilde{\alpha}_{t*}^{-1} F - F\|_{L_2(\mathbf{R}^n;\wedge)} &\longrightarrow 0, && t \longrightarrow 0, \quad F \in L_2(\mathbf{R}^n;\wedge). \end{aligned}$$

Proof. Let $\overline{\Omega} \subset \cup_{j=0}^N V_j$ and $\rho_j : \mathbf{R}^n \to \mathbf{R}^n$, $j = 1 \ldots N$, be the local bilipschitz parametrisations from Definition 2.1 and let $V_0 \subset\subset \Omega$ be contained in the interior. Let $B_1 := B(0,1) \subset B_2 := B(0,2)$ be concentric balls. We may assume that $\Sigma \subset \cup_1^N \rho_j B_1$ and that $\rho_j B_2 \subset\subset V_j$, $j = 1 \ldots N$. Let $\eta \in C_0^\infty(\mathbf{R}^n)$ be such that $0 \le \eta \le 1$, $\eta|_{B_1} = 1$ and $\eta|_{\mathbf{R}^n \setminus B_2} = 0$. Define for $|t| \le T$ the translation map

$$\beta_t : \mathbf{R}^n \longrightarrow \mathbf{R}^n : x \longmapsto x + \eta(|x|) t e_n,$$

let $\alpha_t^j := \rho_j \circ \beta_t \circ \rho_j^{-1}$ and extend to identity outside V_j and let $\alpha_t := \alpha_t^N \circ \ldots \circ \alpha_t^1$. Obviously the required mapping properties for α_t holds.

For the Lie pullback flow, if $\alpha_t^* F = (\alpha_t^1)^* G$ where $G := (\alpha_t^2)^* \ldots (\alpha_t^N)^* F$ then

$$\begin{aligned} \|\alpha_t^* F - F\|_{L_2(\mathbf{R}^n;\wedge)} &\le \|(\alpha_t^1)^* G - F\|_{L_2(V_1;\wedge)} + \|(\alpha_t^1)^* G - F\|_{L_2(V_1^c;\wedge)} \\ &\le \|(\alpha_t^1)^* G - G\|_{L_2(V_1;\wedge)} + \|G - F\|_{L_2(V_1;\wedge)} + \|G - F\|_{L_2(V_1^c;\wedge)}. \end{aligned}$$

Thus it suffices to show that $\|(\alpha_t^j)^* F - F\|_{L_2(V_j;\wedge)} \to 0$. But (essentially) since translation is L_2 continuous, this follows from

$$\|(\alpha_t^j)^* F - F\|_{L_2(V_j;\wedge)} \lesssim \|(\beta_t)^*(\rho_j^* F) - (\rho_j^* F)\|_{L_2(U_j;\wedge)} \longrightarrow 0.$$

The proof of the L_2-continuity of the reduced push forward flow is similar. □

Proof of Proposition 4.3. Let $\eta_s(x) := s^{-n}\eta(x/s)$ be a mollifier. For $F \in \mathsf{D}(d_\Omega)$, let $0 < s << t$ and define approximating fields

$$F_{s,t} := \eta_s * (\alpha_t^* F_z).$$

Then $\|F_{s,t} - F\|_{\mathsf{D}(d_\Omega)} \approx \|F_{s,t} - F\|_{L_2} + \|(dF)_{s,t} - dF\|_{L_2} \to 0$ as $s, t \to 0$. Furthermore, if $d_\Omega F = 0$ and $0 < s << t$, then $dF_{s,t} = 0$ in a neighbourhood of $\overline{\Omega}$, due to Proposition 2.6.

On the other hand for $F \in \mathsf{D}(d_{\overline{\Omega}})$, let $0 < s << -t$ and define approximating fields $F_{s,t}$ as above. Then $F_{s,t} \in C_0^\infty(\Omega; \wedge)$ if $0 < s << -t$ and $\|F_{s,t} - F\|_{\mathsf{D}(d_{\overline{\Omega}})} \to 0$ as $s, t \to 0$. Furthermore, if $d_{\overline{\Omega}} F = 0$ then $d_{\overline{\Omega}} F_{s,t} = 0$. □

Example 4.6. Using Proposition 3.5, we form the corresponding swapping operators Π_{-1}.

(i) If $\Gamma = d_{\overline{\Omega}}$, then the (Hodge–)Dirac operator on Ω with *normal boundary conditions* is

$$\mathbf{D}_{\Omega^\perp} := d_{\overline{\Omega}} + \delta_\Omega.$$

Note that $\mathbf{D}^2_{\Omega^\perp} = d_{\overline{\Omega}}\delta_\Omega + \delta_\Omega d_{\overline{\Omega}}$ is the Hodge–Laplace operator with relative (generalised Dirichlet) boundary conditions. For a scalar function $U : \Omega \to \wedge^0$, we have $\mathbf{D}^2_{\Omega^\perp} U = \delta_\Omega d_{\overline{\Omega}} U$, and $U \in \mathsf{D}(d_{\overline{\Omega}})$ incorporates the boundary condition $U|_\Sigma = 0$ since all scalars are tangential.

(ii) If $\Gamma = d_\Omega$, then the (Hodge–)Dirac operator on Ω with *tangential boundary conditions* is

$$\mathbf{D}_{\Omega^\parallel} := d_\Omega + \delta_{\overline{\Omega}}.$$

Here the Hodge–Laplace operator with absolute (generalised Neumann) boundary conditions is $\mathbf{D}^2_{\Omega^\parallel} = d_\Omega \delta_{\overline{\Omega}} + \delta_{\overline{\Omega}} d_\Omega$. Note that for a scalar function $U : \Omega \to \wedge^0$, we have $\mathbf{D}^2_{\Omega^\perp} U = \delta_{\overline{\Omega}} d_\Omega U$, and $d_\Omega U \in \mathsf{D}(\delta_{\overline{\Omega}})$ incorporates the boundary condition $\frac{\partial U}{\partial \nu} = \langle \nu, \nabla U|_\Sigma \rangle = 0$.

(iii) As is well known, the null spaces

$$\mathsf{N}(\mathbf{D}_{\Omega^\perp}) = \mathsf{N}(d_{\overline{\Omega}}) \cap \mathsf{N}(\delta_\Omega) = \mathsf{N}(\mathbf{D}^2_{\Omega^\perp}),$$
$$\mathsf{N}(\mathbf{D}_{\Omega^\parallel}) = \mathsf{N}(d_\Omega) \cap \mathsf{N}(\delta_{\overline{\Omega}}) = \mathsf{N}(\mathbf{D}^2_{\Omega^\parallel})$$

of the Dirac operators $\mathbf{D}_{\Omega^\perp}$ and $\mathbf{D}_{\Omega^\parallel}$ can be identified with the de Rham cohomology spaces of Ω with normal (relative) and tangential (absolute) boundary conditions, and are thus determined by the global topology of Ω. However, we are here mainly interested in how the local regularity of the boundary Σ influences the Fredholm properties of the Dirac operators $\mathbf{D}_{\Omega^\perp}$ and $\mathbf{D}_{\Omega^\parallel}$.

We now turn to the proof of Theorem 1.3 and show how to reduce to the case $\Omega = B$, where B is the unit ball in $\mathbf{R}^n$. The proof that $\mathbf{D}_{B^\perp}$ and $\mathbf{D}_{\Omega^\parallel}$ are diffuse Fredholm operators is deferred to the end of this section. It follows from either Theorem 4.10, Theorem 4.13 (combined with Lemma 3.14) or Theorem 4.17 (combined with Proposition 3.11).

Proof of Theorem 1.3. (i) We first consider the unperturbed case $k_1 = k_2 = 0$. By Proposition 3.11 it suffices to show that $\mathbf{D}_{\Omega^\perp}$ is a diffuse Fredholm operator. Using Definition 2.1 we see that there exist bilipschitz maps $\rho_j : B \to \Omega_j$, $j = 1, \ldots, N$, where B denotes the open unit ball in $\mathbf{R}^n$, such that $\Omega = \bigcup_{j=1}^N \Omega_j$. Furthermore we may assume that ρ_j extends to a bilipschitz map between slightly larger open sets. Choose a smooth partition of unity $\{\eta_j\}$ such that $\operatorname{supp}\eta_j \subset\subset \mathbf{R}^n \setminus \overline{(\Omega\setminus\Omega_j)}$ and $\sum \eta_j^2 = 1$ on Ω.

Assuming that $\mathbf{D}_{B^\perp}$ is a diffuse Fredholm operator, it follows from Proposition 3.11 that $d_{\overline{B}}$ is a diffuse Fredholm-nilpotent operator. We may now apply Lemma 3.13 with $\mathcal{H} = L_2(\Omega_j;\wedge)$, $\mathcal{H}_0 = L_2(B;\wedge)$, $A = d_{\overline{\Omega}_j}$, $A_0 = d_{\overline{B}}$, $T = (\rho_j^{-1})^*$ and $S = \rho_j^* = T^{-1}$, since Proposition 2.6 proves that T and S intertwine A and A_0. This shows that $d_{\overline{\Omega}_j}$ is a diffuse Fredholm-nilpotent operator.

Applying Proposition 3.11 again with $\Gamma = d_{\overline{\Omega}_j}$ shows that $\mathbf{D}_{\Omega_j^\perp}$ is a diffuse Fredholm operator. Localising, we can now prove that the Dirac operator $\mathbf{D}_{\Omega^\perp}$ is a diffuse Fredholm operator. Indeed, if T_j are compact Fredholm inverses to $\mathbf{D}_{\Omega_j^\perp}$ respectively as in Proposition 3.8, then a compact Fredholm inverse to $\mathbf{D}_{\Omega^\perp}$ is

$$T(F) := \sum_j \eta_j T_j(\eta_j F).$$

Similarly one can show that the Dirac operator $\mathbf{D}_{\Omega^\parallel}$ is a diffuse Fredholm operator.

(ii) To prove that the map (2) is a diffuse Fredholm map for general k_1 and k_2, note that (i) above and Proposition 3.11(iv) with $\Gamma + \Gamma_0 = d_{-k_2^c,\overline{\Omega}}$ and $\Gamma^* + \Gamma_0^* = -\delta_{k_2,\Omega}$ shows that we have a Hodge decomposition

$$L_2(\Omega;\wedge) = \mathsf{R}(\delta_{k_2,\Omega}) \oplus \big(\mathsf{N}(\delta_{k_2,\Omega}) \cap \mathsf{N}(d_{-k_2^c,\overline{\Omega}})\big) \oplus \mathsf{R}(d_{-k_2^c,\overline{\Omega}})$$

and that $\delta_{k_2,\Omega} : \mathsf{N}(d_{-k_2^c,\overline{\Omega}}) \to \mathsf{N}(\delta_{k_2,\Omega})$ is a diffuse Fredholm map. In particular $\mathsf{N}(\delta_{k_2,\Omega})/\mathsf{R}(\delta_{k_2,\Omega})$ is finite dimensional, and similarly $\mathsf{N}(d_{k_1,\overline{\Omega}})/\mathsf{R}(d_{k_1,\overline{\Omega}})$ is finite dimensional. Thus it suffices to prove that

$$\delta_{k_2,\Omega} : \mathsf{R}(d_{k_1,\overline{\Omega}}) \longrightarrow \mathsf{R}(\delta_{k_2,\Omega})$$

is a diffuse Fredholm map. Consider the following diagram

$$\begin{array}{ccccc}
\mathsf{R}(d_{k_1,\overline{\Omega}}) & \xrightarrow{P_1} & \mathsf{R}(d_{-k_2^c,\overline{\Omega}}) & \xrightarrow{\delta_{k_2,\Omega}} & \mathsf{R}(\delta_{k_2,\Omega}) \\
& \searrow^{P_2} & \oplus & & \\
& & \mathsf{N}(\delta_{k_2,\Omega}) & \xrightarrow{\delta_{k_2,\Omega}} & 0,
\end{array}$$

where P_1 and P_2 denotes the associated orthogonal projections. To show that $\delta_{k_2,\Omega}P_1|_{\mathsf{R}(d_{k_1,\overline{\Omega}})} = \delta_{k_2,\Omega}|_{\mathsf{R}(d_{k_1,\overline{\Omega}})}$ is a diffuse Fredholm map, we first prove a priori estimates for $P_1|_{\mathsf{R}(d_{k_1,\overline{\Omega}})}$. Note that (i) above and Proposition 3.11(iv) show that any $F \in \mathsf{R}(d_{k_1,\overline{\Omega}})$ has a potential $F = d_{k_1,\overline{\Omega}}U$ where the map $F \mapsto U$ is compact. This gives

$$\|F\|^2 = \int_\Omega \langle P_1F, F^c\rangle + \langle P_2F, (d_{k_1,\overline{\Omega}}U)^c\rangle = \int_\Omega \langle P_1F, F^c\rangle - (k_1^c + k_2)\langle e_0 \lrcorner F, U^c\rangle.$$

Dividing by $\|F\|$ gives the a priori estimate $\|F\| \lesssim \|P_1F\| + \|U\|$. This shows that $\delta_{k_2,\Omega}(\mathsf{N}(d_{k_1,\overline{\Omega}}))$ is closed and that $\mathsf{N}(\delta_{k_2,\Omega}|_{\mathsf{N}(d_{k_1,\overline{\Omega}})})$ is finite dimensional. Now

Lemma 4.7 below shows that the cokernel $\mathsf{N}(\delta_{k_2,\Omega})\ominus\delta_{k_2,\Omega}\mathsf{N}(d_{k_1,\overline{\Omega}})$ is finite dimensional, which completes the proof. □

Lemma 4.7. *The deficiency indices of the maps* (2) *and* (3) *are*

$$\alpha(\delta_{k_2,\Omega}|_{\mathsf{N}(d_{k_1,\overline{\Omega}})}) = \dim(\mathsf{N}(\delta_{k_2,\Omega})\cap\mathsf{N}(d_{k_1,\overline{\Omega}}))$$
$$\beta(\delta_{k_2,\Omega}|_{\mathsf{N}(d_{k_1,\overline{\Omega}})}) = \dim(\mathsf{N}(\delta_{k_2,\Omega})\cap\mathsf{N}(d_{-k_2^c,\overline{\Omega}})) + \dim(\mathsf{R}(\delta_{-k_1^c,\Omega})\cap\mathsf{R}(d_{-k_2^c,\overline{\Omega}}))$$
$$\alpha(d_{k_1,\overline{\Omega}}|_{\mathsf{N}(\delta_{k_2,\Omega})}) = \dim(\mathsf{N}(\delta_{k_2,\Omega})\cap\mathsf{N}(d_{k_1,\overline{\Omega}}))$$
$$\beta(d_{k_1,\overline{\Omega}}|_{\mathsf{N}(\delta_{k_2,\Omega})}) = \dim(\mathsf{N}(\delta_{-k_2^c,\Omega})\cap\mathsf{N}(d_{k_1,\overline{\Omega}})) + \dim(\mathsf{R}(\delta_{-k_1^c,\Omega})\cap\mathsf{R}(d_{-k_2^c,\overline{\Omega}})).$$

For any k_1 and k_2 these indices are finite. Moreover, if the wave numbers are non zero and $\arg(k_1)+\arg(k_2)\neq 0$ *mod* 2π, *then* $\mathsf{N}(\delta_{k_2,\Omega})\cap\mathsf{N}(d_{k_1,\overline{\Omega}})=\{0\}$.

Proof. (i) Using Theorem 3.3 we get identities

$$\mathsf{N}(\delta_{k_2,\Omega}|_{\mathsf{N}(d_{k_1,\overline{\Omega}})}) = \mathsf{N}(\delta_{k_2,\Omega})\cap\mathsf{N}(d_{k_1,\overline{\Omega}})$$
$$\mathsf{N}(\delta_{k_2,\Omega})\ominus\delta_{k_2,\Omega}\mathsf{N}(d_{k_1,\overline{\Omega}}) = \mathsf{N}(\delta_{k_2,\Omega})\cap d^{-1}_{-k_2^c,\overline{\Omega}}\mathsf{R}(\delta_{-k_1^c,\Omega})$$
$$= \mathsf{N}(\delta_{k_2,\Omega})\cap\mathsf{N}(d_{-k_2^c,\overline{\Omega}})\oplus\mathsf{R}(\delta_{k_2,\Omega})\cap d^{-1}_{-k_2^c,\overline{\Omega}}\mathsf{R}(\delta_{-k_1^c,\Omega}),$$

which gives the deficiency indices for $\delta_{k_2,\Omega}|_{\mathsf{N}(d_{k_1,\overline{\Omega}})}$. Similarly for $d_{k_1,\overline{\Omega}}|_{\mathsf{N}(\delta_{k_2,\Omega})}$.

(ii) The a priori estimate in (ii) in the proof of Theorem 1.3 above shows that $\dim(\mathsf{N}(\delta_{k_2,\Omega})\cap\mathsf{N}(d_{k_1,\overline{\Omega}}))<\infty$ for all k_1, $k_2\in\mathbf{C}$. To prove that the space $\mathsf{N}(\delta_{k_2,\Omega})\cap\mathsf{N}(d_{k_1,\overline{\Omega}})$ vanishes unless k_1 and k_2^c have the same direction, write $\Gamma=d_{\overline{\Omega}}$, $\Gamma^*=-\delta_\Omega$, $\Gamma_0=e_0\wedge(\cdot)$ and $\Gamma_0^*=e_0\lrcorner(\cdot)$. The algebraic property we use here is that not only is $\Gamma\Gamma_0+\Gamma_0\Gamma=0$ but also $\Gamma^*\Gamma_0+\Gamma_0\Gamma^*=0$, which follows from the derivation property (9). Assuming $(\Gamma+k_1\Gamma_0)F=(-\Gamma^*+k_2\Gamma_0^*)F=0$, we calculate

$$0=(F,(\Gamma^*\Gamma_0+\Gamma_0\Gamma^*)F)=(\Gamma F,\Gamma_0F)+(\Gamma_0^*F,\Gamma^*F)$$
$$=-k_1\|\Gamma_0F\|^2+k_2^c\|\Gamma_0^*F\|^2.$$

This shows that $F=0$ under the assumptions on k_1 and k_2 since Γ_0 induces an exact Hodge decomposition. □

Another way of reducing to the case of a smooth domain with Lemma 3.13 is to use the extension maps constructed in Proposition 4.8 below. Let $B\subset\mathbf{R}^n$ be a ball containing Ω, let $\chi_\Omega: L_2(B;\wedge)\to L_2(\Omega;\wedge)$ be the restriction map and pick a δ-extension map $\varepsilon_\Omega: L_2(\Omega;\wedge)\to L_2(B;\wedge)$ as in Proposition 4.8 below. Then, modulo a partition of unity, Lemma 3.13 applies with $\mathcal{H}=L_2(\Omega;\wedge)$, $\mathcal{H}_0=L_2(B;\wedge)$, $A=\delta_\Omega$, $A_0=\delta_B$, $T=\chi_\Omega$ and $S=\varepsilon_\Omega$.

Proposition 4.8. *Let $\chi_\Omega: L_2(\mathbf{R}^n;\wedge)\to L_2(\Omega;\wedge)$ be the restriction operator and let $K\supset\Omega$ be a compact set. Then there exists a bounded extension operator $\varepsilon_\Omega: L_2(\Omega;\wedge)\to L_2(\mathbf{R}^n;\wedge)$ such that*

(i) $\chi_\Omega\varepsilon_\Omega=$ *identity on* $L_2(\Omega;\wedge)$.
(ii) *supp*$(\varepsilon_\Omega F)\subset K$ *for all* $F\in L_2(\Omega;\wedge)$.

(iii) $d_{\mathbf{R}^n}\varepsilon_\Omega - \varepsilon_\Omega d_\Omega$ *extends to an* $L_2(\Omega;\wedge) \to L_2(\mathbf{R}^n;\wedge)$ *bounded map. In particular,* ε_Ω *restricts to a bounded map*

$$\varepsilon_\Omega : \mathsf{D}(d_\Omega) \to \mathsf{D}(d_{\mathbf{R}^n}).$$

(iv) $\varepsilon_\Omega(F_1 + e_0 \wedge F_2) = \varepsilon_\Omega F_1 + e_0 \wedge \varepsilon_\Omega F_2$.

The same holds true when d is replaced by δ.

Proof. Let $\overline{\Omega} \subset \cup_{j=0}^N V_j$, $\eta_j \in C_0^\infty(V_j)$, $\sum_0^N \eta_j|_\Omega = 1$ and let $\rho_j : \mathbf{R}^n \to \mathbf{R}^n$, $j = 1,\dots,N$, be the local bilipschitz parametrisations from Definition 2.1 and $V_0 \subset\subset \Omega$ be contained in the interior.

(i) We first note that it suffices to construct an extension map $\varepsilon : L_2(\mathbf{R}^n_+;\wedge) \to L_2(\mathbf{R}^n;\wedge)$ acting on fields supported in $(\rho_j^{-1}\mathrm{supp}\,\eta_j) \cap \overline{\mathbf{R}^n_+}$. Indeed, this gives local extension maps $\varepsilon_j := (\rho_j^{-1})^*\varepsilon\rho_j^* : L_2(V_j \cap \Omega;\wedge) \to L_2(V_j;\wedge)$, extending fields supported in $\mathrm{supp}\,\eta_j \cap \overline{\Omega}$ to fields compactly supported in V_j, $j = 1,\dots,N$. Then we can construct ε_Ω as

$$\varepsilon_\Omega F := \eta_0 F + \sum_{j=1}^N \varepsilon_j(\eta_j F). \tag{19}$$

Moreover, from the construction of ε below and Proposition 2.6, d commutes with ε_j and thus

$$(d_{\mathbf{R}^n}\varepsilon_\Omega - \varepsilon_\Omega d_\Omega)F = (d\eta_0) \wedge F + \sum_{j=1}^N \varepsilon_j((d\eta_j) \wedge F). \tag{20}$$

Clearly, both (19) and (20) define L_2-bounded operators.

(ii) To construct the extension map $\varepsilon : L_2(\mathbf{R}^n_+;\wedge) \to L_2(\mathbf{R}^n;\wedge)$, consider the stretched reflections

$$r_k : (x', -x_n) \longmapsto (x', kx_n).$$

By Proposition 4.3, it suffices to consider $G \in C_0^\infty(\mathbf{R}^n;\wedge)|_{\mathbf{R}^n_+}$. If we decompose $G(x) = G_1(x) + e_n \wedge G_2(x)$, $e_n \lrcorner G_i = 0$, into parts tangential and normal to $\mathbf{R}^{n-1}$, then the pullbacks are given by $r_k^* G(x', -x_n) = G_1(x', kx_n) - ke_n \wedge G_2(x', kx_n)$, and we see that both tangential and normal parts of the field

$$\widetilde{G} := \begin{cases} G, & x_n > 0, \\ 3r_1^*G - 2r_2^*G, & x_n < 0, \end{cases}$$

are continuous across Σ. We can assume that $\mathrm{supp}\,\eta_j$ is small enough so that $\mathrm{supp}\,\widetilde{G} \subset \rho_j^{-1}V_j$ if $\mathrm{supp}\,G \subset (\rho_j^{-1}\mathrm{supp}\,\eta_j) \cap \overline{\mathbf{R}^n_+}$. Now define $\varepsilon := 3r_1^* - 2r_2^*$.

The proof for δ is analogous. We here use the reduced pushforwards $(\tilde{r}_{k*}^{-1})F = -kF_1 + e_n \wedge F_2$. □

Remark 4.9. (i) We see that in a natural way $\mathsf{D}(d_{\overline{\Omega}}) \subset \mathsf{D}(d_{\mathbf{R}^n})$ and

$$\mathsf{D}(d_\Omega) = \mathsf{D}(d_{\mathbf{R}^n})/\mathsf{N}(\chi_\Omega).$$

Proposition 4.8 shows that $\chi_\Omega : \mathsf{D}(d_{\mathbf{R}^n}) \to \mathsf{D}(d_\Omega)$ is surjective and that $\varepsilon_\Omega : \mathsf{D}(d_\Omega) \to \mathsf{D}(d_{\mathbf{R}^n})$ embeds $\mathsf{D}(d_\Omega)$ as a complement of $\mathsf{N}(\chi_\Omega)$ in $\mathsf{D}(d_{\mathbf{R}^n})$.

(ii) From expressions (19) and (20) we obtain norm estimates

$$\|\varepsilon_\Omega\|_{L_2(\Omega;\wedge)\to L_2(\mathbf{R}^n;\wedge)} \lesssim 1 + \sum_{j=1}^N \left(\sup_{x\in\rho_j^{-1}V_j} \frac{\|\underline{\rho}_j(x)\|_{\text{op}}}{\sqrt{J(\rho_j)(x)}} \right) \left(\sup_{y\in V_j} \frac{\|\underline{\rho}_j^{-1}(y)\|_{\text{op}}}{\sqrt{J(\rho_j^{-1})(y)}} \right),$$

$$\|[d,\varepsilon_\Omega]\|_{L_2(\Omega;\wedge)\to L_2(\mathbf{R}^n;\wedge)} \lesssim \|\nabla\eta_0\|_\infty + \sum_{j=1}^N \left(\sup_{x\in\rho_j^{-1}V_j} \frac{\|\underline{\rho}_j(x)\|_{\text{op}}}{\sqrt{J(\rho_j)(x)}} \right) \left(\sup_{y\in V_j} \frac{\|\underline{\rho}_j^{-1}(y)\|_{\text{op}}}{\sqrt{J(\rho_j^{-1})(y)}} \right) \|\nabla\eta_j\|_\infty,$$

$$\|\varepsilon_\Omega\|_{\mathsf{D}(d_\Omega)\to\mathsf{D}(d_{\mathbf{R}^n})} \le \|\varepsilon_\Omega\|_{L_2(\Omega;\wedge)\to L_2(\mathbf{R}^n;\wedge)} + \|[d,\varepsilon_\Omega]\|_{L_2(\Omega;\wedge)\to L_2(\mathbf{R}^n;\wedge)},$$

where $\underline{\rho}_j(x) : \wedge \to \wedge$ denotes the $\wedge$-homomorphism which extends the Jacobian matrix and $J(\rho_j)$ is the Jacobian determinant.

We end with a discussion of various ways to prove that the Dirac operators $\mathbf{D}_{\Omega^\perp}$ and $\mathbf{D}_{\Omega^\parallel}$ are diffuse Fredholm operators under certain additional regularity and topological assumptions on Σ. First we recall the standard proof in the smooth case. Both here and in Theorem 4.13 we use Lemma 3.14, which shows that it suffices to prove that $\mathsf{D}(d_{\overline{\Omega}}) \cap \mathsf{D}(\delta_\Omega)$ and $\mathsf{D}(d_\Omega) \cap \mathsf{D}(\delta_{\overline{\Omega}})$ are compactly embedded in $L_2(\Omega;\wedge)$.

Theorem 4.10. *Assume that Ω is a bounded open set with C^2-regular boundary Σ. Then*

$$\mathsf{D}(\mathbf{D}_{\Omega^\perp}) = \mathsf{D}(d_{\overline{\Omega}}) \cap \mathsf{D}(\delta_\Omega) = W_2^1(\Omega^\perp;\wedge) := \{F \in W_2^1(\Omega;\wedge) \; ; \; \nu \wedge f = 0\},$$
$$\mathsf{D}(\mathbf{D}_{\Omega^\parallel}) = \mathsf{D}(d_\Omega) \cap \mathsf{D}(\delta_{\overline{\Omega}}) = W_2^1(\Omega^\parallel;\wedge) := \{F \in W_2^1(\Omega;\wedge) \; ; \; \nu \lrcorner f = 0\}.$$

In particular, $\mathsf{D}(d_{\overline{\Omega}}) \cap \mathsf{D}(\delta_\Omega)$ and $\mathsf{D}(d_\Omega) \cap \mathsf{D}(\delta_{\overline{\Omega}})$ are compactly embedded in $L_2(\Omega;\wedge)$.

Moreover, if $\{v_1, \ldots, v_{n-1}\}$ is an ON-frame on Σ of directions of principal inward curvatures κ_i, then we have the Weitzenböck formulae

$$\int_\Omega |\nabla \otimes F(x)|^2 = \int_\Omega |dF(x)|^2 + |\delta F(x)|^2 - \begin{cases} \sum \int_\Sigma \kappa_i(y)|v_i(y) \wedge f(y)|^2, & F \in \mathsf{D}(\mathbf{D}_{\Omega^\perp}), \\ \sum \int_\Sigma \kappa_i(y)|v_i(y) \lrcorner f(y)|^2, & F \in \mathsf{D}(\mathbf{D}_{\Omega^\parallel}). \end{cases} \tag{21}$$

Remark 4.11. (i) Note that when Σ is convex, but not necessarily C^2, then $\kappa_i \ge 0$ and we obtain the inequality $\int_\Omega |\nabla \otimes F(x)|^2 \le \int_\Omega |dF(x)|^2 + |\delta F(x)|^2$ if either $\nu \wedge f = 0$ or $\nu \lrcorner f = 0$. See Mitrea [18] for generalisations of this result.

(ii) Consider also the special case of the Laplace equation as explained in Example 1.1. If U is in the domain $\mathsf{D}(\Delta_D)$ of the Dirichlet Laplacian in Ω, then the gradient $F := \nabla U \in \mathsf{D}(\mathbf{D}_{\Omega^\perp})$. The Weitzenböck formula now reads

$$\int_\Omega |\nabla \otimes \nabla U(x)|^2 = \int_\Omega |\Delta U|^2 - (n-1) \int_\Sigma H(y) \left|\tfrac{\partial u}{\partial \nu}(y)\right|^2,$$

where H is the (inward) mean curvature of Σ, since for normal vector fields $|v_i \wedge f| = |v_i||f| = |f|$. This formula is known as Kadlec's formula, see p.341 in Taylor [24].

Proof. We here only give the proof for $\mathbf{D}_{\Omega^\perp}$, since that for $\mathbf{D}_{\Omega^\parallel}$ is similar.

(i) Assume that $F \in W_2^1(\Omega^\perp; \wedge) \subset \mathsf{D}(\mathbf{D}_{\Omega^\perp})$. Using the boundary theorem 2.7, we obtain identities

$$\int_\Omega |\nabla \otimes F|^2 + \langle F, \Delta F^c\rangle = \int_\Sigma \langle f, (\nu, \nabla)\dot{F}^c|_\Sigma\rangle,$$
$$\int_\Omega |dF|^2 + \langle F, \delta dF^c\rangle = \int_\Sigma \langle f, \nu \lrcorner dF^c|_\Sigma\rangle = 0,$$
$$\int_\Omega |\delta F|^2 + \langle F, d\delta F^c\rangle = \int_\Sigma \langle f, \nu \wedge \delta F^c|_\Sigma\rangle,$$

where $\dot{F}$ denotes the function on which the differential operator ∇ is acting. Thus, subtracting the last two equations from the first gives

$$\int_\Omega |\nabla \otimes F|^2 = \int_\Omega |dF|^2 + |\delta F|^2 - \int_\Sigma \langle f, \nu \wedge \delta F^c|_\Sigma\rangle + \langle f, (\nu, \nabla)\dot{F}^c|_\Sigma\rangle.$$

Using the derivation property (9) and that $\nu \wedge f = 0$ and $\partial_{v_i}\nu = \kappa_i v_i$ we rewrite the boundary integral as

$$\int_\Sigma \langle f, \nu \wedge \delta F^c|_\Sigma\rangle - \langle f, (\nu, \nabla)\dot{F}^c|_\Sigma\rangle = -\int_\Sigma \langle f, \nabla \lrcorner (\nu \wedge \dot{F}^c)\rangle$$
$$= -\sum_{i=1}^{n-1}\int_\Sigma \langle v_i \wedge f, \nu \wedge \partial_{v_i} f^c\rangle = \sum_{i=1}^{n-1}\int_\Sigma \langle v_i \wedge f, (\partial_{v_i}\nu) \wedge f^c\rangle = \sum_{i=1}^{n-1}\int_\Sigma \kappa_i |v_i \wedge f|^2.$$

Since Σ is of regularity C^2, κ_i are continuous on Σ and thus the Sobolev trace theorem shows that the inclusion $i_\Omega : W_2^1(\Omega^\perp; \wedge) \hookrightarrow \mathsf{D}(\mathbf{D}_{\Omega^\perp})$ is a bounded semi-Fredholm map.

(ii) What is left to prove is that the inclusion is surjective. Note that since $e_0\mathbf{D}_{\Omega^\perp}$ is a self-adjoint operator by Proposition 3.5, we have that $\mathbf{D}_{\Omega^\perp} + ie_0 : \mathsf{D}(\mathbf{D}_{\Omega^\perp}) \to L_2(\Omega; \wedge)$ is an isomorphism (for any weakly Lipschitz domain). Thus it suffices to prove that $\mathbf{D}_{\Omega^\perp} + ie_0 : W_2^1(\Omega^\perp; \wedge) \to L_2$ is surjective.

One way to prove this is to perturb the given domain to a domain with an isometric double, e.g. the upper half $T_+^n := \{x \in \mathbf{R}^n \ ; \ 0 < x_n < 1\}/(2\mathbf{Z}+1)^n$ of the flat n-torus $T^n := \mathbf{R}^n/(2\mathbf{Z}+1)^n$ as in Taylor [24]. Since the problem is local, it suffices to prove that if $\rho_t : \Omega = \Omega_0 \to \Omega_t$ is a continuous family of C^2 diffeomorphisms, where Ω_t is a C^2 domain in T^n for $t \in [0,1]$ and $\Omega_1 = T_+^n$, then $\mathbf{D}_{\Omega_0^\perp} + ie_0 : W_2^1(\Omega_0^\perp; \wedge) \to L_2(\Omega_0; \wedge)$ is an isomorphism. From (i) we have a continuous family of semi-Fredholm maps

$$\rho_t^*(\mathbf{D}_{\Omega_t^\perp} + ie_0)(\rho_t^{-1})^* = d_{\overline{\Omega}_0} + \rho_t^*\delta_{\Omega_t}(\rho_t^{-1})^* + ie_0 : W_2^1(\Omega_0^\perp; \wedge) \longrightarrow L_2(\Omega_0; \wedge),$$

since pullbacks preserves normal boundary conditions, and since $[\rho_t^*, \delta] : W_2^1 \to L_2$ depends continuously on t. Perturbation theory [13] now shows that it suffices to prove that $\mathbf{D}_{(T_+^n)^\perp} + ie_0 : W_2^1((T_+^n)^\perp; \wedge) \to L_2(T_+^n; \wedge)$ is surjective. Note that $\mathbf{D}_{T^n} + ie_0 : W_2^1(T^n;\wedge) \to L_2(T^n;\wedge)$ is an isomorphism. We see that, given any $G \in L_2(T^n;\wedge)$ with $\operatorname{supp} G \subset \overline{T^n_+}$, there exists $F \in W_2^1(T^n;\wedge)$ such that $(\mathbf{D}_{T^n} + ie_0)F = G$. Now the anti symmetrised field $F - r^*F$, where $r : T^n_\pm \to T^n_\mp$ is the isometric reflection, belongs to $W_2^1((T^n_+)^\perp;\wedge)$ and $(d+\delta+ie_0)(F-r^*F) = G - r^*G = G$ in T^n_+ since d commutes with r^* and δ commutes with $\tilde{r}_*^{-1} = r^*$ by Proposition 2.6. This finishes the proof. □

For non-smooth Σ, not only the source function $F := \mathbf{D}_{\Omega^\perp}U$ influences the regularity of $U \in \mathsf{D}(\mathbf{D}_{\Omega^\perp})$, but also Σ. A standard example, see e.g. Grisvard [11], is the following.

Example 4.12. Consider a bounded domain $\Omega \subset \mathbf{R}^2$ whose boundary Σ is smooth except at 0 where it coincides with $\overline{\mathbf{R}}_+ \cup e^{i\alpha}\mathbf{R}_+$. Let $U : \mathbf{R}^2 \to \wedge^0$ be a scalar function in Ω, smooth up to the boundary except at 0, such that $U(x) = r^{\frac{\pi}{\alpha}} \sin(\frac{\pi}{\alpha}\theta)$ around 0 and $U|_\Sigma = 0$. Define

$$F(x) := \nabla U(x) = \tfrac{\pi}{\alpha} r^{\frac{\pi}{\alpha}-1}(\sin(\tfrac{\pi}{\alpha}\theta)\hat{r} + \cos(\tfrac{\pi}{\alpha}\theta)\hat{\theta}),$$

for x around 0, where $\hat{r}$ and $\hat{\theta}$ denotes the radial and angular unit vector fields. Then the estimate $|F| \lesssim r^{\frac{\pi}{\alpha}-1}$ shows that $F \in \mathsf{D}(d_{\overline{\Omega}}) \cap \mathsf{D}(\delta_\Omega)$, whereas the estimate $|\frac{\partial F}{\partial r}| \approx r^{\frac{\pi}{\alpha}-2}$ shows that

$$\|F\|^2_{W_2^1(\Omega)} \geq \int_\Omega \left|\frac{\partial F}{\partial r}\right|^2 \gtrsim \int_0^1 r^{2(\frac{\pi}{\alpha}-2)} r\, dr.$$

But in the non-convex case $\alpha > \pi$ the right hand side is infinite so that $F \notin W_2^1(\Omega;\wedge)$. However, one can verify that $\|F\|_{W_2^{1/2}} < \infty$ for any $0 < \alpha < 2\pi$.

For a strongly Lipschitz domain, we use the $L_2(\Sigma;\wedge)$ theory of boundary value problems. This uses the Rellich estimate technique, which was first applied by Verchota [25] to the Laplace equation. This technique was later extended to the full Dirac operator by McIntosh–Mitrea [16] and McIntosh–Mitrea–Mitrea [15].

Theorem 4.13. *Assume that Ω is a bounded, strongly Lipschitz domain. Then we have continuous inclusions*

$$\mathsf{D}(\mathbf{D}_{\Omega^\perp}),\ \mathsf{D}(\mathbf{D}_{\Omega^\parallel}) \subset W_2^{1/2}(\Omega;\wedge).$$

In particular, $\mathsf{D}(d_{\overline{\Omega}}) \cap \mathsf{D}(\delta_\Omega)$ and $\mathsf{D}(d_\Omega) \cap \mathsf{D}(\delta_{\overline{\Omega}})$ are compactly embedded in $L_2(\Omega;\wedge)$.

Proof. Consider the map $\mathbf{D}_{\Omega^\perp} + ie_0 : \mathsf{D}(\mathbf{D}_{\Omega^\perp}) \longrightarrow L_2(\Omega;\wedge)$, which is an isomorphism since $e_0\mathbf{D}_{\Omega^\perp}$ is self-adjoint, and the dense subset

$$S := \{F \in \mathsf{D}(\mathbf{D}_{\Omega^\perp})\ ;\ (\mathbf{D}_{\Omega^\perp} + ie_0)F \in C_0^\infty(\Omega;\wedge)\} \subset \mathsf{D}(\mathbf{D}_{\Omega^\perp}).$$

It suffices to show that we have a continuous inclusion $S \hookrightarrow W_2^{1/2}(\Omega;\wedge)$. Given $G = (\mathbf{D}_{\Omega^\perp} + ie_0)F \in C_0^\infty(\Omega;\wedge)$, let $F_0 := (\mathbf{D}_{\mathbf{R}^n} + ie_0)^{-1}G \in C^\infty(\mathbf{R}^n;\wedge)$ and form its tangential trace $\nu \wedge f_0 \in L_2(\Sigma;\wedge)$. We now apply the Rellich $L_2(\Sigma;\wedge)$ theory of boundary value problems on strongly Lipschitz domains, see [1] for more details, which shows the existence of a field $F_1 : \Omega \to \wedge$ such that

$$\begin{aligned} (\mathbf{D} + ie_0)F_1 &= 0 \qquad \text{in } \Omega \\ \nu \wedge f_1 &= \nu \wedge f_0 \qquad \text{on } \Sigma \\ \|F_1\|_{W_2^{1/2}(\Omega;\wedge)} &\approx \|\nu \wedge f_0\|_{L_2(\Sigma;\wedge)} \\ dF_1, \delta F_1 &\in W_2^{1/2}(\Omega;\wedge) \subset L_2(\Omega;\wedge). \end{aligned}$$

Now let $F' := F_0 - F_1$. We see that

$$\begin{aligned} \|F_0\|_{W_2^{1/2}(\Omega;\wedge)} &\lesssim \|F_0\|_{W_2^1(\Omega;\wedge)} \lesssim \|G\|_{L_2(\Omega;\wedge)} \approx \|F\|_{\mathsf{D}(\mathbf{D}_{\Omega^\perp})} \\ \|F_1\|_{W_2^{1/2}(\Omega;\wedge)} &\approx \|\nu \wedge f_0\|_{L_2(\Sigma;\wedge)} \lesssim \|F_0\|_{W_2^1(\Omega;\wedge)} \lesssim \|F\|_{\mathsf{D}(\mathbf{D}_{\Omega^\perp})}. \end{aligned}$$

Moreover dF', $\delta F' \in L_2(\Omega;\wedge)$ and $\nu \wedge f' = \nu \wedge f_0 - \nu \wedge f_1 = 0$. Thus $F' \in \mathsf{D}(\mathbf{D}_{\Omega^\perp}) \cap W_2^{1/2}(\Omega;\wedge)$ and $(\mathbf{D}_{\Omega^\perp} + ie_0)F' = G = (\mathbf{D}_{\Omega^\perp} + ie_0)F$. Thus $F = F' \in W_2^{1/2}(\Omega;\wedge)$ with $\|F\|_{W_2^{1/2}(\Omega;\wedge)} \lesssim \|F\|_{\mathsf{D}(\mathbf{D}_{\Omega^\perp})}$. □

Remark 4.14. In the proof above we used the fact that

$$\|F_1\|_{W_2^{1/2}(\Omega;\wedge)} \approx \|f_1\|_{L_2(\Sigma;\wedge)}$$

when $(\mathbf{D} + ie_0)F_1 = 0$ in Ω. This result is presented in Fabes [7] with an incomplete proof which is corrected in Mitrea–Mitrea–Pipher [19]. See also [3] for an alternative correction.

Note that Theorem 4.13 is more constructive than Theorem 4.10 in the sense that F' is found by solving the boundary equation $\nu \wedge f_1 = \nu \wedge f_0$. However, if one is only interested in solving for example $d_\Omega U = F$, where $F \in \mathsf{N}(d_\Omega)$, with a "good inverse" in the sense that $F \mapsto U$ is an L_2 compact map, then this can be done much more explicitly using path integrals as we now explain.

Lemma 4.15. *Let $\Omega \subset \mathbf{R}^n$ be an open set with a smooth retraction $\mathcal{F}_t : \Omega \to \mathcal{F}_t(\Omega) \subset \Omega$ to $p \in \Omega$ such that $\mathcal{F}_1 = I$, $\mathcal{F}_t\mathcal{F}_s = \mathcal{F}_{ts}$ for $0 \le t, s \le 1$ and $\mathcal{F}_0 = p$. If $\theta = d\mathcal{F}_t/dt|_{t=1}$ is the vector field with flow $\mathcal{F}_t$, then for smooth fields F in Ω we have the path integral formulae*

$$F(x) = \nabla \wedge \Big(\int_0^1 \theta(x) \lrcorner \mathcal{F}_t^* F(x) \tfrac{dt}{t} \Big), \qquad \text{if } \nabla \wedge F = 0 \text{ and } F|_{\wedge^0} = 0,$$

$$F(x) = \nabla \lrcorner \Big(\int_0^1 \theta(x) \wedge \widetilde{\mathcal{F}}_{t*}^{-1} F(x) \tfrac{dt}{t} \Big), \qquad \text{if } \nabla \lrcorner F = 0 \text{ and } F|_{\wedge^n} = 0.$$

One can prove this lemma by using *Cartan's formula*

$$\mathcal{L}_\theta F = \tfrac{d}{dt}(\mathcal{F}_t^* F(x))|_{t=1} = \nabla \wedge (\theta \lrcorner F) + \theta \lrcorner (\nabla \wedge F)$$

for the Lie derivative of the differential form F and using the homomorphism formulae from Proposition 2.6. For more details, see for example Taylor [24].

Example 4.16. Let Ω be star shaped with respect to 0 and $\mathcal{F}_t(x) := tx$. Then for a smooth j-vector field $F : \Omega \to \wedge^j$ we have the path integrals

$$F(x) = \nabla \wedge \Big(\int_0^1 t^{j-1} x \lrcorner F(tx)\, dt \Big), \qquad \text{if } \nabla \wedge F = 0 \text{ and } j \geq 1,$$

$$F(x) = \nabla \lrcorner \Big(\int_0^1 t^{n-j-1} x \wedge F(tx)\, dt \Big), \qquad \text{if } \nabla \lrcorner F = 0 \text{ and } j \leq n-1.$$

Indeed, using the derivation formula (9) and that $\sum_i e_i \wedge (e_i \lrcorner F(tx)) = jF(tx)$, one can directly verify that $\nabla \wedge (x \lrcorner F(tx)) = jF(tx) + t\frac{d}{dt}(F(tx))$.

In particular, a curl free vector field F has a scalar potential given by the path integral $U(x) = \int_0^1 (x, F(tx))\, dt$ and a divergence free vector field F has a bivector potential $U(x) := \int_0^1 t^{n-2} x \wedge F(tx)\, dt \in \wedge^2$. In classical notation in $\mathbf{R}^3$, the latter translates to $F = -\nabla \times U^\perp$ if $U(x)^\perp := \int_0^1 tx \times F(tx)\, dt$.

A third way to prove that the Dirac operators are diffuse Fredholm operators uses an L_2 version of the classical Poincaré lemma. We here only consider fields with values in $\wedge \mathbf{R}^n = \wedge^0 \oplus \ldots \oplus \wedge^n$. The extension to spacetime setting is straightforward.

Theorem 4.17. *Let $\Omega \subset \mathbf{R}^n$ be a star shaped strongly Lipschitz domain. Let $0 < \epsilon < R < \infty$ be such that $B(0,\epsilon) \subset \Omega \subset B(0,R)$ and such that Ω is star shaped with respect to each $p \in B(0,\epsilon)$. For $1 \leq j \leq n$, let $T^{(j)}$ denote the integral operator*

$$T^{(j)} F(x) := \int_\Omega (x-y) \lrcorner F(y)\, k_j(x,y) dy, \qquad F \in L_2(\Omega; \wedge^j), \tag{22}$$

where k_j denotes the kernel

$$k_j(x,y) := \int_0^1 \eta\Big(x + \frac{y-x}{1-t}\Big) \frac{t^{j-1} dt}{(1-t)^{n+1}}$$

for some fixed $\eta \in C_0^\infty(B(0,\epsilon))$ with $\int \eta = 1$. In particular

$$\operatorname{supp} k_j \subset \{(x,y)\, ;\, y \in \overline{\operatorname{conv}}\,(B(0,\epsilon), x)\},$$

where $\overline{\operatorname{conv}}$ denotes the closed convex hull, k_j is smooth off the diagonal $\{x = y\}$ with estimates

$$|k_j(x,y)| \leq \tfrac{1}{n} \|\eta\|_\infty (R+\epsilon)^n \frac{1}{|x-y|^n}$$

and $T^{(j)}$ defines a compact operator $L_2(\Omega; \wedge^j) \to L_2(\Omega; \wedge^{j-1})$. Then $T := 0 \oplus T^{(1)} \oplus \ldots \oplus T^{(n)} : \mathsf{N}(d_\Omega) \to L_2(\Omega; \wedge \mathbf{R}^n)/\mathsf{N}(d_\Omega)$ is a compact Fredholm inverse to d_Ω. Thus d_Ω is a diffuse Fredholm-nilpotent operator.

The corresponding result for δ_Ω holds true as well.

Proof. Assume that $F \in C_0^\infty(\mathbf{R}^n; \wedge^j)$ and $\operatorname{supp}(\nabla \wedge F) \cap \overline{\Omega} = \emptyset$ as in Proposition 4.3. We define

$$T^{(j)} F(x) = \int \eta(p) dp \Big(\int_0^1 t^{j-1} (x-p) \lrcorner F(p + t(x-p))\, dt \Big),$$

which by using Fubini's theorem and the change of variables $y = p + t(x - p)$ becomes (22). Since $\int \eta = 1$, it follows from Example 4.16 that $d_\Omega T^{(j)} F = F$.

The estimates off $\operatorname{supp} k_j$ and $|k_j(x, y)|$ are straightforward to verify. Since the full kernel for $T^{(j)}$ has the estimate $\lesssim 1/|x-y|^{n-1}$, Schur's lemma shows that $T^{(j)}$ defines a compact operator $L_2(\Omega; \wedge^j) \to L_2(\Omega; \wedge^{j-1})$.

We can now apply Proposition 3.8 with $\mathcal{H}_1 = L_2(\Omega; \wedge)/\mathsf{N}(d_\Omega)$, $\mathcal{H}_2 = \mathsf{N}(d_\Omega)$, $A = d_\Omega$, $T_1 = T_2 = T = 0 \oplus T^{(1)} \oplus \ldots \oplus T^{(n)}$, $K_1 = 0$ and $K_2 =$ orthogonal projection onto scalar constants, which shows that d_Ω is a diffuse Fredholm-nilpotent operator. □

References

[1] A. Axelsson, Oblique and normal transmission problems for Dirac operators with strongly Lipschitz interfaces. To appear in *Comm. Partial Differential Equations* (2002).

[2] A. Axelsson, Transmission problems and boundary operator algebras. To appear in *Integral Equations Operator Theory* (2004), DOI 0.10007/s00020-003-1225-0.

[3] A. Axelsson, *Transmission problems for Dirac's and Maxwell's equations with Lipschitz interfaces.* PhD thesis, Australian National University, submitted 2002.

[4] A. Axelsson, R. Grognard, J. Hogan, A. McIntosh, Harmonic analysis of Dirac operators on Lipschitz domains. Clifford analysis and its applications (Prague, 2000), 231–246, *NATO Sci. Ser. II Math. Phys. Chem.* **25**, Kluwer Acad. Publ., Dordrecht, 2001.

[5] M.S. Birman, M.Z. Solomyak, L_2-theory of the Maxwell operator in arbitrary domains. *Uspekhi Mat. Nauk* **42** (1987), no. 6 (258), 61–76, 247.

[6] M.S. Birman, M.Z. Solomyak, The selfadjoint Maxwell operator in arbitrary domains. *Leningrad Math. J.* **1** (1990), no. 1, 99–115.

[7] E. Fabes, Layer potential methods for boundary value problems on Lipschitz domains. in: Potential theory, surveys and problems (Prague, 1987), 55–80, *Lecture Notes in Math.* **1344**, Springer, Berlin, 1988.

[8] K.O. Friedrichs, Differential forms on Riemannian manifolds. *Comm. Pure Appl. Math.* **8** (1955), 551–590.

[9] M.P. Gaffney, The harmonic operator for exterior differential forms. *Proc. Nat. Acad. Sci. U. S. A.* **37** (1951), 48–50.

[10] M.P. Gaffney, Hilbert space methods in the theory of harmonic integrals. *Trans. Amer. Math. Soc.* **78** (1955), 426–444.

[11] P. Grisvard, *Elliptic problems in nonsmooth domains.* Monographs and Studies in Mathematics, 24. Pitman (Advanced Publishing Program), Boston, MA, 1985.

[12] W.V.D. Hodge, *The Theory and Applications of Harmonic Integrals.* Cambridge University Press, Cambridge, England; Macmillan Company, New York, 1941.

[13] T. Kato, *Perturbation theory for linear operators.* Second edition. Grundlehren der Mathematischen Wissenschaften, Band 132. Springer-Verlag, Berlin, New York, 1976.

[14] K. Kodaira, Harmonic fields in Riemannian manifolds (generalized potential theory). *Ann. of Math.* (2) **50** (1949), 587–665.

[15] A. McIntosh, D. Mitrea, M. Mitrea, Rellich type estimates for one-sided monogenic functions in Lipschitz domains and applications. in: *Analytical and Numerical Methods in Quaternionic and Clifford Algebras,* K. Gürlebeck and W. Sprössig, eds., Proceedings of the Seiffen Conference, Germany, 1996, 135–143.

[16] A. McIntosh, M. Mitrea, Clifford algebras and Maxwell's equations in Lipschitz domains. *Math. Meth. Appl. Sci.* **22** (1999), 1599–1620.

[17] D. Mitrea, M. Mitrea, Finite energy solutions of Maxwell's equations and constructive Hodge decompositions on nonsmooth Riemannian manifolds. *J. Funct. Anal.* **190** (2002), no. 2, 339–417.

[18] M. Mitrea, Dirichlet integrals and Gaffney-Friedrichs inequalities in convex domains. *Forum Math.* **13** (2001), no. 4, 531–567.

[19] D. Mitrea, M. Mitrea, J. Pipher, Vector potential theory on nonsmooth domains in $\mathbf{R}^3$ and applications to electromagnetic scattering. *J. Fourier Anal. Appl.* **3** (1997), no. 2, 131–192.

[20] C.B. Morrey Jr., *Multiple integrals in the calculus of variations.* Die Grundlehren der mathematischen Wissenschaften, Band 130, Springer-Verlag New York, Inc., New York 1966.

[21] R. Picard, An elementary proof for a compact imbedding result in generalized electromagnetic theory. *Math. Z.* **187** (1984), no. 2, 151–164.

[22] M. Schechter, Basic theory of Fredholm operators. *Ann. Scuola Norm. Sup. Pisa* (3) **21** (1967), 261–280.

[23] G. Schwarz, *Hodge decomposition, a method for solving boundary value problems.* Lecture Notes in Mathematics, 1607, Springer-Verlag, Berlin, 1995.

[24] M.E. Taylor, *Partial differential equations. I. Basic theory.* Applied Mathematical Sciences, 115. Springer-Verlag, New York, 1996.

[25] G. Verchota, Layer potentials and boundary value problems for Laplace's equation in Lipschitz domains. *J. Funct. Anal.* **59** (1984), 572–611.

[26] N. Weck, Maxwell's boundary value problem on Riemannian manifolds with nonsmooth boundaries. *J. Math. Anal. Appl.* **46** (1974), 410–437.

[27] H. Weyl, The method of orthogonal projection in potential theory. *Duke Math. J.* **7** (1940), 411–444.

Andreas Axelsson, Alan McIntosh
Centre for Mathematics and its Applications, Mathematical Sciences Institute
Australian National University
Canberra, ACT 0200
Australia
e-mail: `Alan.McIntosh@maths.anu.edu.au`

Trends in Mathematics: *Advances in Analysis and Geometry*, 31–50

Monogenic Functions of Bounded Mean Oscillation in the Unit Ball

Swanhild Bernstein

Abstract. Our goal is to generalize results for analytic functions of bounded mean oscillation (BMOA) in the unit circle to monogenic functions of bounded mean oscillation (BMOM) in the unit ball of $\mathbb{R}^n$. In particular we obtain the duality of BMOM^l and the Hardy space of right monogenic functions $H^{1,r}$.

Mathematics Subject Classification (2000). Primary 30G35; Secondary 31B10, 32A37.

Keywords. Monogenic functions of bounded mean oscillation, monogenic functions, harmonic analysis .

1. Introduction

Both for complex and for harmonic analysis functions of bounded mean oscillation are of fundamental importance. Since monogenic functions represent a possible generalization of analytic functions into higher dimensions and likewise are connected with harmonic functions, we want to regard monogenic functions of bounded mean oscillation.

In [3] important and substantial characteristics of analytic functions on the unit circle and in the unit disc were summarized. In particular the duality of BMOA and H^1 was shown, which is a substantial result of C. Fefferman. We will be able to generalize actually all results for BMOA to monogenic functions on the sphere and in the unit ball of $\mathbb{R}^n$. The substantial difference consists of the fact that we cannot use the harmonic measure, which only for $n = 2$ behaves like the Poisson kernel. Usually monogenic functions are represented by Cauchy integrals which contain the Cauchy kernel. But the Cauchy kernel is not positive. This implies a delicate behavior by using the absolute value of the kernel instead of the kernel itself because

$$\int_S \left| \frac{\bar{y} - \bar{x}}{|y - x|^n} \right| dS_y \to \infty$$

as x approaches any boundary point of the unit ball. Analytic and also monogenic functions are harmonic. Harmonic functions are represented by Poisson integrals. The advantage is that the Poisson kernel is positive! Because both the extension of a monogenic as well as a harmonic function from the unit sphere into the unit ball is unique, the Cauchy integral and the Poisson integral represent the same function.

To start, let us denote by $\mathcal{B} := \{x \in \mathbb{R}^n : |x| < 1\}$ the unit ball in $\mathbb{R}^n$ and by $\mathcal{S} := \{x \in \mathbb{R}^n : |x| = 1\}$ the unit sphere in $\mathbb{R}^n$.

For any left monogenic F in $\mathcal{B}$ with boundary values f on $\mathcal{S}$ the following properties are equivalent:

1. $f \in BMOM^l(\mathcal{S})$, i.e. $f \in BMO(\mathcal{S})$ and the Poisson extension $F = Pf$ of f into the unit ball $\mathcal{B}$ is left monogenic. The upper index "l" shows the space of left monogenic functions, which is right-linear.
2. $(1-|x|)|\nabla F|^2\,dx$ is a Carleson measure on $\mathcal{B}$.
3. Let h be the boundary values of a right monogenic function in $\mathcal{B}$. Then there exists a constant $C(f)$, depending only on f, such that
$$\left|\int_{\mathcal{S}} hf\,d\mathcal{S}\right| \le C(f)||h||_{H^1}.$$
4. There exists a $b \in L^\infty(\mathcal{S})$ such that
$$F(x) = \int_{\mathcal{S}} e(x-\omega)\omega b(\omega)\,d\mathcal{S}.$$
5. There exist left monogenic functions $G_\alpha(x)$, $x \in \mathcal{B}$, such that $\operatorname{Sc} G_\alpha(x) \in L^\infty(\mathcal{B})$ and $F(x) = \sum_\alpha G_\alpha(x)$.

The most important consequence is that by proving 3. we established that the right-linear space of left monogenic functions of bounded mean oscillation $BMOM^l$ is a subset of $(H^{1,r})^*$ the dual space of the left-linear Hardy space of right monogenic functions. On the other hand, if we start with $\Phi \in (H^{1,r})^*$ then a Hahn-Banach extension of Φ to L^1 is represented by some $b \in L^\infty$ whose left monogenic projection f also represents Φ. From 4. $\to$ 1. we get that $f \in BMOM^l$, i.e. $BMOM^l = \left(H^{1,r}\right)^*$. Thus we generalized Fefferman's famous result for analytic functions to monogenic functions.

2. Preliminaries

2.1. Clifford algebras and Clifford analysis

We begin by considering the real Clifford algebra $\mathcal{C}\ell_{n-1}$ generated from $\mathbb{R}^{n-1}$ equipped with a negative inner product, i.e.,
$$x^2 = -\langle x, x\rangle.$$
If $e_1, e_2, \ldots, e_{n-1}$ is an orthonormal basis for $\mathbb{R}^{n-1}$, then
$$e_ie_j + e_je_i = -2\delta_{ij}.$$

There is an anti-automorphism called conjugation or main anti-involution,

$$^{-} : \mathcal{C}\ell_{n-1} \to \mathcal{C}\ell_{n-1} : \overline{(e_{i_1} e_{i_2} \dots e_{i_l})} = \bar{e}_{i_l} \dots \bar{e}_{i_2} \bar{e}_{i_1}, \quad \bar{e}_j = -e_j, \; j = 1, 2, \dots, n-1.$$

Clifford analysis offers a function theory which is a higher dimensional analog of the function theory of the holomorphic functions of one complex variable.

A Clifford-valued function f is called left monogenic (right monogenic, or two-sided monogenic) in an open region Ω of $\mathbb{R}^{n-1}$ if $Df = 0$ ($fD = 0$, or $Df = fD = 0$, respectively). Here, D is the Dirac operator

$$D = \sum_{j=0}^{n-1} e_j \frac{\partial}{\partial x_j}.$$

The most important function which is both left and right monogenic is the Cauchy kernel

$$e(x-y) = \frac{1}{m(\mathcal{S})} \frac{\overline{x-y}}{|x-y|^n},$$

where $m(\mathcal{S})$ is the surface area of the unit sphere in $\mathbb{R}^{n-1}$.

Cauchy's theorem and the Cauchy integral formula follow:

Corollary 2.1 (see [20]). *Let Ω be a bounded Lipschitz domain in $\mathbb{R}^n$ with boundary $\partial\Omega$ and exterior unit normal $n(y)$ defined for almost all $y \in \partial\Omega$. Suppose that f is a right monogenic function in a neighborhood of $\overline{\Omega} = \Omega \cap \partial\Omega$ and g is a left monogenic function in a neighborhood of $\overline{\Omega}$. Then*

$$\int_{\partial\Omega} f(y) n(y) g(y) d\Sigma_y = 0,$$

$$C_r f(x) = \int_{\partial\Omega} f(y) n(y) e(x-y) d\Sigma_y = \begin{cases} f(x), & x \in \Omega, \\ 0, & x \notin \overline{\Omega}, \end{cases}$$

$$C_l g(x) = \int_{\partial\Omega} e(x-y) n(y) g(y) d\Sigma_y = \begin{cases} g(x), & x \in \Omega, \\ 0, & x \notin \overline{\Omega}. \end{cases}$$

As an immediate consequence we obtain that left as well as right monogenic functions can be represented by Cauchy integrals.

The domain we are interested in is the unit ball $\mathcal{B}$ in $\mathbb{R}^n$ and its boundary is the unit sphere $\mathcal{S}$.

2.2. Spherical Harmonics and Spherical Monogenics

In classical complex analysis polynomials play an important role. They are easy to handle and they are also analytic. Obviously also in higher dimensions polynomials are easy to handle but they need not to be monogenic. For example

$$x = x_1 e_1 + x_2 e_2 + \dots + x_{n-1} e_{n-1}$$

is a nice polynomial but

$$Dx = e_1^2 + e_2^2 + \dots + e_{n-1}^2 = -(n-1) \neq 0$$

and thus x is not monogenic.

On the other hand monogenic functions have harmonic components. Harmonic functions have also very nice properties. Let us denote by $\mathcal{H}_k(\mathbb{R}^n)$ the set of all harmonic polynomials in $\mathbb{R}^n$ of degree k. Then a *spherical harmonic* of *degree* k is the restriction to $\mathcal{S}$ of an element of $\mathcal{H}_k(\mathbb{R}^n)$. The collection of all spherical harmonics of degree k will be denoted by $\mathcal{H}_k(\mathcal{S})$; thus

$$\mathcal{H}_k(\mathcal{S}) = \{p\,|_{\mathcal{S}} \colon p \in \mathcal{H}_k(\mathbb{R}^n)\}.$$

One main result of spherical harmonics is now (see for example [1])

Proposition 2.2.

$$L^2(\mathcal{S}) = \bigoplus_{k=0}^{\infty} \mathcal{H}_k(\mathcal{S}),$$

i.e., any function $f \in L^2(\mathcal{S})$ can be uniquely expanded into a series of spherical harmonics.

We will use the following denotations:

Definition 2.3 ([11])**.** Call $M^+(k)$ the space of homogeneous monogenic Clifford-valued polynomials of degree k in $\mathbb{R}^n$. An arbitrary element of $M^+(k)$, denoted by P_k, is called an *inner spherical monogenic of order k.*
Call $M^-(k)$ the space of homogeneous monogenic Clifford-valued polynomials of degree $-(k+n-1)$ in $\mathbb{R}^n\backslash\{0\}$. An arbitrary element of $M^-(k)$, denoted by Q_k, is called an *outer spherical monogenic of order k.*
Call $\mathcal{M}^+(k)$ and $\mathcal{M}^-(k)$ the spaces consisting of the restrictions to the unit sphere $\mathcal{S}$ of the elements belonging to $M^+(k)$ and $M^-(k)$, respectively, and put

$$\mathcal{M}(k) = \mathcal{M}^+(k) + \mathcal{M}^-(k).$$

The elements of $\mathcal{M}(k)$ are called *spherical monogenics of degree k.* Arbitrary elements of $\mathcal{M}^+(k)$ and $\mathcal{M}^-(k)$ will still be denoted by P_k and Q_k, respectively.

For the construction of spherical monogenics we need a representation of the Dirac operator in spherical coordinates. Using polar coordinates $x = r\xi$, with $r = |x|$ and $\xi \in \mathcal{S}$, we may write

$$D_x = \xi\partial_r + \tfrac{1}{r}\partial_\xi,$$

where ∂_ξ is a differential operator depending only on the angular coordinates. Then the *spherical Dirac operator* is

$$\Gamma_\xi = \overline{\xi}\partial_\xi$$

and we have $D_x = \xi\left(\partial_r + \tfrac{1}{r}\Gamma_\xi\right)$. If we express the Laplacian Δ_x in $\mathbb{R}^n$ in polar coordinates we have

$$\Delta_x = \partial_r^2 + \tfrac{n-1}{r}\partial_r + \tfrac{1}{r^2}\Delta_\xi,$$

where $-\Delta_\xi$ is the Laplace-Beltrami operator on $\mathcal{S}$. With the aid of the spherical Dirac operator Γ_ξ we can factorize the Laplace-Beltrami operator in the following way:

$$\Delta_\xi = ((n-2)\mathbf{1} - \Gamma_\xi)\,\Gamma_\xi.$$

The space $\mathcal{H}_k(S^1) = \mathcal{O}^+(k) \oplus \mathcal{O}^-(k)$, where $\mathcal{O}^\pm(k)$ are exactly the eigenspaces of the Cauchy-Riemann operator restricted to the unit circle S^1. In [11] it is proven that the same is true in higher dimensions. Especially we have

Proposition 2.4.

$$\mathcal{H}_k(\mathcal{S}) = \mathcal{M}^+(k) \oplus \mathcal{M}^-(k),$$

where $\mathcal{M}^\pm(k)$ are the eigenspaces of the spherical Dirac operator.

Because of this decomposition we have

Theorem 2.5 ([11]). *Let $f \in L^2(\mathcal{S})$. Then there exist unique sequences of spherical monogenics $\{P(k)f\}_{k\in\mathbb{N}}$, $P(k)f \in \mathcal{M}^+(k)$ and $\{Q(k)f\}_{k\in\mathbb{N}}$, $Q(k)f \in \mathcal{M}^-(k)$ such that*

$$f = \sum_{k=0}^{\infty}(P(k)f + Q(k)f).$$

Here $P(k)$ denotes the projection operator $P(k) : L^2(\mathcal{S}) \to \mathcal{M}^+(k)$ given by

$$P(k)f(\xi) = \frac{1}{m(\mathcal{S})}\int_{\mathcal{S}} C^+_{n,k}(\xi,\omega)f(\omega)d\mathcal{S}$$

and $Q(k)$ denotes the projection operator $Q(k) : L^2(\mathcal{S}) \to \mathcal{M}^-(k)$ given by

$$Q(k)f(\xi) = \frac{1}{m(\mathcal{S})}\int_{\mathcal{S}} C^-_{n,k}(\xi,\omega)f(\omega)d\mathcal{S}.$$

Moreover,

$$||f||^2_{L^2} = \sum_{k=0}^{\infty}(||P(k)f||^2_{L^2} + ||Q(k)f||^2_{L^2}).$$

We will use Gegenbauer polynomials, for which we refer to [8] and [15]. For $\nu \in \mathbb{C}$ with $\operatorname{Re}\nu > -\frac{1}{2}$ we have that for all $x, t \in \mathbb{R}$

$$(1 - 2tx + x^2)^{-\nu} = \sum_{k=0}^{\infty} C^\nu_k(t)x^k,$$

where C^ν_k is the Gegenbauer polynomial of degree k associated to ν. Let $t = \langle\xi, \omega\rangle$, $\xi, \omega \in \mathcal{S}$- Then we get

$$\begin{aligned}
&C^+_{n,k}(\xi,\omega)\\
&= \frac{1}{2-n}\left[-(m+k-2)C_k^{(n-2)/2}(t) + (2-n)C_{k-1}^{n/2}(t)((\omega_0\underline{\xi} - \xi_0\underline{\omega})e_0 + \underline{\xi}\wedge\underline{\omega})\right],\\
&C^-_{n,k}(\xi,\omega) = \frac{1}{2-n}\left[(k+1)C_k^{(n-2)/2}(t) + (2-n)C_{k-1}^{n/2}(t)((\omega_0\underline{\xi} - \xi_0\underline{\omega})e_0 + \underline{\xi}\wedge\underline{\omega})\right]
\end{aligned} \tag{2.1}$$

The most important (two-sided) monogenic function is the Cauchy kernel for which we have the following expansions. Let $x = r\xi \in \mathbb{R}^n$ and $y = r\omega \in \mathbb{R}^n\backslash\{0\}$. Then

$$\frac{\overline{y} - \overline{x}}{|x-y|^n} = \sum_{k=0}^{\infty}\frac{|x|^k}{|y|^{n+k-1}}C^+_{n,k}(\xi,\omega)\overline{\omega} = \sum_{k=0}^{\infty}\frac{|x|^k}{|y|^{n+k-1}}C^-_{n,k}(\omega,\xi)\overline{\xi}.$$

Following [11] let us denote by $\mathcal{M}_2(\mathcal{S})$ the space of those elements f in $L^2(\mathcal{S})$ such that their decomposition into spherical monogenics is of the form

$$f(\omega) = \sum_{k=0}^{\infty} P(k) f(\omega),$$

i.e., $Q(k)f(\omega) = 0$ for all $k \in \mathbb{N}$. $\mathcal{M}_2(\mathcal{S})$ is obviously a closed subspace of $L^2(\mathcal{S})$. The direct complement of $\mathcal{M}_2(\mathcal{S})$ will be denoted by $\mathcal{M}_{0,2}(\mathcal{S})$, thus

$$L^2(\mathcal{S}) = \mathcal{M}_2(\mathcal{S}) \oplus \mathcal{M}_{0,2}(\mathcal{S}).$$

It should be mentioned that the sum is not only direct but also orthogonal w.r.t. the usual $L^2(\mathcal{S})$ inner product $\int_{\mathcal{S}} \bar{g}(\omega) f(\omega) dS$.

The connection between inner spherical monogenics and the Hardy space $H^2(\mathcal{S})$ is given by the following theorem:

Theorem 2.6 ([11]). *A function f belongs to $\mathcal{M}_2(\mathcal{S})$ if and only if there exists a function F which is monogenic in the open ball $\mathcal{B}$ such that*

$$f(\xi) = \lim_{r \to 1-0} F(r\xi),$$

the convergence being taken in $L^2(\mathcal{S})$.

2.3. Monogenic Hardy spaces

Definition 2.7. If F is any function with domain $\mathcal{B}$, and $0 < r < 1$, then F_r denotes the dilated function defined for $|x| < \frac{1}{r}$ by $F_r(x) = F(rx)$. A monogenic function F with domain $\mathcal{B}$ is in the *monogenic Hardy space* $H^p(\mathcal{B})$ $(0 < p < \infty)$ if

$$\sup_{0<r<1} \int_{\mathcal{S}} |F_r|^p dS < \infty.$$

In the latter case, we write $||F||_p$ for the pth root of the left-hand side. This is a norm if $p \geq 1$.

Definition 2.8. Let $M(\mathcal{B})$ denote the class of continuous functions $F : \mathcal{B} \to \mathcal{C}\ell_{0,n-1}$ which are continuous on the closed ball $\overline{\mathcal{B}}$ and which are left monogenic in its interior $\mathcal{B}$. Further, let $M(\mathcal{S})$ denote the class of all continuous functions f defined on $\mathcal{S}$ which are restrictions to $\mathcal{S}$ of members of $M(\mathcal{B})$.
For $0 < p < \infty$, let $H^p(\mathcal{S})$ be the L^p-closure of $M(\mathcal{S})$.

In the following proposition we summarize relations between monogenic functions and their boundary values. In [7] these results are proven for the unit disk (Theorem 3.1 and Corollary 2) as well as for Jordan domains with rectifiable boundary (Theorem 10.4).

Proposition 2.9. *If $F \in H^p(\mathcal{B})$, then $F^+ \in H^p(\mathcal{S})$ and the norms of F and F^+ are equivalent, i.e., $||F^+||_p \approx ||F||_p$.*
If $p \geq 1$ and $f \in H^p(\mathcal{S})$, then $Pf = C_l f \in H^p(\mathcal{B})$, and $f = (C_l f)^+$, where F^+ is the non-tangential boundary trace of F.

Proof. For $f \in M(\mathcal{S})$ we have

$$\begin{aligned} Pf(a) &= \frac{1}{m(\mathcal{S})}\int_{\mathcal{S}} \frac{1-|a|^2}{|x-a|^n} f(x)\,d\mathcal{S} = \frac{1}{m(\mathcal{S})}\int_{\mathcal{S}} \frac{(\bar{x}-\bar{a})x+\bar{a}(x-a)}{|x-a|^n} f(x)\,d\mathcal{S} \\ &= \frac{1}{m(\mathcal{S})}\int_{\mathcal{S}} \frac{(\bar{x}-\bar{a})}{|x-a|^n}\, x\, f(x)\,d\mathcal{S} + \frac{\bar{a}}{m(\mathcal{S})}\int_{\mathcal{S}} \overline{\frac{(\bar{x}-\bar{a})}{|x-a|^n}} f(x)\,d\mathcal{S} = J_1 + J_2 \\ &= \frac{1}{m(\mathcal{S})}\int_{\mathcal{S}} \frac{(\bar{x}-\bar{a})}{|x-a|^n}\, x\, f(x)\,d\mathcal{S} = C_l f(a). \end{aligned}$$

The integral J_2 vanishes because $\frac{(\bar{x}-\bar{a})}{|x-a|^n}$ is a right monogenic function in $\overline{\mathcal{B}}$ and f is the non-tangential limit of a left monogenic function in $\mathcal{B}$. Hence $Pf = C_l f$ for any $f \in H^p(\mathcal{S})$.

To finish the proof we use harmonic analysis. Corollary (7.21) (on page 139) in [13] applies to the case of the unit sphere and the unit ball. □

Remark 2.10. Thus, the map $F \to F^+$ is a linear isometry of $H^p(\mathcal{B})$ onto $H^p(\mathcal{S})$. $H^1(\mathcal{S})$ is a closed subspace of $L^1(\mathcal{S})$. A similar definition and proposition applies to right monogenic functions.

2.4. Green's function for the unit ball

The Green's function for the unit ball is a composition of the fundamental solution of the Laplace operator Δ and a Möbius transformation $\varphi_a(x)$. We have

$$g(x,a) = \left(\frac{1}{|\varphi_a(x)|^{n-2}} - 1\right), \quad \text{where} \quad \varphi_a(x) = (a-x)(1-\bar{a}x)^{-1}.$$

Especially for $a = 0$ we have

$$g(x,0) = \left(\frac{1}{|x|^{n-2}} - 1\right).$$

Proposition 2.11. *Let $\Phi \in C^2(\overline{\mathcal{B}})$. Then*

$$\frac{1}{m(\mathcal{S})}\int_{\mathcal{S}} \Phi\,d\mathcal{S} = \frac{1}{(n-2)m(\mathcal{S})}\int_{\mathcal{B}} g(x,0)\Delta\Phi\,d\mathcal{B} + \Phi(0). \tag{2.2}$$

Proof. This is an immediate consequence of Green's theorem for the Laplacian and the fact that $g(x,0)\,|_{\mathcal{S}} = 0$. □

Remark 2.12. More general holds for $\Phi \in C^2(\overline{\mathcal{B}})$ and $a \in \mathcal{B}$:

$$\int_{\mathcal{S}} \Phi\,d\mu_a = \frac{1}{(n-2)m(\mathcal{S})}\int_{\mathcal{B}} G(x,a)\Delta\Phi\,d\mathcal{B} + \Phi(a),$$

where $d\mu_a = \frac{1}{m(\mathcal{S})}\frac{1-|a|^2}{|x-a|^n}d\mathcal{S}$ and $G(x,a) = \frac{1}{|1-\bar{a}x|^{n-2}}g(x,a)$, i.e., Green's function for the unit ball. This can be seen from the fact that the right-hand side as well as the left-hand side of the equation are harmonic functions which have the same boundary values.

Proposition 2.13. *Let F be a harmonic or monogenic function on the unit ball $\mathcal{B}$ which is continuous on $\overline{\mathcal{B}}$. Denote by f the restriction of F to the boundary. Then*

$$\frac{1}{m(\mathcal{S})}\int_{\mathcal{S}}|f(x)-F(0)|^2 d\mathcal{S} = \frac{2}{(n-2)m(\mathcal{S})}\int_{\mathcal{B}} g(x,0)|\nabla F|^2\, d\mathcal{B}.$$

Proof. Under the assumption that F is harmonic or monogenic the compounds F_k of F are harmonic. A simple computation shows now that $\Delta|F(x)-F(0)|^2 = 2|\nabla F|^2$. □

3. The space BMOM

An integrable function g belongs to the space of functions of bounded mean oscillation on the unit sphere $\mathcal{S}$ if there exists a positive constant $C>0$ such that

$$\frac{1}{m(E)}\int_E |g-g_E|\, d\mathcal{S} \le C$$

for all surface balls $E\subseteq\mathcal{S}$. Here $g_E := \frac{1}{m(E)}\int_E g\, d\mathcal{S}$ denotes the mean value of g over E. The space BMO is usually equipped with the norm

$$||g||_{**} = \sup_{\text{surface ball } E\subseteq\mathcal{S}} \frac{1}{m(E)}\int_E |g-g_E|\, d\mathcal{S}. \tag{3.1}$$

It should be mentioned that $||c||_{**}=0$ for constants. Therefore we identify functions in BMO which differ by a constant.

We use the Poisson kernel to represent monogenic functions. Let $a\in\mathcal{B}$ and $x\in\mathcal{S}$. Then the Poisson kernel for the unit ball is given by

$$P_a(x) = \frac{1}{m(\mathcal{S})}\frac{1-|a|^2}{|x-a|^n} = \frac{1}{m(\mathcal{S})}\frac{1-|a|^2}{(1+|a|^2-2|a|\cos\angle(x,a))^{n/2}},$$

where $m(\mathcal{S})$ denotes the Lebesgue measure of $\mathcal{S}$. In an analogous way to [3] for analytic functions we define the space of monogenic BMO-functions $BMOM(\mathcal{S})$:

Definition 3.1. An integrable function f on $\mathcal{S}$ belongs to $BMOM(\mathcal{S})$ if the Poisson integral of f

$$(Pf)(a) = \int_{\mathcal{S}} P_a(x)f(x)d\mathcal{S}_x$$

is a monogenic function in the unit ball $\mathcal{B}$ and $||f||_* < \infty$, where

$$||f||_* := \sup_{a\in\mathcal{B}} P|f-(Pf)(a)|(a) = \sup_{a\in\mathcal{B}}\int_{\mathcal{S}}|f-(Pf)(a)|d\mu_a, \tag{3.2}$$

$$\text{with } d\mu_a = \frac{1}{m(\mathcal{S})}\frac{1-|a|^2}{|x-a|^n}d\mathcal{S}.$$

For $n=2$ the measure $d\mu_a$ is the harmonic measure. In higher dimensions the harmonic measure has not the right behavior for creating an equivalent norm to the standard BMO-norm. A John-Nirenberg-type inequality using the measure μ_a is given in [5] and [6].

4. Equivalent norms of BMOM

To demonstrate that the new norm is indeed a norm for functions of bounded mean oscillation on the unit sphere we will demonstrate that the norm $||.||_*$ (3.2) is equivalent to the standard norm $||.||_{**}$ (3.1).

Theorem 4.1. *For functions $f \in BMOM(\mathcal{S})$ the norms $||f||_*$ and $||f||_{**}$ are equivalent.*

Proof. Let us choose an arbitrary point $a \in \mathcal{B}$ and consider the surface ball E with the following properties:

1. The center of E is $x_0 = \frac{a}{|a|}$.
2. The radius of E is $r = 1 - |a|$.

Under these assumptions we get for $x \in \mathcal{S}$:

$$\left| x - \frac{a}{|a|} \right|^2 \leq r^2 \iff \sum_{i=1}^{n} \left(x_i^2 + \frac{a_i^2}{|a|^2} - 2\cos\angle(x,a) \right) \leq r^2$$

$$\iff 2(1 - \cos\angle(x,a)) \leq r^2 = (1 - |a|)^2$$

and we can estimate

$$|x - a|^2 = 1 + |a|^2 - 2|a|\cos\angle(x,a) \leq (1 - |a|)^2(1 + |a|)$$

or equivalently

$$|x - a|^n \leq (1 - |a|)^n (1 + |a|)^{n/2}.$$

This allows us to get a lower bound for the Poisson kernel. We have

$$\begin{aligned} P_a(x) = \frac{1}{m(\mathcal{S})} \frac{1 - |a|^2}{|x - a|^n} &\geq \frac{1}{m(\mathcal{S})} \frac{1 - |a|^2}{(1 - |a|)^n (1 + |a|)^{n/2}} \\ &\geq \frac{(1 + |a|)^{1-n/2} 2^{n-1} C(n)}{m(\mathcal{S}) 2\pi^{n-2}(n-1)} \frac{1}{m(E)} \geq \frac{K(n)}{m(E)}, \end{aligned}$$

where an estimate of the constant $C(n)$ which depends only on n is given in the appendix. Hence,

$$\int_{\mathcal{S}} |f - f(a)| d\mu_a \geq \int_E |f - f(a)| d\mu_a \geq \frac{K(n)}{m(E)}$$

which gives raise to the final estimation

$$\begin{aligned} \frac{1}{m(E)} \int_E |f - f_E| d\mathcal{S} &\leq \frac{1}{m(E)} \int_E |f - f(a) + f(a) - f_E| d\mathcal{S} \\ &\leq \frac{1}{m(E)} \int_E |f - f(a)| d\mathcal{S} + \frac{1}{m(E)} \int_E d\mathcal{S} |f(a) - f_E| \\ &\leq \frac{2}{m(E)} \int_E |f - f(a)| d\mathcal{S} \leq \frac{2}{K(n)} \int_{\mathcal{S}} |f - f(a)| d\mu_a \end{aligned}$$

and thus

$$||f||_{**} \leq \frac{2}{K(n)} ||f||_* \ .$$

To prove the opposite inequality we construct a sequence of surface balls $E^m := 2^m E$, $m = 0, 1, \ldots, M-1$, i.e., $E^0 = E$, with the following properties. Let a be again an arbitrary point in the unit ball $\mathcal{B}$.

1. Each surface ball E^m has the center $\frac{a}{|a|}$ as E.
2. The radius of E^m is $r = 2^m(1-|a|)$.

Let M be the smallest natural number such that $2^M(1-|a|) \geq 2$. Then we set $E^M := \mathcal{S}$. For all $x \in \mathcal{S}$ it holds

$$\frac{1-|a|^2}{(1+|a|^2-2|a|\cos\angle(x,a))^{n/2}} \leq \frac{1-|a|^2}{(1+|a|^2-2|a|)^{n/2}} \leq \frac{2}{(1-|a|)^{n-1}}.$$

We can improve the last estimation on $\mathcal{S}\backslash E^m$. Because of

$$\left| x - \frac{a}{|a|} \right|^2 \geq (2^m(1-|a|))^2 \iff 2(1-\cos\angle(x,a)) \geq 2^{2m}(1-|a|)^2$$

we get

$$\begin{aligned} \frac{1-|a|^2}{(1+|a|^2-2|a|\cos\angle(x,a))^{n/2})} &\leq \frac{(1+|a|)(1-|a|)}{(1+|a|^2-2|a|+2|a|(1-\cos\angle(x,a)))^{n/2}} \\ \leq \frac{(1+|a|)(1-|a|)}{(1-|a|)^2+|a|2^{2m}(1-|a|)^2)^{n/2}} &\leq \frac{1+|a|}{(1-|a|)^{n-1}2^{nm}|a|^{n/2}} \leq \frac{2^{1-(2m-1)n/2}}{(1-|a|)^{n-1}}. \end{aligned}$$

To get the last estimation we have incorporated that if $m \geq 1$ then $M \geq 2$ and thus $2^2(1-|a|) \leq 2 \iff (1-|a|) \leq \frac{1}{2} \iff |a| \geq \frac{1}{2} \iff |a|^{-n/2} \leq 2^{n/2}$. □

5. A Carleson measure characterization

Definition 5.1. A positive measure ν defined on the unit ball $\mathcal{B}$ is a *Carleson measure* if there exists a constant $A > 0$ such that

$$\nu(Q_h(\xi)) \leq A h^{n-1}$$

for any subset

$$Q_h(\xi) := \{x \in \mathcal{B} : |x-\xi| < h, \}$$

where $h > 0$ and $\xi \in \mathcal{S}$.

Theorem 5.2. *Let be $F(x) = (Pf)(x)$. For $f \in BMOM(\mathcal{S})$ the measure*

$$(1-|x|)|\nabla F|^2\, dx$$

is a Carleson measure.

Proof. We choose an arbitrary $x_0 \in \mathcal{S}$ and set $B_r(x_0) = \{x \in \mathcal{B} : |x-x_0| < r\}$, $E_r = E(x_0, r)$, $r > 0$, where E_r is the surface ball with radius r and center x_0 on the unit sphere $\mathcal{S}$. Now, we decompose f in the usual way:

$$f = (f - f_{E_r})\chi_{E_r} + (f - f_{E_r})\chi_{E_r^c} + f_{E_r} = f_1 + f_2 + f_{E_r}.$$

Hence

$$Pf = Pf_1 + Pf_2 + Pf_{E_r} = Pf_1 + Pf_2 + f_{E_r} = F_1 + F_2 + f_{E_r}$$

and

$$\nabla F = \nabla(Pf) = \nabla F_1 + \nabla F_2.$$

Because of $|\nabla F|^2 \leq c(|\nabla F_1|^2 + |\nabla F_2|^2)$ we only need to prove that $(1-|x|)|\nabla F_1|^2$ as well as $(1-|x|)|\nabla F_2|^2$ are Carleson measures on $\mathcal{B}$. First, we consider F_1.

$$\begin{aligned}
\int_{\mathcal{B}} (1-|x|)|\nabla F_1|^2\,dx &\leq \tfrac{(n-2)}{2} \int_{\mathcal{B}} g(x,0)|\nabla F_1|^2\,dx = \int_{\mathcal{S}} |f_1(x) - F_1(0)|^2\,d\mathcal{S} \\
&= \tfrac{(n-2)}{2} \int_{\mathcal{S}} \left| (f - f_{E_r})\chi_{E_r} - \int_{\mathcal{S}} (f - f_{E_r})\chi_{E_r}\,d\mathcal{S} \right|^2 d\mathcal{S} \\
&\leq (n-2)\left(\int_{\mathcal{S}} |(f - f_{E_r})\chi_{E_r})|^2\,d\mathcal{S} + \int_{\mathcal{S}} \left| \int_{\mathcal{S}} (f - f_{E_r})\chi_{E_r})\,d\mathcal{S} \right|^2 d\mathcal{S} \right) \\
&\leq (n-2)\int_{E_r} |f - f_{E_r}|^2\,d\mathcal{S} + 4(n-2)\pi \left(\int_{E_r} |f - f_{E_r}|\,d\mathcal{S} \right)^2 \\
&\leq (n-2)\frac{m(E_r)}{m(E_r)} \int_{E_r} |f - f_{E_r}|^2\,d\mathcal{S} + 4(n-2)\pi \frac{(m(E_r))^2}{m(E_r)} \int_{E_r} |f - f_{E_r}|^2\,d\mathcal{S} \\
&\leq \mathit{const} \cdot ||f||_{BMO(\mathcal{S})} m(E_r) \leq \mathit{const} \cdot ||f||_{BMO(\mathcal{S})} \cdot r^{(n-1)}
\end{aligned}$$

due to $m(E_r) \leq m(\mathcal{S})$. To get the estimation for F_2 we consider the partial derivatives

$$F_2(x) = \int_{\mathcal{S}} P_x(y) f_2(y)\,d\mathcal{S}_y = \frac{1}{m(\mathcal{S})} \int_{E_r^c} \frac{1-|x|^2}{|x-y|^n} (f(y) - f_{E_r})\,d\mathcal{S}_y,$$

the integral is non-singular and we can take the derivative straightforward manner

$$\frac{\partial}{\partial x_j}(F_2(x)) = \frac{1}{m(\mathcal{S})} \int_{E_r^c} \left(\frac{-2x_j}{|x-y|^n} - n\frac{(1-|x|^2)(y_j - x_j)}{|x-y|^{n+2}} \right) (f(y) - f_{E_r})\,d\mathcal{S}_y.$$

Because of $y \in \mathcal{S}$ we have $|1-|x|| = ||y| - |x|| \leq |y-x|$, and if $x \in B_r(x_0)$ we get $\frac{1}{2}|x_0 - y| \leq |x-y| \leq 2|x_0 - y|$.

$$\begin{aligned}
\left| \frac{\partial}{\partial x_j}(F_2(x)) \right| &\leq \mathit{const} \cdot \int_{E_r^c} \frac{|f - f_{E_r}|}{|x-y|^n}\,d\mathcal{S} \leq \mathit{const} \cdot \int_{E_r^c} \frac{|f - f_{E_r}|}{|x_0 - y|^n}\,d\mathcal{S} \\
&= \mathit{const} \cdot \frac{1}{r} \int_{E_r^c} |f - f_{E_r}| \frac{r}{|x_0 - y|^n}\,d\mathcal{S}.
\end{aligned}$$

We rewrite the last integral as an infinite series that converges with respect to r.

$$\sum_{k=0}^{\infty} \int_{E_{2^{k+1}r} \setminus E_{2^k r}} |f - f_{E_r}| \frac{r}{|x_0 - y|^n}\,d\mathcal{S} \leq \sum_{k=0}^{\infty} 2^{-nk} r^{-(n-1)} \int_{E_{2^{k+1}r}} |f - f_{E_r}|\,d\mathcal{S}.$$

Taking into account

$$\begin{aligned}
|f_{E_{2^{k+1}r}} - f_{E_r}| &= |f_{E_{2^{k+1}r}} - f_{E_{2^k r}} + f_{E_{2^k r}} - f_{E_{2^{k-1}r}} + \ldots - f_{E_r}| \\
&\leq |f_{E_{2^{k+1}r}} - f_{E_{2^k r}}| + |f_{E_{2^k r}} - f_{E_{2^{k-1}r}}| + \ldots + |f_{E_{2r}} - f_{E_r}|
\end{aligned}$$

$$\leq \frac{m(E_{2^{k+1}r})}{m(E_{2^k r})} \frac{1}{m(E_{2^{k+1}r})} \int_{E_{2^{k+1}r}} |f - f_{E_{2^{k+1}}}|\, dS + \frac{m(E_{2^k r})}{m(E_{2^{k-1}r})} \frac{1}{m(E_{2^k r})} \int_{E_{2^k r}} |f - f_{E_{2^k}}|\, dS + + \ldots + \frac{m(E_{2r})}{m(E_r)} \frac{1}{m(E_{2r})} \int_{E_{2r}} |f - f_{E_{2r}}|\, dS$$

$$\leq \frac{(m(E_{2^{k+1}r}))^2}{m(E_{2^k r})} \left(\frac{1}{m(E_{2^{k+1}r})} \int_{E_{2^{k+1}r}} |f - f_{E_{2^{k+1}}}|^2\, dS \right)^{1/2} + \frac{(m(E_{2^k r}))^2}{m(E_{2^{k-1}r})} \left(\frac{1}{m(E_{2^k r})} \int_{E_{2^k r}} |f - f_{E_{2^k}}|^2\, dS \right)^{1/2} + + \ldots + \frac{(m(E_{2r}))^2}{m(E_r)} \left(\frac{1}{m(E_{2r})} \int_{E_{2r}} |f - f_{E_{2r}}|^2\, dS \right)^{1/2}$$

$$\leq const \cdot k\, ||f||^{1/2}_{BMO(\mathcal{S})}$$

we estimate

$$\sum_{k=0}^{\infty} 2^{-nk} r^{-2} \int_{E_{2^{k+1}r}} |f - f_{E_r}|\, dS \leq const \cdot \sum_{k=0}^{\infty} 2^{n-1-k} k\, ||f||^{1/2}_{BMO(\mathcal{S})} \leq const \cdot ||f||^{1/2}_{BMO(\mathcal{S})}.$$

Finally, we obtain

$$\begin{aligned} \int_{B_r(x_0)} (1 - |x|) |\nabla F|^2\, dx &\leq const \cdot \int_{B_r(x_0)} (1 - |x|) \left(|\nabla F_1|^2 + |\nabla F_2|^2 \right) dx \\ &\leq const \cdot \int_{\mathcal{B}} (1 - |x|) |\nabla F_1|^2\, dx + \int_{B_r(x_0)} |\nabla F_2|^2\, dx \\ &\leq const \cdot ||f||_{BMO(\mathcal{S})} \cdot r^{(n-1)} + r^n \frac{1}{r} ||f||_{BMO(\mathcal{S})} \\ &\leq const \cdot ||f||_{BMO(\mathcal{S})} \cdot r^{(n-1)}. \end{aligned}$$

□

6. Duality and representation

The bilinear form

$$f \mapsto \int_{\mathcal{S}} hf\, dS$$

is well-defined for left-linear h and right-linear f, i.e., it defines left-linear functional over a right-linear space.

Theorem 6.1. *Let $f \in BMOM^l(\mathcal{S})$ (the right-linear space of boundary values of left monogenic functions in $\mathcal{B}$) and $h \in H^{1,r}(\mathcal{S})$ (the left-linear space of boundary values of right monogenic Hardy space $H^1(\mathcal{B})$. Then*

$$\left|\int_{\mathcal{S}} hf\, d\mathcal{S}\right| \leq C(f)||h||_{H^{1,r}} \ .$$

Proof. Let H be the right monogenic extension of h and let F be the left monogenic extension of f into $\mathcal{B}$. We assume that $H(0) = 0$. Using Green's formula for harmonic functions in the unit ball (see (2.2)) we get

$$\begin{aligned}\left|\int_{\mathcal{S}} hf\, d\mathcal{S}\right| &= \left|\tfrac{1}{n-2}\int_{\mathcal{B}} g(x,0)\Delta(HF)\, dx + \tfrac{1}{m(\mathcal{S})}(HF)(0)\right| \\ &= \left|\tfrac{1}{n-2}\int_{\mathcal{B}} g(x,0)\Delta(HF)\, dx\right| = \left|\tfrac{2}{n-2}\sum_{j=1}^{n}\int_{\mathcal{B}} \frac{\partial H}{\partial x_j}\frac{\partial F}{\partial x_j} g(x,0)\, dx\right|\end{aligned}$$

due to $H(0) = 0$, and

$$\begin{aligned}\Delta(HF) &= \sum_{j=0}^{n}\frac{\partial^2}{\partial x_j^2}(HF) = 2\sum_{j=0}^{n}\frac{\partial^2 H}{\partial x_j^2}F + \frac{\partial H}{\partial x_j}\frac{\partial F}{\partial x_j} + H\frac{\partial^2 F}{\partial x_j^2} = \\ &= 2\left[(\Delta H)F + H(\Delta F) + \sum_{j=0}^{n}\frac{\partial H}{\partial x_j}\frac{\partial F}{\partial x_j}\right] = 2\sum_{j=0}^{n}\frac{\partial H}{\partial x_j}\frac{\partial F}{\partial x_j},\end{aligned}$$

because any right or left monogenic function is harmonic. We get using Fefferman's trick:

$$\begin{aligned}\left|\int_{\mathcal{S}} hf\, d\mathcal{S}\right| &\leq \tfrac{2}{n-2}\int_{\mathcal{B}} |\nabla F|\cdot|\nabla H|\, g(x,0)\, dx \\ &\leq \tfrac{2}{n-2}\left(\int_{\mathcal{B}} |\nabla F|^2\cdot|H|\, g(x,0)\, dx\right)^{1/2}\left(\int_{\mathcal{B}} |\nabla H|^2\cdot|H|^{-1} g(x,0)\, dx\right)^{1/2} \\ &= \tfrac{2}{n-2}J_1^{1/2}J_2^{1/2}.\end{aligned}$$

At first we consider the integral J_1. If $|x| < \frac{1}{2}$, the integral exists because

$$\begin{aligned}\int_{\{x\in\mathcal{B}:|x|<\frac{1}{2}\}} |\nabla F|^2\cdot|H|\, g(x,0)\, dx &\leq \sup_{\{x\in\mathcal{B}:|x|<\frac{1}{2}\}}|H|\cdot\int_{\{x\in\mathcal{B}:|x|<\frac{1}{2}\}} |\nabla F|^2\, g(x,0)\, dx \\ &\leq ||h||_{H^1}\int_{\mathcal{B}} |\nabla F|^2 g(x,0)\, dx = ||h||_{H^1}\tfrac{(n-2)}{2}\int_{\mathcal{S}} |f - F(0)|^2 d\mu \\ &\leq C||f||^2_{BMO}\cdot||h||_{H^1}.\end{aligned}$$

From the estimation $|g(x,0)| \leq 2^{n-2}(n-2)(1-|x|)$ on $|x| \geq \frac{1}{2}$ and the fact that $|\nabla F|^2 (1-|x|)$ is a Carleson measure we obtain

$$\int_{\{x\in\mathcal{B}:|x|\geq\frac{1}{2}\}} |\nabla F|^2 \cdot |H|\, g(x,0)\, dx \leq 2^{n-2}(n-2) \int_{\mathcal{B}} |\nabla F|^2 \cdot |H|\,(1-|x|)\, dx$$
$$\leq C(f) \cdot ||h||_{H^1}.$$

and thus $|J_1| \leq c(f) \cdot ||h||_{H^1}$. Next, we estimate the integral J_2:

$$\int_{\mathcal{B}} |\nabla H|^2 \cdot |H|^{-1} g(x,0)\, dx \leq n \int_{\mathcal{B}} \Delta |H| g(x,0) dx$$
$$= n \int_{\mathcal{S}} |h| d\mathcal{S} - H(0) = n \int_{\mathcal{S}} |h| d\mathcal{S} \leq n ||h||_{H^1}.$$

Here, the subharmonicity of the right monogenic function H is used.

Finally, we obtain the desired estimation

$$\left| \int_{\mathcal{S}} hf\, d\mathcal{S} \right| \leq C(f) \cdot ||h||_{H^1}.$$

If $H(0) = H_0 \neq 0$, we decompose

$$\begin{aligned}\left|\int_{\mathcal{S}} hf\, d\mathcal{S}\right| &= \left|\int_{\mathcal{S}} (h-H_0) f\, d\mathcal{S} + \int_{\mathcal{S}} H_0 f\, d\mathcal{S}\right| \\ &\leq \left|\int_{\mathcal{S}} (h-H_0) f\, d\mathcal{S}\right| + |H_0| \left|\int_{\mathcal{S}} f\, d\mathcal{S}\right| \\ &\leq \left|\int_{\mathcal{S}} (h-H_0) f\, d\mathcal{S}\right| + ||f||_{L^1} ||h||_{H^1} \\ &\leq C(f) ||h-H_0||_{H^1} + ||f||_{BMO} ||h||_{H^1} \leq \tilde{C}(f) ||h||_{H^1}.\end{aligned}$$

This completes the proof. □

Remark 6.2. Right monogenic functions are harmonic and fulfill also the subharmonicity property (see [13], page 106), and they form a left-(Clifford)-linear space.

Now, the space $H^1(\mathcal{S})$ is a closed subspace of $L^1(\mathcal{S})$ and therefore

$$f \mapsto \int_{\mathcal{S}} hf\, d\mathcal{S}$$

extends to a functional over $L^1(\mathcal{S})$. Hence, there exists a $b \in L^\infty(\mathcal{S})$ such that

$$\int_{\mathcal{S}} hf\, d\mathcal{S} = \int_{\mathcal{S}} hb\, d\mathcal{S}$$

for all $h \in L^1(\mathcal{S})$. Let us take the functions $C^+_{n,k}(\omega,\xi) \in L^1(\mathcal{S})$ (cf. (2.1)). Then

$$\int_{\mathcal{S}} C^+_{n,k}(\omega,\xi) f\, d\mathcal{S} = \int_{\mathcal{S}} C^+_{n,k}(\omega,\xi) b\, d\mathcal{S},$$

i.e., the inner monogenic projections are equal to $P_k f = P_k b$ and thus

$$F(x) = \sum_{k=0}^{\infty} \frac{|x|^k}{m(\mathcal{S})} \int_{\mathcal{S}} C^+_{n,k}(\omega, \xi) f \, d\mathcal{S} = \sum_{k=0}^{\infty} \frac{|x|^k}{m(\mathcal{S})} \int_{\mathcal{S}} C^+_{n,k}(\omega, \xi) b \, d\mathcal{S}$$

$$= \sum_{k=0}^{\infty} \frac{|x|^k}{m(\mathcal{S})} \int_{\mathcal{S}} C^+_{n,k}(\omega, \xi) \overline{\omega} \omega b \, d\mathcal{S} = \int_{\mathcal{S}} e(x - \omega) \omega b \, d\mathcal{S} = C_l b(x).$$

Hence,

Theorem 6.3. *A left monogenic function F in $\mathcal{B}$ with boundary values $f \in BMOM^l(\mathcal{S})$ can be represented as a Cauchy integral of a function $b \in L^{\infty,l}(\mathcal{S})$:*

$$F(x) = \int_{\mathcal{S}} e(x - \omega) \omega b(\omega) \, d\mathcal{S}.$$

Proof. The Clifford-valued function b may be written as

$$b(x) = \sum_{\alpha} b_\alpha e_\alpha,$$

and we define

$$F_\alpha(x) := \int_{\mathcal{S}} e(x - \omega) \omega b_\alpha(\omega) \, d\mathcal{S}.$$

Then $F(x) = F_\alpha(x) e_\alpha$. Now, we consider the scalar part of each F_α. We have

$$\operatorname{Sc} F_\alpha(x) = \operatorname{Sc} \int_{\mathcal{S}} e(x - \omega) \omega b_\alpha(\omega) \, d\mathcal{S} = \int_{\mathcal{S}} \operatorname{Sc} \left(e(x - \omega) \omega \right) b_\alpha(\omega) \, d\mathcal{S}$$

$$= \int_{\mathcal{S}} \left(e(x - \omega) \cdot \omega \right) b_\alpha(\omega) \, d\mathcal{S}.$$

Hence $\operatorname{Sc} F_\alpha$ is the double layer potential of $b_\alpha \in L^\infty(\mathcal{S})$ and therefore $\operatorname{Sc} F_\alpha \in L^\infty(\mathcal{B})$. □

Theorem 6.4. *A left monogenic function F in $\mathcal{B}$ with boundary values $f \in BMOM^l(\mathcal{S})$ can be represented as*

$$F = \sum_{\alpha} F_\alpha e_\alpha,$$

where $\operatorname{Sc} F_\alpha \in L^\infty$.

7. Remark

We got a relation among functions from a Hardy space and functions of bounded mean oscillation both representing boundary values of functions monogenic inside the unit ball $\mathcal{B}$. But we can also obtain similar relations among monogenic functions inside the unit ball and outside. This comes from the fact that the inversion operator I^* transforms a function f monogenic inside the unit ball into a function

I^*f monogenic outside the unit ball ([11]). If we extend the operator I^* to $L_2(\mathcal{S})$ by putting for $f \in L_2(\mathcal{S})$

$$I^*f(\omega) = \omega f(\omega)$$

we get that

$$\int_{\mathcal{S}} hf\, d\mathcal{S} = \int_{\mathcal{S}} h\overline{\omega} g\, d\mathcal{S},$$

where $g = \omega f$ are the boundary values of a monogenic function outside the unit ball.

In $\mathbb{R}^2 \sim \mathbb{C}^1$ there is only one Poisson kernel. This it no longer true in higher complex dimensions. Here, there must be distinguished between the Poisson kernel and the Poisson-Szegö kernel. The same situation occurs in higher real dimensions. We used the Poisson kernel for the unit ball in $\mathbb{R}^n$,

$$P_a(x) = \frac{1}{m(\mathcal{S})} \frac{1-|a|^2}{|x-a|^n}.$$

This kernel is also used in harmonic analysis. But there is also a Poisson-Szegö kernel (introduced in [10])

$$PS_a(x) = \frac{1}{m(\mathcal{S})} \left(\frac{1-|a|^2}{|x-a|^2} \right)^{n-1}.$$

The basic feature that allows to prove the equivalence of the standard BMO-norm with $||.||_*$ is that

$$P_a(x) \sim \frac{1}{(1-|a|)^{n-1}} \sim \frac{1}{m(E)}.$$

But this is also fulfilled for $PS_a(x)$ because from

$$\frac{2}{(1-|a|)} \geq \frac{1-|a|^2}{|x-a|^2} \geq \frac{(1+|a|)^{1/2}}{(1-|a|)} \geq \frac{1}{(1-|a|)}$$

we also get

$$PS_a(x) \sim \frac{1}{(1-|a|)^{n-1}} \sim \frac{1}{m(E)}.$$

Especially $PS_a(x)$ is much more connected to conformal invariance than $P_a(x)$. But as in several complex variables the Poisson-Szegö kernel is *not* harmonic but *M-harmonic* (see [9]).

In classical complex analysis the Dirichlet and the Bloch space can be connected by the scale of Q_p-spaces and $Q_1 = BMOA$ (space of analytic functions of bounded mean oscillation), see [2]. The Q_p-spaces are generalized to higher dimensions in a different way in [14] and in [10]. But it could not be proven that $Q_1 = BMOM$ or at least $Q_p = BMOM$ for some p.

References

[1] S. Axler, P. Bourdon, W. Ramey: *Harmonic Function Theory,* 2. Ed., Graduate Texts in Mathematics 137, Springer-Verlag, New York, Berlin, Heidelberg, 2001.

[2] R. Auleskari, J. Xiao and R. Zhao: *On subspaces of BMOA and UBC*, Analysis **15** (1995), 101–121.

[3] A. Baernstein II.: Analytic functions of Bounded Mean Oscillation, in: Aspects of contemporary analysis, Proc. instr. Conf. Durham/England 1979, 3–36, 1980.

[4] C.A. Berenstein and R. Gay: *Complex Analysis and Special Topics in Harmonic Analysis*, Springer-Verlag, New York, Berlin, Heidelberg, 1995.

[5] S. Bernstein: *Integralgleichungen und Funktionenräume für Randwerte monogener Funktionen,* Habilitationsschrift, Fakultät für Mathematik und Informatik, TU Bergakademie Freiberg, 2001.

[6] S. Bernstein: *The space of monogenic BMO-functions and a John-Nirenberg inequality,* 3rd international ISAAC congress, Berlin, 2001, to appear.

[7] P.L. Duren: *Theory of H^p spaces,* Pure and Applied Mathematics, Vol. 38, Academic Press, 1970.

[8] A. Erdélyi, W. Magnus, F. Oberhettinger and F. Tricomi: *Higher Transcendental Functions,* Mc Graw-Hill, New York, 1953.

[9] P. Cerejeiras: *Reproducing Properties of Hyperbolically-Harmonic and Monogenic Functions,* Adv. in Appl. Clifford Algebras **11(S2)**, (2001), 61–66.

[10] J. Cnops, R. Delanghe: *Möbius invariant spaces in the unit ball*, Appl. Anal. **73**, No. 1–2, (1999), 45–64.

[11] R. Delanghe, F. Sommen and V. Souček: *Clifford Algebra and Spinor-Valued Functions, A Function Theory for the Dirac Operator,* Kluwer Academic Publishers, Dordrecht, Boston, London, 1992.

[12] C. Fefferman, E. M. Stein: *H^p spaces of several variables,* Acta Math. **129**, (1972), 137–193.

[13] J. E. Gilbert, M. A. M. Murray: *Clifford algebras and Dirac operators in harmonic analysis,* (Cambridge studies in advances mathematics 26), Cambridge University Press, 1991.

[14] K. Gürlebeck, U. Kähler, M.V. Shapiro and L.M. Tovar: *On Q_p-Spaces of Quaternion-Valued Functions*, Complex Variables **39** (1999), 115–135.

[15] H. Hochstadt: *The Functions of Mathematical Physics,* Wiley, New York, 1971.

[16] K. Hoffman: *Banach Spaces of Analytic Functions,* Dover Publications, Inc., 1988, (reprint of the work first published by Prentice-Hall, Inc., Englewood Cliffs, N.J., in 1962).

[17] F. John and L. Nirenberg: *On functions of bounded mean oscillation,* Comm. Pure Appl. Math. **14**, (1960), 415–426.

[18] H. Leutwiler: *Lectures on BMO*, Universität Erlangen-Nürnberg, Mathematisches Institut, 1988.

[19] H. Leutwiler: *Harmonic functions of bounded mean oscillation*, Mathematische Annalen **244**, (1979), 167–183.

[20] A. McIntosh: *Clifford algebras, Fourier theory, singular integrals, and harmonic functions on Lipschitz domains.* in: Ryan, John (ed.), Clifford algebras in analysis and related topics. Based on a conference, Fayetteville, AR, USA, April 8-10, 1993. Boca Raton, FL: CRC Press. Studies in Advanced Mathematics, 33–87, (1996).

[21] M. Mitrea: *Clifford Waveletts, Singular Integrals and Hardy Spaces,* Lecture Notes in Mathematics, vol. 1575, Springer-Verlag, New York, Berlin, Heidelberg, 1994.

[22] G.C. Moisil, N. Teodorescu, *Fonctions holomorphes dans l'espace,* Mathematicae Cluj **5**, (1931), 142–150.

[23] W. Rudin: *Function theory in the Unit Ball of* $\mathbb{C}^n$, Springer-Verlag, New York, Heidelberg, Berlin, 1980.

Appendix

The unit sphere is a very special manifold and has a lot of useful properties. As a compact space of homogenous type the unit sphere fulfills the doubling condition, i.e., there exist positive constants $b, c > 0$ such that for all $x \in \mathcal{S}$, $t \geq 1$ and all $r > 0$

$$m(E(x,tr)) \leq ct^b\, m(E(x,r)),$$

where $E(x,r)$ denotes the surfaces ball with center $x \in \mathcal{S}$ and radius $r > 0$. But to make it clear we will give a simple proof of this fact and fix some constants.

First of all, we want to calculate the area measure of the set

$$E(x_0,r) := \left\{x \in \mathcal{S} : \left|x - x_0\right| < r\right\},$$

for an arbitrarily chosen $x_0 \in \mathcal{S}$. Without loss of generality we can assume

$$x_0 = (1,0,\dots,0)$$

and thus $|x - (1,0,\dots,0)|^2 < r^2$, and

$$\sum_{j=1}^{n} x_j^2 + 1 - 2x_1 = 2(1 - x_1) < r^2.$$

Using spherical coordinates we have $x_1 = \cos\vartheta_1$ and

$$2(1-x_1) = 2(1-\cos\theta_1) < r^2 \iff \sin\frac{\vartheta_1}{2} < \frac{r}{2} \iff \vartheta_1 < 2\arcsin\frac{r}{2}.$$

Hence,

$$\begin{aligned} m(E(x_0,r)) &= \int_{\mathcal{S}\cap E(x_0,r)} d\mathcal{S} \\ &= \int_0^{2\arcsin\frac{r}{2}} \sin^{n-2}\vartheta_1\, d\vartheta_1 \int_0^{\pi} \sin^{n-3}\vartheta_2\, d\vartheta_2 \dots \int_0^{\pi} \sin\vartheta_{n-2}\, d\vartheta_{n-2} \int_{-\pi}^{\pi} d\vartheta_{n-1} \\ &= C(n) \int_0^{2\arcsin\frac{r}{2}} \sin^{n-2}\vartheta_1\, d\vartheta_1, \end{aligned}$$

where $C(n) = \int_0^{\pi} \sin^{n-3}\vartheta_2\, d\vartheta_2 \dots \int_0^{\pi} \sin\vartheta_{n-2}\, d\vartheta_{n-2} \int_{-\pi}^{\pi} d\vartheta_{n-1}$.

In case of

$$0 < 2\arcsin\frac{r}{2} \leq \frac{\pi}{2} \iff 0 < \frac{r}{2} \leq \frac{\sqrt{2}}{2} \iff 0 < r \leq \sqrt{2},$$

we get

$$\arcsin\frac{r}{2} = \frac{r}{2} + \frac{1}{2}\left(\frac{r}{2}\right)^3 + \frac{1}{2}\cdot\frac{3}{4}\left(\frac{r}{2}\right)^5 + \ldots = \frac{r}{2}\left(1 + \frac{1}{2}\left(\frac{r}{2}\right)^2 + \frac{1}{2}\cdot\frac{3}{4}\left(\frac{r}{2}\right)^4 + \ldots\right)$$

and

$$1+\frac{1}{2}\left(\frac{r}{2}\right)^2+\frac{1}{2}\cdot\frac{3}{4}\left(\frac{r}{2}\right)^4+\ldots \leq \sum_{n=0}^{\infty}\left(\frac{r}{2}\right)^{2n} \leq \sum_{n=0}^{\infty}\left[\left(\frac{r}{2}\right)^2\right]^n = \frac{1}{1-\left(\frac{r}{2}\right)^2} \leq \frac{1}{1-\frac{1}{2}} = 2,$$

which leads to the estimation

$$r = 2\cdot\frac{r}{2} \geq \arcsin\frac{r}{2} \geq \frac{r}{2}.$$

If $0 \leq \vartheta_1 \leq \frac{\pi}{2}$ we have

$$1 \geq \frac{\sin\vartheta_1}{\vartheta_1} \geq \frac{2}{\pi},$$

and thus

$$\vartheta_1 \geq \sin\vartheta_1 \geq \frac{2}{\pi}\vartheta_1 \iff \vartheta_1^{n-2} \geq \sin^{n-2}\vartheta_1 \geq \left(\frac{2}{\pi}\right)^{n-2}\vartheta_1^{n-2}.$$

Now, we are able to estimate the expression

$$m(E(x_0,r)) = C(n)\int_0^{2\arcsin\frac{r}{2}} \sin^{n-2}\vartheta_1 d\vartheta_1.$$

We have

$$\begin{aligned} C(n)\int_0^{2\arcsin\frac{r}{2}} \vartheta_1^{n-2}\,d\vartheta_1 \geq m(E(x_0,r)) \geq C(n)\left(\frac{2}{\pi}\right)^{n-2}\int_0^{2\arcsin\frac{r}{2}} \vartheta_1^{n-2}\,d\vartheta_1 \\ \iff \frac{C(n)}{n-1}\left(2\arcsin\frac{r}{2}\right)^{n-1} \geq m(E(x_0,r)) \geq \frac{C(n)}{n-1}\left(\frac{2}{\pi}\right)^{n-2}\left(2\arcsin\frac{r}{2}\right)^{n-1} \\ \iff \frac{C(n)2^{n-1}}{n-1}\,r^{n-1} \geq m(E(x_0,r)) \geq \frac{C(n)2^{n-1}}{(n-1)2\pi^{n-2}}\,r^{n-1}, \\ 0 < r \leq \sqrt{2}. \end{aligned}$$

Let us now consider the case $\frac{m(E(x_0,tr))}{m(E(x_0,r)}$ where $t \geq 1$. In case of $tr < \sqrt{2} \iff r < \frac{\sqrt{2}}{t}$ we already obtained that

$$\frac{m(E(x_0,tr))}{m(E(x_0,r))} \leq \frac{C(n)2^{n-1}(tr)^{n-1}2\pi^{n-2}(n-1)}{(n-1)C(n)2^{n-1}r^{n-1}} \leq 2\pi^{n-2}t^{n-1}.$$

But if $r \geq \frac{\sqrt{2}}{t}$, then $m(E(x_0, tr)) \leq m(\mathcal{S})$ and $m(E(x_0, r)) \geq m(E_{\frac{\sqrt{2}}{t}})$, i.e.,

$$\frac{m(E(x_0, tr))}{m(E(x_0, r))} \leq \frac{m(\mathcal{S})(n-1)2\pi^{n-2}t^{n-1}}{\sqrt{2}^{n-1}C(n)2^{n-1}}.$$

Therefore we obtain

$$m(E(x_0, tr)) \leq c\, t^{n-1}\, m(E(x_0, r)).$$

Swanhild Bernstein
Bauhaus-Universität Weimar
Mathematische Optimierung
Coudraystr. 13B
99421 Weimar
Germany
e-mail: swanhild.bernstein@fossi.uni-weimar.de

Trends in Mathematics: *Advances in Analysis and Geometry*, 51–63

$B^{p,q}$-Functions and their Harmonic Majorants

S. Bernstein, K. Gürlebeck, L. F. Reséndis, and Luis M. Tovar S.

Abstract. The aim of this work is to characterize hyperholomorphic $\mathbf{B^{p,q}}$-functions in terms of harmonic majorants. In addition we point out how some important relations between $\mathbf{B^{p,q}}$, Bloch and $\mathbf{Q_p}$-spaces can be expressed in terms of their corresponding harmonic majorants.

Mathematics Subject Classification (2000). Primary 30G35; Secondary 31B05.

Keywords. Hyperholomorphic $B^{p,q}$-functions, harmonic majorants.

1. Introduction

Let $\Delta = \{z \in \mathbb{C} : |z| < 1\}$ be the complex unit disk and let $\mathcal{A}$ be the family of analytic functions in Δ. Consider the well-known BMOA space

$$BMOA = \{f \in \mathcal{A} : \sup_{a\in\Delta} \int\int_{\Delta} |f'(z)|^2(1 - |\varphi_a(z)|^2)dxdy < \infty\},$$

where $\varphi_a(z)$ is the Möbius transformation

$$\varphi_a(z) = \frac{a - z}{1 - \bar{a}z}, \quad |a| < 1.$$

This function maps the unit disc onto the unit disc, $\varphi_a(\Delta) = \Delta$, and has the properties $\varphi_a(\varphi_a(z)) = z, \varphi_a(0) = a, \varphi_a(a) = 0$ and $\lim_{|z|\mapsto 1} |\varphi_a(z)| = 1 \quad \forall a \in \Delta$.

In 1984, Kobayashi [11] gave a characterization of a function f in the space BMOA in terms of the least harmonic majorants of the subharmonic functions

$$u_a(z) = |f(z) - f(a)|^2 \quad \text{for } a \in \Delta.$$

In 1993, R. Aulaskari and P. Lappan [2] introduced the $\mathbf{Q_p}$-spaces for any $p > 0$ as the family of functions in $\mathcal{A}$ that satisfies

$$\mathbf{Q_p} = \{f \in \mathcal{A} : \sup_{a\in\Delta} \int\int_{\Delta} |f'(z)|^2(1 - |\varphi_a(z)|^2)^p dxdy < \infty\}.$$

Thus for $p = 1$ we obtain the BMOA space and for $p = 0$ we obtain the Dirichlet space

$$\mathbf{D} = \{f \in \mathcal{A} : \int\int_{\Delta} |f'(z)|^2 dxdy < \infty\}.$$

Besides, Aulaskari and Lappan proved that for $p > 1$, $\mathbf{Q_p}$ coincides always with the well studied Bloch space.

Several difficulties occur if we try to generalize Kobayashi's tools, used in [11] for functions in $\mathbf{Q_p}$. So Aulaskari et al in [3] introduced a new idea to characterize $\mathbf{Q_p}$-functions in terms of harmonic majorants in such a way that the main properties of the $\mathbf{Q_p}$-theory have their correspondent one with harmonic majorants.

In 1999 Gürlebeck et al [10] defined $\mathbf{Q_p}$-spaces of hyperholomorphic functions defined in the unit ball of $\mathbb{R}^3$. Recently in [1] the authors introduced the $\mathbf{B^{p,q}}$-spaces of hyperholomorphic functions that generalize in several ways the $\mathbf{Q_p}$-spaces. These hyperholomorphic $\mathbf{Q_p}$-spaces are defined with the help of a quaternionic derivative and the quaternionic Möbius transform. The quaternionic Möbius transform is needed to define a weight function that concentrates the mass of the weight at the point a. Doing this the new definition extends the complex one-dimensional case in such a way that the restriction of the new model to the two-dimensional case reproduces the original complex function spaces. Moreover, to keep the results of this paper compatible to Kobayashi's results, we should work with weight functions of the same type. It should be mentioned that for other purposes one can work also with equivalent weight functions based on the distance of the point to the boundary only.

In this work we will extend the techniques used in [3] for the complex case to obtain the corresponding characterizations of hyperholomorphic $\mathbf{B^{p,q}}$-functions in terms of harmonic majorants. Besides, we present, too, how several important relationships among hyperholomorphic $\mathbf{B^{p,q}}$, Bloch, and $\mathbf{Q_p}$-spaces can be translated in terms of its corresponding harmonic majorants.

To introduce the meaning of hyperholomorphic functions let $\mathbb{H}$ be the set of real quaternions. This means that elements of $\mathbb{H}$ are of the form

$$a = \sum_{k=0}^{3} a_k e_k, \quad \text{where} \quad \{a_k | k \in \mathbb{N}_3^0 := \mathbb{N}_3 \cup \{0\}; \ \mathbb{N}_3 := \{1,2,3\}\} \subset \mathbb{R};$$

$e_0 = 1$ the unit; e_1, e_2, e_3 are called imaginary units. They define arithmetic rules in $\mathbb{H}$; by definition $e_k^2 = -e_0$, $k \in \mathbb{N}_3$; $e_1e_2 = -e_2e_1 = e_3$; $e_2e_3 = -e_3e_2 = e_1$; $e_3e_1 = -e_1e_3 = e_2$.

Natural operations of addition and multiplication in $\mathbb{H}$ turn it into a skew-field. The main involution in $\mathbb{H}$, the quaternionic conjugation, is defined by

$$\bar{e}_0 := e_0; \quad \bar{e}_k := -e_k; \quad \text{for} \quad k \in \mathbb{N}_3,$$

and it extends onto $\mathbb{H}$ by $\mathbb{R}$-linearity, i.e., for $a \in \mathbb{H}$,

$$\bar{a} := \sum_{k=0}^{3} \overline{a_k e_k} = \sum_{k=0}^{3} a_k \bar{e}_k = a_0 - \sum_{k=1}^{3} a_k e_k.$$

Note that

$$\bar{a}a = a\bar{a} = \sum_{k=0}^{3} a_k^2 = |a|_{\mathbb{R}^4}^2 =: |a|_{\mathbb{H}}^2.$$

Therefore, for $a \in \mathbb{H} \setminus \{0\}$ the quaternion

$$a^{-1} := \frac{1}{|a|^2}\bar{a}$$

is an inverse to a.

Let Ω be a domain in $\mathbb{R}^3$. Then we consider $\mathbb{H}$-valued functions defined in Ω (depending on $x = (x_0, x_1, x_2)$):

$$f \,:\, \Omega \longrightarrow \mathbb{H}.$$

The notation $C^p(\Omega; \mathbb{H})$, $p \in \mathbb{N} \cup \{0\}$, has the usual component-wise meaning. On $C^1(\Omega; \mathbb{H})$ we define a generalized Cauchy-Riemann operator D by

$$D(f) := \sum_{k=0}^{2} e_k \frac{\partial f}{\partial x_k} =: \sum_{k=0}^{2} e_k \, \partial_k f.$$

D is a right-linear operator with respect to scalars from $\mathbb{H}$. The operator $\overline{D}$ defined by

$$\overline{D}(f) := \sum_{k=0}^{2} \bar{e}_k \frac{\partial f}{\partial x_k} =: \sum_{k=0}^{2} \bar{e}_k \, \partial_k f$$

is the adjoint Cauchy-Riemann operator. The solutions of $Df = 0$, $x \in \Omega$, are called (left) hyperholomorphic (or monogenic) functions and generalize the class of holomorphic functions from the one-dimensional complex function theory. Let $\triangle_3$ be the three-dimensional Laplace operator, $\triangle_3 := \sum_{k=0}^{2} \partial_k^2$. Then on $C^2(\Omega; \mathbb{H})$ analogously to the complex case the following equalities hold:

$$\triangle_3 = D\overline{D} = \overline{D}D.$$

This property means, in particular, that hyperholomorphic functions are harmonic in all their coordinates and we can use results from classical harmonic analysis also for our hyperholomorphic functions. Using the adjoint generalized Cauchy-Riemann operator $\overline{D}$ instead of the derivative $f'(z)$, the quaternionic Möbius transformation $\varphi_a(x) = (a - x)(1 - \bar{a}x)^{-1}$, and the modified fundamental solution $g(x) = \frac{1}{4\pi}\left(\frac{1}{|x|} - 1\right)$ of the real Laplacian, in [10] generalized $\mathbf{Q_p}$-spaces are defined by

$$\mathbf{Q_p} = \left\{ f \in \ker D : \; Q_p(f) := \sup_{a \in B_1(0)} \int_{B_1(0)} \left|\overline{D}f(x)\right|^2 \left(g(\varphi_a(x))\right)^p dB_x < \infty \right\},$$

where $B_1(0)$ stands for the unit ball in $\mathbb{R}^3$. Here, the generalizations of the Green function and of the higher dimensional Möbius transformation seem to be natural; that $-\frac{1}{2}\overline{D}$ plays the role of a derivative is shown in [9] for arbitrary dimensions and in [12] and [14] for dimension four.

From the consideration of $\mathbf{Q_p}$ as $p \to \infty$ in [10] the following definition of the Bloch seminorm in the three dimensional case is introduced:

$$\mathcal{B}(f) = \sup_{x \in B_1(0)} (1 - |x|^2)^{\frac{3}{2}} |\overline{D}f(x)|.$$

Based on these definitions it is proved in [10] that

$$\mathcal{D} \subset \mathbf{Q_p} \subset \mathbf{Q_q} \subset \mathcal{B} \text{ for } 0 < p < q \leq 2 \quad \text{and} \quad \mathbf{Q_q} = \mathcal{B} \text{ for } q > 2,$$

where $\mathcal{D}$ is the hyperholomorphic Dirichlet space given by (see [10]):

$$\mathcal{D} = \left\{ f : f \in ker D \quad \text{and} \quad \int_{B_1(0)} |\overline{D}f(x)|^2 dB_x < \infty \right\}.$$

For more information about the study of $\mathbf{Q_p}$ spaces of hyperholomorphic functions, we refer to [5], [8], [10], and [9].

Recently, $\mathbf{B^q}$-spaces of quaternion-valued functions were defined in [7] as follows:

$$\mathbf{B^q} = \{ f \in \ker D : B^q(f) < \infty \},$$

with

$$B^q(f) := \sup_{a \in B_1(0)} \int_{B_1(0)} |\overline{D}f(x)|^q \left(1 - |x|^2\right)^{\frac{3}{2}q - 3} \left(1 - |\varphi_a(x)|^3\right)^3 dB_x,$$

where $0 < q < \infty$. It is shown in [7] that all these spaces are equivalent to the quaternionic Bloch space. This was the reason to work in [8] with another exponent of the weight function

$$B^q(f) := \sup_{a \in B_1(0)} \int_{B_1(0)} |\overline{D}f(x)|^q \left(1 - |x|^2\right)^{\frac{3}{2}q - 3} \left(1 - |\varphi_a(x)|^3\right)^2 dB_x,$$

where again $0 < q < \infty$ and the exponent 2=3-1 is related to the real space dimension. In this case it is known that for $q < 2$ these $\mathbf{B^q}$-spaces form a proper scale and all spaces are included in the Bloch space.

Using the same definition, Reséndis and Tovar [13] obtained a Besov-type characterization for quaternionic Bloch functions that generalizes a Stroethoff theorem (see [15]) in $\mathbb{R}^3$.

In the present paper we define $\mathbf{B^{p,q}}$-spaces of quaternion-valued functions as follows:

$$\mathbf{B^{p,q}} = \left\{ f \in \ker D : B^{p,q}(f) < \infty \right\},$$

with

$$B^{p,q}(f) := \sup_{a\in B_1(0)} \int_{B_1(0)} |\overline{D}f(x)|^q (1-|x|^2)^{\frac{3}{2}q-p} (1-|\varphi_a(x)|^2)^p dB_x,$$

where $0 < q < \infty$ and $0 < p < \infty$. This definition refines the scale by a better controlling of the weight function. If $p = 3$, we will get the space $\mathbf{B^q}$ of hyperholomorphic functions as defined in [7]. Also, if $q = 2$ and $p = 3$ we will get the space $\mathbf{Q_2}$ of hyperholomorphic functions as studied in [10].

Remark 1.1. The functionals $B(f)$, $Q_p(f)$, $B^q(f)$ and $B^{p,q}(f)$ define only seminorms. Nevertheless, if U_ε is a small neighbourhood of the origin, with an arbitrary but fixed $\varepsilon > 0$, then we can add the L_1-norm of f over U_ε to our seminorms and $\mathbf{B}$ as well as $\mathbf{Q_p}$, $\mathbf{B^p}$, $\mathbf{B^{p,q}}$ will become Banach spaces. For details see [9].

Remark 1.2. We restrict ourselves to the study of hyperholomorphic functions in $\mathbb{R}^3$ with values in the quaternions even though some of the results on harmonic functions are stated in $\mathbb{R}^n$, and hyperholomorphic functions can be defined also as functions from $\mathbb{R}^{n+1}$ with values in Clifford algebras $\mathcal{Cl}_{0,n}$. One reason is that the used hyperholomorphic $B^{p,q}$-spaces are studied only in the three-dimensional case. The seminorms introduced by the hypercomplex derivative $\bar{D}$ for functions defined in $\mathbb{R}^{n+1}$ with values in $\mathcal{Cl}_{0,n}$ become weaker with increasing dimension because the hyperholomorphic constants (functions belonging to $\ker D \cap \ker \bar{D})$) are isomorphic to the Clifford algebra $\mathcal{Cl}_{0,n-1}$.

2. Classical Dirichlet Problem

This section is devoted to present definitions and results related to the classical Dirichlet problem for real valued functions defined in some subset of $\mathbb{R}^n$. For Proposition 2.1, Theorem 2.2 and Corollary 2.3 and the basic definition we refer to [6].

Let $\Omega \subset \mathbb{R}^n (n \geq 2)$ be a bounded open set, with boundary Γ and $h \in C^0(\Gamma)$. We call a function $H : \Omega \to \mathbb{R}^n$ a classical Dirichlet solution if H is harmonic in Ω, i.e., $\Delta_n(H(x)) = 0$ for all $x \in \Omega$, and such that

$$\lim_{\substack{x\to z\\ x\in\Omega}} H(x) = h(z) \qquad \text{for all} \quad z \in \Gamma.$$

We denote by $\mathfrak{H}(\Omega)$ the family of harmonic functions in Ω. If H is the classical Dirichlet solution of h in Ω, then $H \in \mathfrak{H}(\Omega) \cap C^0(\bar{\Omega})$ and h is the trace of H on Γ. From the maximum principle for harmonic functions we have that the linear map that takes H to its trace h on Γ is an isometry with the uniform convergence norm on Ω given by

$$||H|| = \max_{x\in\Gamma} |H(x)|$$

and the uniform convergence norm on Γ given by

$$||h|| = \max_{x\in\Gamma} |h(x)|.$$

Thus we obtain that every uniform limit of harmonic functions on Ω results in a harmonic function on Ω, and we have

Proposition 2.1. [6] *Let Ω be a bounded open set of $\mathbb{R}^n$ with boundary Γ. The set $E(\Omega)$ of the $h \in C^0(\Gamma)$ such that the Dirichlet problem admits a classical solution is a closed vector subspace of $C^0(\Gamma)$.*

For the case $n = 3$ and $\Omega = B_{r_0}(x_0)$, the ball centered in x_0 of radius r_0, Poisson's integral formula gives us the classical Dirichlet solution for $h \in C^0(\Gamma)$, where $\Gamma = \partial B_{r_0}(x_0)$:

$$H(x) = \frac{1}{r_0^2 \sigma_3} \int_\Gamma \frac{r_0^2 - |x - x_0|^2}{|t - x|^3} h(t)\, d\Gamma_t.$$

Here σ_3 is the total surface area of the unit sphere in $\mathbb{R}^3$. Consider now the next immediate consequence of Harnack's theorem:

Theorem 2.2. [6] *Let $\Omega \subset \mathbb{R}^n$ be a connected open set and let $\{H_m\}$ be a monotonic sequence of harmonic functions on Ω. Suppose that there exists an $x_0 \in \Omega$ such that $\{H_m(x_0)\}$ is bounded. Then the sequence $\{H_m\}$ is uniformly convergent to a harmonic function in Ω on every compact subset of Ω.*

Corollary 2.3. [6] *Let $\Omega \subset \mathbb{R}^n$ be a bounded open set and let $\{H_m\}$ be a sequence of functions, harmonic on Ω and continuous on $\bar{\Omega}$, satisfying for some $M > 0$, $|H_m(x)| \leq M$ for all $x \in \Gamma = \partial\Omega$ and all $m \in \mathbb{N}$. Then there exists a subsequence of $\{H_m\}$ converging to a harmonic function on Ω.*

Now we will prove some facts about continuous functions on the unit ball $B_1(0)$ and its boundary $\mathbb{T} = \partial B_1(0)$ in $\mathbb{R}^3$.

Proposition 2.4. *Let $h : B_1(0) \to \mathbb{R}^+ \cup \{0\}$ be a continuous function and $0 < R < 1$. Let $h_R : \mathbb{T} \to \mathbb{R}$ be a function defined in spherical coordinates (r, φ, θ) by*

$$h_R(1, \varphi, \theta) = \max_{0 \leq r \leq R} h(r, \varphi, \theta), \quad 0 \leq \varphi \leq \pi, \quad 0 \leq \theta \leq 2\pi.$$

Then h_R is a continuous function on $\mathbb{T}$. Moreover, if $0 < R_1 < R_2 < 1$, then $h_{R_1}(1, \varphi, \theta) \leq h_{R_2}(1, \varphi, \theta)$ for every $(\varphi, \theta) \in [0, \pi] \times [0, 2\pi]$.

Proof. Let $\varphi_0 \in [0, \pi]$ and $\theta_0 \in [0, 2\pi]$ be fixed. By compactness of the segment $[0, (R_1, \varphi, \theta)]$ in $B_1(0)$, the maximum of h in this segment is attained. We proceed by contradiction: Suppose that h_R is not continuous at $(1, \varphi_0, \theta_0)$, then if $h_R(1, \varphi_0, \theta_0) = h(r_0, \varphi_0, \theta_0)$ for some $r_0 \in [0, R]$, there exists $\varepsilon > 0$, such that for each $n \in \mathbb{N}$, there exist φ_n and θ_n, with $|\varphi_n - \varphi_0| < \frac{1}{n}$, and $|\theta_n - \theta_0| < \frac{1}{n}$, but $|h_R(1, \varphi_0, \theta_0) - h_R(1, \varphi_n, \theta_n)| \geq \varepsilon$. In particular there exists a sequence $\{r_n\} \subset [0, R]$, with $h_R(1, \varphi_n, \theta_n) = h(r_n, \varphi_n, \theta_n)$. Besides, we can suppose that r_n converges to $r^* \in [0, R]$, with $r^* \neq r_0$ and by continuity of the function h at $(r_0, \varphi_0, \theta_0)$, necessarily we have that $h_R(1, \varphi_0, \theta_0) + \varepsilon \leq h_R(1, \varphi_n, \theta_n) = h(r_n, \varphi_n, \theta_n)$. Taking the limit we get a contradiction. □

Corollary 2.5. *Let $h : B_1(0) \to \mathbb{R}$ be a continuous function. Let $h_1 : \mathbb{T} \to \mathbb{R}$ be defined by*

$$h_1(1,\varphi,\theta) = \sup_{0 \le R < 1} h_R(r,\varphi,\theta), \quad 0 \le \varphi \le \pi, \quad 0 \le \theta < 2\pi.$$

If $h_1(1,\varphi,\theta) < +\infty$, then $\lim_{R\to 1} h_R(1,\varphi,\theta) = h_1(1,\varphi,\theta)$.

h_1 is not necessarily a continuous function on $\mathbb{T}$ (see [3]). Let $h_R : \mathbb{T} \to \mathbb{R}$ be a continuous function and $0 < R < 1$. Let $H_R : B_1(0) \to \mathbb{R}$ be a harmonic function such that $H_R|_{\mathbb{T}} = h_R$, i.e., H_R is the classical solution of the Dirichlet problem with boundary values h_R on $\mathbb{T}$. Then we have

Theorem 2.6. *Let $h : B_1(0) \to \mathbb{R}^+ \cup \{0\}$ be a continuous function. Let $h_1 : \mathbb{T} \to \mathbb{R}$ be defined as in the previous corollary and let $H_1 : B_1(0) \to \mathbb{R}$ be defined by*

$$H_1(x) = \sup_{0 \le R < 1} H_R(x), \quad \textit{for every } x \in B_1(0).$$

Then H_1 is a harmonic function on $B_1(0)$ or $H_1(x) \equiv \infty$ on $B_1(0)$. In the first case H_1 can be obtained through the application of the Poisson kernel to h_1; moreover if H_R is positive then

$$\lim_{R\to 1} H_R(1,\varphi,\theta) = H_1(1,\varphi,\theta) \quad a.e.$$

and $\lim_{r\to 1} H_1(1,\varphi,\theta)$ exists a.e. on $[0,\pi] \times [0,2\pi]$.

Proof. Let $\{R_n\}$ be an increasing sequence of positive numbers on $(0,1)$ with $\lim_{n\to\infty} R_n = 1$. By Proposition 2.4 and the formula for the classical solution of the Dirichlet problem, the sequence $\{H_{R_n}\}$ satisfies

$$H_{R_n}(x) \le H_{R_m}(x) \le \sup_{y \in B_1(0)} h(y), \text{ for all } x \in B_1(0) \text{ and } n \le m.$$

Thus $\{H_{R_n}\}$ is an increasing sequence such that

$$\lim_{n\to\infty} H_{R_n}(x) = H_1(x) \text{ for every } x \in B_1(0).$$

By Harnack's theorem $H_1 \equiv \infty$ or H_1 is a harmonic function. By Corollary 2.3 H_1 is a harmonic function. It is easy to see the that limit does not depend on the sequence. Besides, since

$$H_R(1,\varphi,\theta) = h_R(1,\varphi,\theta) \le h_1(1,\varphi,\theta), \quad (\varphi,\theta) \in [0,\pi] \times [0,2\pi]$$

we have that $H_R(1,\varphi,\theta) \to h_1(1,\varphi,\theta)$ if $R \to 1^-$ and $\lim_{R\to 1^-} H_1(1,\varphi,\theta)$ exists a.e. (see [4] p.55).

For every fixed $x \in B_1(0)$, we consider the sequence of functions

$$\left\{\frac{1}{\sigma_3}\frac{1-|x|^2}{|t-x|^2}\, h_{R_n}(t)\right\}.$$

So there is a sequence of positive Lebesgue integrable functions such that for $R_{n_1} \leq R_{n_2}$

$$0 \leq \frac{1}{\sigma_3}\frac{1-|x|^2}{|t-x|^2}\, h_{R_{n_1}}(t) \leq \frac{1}{\sigma_3}\frac{1-|x|^2}{|t-x|^2}\, h_{R_{n_2}}(t)$$

and

$$\frac{1}{\sigma_3}\frac{1-|x|^2}{|t-x|^2}\, h_{R_{n_m}}(t) \to \frac{1}{\sigma_3}\frac{1-|x|^2}{|t-x|^2}\, h_1(t) \quad as \quad m \mapsto \infty.$$

Thus for the classical theorem on a monotonic sequence of Lebesgue integrable functions we have that

$$\frac{1}{\sigma_3}\frac{1-|x|^2}{|t-x|^2}\, h_1(t)$$

is an integrable function such that

$$\frac{1}{\sigma_3}\int_{\mathbb{T}}\frac{1-|x|^2}{|t-x|^2}\, h_{R_n}(t)\, d\Gamma_t \to \frac{1}{\sigma_3}\int_{\mathbb{T}}\frac{1-|x|^2}{|t-x|^2}\, h_1(t)\, d\Gamma_t \quad as \quad n \mapsto \infty. \qquad \square$$

Corollary 2.7. *Let $G : B_1(0) \to \mathbb{R}$ be a continuous radially increasing function, that is, for $0 < R_1 < R_2 < 1$,*

$$G(R_1, \varphi, \theta) \leq G(R_2, \varphi, \theta), \text{ for every } \varphi \in [0, \pi], \quad \theta \in [0, 2\pi].$$

Then $h_1 : \mathbb{T} \to \mathbb{R}$, constructed in Corollary 2.5, defines through the Poisson kernel a harmonic function H_1 on $B_1(0)$ with boundary values given by h_1 a.e. on $\mathbb{T}$.

Corollary 2.8. *Under the hypotheses of the previous theorem, we have*

$$H_R(x) \leq H_1(x) \leq \sup_{y \in B_1(0)} H_1(y) = \sup_{w \in T} h(w).$$

Let $\Omega \subset \mathbb{R}^3$ be an arbitrary set and let $G : \Omega \to \mathbb{R}$ be a function that is not identically $-\infty$. We say that $H : \Omega \to \mathbb{R}$, a harmonic function on Ω, is the least harmonic majorant of G if it satisfies:

a) $G(x) \leq H(x)$ for each $x \in \Omega$,
b) If I is any harmonic function on Ω such that $G(x) \leq I(x)$ for each $x \in \Omega$, then $H(x) \leq I(x)$ for each $x \in \Omega$.

The least harmonic majorant of a bounded and radially increasing continuous function is not necessarily the Poisson extension of its boundary value (proved in [3]). Remark that the proof is given there for the case of the unit disc in $\mathbb{C}$. It is easy to get the corresponding result for functions defined on $B_1(0)$.

3. The $\mathbf{B}^{\mathbf{p},\mathbf{q}}$-Harmonic Majorant

Proposition 3.1. *For $0 < q < \infty$ and $0 < p < \infty$, let $f \in \mathbf{B}^{\mathbf{p},\mathbf{q}}$ and $F : B_1(0) \to \mathbb{R}$ be defined by*

$$F(a) = \int_{B_1(0)} |\bar{D}f(x)|^q \left(1 - |x|^2\right)^{\frac{3}{2}q-p} \left(1 - |\varphi_a(x)|^2\right)^p \, dB_x.$$

Then F is a bounded continuous function on $B_1(0)$.

Proof. Let $a \in B_1(0)$ be fixed and let $0 < \delta_a < 1 - |a|$ be such that the closed ball $\overline{B_{\delta_a}(a)} \subset B_1(0)$. The function $l : B_1(0) \times \overline{B_{\delta_a}(a)} \to \mathbb{R}$ defined by

$$(x, \xi) \to \frac{(1 - |\xi|)^p}{|1 - \bar{\xi}x|^{2p}}$$

is uniformly continuous on $B_1(0) \times \overline{B_{\delta_a}(a)}$. Then for given $\varepsilon > 0$ there exists $\delta > 0$ such that if $|x - x'| < \delta$ and $|\xi - \xi'| < \delta$, then

$$|l(x', \xi') - l(x, \xi)| < \frac{\varepsilon}{\int_{B_1(0)} |\bar{D}f(x)|^q (1 - |x|^2)^{\frac{3}{2}q} \, dB_x}.$$

Therefore

$$|F(a) - F(b)| \le \int_{B_1(0} |\bar{D}f(x)|^q \left(1 - |x|^2\right)^{\frac{3}{2}q} |l(x, a) - l(x, b)| \, dB_x.$$

To prove this we use the equality

$$\frac{1 - |\varphi_a(x)|^2}{1 - |x|^2} = \frac{1 - |a|^2}{|1 - \bar{a}x|^2}$$

that follows directly from the properties of the Möbius transform. □

Then for $f \in \mathbf{B}^{\mathbf{p},\mathbf{q}}$, $a = (r, \varphi, \theta)$, $0 \le R < 1$, $(\varphi, \theta) \in [0, \pi] \times [0, 2\pi]$ define

$$h_R(1, \varphi, \theta) = \sup_{\substack{a=(r,\varphi,\theta) \\ 0<r\le R}} \int_{B_1(0)} |\bar{D}f(x)|^q \left(1 - |x|^2\right)^{\frac{3}{2}q-p} (1 - |\varphi_a(x)|^2)^p \, dB_x < +\infty.$$

By Proposition 2.4, each function h_R is continuous on $\mathbb{T}$. So we can associate through the classical Dirichlet solution a harmonic function $H_R(x)$ on $B_1(0)$ given by

$$H_R(x) = \frac{1}{R\sigma_3} \int_{\partial B_1(0)} \frac{1 - |x|^2}{|t - x|^3} h_R(t) \, d\Gamma_t.$$

Observe that for $R_1 \le R_2$ we have $h_{R_1}(1, \varphi, \theta) \le h_{R_2}(1, \varphi, \theta)$, thus

$$H_{R_1}(x) \le H_{R_2}(x) < B^{p,q}(f).$$

If we apply Theorem 2.6, then we obtain that the limit

$$\lim_{R \to 1^-} H_R(x) = H_1(x)$$

is harmonic on $B_1(0)$. Besides, $\lim_{R\to 1^-} h_R(x) := h_1(x)$ is a well defined function. Observe that $H_1(x)$ is the least harmonic majorant of the family $\{H_R(x)\}$ on $B_1(0)$. It follows from Theorem 2.6

$$\sup_{y\in B_1(0)} H_R(y) \le \sup_{y\in B_1(0)} H_1(y) \le B^{p,q}(f).$$

In a similar way we obtain

Proposition 3.2. *Let f be a non-constant hyperholomorphic function on $B_1(0)$ and $0 < R < 1$. Then the function $K_R : \mathbb{T} \to \mathbb{R}$ defined by*

$$K_R(1,\varphi,\theta) = \min_{\substack{a=(r,\varphi,\theta)\\ 0\le r<R}} \int_{B_1(0)} |\bar{D}f(x)|^q \left(1-|x|^2\right)^{\frac{3}{2}q-p} \left(1-|\varphi_a(x)|^2\right)^p dB_x$$

is continuous and positive.

Now we are in the position to obtain a characterization of hyperholomorphic $\mathbf{B^{p,q}}$-functions in terms of a very natural least harmonic majorant associated to f.

Theorem 3.3. *Let $f : B_1(0) \to \mathbb{H}(\mathbb{R})$ be a hyperholomorphic function and $0 < p < \infty$, $0 < q < \infty$. Then f belongs to the class $\mathbf{B^{p,q}}$ if and only if $h_1 : \mathbb{T} \to \mathbb{R}$ defined by*

$$h_1(1,\varphi,\theta) = \sup_{\substack{a=(r,\varphi,\theta)\\ 0\le r<1}} \int_{B_1(0)} |\bar{D}f(x)|^q \left(1-|x|^2\right)^{\frac{3}{2}q-p} \left(1-|\varphi_a(x)|^2\right)^p dB_x$$

is bounded. Because h_1 is a well defined bounded function on $\mathbb{T}$, the function $H_1 : B_1(0) \to \mathbb{H}(\mathbb{R})$ can be obtained from h_1 through the classical Dirichlet solution with the Poisson kernel, and H_1 is the least harmonic majorant of the family of harmonic functions $\{H_R\}$, $0 < R < 1$.

Proof. For the continuous function F of proposition 3.1, consider the associated family of functions h_R introduced in Proposition 2.4. Then associate, for each $0 < R < 1$, the corresponding harmonic Poisson extension H_R and finally apply Theorem 2.6 . □

Corollary 3.4. *Let $H_1 : B_1(0) \to \mathbb{R}^+ \cup \{0\}$ be as above. Then H_1 admits a finite nontangential limit $h_1(1,\varphi,\theta)$ almost everywhere on $\mathbb{T}$.*

Proof. See [4]. □

We will study now some relationships proved by El-Sayed Ahmed et al in [1] and show their corresponding translation in terms of harmonic majorants.

Proposition 3.5. *Let f be a hyperholomorphic function in $B_1(0)$. If $f \in \mathcal{B}$ then for every a in $B_1(0)$, $1 \le p < \infty$ and $0 < q < \infty$ we have*

$$\int_{B_1(0)} |\bar{D}f(x)|^q \left(1-|x|^2\right)^{\frac{3}{2}q-p} \left(1-|\varphi_a(x)|^2\right)^p dB_x \le 4\pi (B(f))^q$$

where $B(f)$ denotes the Bloch seminorm of f.

As a consequence we obtain

Corollary 3.6. *For $1 \le p < \infty$ and $0 < q < \infty$ we have*

$$\mathcal{B} \subset \mathbf{B^{p,q}}.$$

Let now $f \in \mathcal{B}$ and denote by h_1 and $h_1^{p,q}$ the corresponding functions associated through the technique outlined previously. Likewise let H_1 and $H_1^{p,q}$ be the corresponding least harmonic majorants associated to f with boundary limits given a.e. by h_1 and $h_1^{p,q}$, respectively. So we can reformulate the previous proposition in terms of harmonic majorants associated to f as follows.

Theorem 3.7. *Let f be a hyperholomorphic function in $B_1(0)$ and assume that $f \in \mathcal{B}$. For every $0 \le p < \infty$ and $0 < q < \infty$,*

$$\sup_{x\in B_1(0)} H_1^{p,q}(x) = \sup_{x\in \partial B_1(0)} h_1^{p,q}(x) \le \sup_{x\in \partial B_1(0)} h_1(x) = \sup_{x\in B_1(0)} H_1(x).$$

In [1] the authors obtained the next result, too:

Theorem 3.8. *For $1 \le q < \infty$ and $3 \le p < \infty$ we have*

$$\mathbf{B^{p,q}} \subset \mathcal{B}.$$

In the proof they used essentially the inequality

$$\eta^*(R)(1-|a|^2)^{\frac{3}{2}q}|\bar{D}f(a)|^q \le \int_{B_1(0)} |\bar{D}f(x)|^q \left(1-|x|^2\right)^{\frac{3}{2}q-p} \left(1-|\varphi_a(x)|^2\right)^p dB_x.$$

So its follows directly that this statement can be rewritten for harmonic majorants of f as:

Theorem 3.9. *For $1 \le q < \infty$ and $3 \le p < \infty$ we have for f in $\mathbf{B^{p,q}}$ that*

$$\sup_{x\in \partial B_1(0)} h_1(x) = \sup_{x\in B_1(0)} H_1(x) \le \sup_{x\in B_1(0)} H_1^{p,q}(x) = \sup_{x\in \partial B_1(0)} h_1^{p,q}(x).$$

At the end we extend some results by Stroethoff [15] for $\mathbf{B^{p,q}}$-spaces to hyperholomorphic functions and obtain several characterizations of functions in the Bloch space. (See Theorems 3.1 and 1.2 in [1])

Theorem 3.10. *Let f be hyperholomorphic in the unit ball. Then the following conditions are equivalent.*

1. $f \in \mathcal{B}$,
2. $f \in \mathbf{B^{p,q}}$ *for all* $0 < q < \infty$ *and* $0 < p < \infty$,
3. $f \in \mathbf{B^{p,q}}$ *for some* $q \ge 1$ *and* $3 \le p < \infty$.

This result can be translated for H_1 and $H_1^{p,q}$ as

Theorem 3.11. *Let f be hyperholomorphic in the unit ball. Then the following conditions are equivalent*

1. H_1 *is bounded,*
2. $H_1^{p,q}$ *is bounded for all* $0 < q < \infty$ *and* $0 < p < \infty$,
3. $H_1^{p,q}$ *is bounded fore same* $q \ge 1$ *and* $3 \le p < \infty$.

4. Acknowledgement

This work was supported by DFG,CONACYT and IPN.

References

[1] A. El-Sayed Ahmed, K. Gürlebeck, L.F. Resendis, L.M. Tovar, Characterizations for $\mathbf{B}^{\mathbf{p},\mathbf{q}}$ spaces by Bloch spaces in Clifford analysis, preprint Bauhaus Univ. Weimar, Germany 2002.

[2] R. Aulaskari and P. Lappan, Criteria for an analytic function in the Bloch space and harmonic or meromorphic function to be normal, Complex analysis and its application, Pitman Research Notes in Math. 305, Longman Scientific of Technical, Harlow, (1994), 136–146.

[3] R. Aulaskari, L.F. Resendis, L.M. Tovar, Q_p-spaces and Harmonic majorants, to appear in *Complex Variables.*

[4] S.Axler, P. Bourlau, W. Raney, *Harmonic Function Theory*, Grad. Text in Math., Vol. 137, Springer Verlag, 2001.

[5] J. Cnops and R. Delanghe, Möbius invariant spaces in the unit ball, *Appl. Anal.* **73** (2000), 45–64.

[6] R. Dautray, J. L. Lions, *Mathematical analysis and numerical methods for Science and Technology*, Vol. 1, Springer Verlag, 1990.

[7] K. Gürlebeck and A. El-Sayed Ahmed, Integral norms for hyperholomorphic Bloch-functions in the unit ball of $\mathbb{R}^3$, in: *Progress in Analysis*, Proceedings of the 3rd International ISAAC Congress, Volume I, Editors H. Begehr, R. Gilbert, M.W. Wong, World Scientific, New Jersey, London, Singapore, Hong Kong, 2003

[8] K. Gürlebeck and A. El-Sayed Ahmed, On $\mathbf{B}^{\mathbf{q}}$ spaces of hyperholomorphic functions and the Bloch space in $\mathbb{R}^3$, in: *Finite or Infinite Dimensional Complex Analysis and Applications*, Le Hung Son (Ed.), W. Tutschke (Ed.), C.-C. Yang (Ed.), Advances in Complex Analysis and its Applications – 2, Kluwer Boston, 2003

[9] K. Gürlebeck and H. R. Malonek, On strict inclusions of weighted Dirichlet spaces of monogenic functions, *Bull. Austral. Math. Soc.* **64** (2001), 33–50.

[10] K. Gürlebeck, U. Kähler, M. Shapiro, L.M. Tovar, On Q_p-spaces of quaternion-valued functions, *Complex Variables* **39** (1999), 115–135.

[11] S. Kobayashi, Range sets and BMO norms of analytic functions, *Canadian S. Math.* **XXXVI**, No. 4 (1984), 747–755.

[12] I.M. Mitelman and M.V. Shapiro, Differentiation of the Martinelli-Bochner Integrals and the Notion of Hyperderivability, *Math. Nachr.* **172** (1995), 211–238.

[13] L. F. Reséndis and L. M. Tovar, Besov-type Characterizations for Quaternionic Bloch Functions, in: *Finite or Infinite Dimensional Complex Analysis and Applications*, Le Hung Son (Ed.), W. Tutschke (Ed.), C.-C. Yang (Ed.), Advances in complex analysis and its applications – 2, Kluwer Boston, 2003

[14] A. Sudbery, Quaternionic analysis, *Math. Proc. Cambridge Philos. Soc.* **85** (1979), 199–225.

[15] K. Stroethoff, Besov-type characterizations for the Bloch space, *Bull. Austral. Math. Soc.* **39** (1989), 405–420.

S. Bernstein
Bauhaus-Universität Weimar
Institut für Mathematik und Physik
Coudray-Str. 13 B
D-99423 Weimar
Germany
e-mail: swanhild.bernstein@fossi.uni-weimar.de

K. Gürlebeck
Bauhaus-Universität Weimar
Institut für Mathematik und Physik
Coudray-Str. 13 B
D-99423 Weimar
Germany
e-mail: guerlebe@fossi.uni-weimar.de

L. F. Reséndis
Universidad Autónoma Metropolitana, Unidad Azcapotzalco
C.B.I., Apartado Postal 16-306 C.P. 02200 México
16, D.F. Area de Análisis Matemático y sus Aplicaciones
México
e-mail: lfro@correo.azc.uam.mx

Luis M. Tovar S.
Escuela Superior de Física y Matemáticas del IPN Edif. 9
Unidad ALM, Zacatenco del IPN.
C.P. 07300, D.F.
México
e-mail: tovar@esfm.ipn.mx

Trends in Mathematics: *Advances in Analysis and Geometry*, 65–96

Spherical Means and Distributions in Clifford Analysis

Fred Brackx, Richard Delanghe, and Frank Sommen

Abstract. This paper consists of two parts. In the first part we deal with specific higher dimensional distributions within the framework of Clifford analysis. These distributions are "classical" in the sense that they were already introduced, albeit dispersed, in the literature on harmonic analysis and on Clifford analysis. Amongst these classical distributions are the fundamental solutions of the natural powers of the Laplace and the Dirac operators, and the integral kernel of the Hilbert transform. The strength of our approach is its unifying character.
In the second part new higher dimensional distributions are introduced, generalizing the distributions of part I. Crucial to this generalization is the use of so-called surface spherical monogenics. The whole picture thus obtained offers structural clarity and unity.

Mathematics Subject Classification (2000). Primary 30G35; Secondary 46F10.

Keywords. Distributions, spherical means, Clifford analysis.

Introduction

The immediate cause of writing this paper was Delanghe's paper [5] in which he studies a higher dimensional analogue of the Principal Value distribution on the real line. In that paper the integrated approach to all dimensions, offered by Clifford analysis, is fully exploited, as opposed to a traditional tensorial or cartesian approach with a number of copies of one dimensional phenomena. We realized that this higher dimensional Principal Value distribution and the well known fundamental solution of the Dirac operator are but two examples of vector valued Clifford distributions, out of an infinite collection of such kind of distributions, which moreover can be obtained by letting act the Dirac operator on a corresponding infinite set of classical real valued radial distributions. We also realized that by making use of the well known spherical means, which arise naturally by introducing spherical co-ordinates, a simple, powerful and highly efficient technique could be designed allowing us to carry out the explicit calculations on the real line and exporting them to the original setting of Euclidean space. Finally we realized that introducing generalized spherical means involving the so-called spherical monogenics – the well known counterparts in Clifford analysis to the spherical harmonics of harmonic analysis – gave rise to much more general Clifford distributions, encompassing those from part I in the special case where the degree of the spherical monogenic considered is zero.

The aim of this paper is twofold. From the scientific point of view we introduce new Clifford distributions generalizing and extending the existing ones in harmonic analysis and Clifford analysis. From the didactic point of view our unifying approach offers structural clarity and gathers results and formulae spread over the literature on harmonic and Clifford analysis, at the same time proving once more the power and elegance of Clifford analysis.

Part I: Classical Clifford Distributions

1. A fundamental distribution on the real line

In this section we recall the definition and some properties of the classical distribution $Fp\, x_+^\mu$ (see e.g. [11], [7]).
Let μ be a complex parameter and x be a real variable. We consider the function

$$x_+^\mu = \begin{cases} x^\mu & , \quad x > 0 \\ 0 & , \quad x < 0 \end{cases}$$

For $\operatorname{Re}\mu > -1$, this function x_+^μ is a regular distribution given, for any test function ϕ, by

$$\langle\, x_+^\mu, \phi \,\rangle = \int_0^{+\infty} x^\mu \phi(x) dx$$

For each $n \in \mathbb{N}$ and for $\mu \in \mathbb{C}$ such that $-n-1 < \operatorname{Re}\mu < -n$, the distribution $Fp\, x_+^\mu$ – where Fp stands for "finite part" – is defined by

$$\begin{aligned} \langle\, Fp\, x_+^\mu, \phi \,\rangle &= \int_0^{+\infty} x^\mu \left(\phi(x) - \phi(0) - \frac{\phi'(0)}{1!}\, x - \ldots - \frac{\phi^{(n-1)}(0)}{(n-1)!}\, x^{n-1} \right) dx \\ &= \lim_{\substack{\varepsilon \to 0 \\ >}} \left(\int_\varepsilon^{+\infty} x^\mu \phi(x) + \phi(0) \frac{\varepsilon^{\mu+1}}{\mu+1} + \ldots + \frac{\phi^{(n-1)}(0)}{(n-1)!}\, \frac{\varepsilon^{\mu+n}}{\mu+n} \right) \end{aligned}$$

As a function of μ, x_+^μ is holomorphic in $\operatorname{Re}\mu > -1$, and by analytic continuation $Fp\, x_+^\mu$ is holomorphic in $\mathbb{C}\backslash\{-1,-2,-3,\ldots\}$; the singular points $-n, n \in \mathbb{N}$, are simple poles with residue $\frac{(-1)^{n-1}}{(n-1)!}\, \delta_x^{(n-1)}$.

The derivative of $Fp\, x_+^\mu$ is given by

$$\frac{d}{dx}\, Fp\, x_+^\mu = \mu\, Fp\, x_+^{\mu-1}, \qquad \mu \neq 0, -1, -2, -3, \ldots$$

and multiplication with powers of the variable x follows the rule

$$x\, Fp\, x_+^\mu = Fp\, x_+^{\mu+1}, \qquad \mu \neq -1, -2, -3, \ldots$$

The distribution $Fp\, x_+^\mu$ may be normalized by dividing by an appropriate Gamma-function showing the same singularities in the complex μ-plane; the functional

$$\langle \frac{x_+^\mu}{\Gamma(\mu+1)}, \phi \rangle$$

is an entire function, so that the distribution $\frac{x_+^\mu}{\Gamma(\mu+1)}$ is well defined at $\mu = -n$, $n \in \mathbb{N}$, where

$$\left[\frac{x_+^\mu}{\Gamma(\mu+1)}\right]_{\mu=-n} = \delta_x^{(n-1)}$$

However, by slightly changing the above definition, $Fp\, x_+^\mu$ may be defined for negative entire exponents, leading to the so-called *monomial pseudofunctions* $Fp\, x_+^{-n}$, $n \in \mathbb{N}$ (see [11]), given by

$$\langle\, Fp\, x_+^{-n}, \phi(x)\, \rangle =$$

$$\lim_{\substack{\varepsilon \to 0 \\ >}} \left(\int_\varepsilon^{+\infty} x^{-n}\phi(x)\, dx + \phi(0)\frac{\varepsilon^{-n+1}}{-n+1} + \ldots + \frac{\phi^{(n-2)}(0)}{(n-2)!}\frac{\varepsilon^{-1}}{(-1)} + \frac{\phi^{(n-1)}(0)}{(n-1)!}\ln\varepsilon \right).$$

Their derivatives are given by

$$\frac{d}{dx}\, Fp\, x_+^{-n} = (-n) Fp\, x_+^{-n-1} + (-1)^n \frac{1}{n!}\, \delta_x^{(n)}, \qquad n \in \mathbb{N},$$

and they satisfy the multiplication rule

$$\begin{aligned} x\, Fp\, x_+^{-1} &= Y(x) \\ x\, Fp\, x_+^{-n} &= Fp\, x_+^{-n+1}, \qquad n = 2, 3, 4, \ldots \end{aligned}$$

where $Y(x)$ stands for the Heaviside distribution, which is identified with $Fp\, x_+^0$.

In the sequel we will also make use of the following technical lemma on the division of $Fp\, x_+^\mu$ by natural powers of the variable x.

Lemma 1.1. *If the test function ϕ is such that $\phi(0) = \phi'(0) = \ldots = \phi^{(k-1)}(0) = 0$, then*

$$\langle\, Fp\, x_+^\mu, \frac{1}{x^k}\phi(x)\, \rangle = \langle\, Fp\, x_+^{\mu-k}, \phi(x)\, \rangle, \quad \mu \in \mathbb{C}\backslash\{0, 1, 2, \ldots, k-1\}$$

and

$$\langle\, x_+^n, \frac{1}{x^k}\phi(x)\, \rangle = \langle\, Fp\, x_+^{n-k}, \phi(x)\, \rangle, \qquad n = 0, 1, 2, \ldots, k-1.$$

Proof. If we put $\frac{1}{x^k}\phi(x) = \psi(x)$ then

(i) $\phi^{(j)}(0) = 0 \quad \text{for} \quad j = 0, 1, 2, \ldots, k-1$

(ii) $\phi^{(k)}(0) = k!\ \psi(0)$

(iii) $\phi^{(j)}(0) = \dfrac{j!}{(j-k)!}\ \psi^{(j-k)}(0) \quad \text{for} \quad j = k+1, k+2, \ldots$

Choose an arbitrary $\mu \in \mathbb{C}\backslash\{k-1, k-2, \ldots\}$ and take $n \in \mathbb{N}$ such that $-n-1 < \operatorname{Re}\mu < -n$. Then $-n-k-1 < \operatorname{Re}\mu - k < -n-k$ and thus by definition

$$
\begin{aligned}
&\langle\, Fp\, x_+^{\mu-k}, \phi \,\rangle \\
&= \int_0^{+\infty} x^{\mu-k}\left(\phi(x) - \phi(0) - \frac{\phi'(0)}{1!}\,x - \ldots - \frac{\phi^{(n+k-1)}(0)}{(n+k-1)!}\,x^{n+k-1}\right) dx \\
&= \int_0^{+\infty} x^{\mu-k}\left(\phi(x) - \psi(0)x^k - \frac{\psi'(0)}{1!}\,x^{k+1} - \ldots - \frac{\psi^{(n-1)}(0)}{(n-1)!}\,x^{n+k-1}\right) dx \\
&= \langle\, Fp\, x_+^{\mu}\,,\ \psi(x) \,\rangle
\end{aligned}
$$

Next for $n = -1, -2, -3, \ldots$ we have by definition

$$
\begin{aligned}
&\langle\, Fp\, x_+^{n-k}\,,\ \phi(x) \,\rangle \\
&= \lim_{\substack{\varepsilon \to 0 \\ >}} \left(\int_\varepsilon^{+\infty} x^{n-k}\phi(x)\,dx + \frac{\phi^{(k)}(0)}{k!}\,\frac{\varepsilon^{n+1}}{n+1} + \ldots + \frac{\phi^{(k-n-1)}(0)}{(k-n-1)!}\ln\varepsilon\right) \\
&= \lim_{\substack{\varepsilon \to 0 \\ >}} \left(\int_\varepsilon^{+\infty} x^{n}\psi(x)\,dx + \psi(0)\frac{\varepsilon^{n+1}}{n+1} + \ldots + \frac{\psi^{(-n-1)}(0)}{(-n-1)!}\ln\varepsilon\right) \\
&= \langle\, Fp\, x_+^{n}\,,\ \psi(x) \,\rangle
\end{aligned}
$$

Finally for $n = 0, 1, 2, \ldots, k-1$ we get

$$
\begin{aligned}
\langle\, Fp\, x_+^{n-k}\,,\ \phi(x) \,\rangle &= \lim_{\substack{\varepsilon \to 0 \\ >}} \int_\varepsilon^{+\infty} x^{n-k}\phi(x)\,dx \\
&= \int_0^{+\infty} x^{n}\psi(x)\,dx \;=\; \langle\, x_+^{n}\,,\ \psi(x) \,\rangle
\end{aligned}
$$

□

2. Clifford analysis (I)

Clifford analysis (see e.g. [2],[6]) offers a function theory which is a higher dimensional analogue of the theory of holomorphic functions of one complex variable. In this section we only present the necessary basic definitions and results in Clifford analysis needed in the first part of this paper. The necessary material for the second part will be presented there (section 7).

Let $\mathbb{R}^{0,m}$ be the real vector space $\mathbb{R}^m$, endowed with a non-degenerate quadratic form of signature $(0, m)$, let $(e_1, \ldots, e_m)$ be an orthonormal basis for $\mathbb{R}^{0,m}$, and let $\mathbb{R}_{0,m}$ be the universal Clifford algebra constructed over $\mathbb{R}^{0,m}$. The non-commutative multiplication in $\mathbb{R}_{0,m}$ is governed by the rules

$$e_i^2 = -1, \qquad i = 1, 2, \ldots, m$$

and

$$e_i e_j + e_j e_i = 0, \qquad 1 \leq i \neq j \leq m.$$

For a set $A = \{i_1, \ldots, i_h\} \subset \{1, \ldots, m\}$, ordered in the natural way: $1 \leq i_1 < i_2 < \ldots < i_h \leq m$, we put

$$e_A = e_{i_1} e_{i_2} \ldots e_{i_h}$$

and

$$e_\phi = 1,$$

the latter being the identity element. Then $(e_A : A \subset \{1, \ldots, m\})$ is a basis for the Clifford algebra $\mathbb{R}_{0,m}$. Any $a \in \mathbb{R}_{0,m}$ may thus be written as

$$a = \sum_A a_A \, e_A, \qquad a_A \in \mathbb{R}$$

or still as

$$a = \sum_{k=0}^{m} [a]_k$$

where

$$[a]_k = \sum_{|A|=k} a_A \, e_A$$

is a so-called k-vector ($k = 0, 1, \ldots, m$).
If we denote the space of k-vectors by $\mathbb{R}_{0,m}^k$, then

$$\mathbb{R}_{0,m} = \sum_{k=0}^{m} \oplus \mathbb{R}_{0,m}^k$$

leading to the identification of $\mathbb{R}$ and $\mathbb{R}^{0,m}$ with $\mathbb{R}_{0,m}^0$ and $\mathbb{R}_{0,m}^1$ respectively.
We will also identify an element $\underline{x} = (x_1, \ldots, x_m) \in \mathbb{R}^m$ with the one-vector (or vector for short)

$$\underline{x} = \sum_{j=1}^{m} x_j \, e_j.$$

For any two vectors $\underline{x}$ and $\underline{y}$ we have

$$\underline{x}\,\underline{y} = \underline{x} \bullet \underline{y} + \underline{x} \wedge \underline{y}$$

where

$$\underline{x} \bullet \underline{y} = -\langle \, \underline{x}, \underline{y} \, \rangle = -\sum_{j=1}^{m} x_j y_j = \frac{1}{2}(\underline{x}\,\underline{y} + \underline{y}\underline{x})$$

is a scalar, and

$$\underline{x} \wedge \underline{y} = \sum_{i<j} e_{ij}(x_i y_j - x_j y_i) = \frac{1}{2}(\underline{x}\,\underline{y} - \underline{y}\underline{x})$$

is a 2-vector, also called bivector.

In particular

$$\underline{x}^2 = \underline{x} \bullet \underline{x} = -|\underline{x}|^2 = -\sum_{j=1}^{m} x_j^2.$$

Conjugation in $\mathbb{R}_{0,m}$ is defined as the anti-involution for which

$$\bar{e}_j = -e_j, \qquad j = 1, \ldots, m.$$

In particular for a vector $\underline{x}$ we have

$$\overline{\underline{x}} = -\underline{x}.$$

The Dirac operator in $\mathbb{R}^m$ is the first order vector valued differential operator

$$\underline{\partial} = \sum_{j=1}^{m} e_j \partial_{x_j}$$

its fundamental solution being given by

$$E_m(\underline{x}) = \frac{1}{a_m} \frac{\overline{\underline{x}}}{|\underline{x}|^m}$$

with $a_m = \frac{2\pi^{m/2}}{\Gamma(m/2)}$ the area of the unit sphere S^{m-1} in $\mathbb{R}^m$.

Considering functions defined in $\mathbb{R}^m$ and taking values in $\mathbb{R}_{0,m}$, we say that the function f is left-monogenic in the open region Ω of $\mathbb{R}^m$ iff f is continuously differentiable in Ω and satisfies in Ω :

$$\underline{\partial} f = 0.$$

As

$$\overline{\underline{\partial} f} = \overline{f}\, \overline{\underline{\partial}} = -\overline{f} \underline{\partial}$$

a function f is left monogenic in Ω iff $\overline{f}$ is right monogenic in Ω.
As moreover the Dirac operator factorizes the Laplace operator :

$$-\underline{\partial}^2 = \underline{\partial}\, \overline{\underline{\partial}} = \overline{\underline{\partial}}\, \underline{\partial} = \Delta,$$

a monogenic function in Ω is harmonic and hence C_∞ in Ω.
Introducing spherical co-ordinates :

$$\underline{x} = r\underline{\omega}, \quad r = |\underline{x}|, \quad \underline{\omega} \in S^{m-1},$$

the Dirac operator $\underline{\partial}$ may be written as

$$\underline{\partial} = \underline{\omega}\partial_r + \frac{1}{r}\partial_{\underline{\omega}} = \underline{\omega}\left(\partial_r - \frac{1}{r}\,\underline{\omega}\,\partial_{\underline{\omega}}\right),$$

while the Laplace operator takes the form

$$\Delta = \partial_r^2 + \frac{m-1}{r}\,\partial_r + \frac{1}{r^2}\,\Delta^*,$$

Δ^* being the Laplace-Beltrami operator on S^{m-1}.

To illustrate the meaning of the operators $\partial_{\underline{\omega}}$ and $\Gamma_{\underline{\omega}} = -\underline{\omega}\,\partial_{\underline{\omega}}$, consider the traditional case $m = 3$, where

$$x_1 = r\sin\theta\cos\varphi, \quad x_2 = r\sin\theta\sin\varphi, \quad x_3 = r\cos\theta \qquad (0 \le \theta < \pi, -\pi \le \varphi < \pi)$$

and

$$\begin{aligned}
\underline{\omega} &= e_1 \sin\theta\cos\varphi + e_2\sin\theta\sin\varphi + e_3\cos\theta \\
\partial_{\underline{\omega}} &= e_\theta\,\partial_\theta + e_\varphi\,\frac{1}{\sin\theta}\,\partial_\varphi \\
\Gamma_{\underline{\omega}} &= (\cos\varphi\, e_1e_3 + \sin\varphi\, e_2e_3)\partial_\theta \\
&\quad +(-e_1e_2 - \cot\theta\sin\varphi\, e_1e_3 + \cot\theta\cos\varphi\, e_2e_3)\partial_\varphi \\
&= e_1e_2e_3\,(e_\varphi\,\partial_\theta - e_\theta\,\frac{1}{\sin\theta}\,\partial_\varphi)
\end{aligned}$$

where

$$e_\theta = \cos\theta\cos\varphi\, e_1 + \cos\theta\sin\varphi\, e_2 - \sin\theta\, e_3$$

and

$$e_\varphi = -\sin\varphi\, e_1 + \cos\varphi\, e_2$$

are unit vectors ($e_\theta^2 = e_\varphi^2 = -1$) in the tangent plane to S^{m-1} at $\underline{\omega}$.

Denoting by $\mathcal{D}$ and $\mathcal{S}$ the space of the compactly supported, respectively rapidly decreasing, real-valued test functions in $\mathbb{R}^m$, we will consider the modules of testfunctions $\prod_{A\subset\{1,\ldots,m\}} \mathcal{D}$ and $\prod_{A\subset\{1,\ldots,m\}} \mathcal{S}$; any such test function φ may be written as

$$\varphi = \sum_{A\subset\{1,\ldots,m\}} e_A\,\phi_A, \qquad \phi_A \in \mathcal{D} \text{ or } \mathcal{S}.$$

In most cases we will even use real-valued test functions.
A left Clifford distribution $\mathcal{T}^{(l)}$ is then a bounded left $\mathbb{R}_{0,m}$-linear functional for which there exist bounded real-linear functionals $\mathcal{T}_B$ ($B \subset \{1,\ldots,m\}$) such that

$$\langle\, \varphi, \mathcal{T}^{(l)}\,\rangle = \sum_{A,B} e_A\, e_B \langle\, \mathcal{T}_B, \phi_A\,\rangle,$$

and a similar definition for a right Clifford distribution $\mathcal{T}^{(r)}$:

$$\langle\, \mathcal{T}^{(r)}, \varphi\,\rangle = \sum_{A,B} e_B\, e_A \langle\, \mathcal{T}_B, \phi_A\,\rangle.$$

3. Generalized spherical means (I)

The spherical co-ordinates are really fundamental to the approach presented in this paper. They generate in a quite natural way the so-called generalized spherical means (see e.g. [13]).

Let $\phi(\underline{x})$ be a scalar valued test function in $\mathbb{R}^m$; putting $\underline{x} = r\,\underline{\omega}$, $r = |\underline{x}|$, $\underline{\omega} \in S^{m-1}$, we define the generalized spherical means :

$$\Sigma^{(0)}[\phi] = \frac{1}{a_m} \int_{S^{m-1}} \phi(r\,\underline{\omega})\, dS(\underline{\omega})$$

and

$$\Sigma^{(1)}[\phi] = \Sigma^{(0)}[\underline{\omega}\phi] = \frac{1}{a_m} \int_{S^{m-1}} \underline{\omega}\phi(r\underline{\omega})\, dS(\underline{\omega})$$

where $dS(\underline{\omega})$ denotes the Lebesgue measure on S^{m-1}.

Note that $\Sigma^{(0)}[\phi]$ is nothing else but the classical spherical mean introduced by John in [10].

The generalized spherical means enjoy the following properties. Firstly we note that both spherical means are interrelated by the action of the Dirac operator. In accordance with [13] we have, extending the notion of spherical mean to Clifford algebra valued, in casu vector valued, functions :

Proposition 3.1. *For a scalar valued test function ϕ one has*

(i)

$$\Sigma^{(0)}[\underline{\partial}\phi] \;=\; \left(\partial_r + \frac{m-1}{r}\right)\, \Sigma^{(1)}[\phi]$$

(ii)

$$\Sigma^{(1)}[\underline{\partial}\phi] \;=\; -\,\partial_r\, \Sigma^{(0)}[\phi]$$

(iii)

$$\Sigma^{(0)}[\partial_{\underline{\omega}}\phi] \;=\; (m-1)\, \Sigma^{(1)}[\phi]$$

(iv)

$$\Sigma^{(1)}[\partial_{\underline{\omega}}\phi] \;=\; 0.$$

Proof. Consider the ball $B(0,\rho)$ with arbitrary radius $\rho > 0$ and apply Stokes's Theorem (see [2]) to obtain:

$$\int_{B(0,\rho)} \underline{\partial}\phi\, dV(\underline{x}) \;=\; \int_{\partial B(0,\rho)} d\underline{\sigma}\,\phi \;=\; \rho^{m-1} \int_{S^{m-1}} \underline{\omega}\;\phi(\rho\underline{\omega})\, dS(\underline{\omega})$$

For the left hand side we get:

$$\int_0^\rho r^{m-1}\, dr \int_{S^{m-1}} \left(\underline{\omega}\;\partial_r\phi + \frac{1}{r}\;\partial_{\underline{\omega}}\phi\right)\; dS(\underline{\omega})$$

$$= \int_{S^{m-1}} dS(\underline{\omega})\;\underline{\omega} \int_0^\rho r^{m-1}\;\partial_r\phi\, dr + \int_0^\rho r^{m-2}\, dr \int_{S^{m-1}} \partial_{\underline{\omega}}\phi\, dS(\underline{\omega})$$

$$= \rho^{m-1} \int_{S^{m-1}} \underline{\omega}\;\phi(\rho\underline{\omega})\, dS(\underline{\omega}) + \int_0^\rho r^{m-2}\, dr \int_{S^{m-1}} \left(\partial_{\underline{\omega}}\phi - (m-1)\;\underline{\omega}\;\phi\right)\; dS(\underline{\omega})$$

As the first term in the last expression equals the right hand side of the above Stokes formula, it follows that

$$\int_{S^{m-1}} \partial_{\underline{\omega}}\phi\, dS(\underline{\omega}) = (m-1) \int_{S^{m-1}} \underline{\omega}\;\phi\, dS(\underline{\omega})$$

or $\Sigma^{(0)}[\partial_{\underline{\omega}}\phi] = (m-1)\ \Sigma^{(1)}[\phi]$, which proves formula (iii).

Next we have:

$$a_m\ \Sigma^{(0)}[\underline{\partial}\phi] = \int_{S^{m-1}} \underline{\partial}\phi\ dS(\underline{\omega}) = \int_{S^{m-1}} \left(\underline{\omega}\ \partial_r\phi + \frac{1}{r}\ \partial_{\underline{\omega}}\phi\right)\ dS(\underline{\omega})$$

$$= a_m\ \partial_r\ \Sigma^{(1)}[\phi] + a_m\ \frac{1}{r}\ \Sigma^{(0)}[\partial_{\underline{\omega}}\phi] = a_m\ \left(\partial_r + \frac{m-1}{r}\right)\ \Sigma^{(1)}[\phi]$$

which proves formula (i).

Now apply again Stokes's Theorem to obtain

$$\int_{B(0,\rho)} (\underline{x}\ \phi)\ \underline{\partial}\ dV(\underline{x}) = \int_{\partial B(0,\rho)} \underline{x}\ \phi\ d\underline{\sigma} = -\ \rho^m \int_{S^{m-1}} \phi(\rho\underline{\omega})\ dS(\underline{\omega})$$

For the left hand side we get

$$\int_{B(0,\rho)} (\underline{x}\ \underline{\partial}\phi - m\ \phi)\ dV(\underline{x})$$

$$= -m \int_0^\rho r^{m-1}\ \Sigma^{(0)}[\phi]\ dr - \int_0^\rho r^m\ dr \int_{S^{m-1}} \partial_r\phi\ dS(\underline{\omega}) + \int_0^\rho r^{m-1}\Sigma^{(1)}[\partial_{\underline{\omega}}\phi]\ dr$$

$$= -\ \rho^m \int_{S^{m-1}} \phi(\rho\underline{\omega})\ dS(\underline{\omega})\ +\ \int_0^\rho r^{m-1}\ \Sigma^{(1)}[\partial_{\underline{\omega}}\phi]\ dr$$

As the first term in the last expression equals the right hand side of the above Stokes Formula, formula (iv) follows.
Finally,

$$a_m\ \Sigma^{(1)}[\underline{\partial}\phi]\ =\ \int_{S^{m-1}} \underline{\omega}\ \left(\underline{\omega}\ \partial_r\phi + \frac{1}{r}\ \partial_{\underline{\omega}}\phi\right)\ dS(\underline{\omega})$$

$$=\ -\ \Sigma^{(0)}[\partial_r\phi]\ +\ \frac{1}{r}\ \Sigma^{(1)}[\partial_{\underline{\omega}}\phi]\ =\ -\ \partial_r\ \Sigma^{(0)}[\phi]$$

proving formula (ii). □

Proposition 3.2. *If $\phi(\underline{x})$ is a scalar valued test function, then the spherical mean $\Sigma^{(0)}[\phi]$ is an even, scalar valued test function on the real r-axis; its derivatives of odd order vanish at the origin $r = 0$, while for its derivatives of even order one has*

$$\left\{\underline{\partial}^{2l}\phi(\underline{x})\right\}_{\underline{x}=\underline{0}} = (-1)^l C(l)\left\{\partial_r^{2l}\Sigma^{(0)}[\phi]\right\}_{r=0}$$

or, equivalently, in terms of distributions,

$$\langle\ \underline{\partial}^{2l}\delta(\underline{x}), \phi(\underline{x})\ \rangle = (-1)^l\ C(l)\langle\ \partial_r^{2l}\delta(r), \Sigma^{(0)}[\phi]\ \rangle,$$

where

$$C(l) = \frac{2^{2l}l!}{(2l)!}\left(\frac{m}{2}+l-1\right)\cdots\left(\frac{m}{2}\right) = \frac{2^{2l}l!}{(2l)!}\ \frac{\Gamma(\frac{m}{2}+l)}{\Gamma(\frac{m}{2})},\quad l = 0, 1, 2, \ldots$$

Proof. We prove the formula for the derivatives of even order by induction on l. For $l = 0$ we have to prove that

$$\langle\ \delta(\underline{x})\ ,\ \phi(\underline{x})\ \rangle = \langle\ \delta(r)\ ,\ \Sigma^{(0)}[\phi]\ \rangle$$

or

$$\phi(\underline{0}) = \left\{ \frac{1}{a_m} \int_{S^{m-1}} \phi(r\underline{\omega})\ dS(\underline{\omega}) \right\}_{r=0}$$

which is of course valid.
For $l = 1$ we have to prove that

$$\langle\ -\Delta\delta(\underline{x})\ ,\ \phi(\underline{x})\ \rangle = -m\ \langle\ \delta''(r)\ ,\ \Sigma^{(0)}[\phi]\ \rangle.$$

Indeed, we have

$$\begin{aligned} a_m\ \Sigma^{(0)}[\Delta\phi] &= \int_{S^{m-1}} \Delta\phi\ dS(\underline{\omega}) \\ &= \int_{S^{m-1}} \left(\partial_r^2 + \frac{m-1}{r}\,\partial_r + \frac{1}{r^2}\,\Delta^* \right) \phi(r\underline{\omega})\ dS(\underline{\omega}) \\ &= a_m \left(\partial_r^2 + \frac{m-1}{r}\,\partial_r \right) \Sigma^{(0)}[\phi] \end{aligned}$$

since

$$\int_{S^{m-1}} \Delta^*\phi(r\underline{\omega})\ dS(\underline{\omega}) = 0.$$

It thus follows that

$$\left\{ \Sigma^{(0)}[\Delta\phi] \right\}_{r=0} = \lim_{\substack{r \to 0 \\ >}} \left(\partial_r^2 + \frac{m-1}{r}\,\partial_r \right) \Sigma^{(0)}[\phi] = m\ \left\{ \partial_r^2\ \Sigma^{(0)}[\phi] \right\}_{r=0}$$

and hence, in view of the result for $l = 0$,

$$\begin{aligned} \langle\ \Delta\delta(\underline{x})\ ,\ \phi(\underline{x})\ \rangle &= \langle\ \delta(\underline{x})\ ,\ \Delta\phi\ \rangle \\ &= \langle\ \delta(r)\ ,\ \Sigma^{(0)}[\Delta\phi]\ \rangle = m\ \langle\ \delta''(r)\ ,\ \Sigma^{(0)}[\phi]\ \rangle \end{aligned}$$

which proves the formula for $l = 1$.
Now assume the formula to be valid for l; then we have consecutively:

$$\begin{aligned} &\langle\ \underline{\partial}^{2l+2}\delta(\underline{x})\ ,\ \phi(\underline{x})\ \rangle = \langle\ \underline{\partial}^{2l}\delta(\underline{x})\ ,\ \Delta\phi\ \rangle \\ &= (-1)^l\ C(l)\ \langle\ \partial_r^{2l}\delta(r)\ ,\ \Sigma^{(0)}[\Delta\phi]\ \rangle \\ &= (-1)^l\ C(l)\ \langle\ \partial_r^{2l}\delta(r)\ ,\ \left(\partial_r^2 + \frac{m-1}{r}\,\partial_r \right)\ \Sigma^{(0)}[\phi]\ \rangle \\ &= (-1)^l\ C(l)\ \langle\ \partial_r^{2l+2}\delta(r)\ ,\ \Sigma^{(0)}[\phi]\ \rangle + (-1)^l\ C(l)\ \frac{m-1}{2l+1} \langle\ \partial_r^{2l+2}\delta(r)\ ,\ \Sigma^{(0)}[\phi]\ \rangle \\ &= (-1)^{l+1}\ C(l+1)\ \langle\ \partial_r^{2l+2}\delta(r)\ ,\ \Sigma^{(0)}[\phi]\ \rangle \end{aligned}$$

which proves the formula for $l + 1$. □

Proposition 3.3. *If $\phi(\underline{x})$ is a scalar valued test function, then the spherical mean $\Sigma^{(1)}[\phi]$ is an odd, vector valued test function on the real r-axis; its derivatives of even order vanish at the origin $r = 0$, while for the derivatives of odd order one has*

$$\left\{\underline{\partial}^{2l+1}\phi(\underline{x})\right\}_{\underline{x}=\underline{0}} = (-1)^l C(l+1)\left\{\partial_r^{2l+1}\Sigma^{(1)}[\phi]\right\}_{r=0}$$

or, equivalently, in terms of distributions,

$$\langle\ \underline{\partial}^{2l+1}\delta(\underline{x}), \phi(\underline{x})\ \rangle = (-1)^l\ C(l+1)\langle\ \partial_r^{2l+1}\delta(r), \Sigma^{(1)}[\phi]\ \rangle.$$

Proof. We prove the formula for the derivatives of odd order. Let ϕ be a scalar valued test function. Then by Proposition 3.2 and Proposition 3.1, we have:

$$\begin{aligned}
&\langle\ \underline{\partial}^{2l+1}\delta(\underline{x})\ ,\ \phi(\underline{x})\ \rangle = -\langle\ \underline{\partial}^{2l}\delta(\underline{x})\ ,\ \underline{\partial}\phi\ \rangle \\
&= (-1)^{l+1}\ C(l)\ \langle\ \partial_r^{2l}\delta(r)\ ,\ \Sigma^{(0)}[\underline{\partial}\phi]\ \rangle \\
&= (-1)^{l+1}\ C(l)\ \langle\ \partial_r^{2l}\delta(r)\ ,\ \left(\partial_r + \frac{m-1}{r}\right)\ \Sigma^{(1)}[\phi]\ \rangle \\
&= (-1)^l\ C(l)\ \langle\ \partial_r^{2l}\delta(r)\ ,\ \Sigma^{(1)}[\phi]\ \rangle + \frac{m-1}{2l+1}\ (-1)^l\ C(l)\ \langle\ \partial_r^{2l+1}\delta(r)\ ,\ \Sigma^{(1)}[\phi]\ \rangle \\
&= (-1)^{l+1}\ C(l+1)\ \langle\ \partial_r^{2l+1}\delta(r)\ ,\ \Sigma^{(1)}[\phi]\ \rangle
\end{aligned}$$ □

Note that in particular $\{\Sigma^{(0)}[\phi]\}_{r=0} = \phi(\underline{0})$, while $\{\Sigma^{(1)}[\phi]\}_{r=0} = 0$ or

$$\int_{S^{m-1}} \underline{\omega}\ dS(\underline{\omega}) = 0$$

4. Classical and Clifford distributions in Euclidean space

In this section we will define classical and new distributions in the Clifford setting making use of the spherical co-ordinates, the fundamental distribution $Fp\, r_+^\mu$ of section 1 and the spherical means of section 3.

Let us explain the underlying idea by considering the special case of a locally integrable radial function $T(r)$. Its action as a regular distribution on a scalar valued test function $\phi(\underline{x})$ is given by

$$\langle\ T(r), \phi(\underline{x})\ \rangle = \int_{\mathbb{R}^m} T(r)\ \phi(\underline{x})\ dV(\underline{x}),$$

$dV(\underline{x})$ denoting the Lebesgue measure in $\mathbb{R}^m$.
Introducing spherical co-ordinates this integral takes the form

$$\int_0^{+\infty} T(r) r^{m-1} dr \int_{S^{m-1}} \phi(r\underline{\omega}) dS(\underline{\omega}) = \int_0^{+\infty} T(r) r^{m-1} \Sigma^{(0)}[\phi]\ dr.$$

Hence for $T(r) = r^\lambda$ with $\operatorname{Re}\lambda > -m$, we find

$$\langle\ r^\lambda, \phi(\underline{x})\ \rangle = \int_0^{+\infty} r^\mu \Sigma^{(0)}[\phi]\ dr = \langle\ r_+^\mu, \Sigma^{(0)}[\phi]\ \rangle,$$

where we have put $\mu = \lambda + m - 1$. By this procedure the action of the distribution $T(r) = r^\lambda$ in $\mathbb{R}^m$ is converted into an action of the distribution r_+^μ on the real line. It is this conversion which is used in defining our Clifford distributions in $\mathbb{R}^m$.

Let λ be a complex parameter and let ϕ be a scalar valued test function. We define the scalar valued distributions T_λ and the vector valued distributions U_λ by

$$\langle\, T_\lambda, \phi \,\rangle = a_m \langle\, Fp\, r_+^\mu,\ \Sigma^{(0)}[\phi] \,\rangle$$

and

$$\langle\, U_\lambda, \phi \,\rangle = a_m \langle\, Fp\, r_+^\mu,\ \Sigma^{(1)}[\phi] \,\rangle,$$

where we have put

$$\mu = \lambda + m - 1.$$

These distributions are classical, in the sense as explained in the introduction: the T_λ distributions appear in harmonic analysis where they are sometimes denoted by $Fp\, r^\lambda$, while the U_λ distributions – at least some of them – appear in Clifford analysis. More familiar expressions are given in the next proposition.

Proposition 4.1. *For $\mu = \lambda + m - 1$ such that $-2l - 1 < \operatorname{Re}\mu < -2l + 1$ one has*

$$\langle\, T_\lambda, \phi \,\rangle = \lim_{\substack{\varepsilon \to 0\\ >}} \left(\int_{\mathbb{R}^m \setminus B(0,\varepsilon)} r^\lambda \phi(\underline{x})\, dV(\underline{x}) + \sum_{j=0}^{l} \widetilde{C}(j)(\Delta^j \phi)(0) \frac{\varepsilon^{\lambda+m+2j}}{\lambda + m + 2j} \right)$$

while for $l = 0, 1, 2, \ldots$

$\langle\, T_{-m-2l}, \phi \,\rangle =$

$$\lim_{\substack{\varepsilon \to 0\\ >}} \left(\int_{\mathbb{R}^m \setminus B(0,\varepsilon)} r^{-m-2l} \phi(\underline{x})\, dV(\underline{x}) + \sum_{j=0}^{l-1} \widetilde{C}(j)(\Delta^j \phi)(0) \frac{\varepsilon^{2j-2l}}{2j - 2l} + \widetilde{C}(l)(\Delta^l \phi)(0) \ln \varepsilon \right)$$

with

$$\begin{aligned} \widetilde{C}(j) &= a_m \frac{1}{(2j)!} \frac{1}{C(j)} = a_m \frac{1}{2^{2j} j! (\frac{m}{2} + j - 1) \ldots (\frac{m}{2})} \\ &= \pi^{m/2} \frac{1}{2^{2j-1} j! \Gamma(\frac{m}{2} + j)}, \qquad j = 0, 1, 2, \ldots \end{aligned}$$

Note that the first formula is in accordance with [11].
Examples of the T_λ and U_λ distributions are given in the next section.

5. The action of the Dirac operator on the distributions T_λ and U_λ

As the distributions T_λ are scalar valued it is not necessary to distinguish between an action of the Dirac operator from the left or from the right. The results of that action, $\underline{\partial} T_\lambda$, will be vector valued distributions, and as long as the testfunctions ϕ remain scalar valued it is neither necessary to distinguish between $\underline{\partial} T_\lambda$ as to be a left or a right distribution.

Proposition 5.1. *For $\lambda \in \mathbb{C}\backslash\{-m-2l : l = 0,1,2,\ldots\}$ one has*

$$\underline{\partial} T_\lambda = \lambda\, U_{\lambda-1}$$

and in particular $\underline{\partial} T_0 = \underline{\partial} 1_{\underline{x}} = 0$, while for $l = 0,1,2,\ldots$

$$\underline{\partial}\, T_{-m-2l} = -(m+2l)U_{-m-2l-1} + (-1)^{l+1}2(l+1)\,\widetilde{C}(l+1)\,\underline{\partial}^{2l+1}\delta(\underline{x})$$

and

$$\underline{\partial}\,\ln r = U_{-1}.$$

Proof. Firstly we proof the formula in the general case where $\lambda \neq -m-n$, $n = 0,1,2,\ldots$.
For any scalar valued test function ϕ we have by the definition of the derivative of a distribution:

$$\langle\, \underline{\partial}\, T_\lambda\, ,\, \phi\, \rangle = -\,\langle\, T_\lambda\, ,\, \underline{\partial}\, \phi\, \rangle.$$

By the definition of the distributions T_λ, this equals

$$-\,a_m\,\langle\, Fp\, r_+^\mu\, ,\, \Sigma^{(0)}[\underline{\partial}\,\phi]\,\rangle,$$

which, by Proposition 3.1, is turned into

$$\begin{aligned}
& -\,a_m\langle\, Fp\, r_+^\mu\, ,\, \left(\partial_r + \frac{m-1}{r}\right)\Sigma^{(1)}[\phi]\,\rangle \\
= \;& a_m\langle\, \partial_r\, Fp\, r_+^\mu\, ,\, \Sigma^{(1)}[\phi]\,\rangle - a_m\,(m-1)\langle\, Fp\, r_+^{\mu-1}\, ,\, \Sigma^{(1)}[\phi]\,\rangle \\
= \;& a_m\,(\mu-m+1)\,\langle\, Fp\, r_+^{\mu-1}\, ,\, \Sigma^{(1)}[\phi]\,\rangle \\
= \;& \lambda\,\langle\, U_{\lambda-1}\, ,\, \phi\,\rangle
\end{aligned}$$

Now for the exceptional cases where $\lambda = -m-2l$, $l = 0,1,2,\ldots$, we have consecutively, taking Lemma 1.1 into account:

$$\begin{aligned}
& \langle\, \underline{\partial}\, T_{-m-2l}, \phi\,\rangle \;=\; -\,\langle\, T_{-m-2l}, \underline{\partial}\,\phi\,\rangle \\
= \;& -\,a_m\,\langle\, Fp\, r_+^{-2l-1}\, ,\, \Sigma^{(0)}[\underline{\partial}\,\phi]\,\rangle \\
= \;& -\,a_m\,\langle\, Fp\, r_+^{-2l-1}\, ,\, \left(\partial_r + \frac{m-1}{r}\right)\Sigma^{(1)}[\phi]\,\rangle \\
= \;& a_m\,\langle\, \partial_r\, Fp\, r_+^{-2l-1}\, ,\, \Sigma^{(1)}[\phi]\,\rangle - a_m(m-1)\langle\, Fp\, r_+^{-2l-2}\, ,\, \Sigma^{(1)}[\phi]\,\rangle \\
= \;& -(m+2l)\, a_m\langle\, Fp\, r_+^{-2l-2}\, ,\, \Sigma^{(1)}[\phi]\,\rangle \\
& + a_m(-1)^{2l+1}\frac{1}{(2l+1)!}\langle\, \delta_{(r)}^{(2l+1)}\, ,\, \Sigma^{(1)}[\phi]\,\rangle \\
= \;& -(m+2l)a_m\langle\, Fp\, r_+^{-2l-2}\, ,\, \Sigma^{(1)}[\phi]\,\rangle \\
& + a_m(-1)^{2l+1}\frac{1}{(2l+1)!}\,\frac{(-1)^l}{C(l+1)}\langle\, \underline{\partial}^{2l+1}\delta(\underline{x})\, ,\, \phi(\underline{x})\,\rangle
\end{aligned}$$

whence

$$\underline{\partial} T_{-m-2l} = -(m+2l)U_{-m-2l-1} + a_m(-1)^{l+1}\,\frac{\underline{\partial}^{2l+1}\delta(\underline{x})}{2^{2l+1}l!\left(\frac{m}{2}+l\right)\ldots\left(\frac{m}{2}\right)}.$$

The exceptional cases where $\lambda = -m-2l-1,\ l = 0,1,2,\ldots$ are treated in a similar manner; there is however no singular term involving derivatives of the delta distribution, since the even order derivatives of $\Sigma^{(1)}[\phi]$ vanish at $r=0$.
Finally we also have

$$\begin{aligned}
&\langle\, \underline{\partial} \ln r\ ,\ \phi\,\rangle > = \ -\ \langle\, \ln r\ ,\ \underline{\partial}\phi\,\rangle \\
= &\int_0^{+\infty} r^{m-1}\ \ln r\ dr \int_{S^{m-1}} \underline{\partial}\phi\ dS(\underline{\omega}) \\
= &-\int_0^{+\infty} r^{m-1}\ \ln r\ \partial_r \Sigma^{(1)}[\phi]\ dr\ -\ (m-1)\int_0^{+\infty} r^{m-2}\ \ln r\ \Sigma^{(1)}[\phi]\ dr \\
= &\int_0^{+\infty} r^{m-2}\ \Sigma^{(1)}[\phi]\ dr \\
= &\ \langle\, Fp\, r_+^{m-2}\ ,\ \Sigma^{(1)}[\phi]\,\rangle\ =\ \langle\, U_{-1}\ ,\ \phi\,\rangle
\end{aligned}$$

□

As the distributions U_λ are vector valued, we should, a priori, make a distinction between an action from the left or an action from the right of the Dirac operator and at the same time consider $\underline{\partial}U_\lambda$ as a left distribution, respectively $U_\lambda\underline{\partial}$ as a right distribution. Both cases can be treated similarly and it turns out that both actions lead to the same scalar valued distribution.

Proposition 5.2. *For* $\lambda \in \mathbb{C}\backslash\{-m-2l-1 : l = -1,0,1,2,\ldots\}$ *one has*

$$\underline{\partial}U_\lambda = U_\lambda\underline{\partial} = -(\lambda+m-1)T_{\lambda-1}$$

while for $l = 0,1,2,\ldots$

$$\underline{\partial}U_{-m-2l-1} = U_{-m-2l-1}\ \underline{\partial} = (2l+2)T_{-m-2l-2} - \widetilde{C}(l+1)\Delta^{l+1}\delta(\underline{x})$$

and

$$\underline{\partial}U_{-m+1} = U_{-m+1}\ \underline{\partial} = -a_m\delta(\underline{x}).$$

Proof. Firstly we proof the formula in the general case where $\lambda \neq -m-n,\ n = -1,0,1,2,\ldots$.
For any scalar valued test function ϕ one has, considering right distributions,

$$\begin{aligned}
&\langle\, U_\lambda\underline{\partial}\ ,\ \phi\,\rangle = -\langle\, U_\lambda\ ,\ \underline{\partial}\phi\,\rangle \\
= &\ -a_m\ \langle\, Fp\, r_+^\mu\ ,\ \Sigma^{(1)}[\underline{\partial}\phi]\,\rangle = -a_m\ \langle\, Fp\, r_+^\mu\ ,\ -\partial_r\Sigma^{(0)}[\phi]\,\rangle \\
= &\ -a_m\ \langle\, \mu\ Fp\, r_+^{\mu-1}\ ,\ \Sigma^{(0)}[\phi]\,\rangle = -(\lambda+m-1)\ \langle\, T_{\lambda-1}\ ,\ \phi\,\rangle
\end{aligned}$$

Next for the exceptional case where $\lambda = -m+1$ we get

$$\begin{aligned}
&\langle\, U_{-m+1}\underline{\partial}\ ,\ \phi\,\rangle = -\langle\, U_{-m+1}\ ,\ \underline{\partial}\phi\,\rangle \\
= &\ -a_m\ \langle\, Fp\, r_+^0\ ,\ \Sigma^{(1)}[\underline{\partial}\phi]\,\rangle = -a_m\ \langle\, Y(r)\ ,\ -\partial_r\Sigma^{(0)}[\phi]\,\rangle \\
= &\ -a_m\ \langle\, \delta(r)\ ,\ \Sigma^{(0)}[\phi]\,\rangle = -a_m\ \langle\, \delta(\underline{x})\ ,\ \phi(\underline{x})\,\rangle.
\end{aligned}$$

Finally for the exceptional cases where $\lambda = -m - 2l - 1$, $l = 0, 1, 2, \ldots$, we have, taking Lemma 1.1 into account:

$$\begin{aligned}
&\langle\, U_{-m-2l-1}\, \underline{\partial}, \phi \,\rangle = -\langle\, U_{-m-2l-1}, \underline{\partial}\phi \,\rangle = -a_m\langle\, Fp\, r_+^{-2l-2}, \Sigma^{(1)}[\underline{\partial}\,\phi] \,\rangle \\
&= -a_m\langle\, Fp\, r_+^{-2l-2}, -\partial_r \Sigma^{(0)}[\phi] \,\rangle = -a_m\langle\, \partial_r Fp\, r_+^{-2l-2}, \Sigma^{(0)}[\phi] \,\rangle \\
&= -a_m\langle\, (-2l-2) Fp\, r_+^{-2l-3} + \frac{1}{(2l+2)!}\, \delta_{(r)}^{(2l+2)}, \Sigma^{(0)}[\phi] \,\rangle \\
&= a_m (2l+2) \langle\, Fp\, r_+^{-2l-3}, \Sigma^{(0)}[\phi] \,\rangle - a_m\, \frac{1}{(2l+2)!} \langle\, \delta_{(r)}^{(2l+2)}, \Sigma^{(0)}[\phi] \,\rangle
\end{aligned}$$

whence

$$U_{-m-2l-1}\, \underline{\partial} = (2l+2) T_{-m-2l-2} - (-1)^{l+1} \frac{a_m}{(2l+2)! C(l+1)}\, \underline{\partial}^{2l+2} \delta(\underline{x})$$

or still

$$U_{-m-2l-1}\, \underline{\partial} = (2l+2) T_{-m-2l-2} - \widetilde{C}(l+1) \Delta^{l+1} \delta(\underline{x}).$$

The exceptional cases where $\lambda = -m - 2l$, $l = 0, 1, 2, \ldots$ are treated in a similar manner; there is however no singular term involving derivatives of the delta distribution, since the odd order derivatives of $\Sigma^{(0)}[\phi]$ vanish at $r = 0$. □

Remark 5.3. The proofs of Propositions 5.1 and 5.2 are fundamental; they nicely demonstrate the technique as explained at the beginning of this section. The proofs of the similar propositions in the rest of the paper run along the same lines and will not be given anymore.

6. Some specific distributions in Euclidean space

In this section we give some explicit examples of the distributions T_λ and U_λ introduced in the foregoing sections. It is shown that for specific values of the parameter λ, distributions are obtained which were already known in Clifford analysis, thus illustrating the unifying character of our approach.

6.1.

For $\lambda = 0$ we have

$$\begin{aligned}
\langle\, U_0, \phi \,\rangle &= a_m \langle\, Fp\, r_+^{m-1}, \Sigma^{(1)}[\phi] \,\rangle = \int_0^{+\infty} r^{m-1}\, dr \int_{S^{m-1}} \underline{\omega} \phi(r\underline{\omega})\, dS(\underline{\omega}) \\
&= \int_{\mathbb{R}^m} \underline{\omega}\, \phi(\underline{x})\, dV(\underline{x}),
\end{aligned}$$

in other words, $U_0(\underline{x})$ is the locally integrable function $\underline{\omega}$.
We also have

$$\underline{\partial}\, T_1 = \underline{\partial} r = U_0 = \underline{\omega},$$

and

$$\underline{\partial}\, \underline{\omega} = \underline{\omega}\, \underline{\partial} = -(m-1) T_{-1} = -(m-1) \frac{1}{r} = -\Delta r$$

and also

$$\Delta \underline{\omega} = (m-1)\underline{\partial}\, \frac{1}{r} = -(m-1)U_2 = (m-1)\, \frac{\overline{\underline{\omega}}}{r^2}.$$

The distribution $\underline{\omega}$ is the higher dimensional analogue of the one dimensional "signum"-distribution

$$\operatorname{sgn}(x) = \begin{cases} -1 & , \quad x < 0 \\ 1 & , \quad x > 0. \end{cases}$$

6.2.

For $\lambda = -m$ we have

$$\begin{aligned} \langle\, U_{-m}, \phi \,\rangle &= a_m \langle\, Fp\, r_+^{-1}, \Sigma^{(1)}[\phi] \,\rangle = \int_0^{+\infty} \frac{1}{r}\, dr \int_{S^{m-1}} \underline{\omega}\phi\, dS(\underline{\omega}) \\ &= \int_{\mathbb{R}^m} \frac{\underline{\omega}}{r^m} (\phi(\underline{x}) - \phi(\underline{0}))dV(\underline{x}) = \lim_{\substack{\varepsilon \to 0 \\ >}} \int_{\mathbb{R}^m \setminus B(0,\varepsilon)} \frac{\underline{\omega}}{r^m}\, \phi(\underline{x}) dV(\underline{x}) \end{aligned}$$

so that (see e.g. [5])

$$U_{-m} = -Pv\, \frac{\overline{\underline{\omega}}}{r^m}$$

where Pv stands for "principal value". It is the higher dimensional analogue of the one dimensional $Pv\, \frac{1}{x}$ distribution given by

$$\langle\, Pv\, \frac{1}{x}, \phi(x) \,\rangle = \lim_{\substack{\varepsilon \to 0 \\ >}} \left(\int_{-\infty}^{-\varepsilon} + \int_{\varepsilon}^{+\infty} \frac{\phi(x)}{x}\, dx \right) = \int_{-\infty}^{+\infty} \frac{\phi(x) - \phi(o)}{x}\, dx.$$

For this distribution U_{-m} we have the following formulae:

(i) $\quad r^m \underline{\omega}\, Pv\, \dfrac{\overline{\underline{\omega}}}{r^m} = 1$

(ii) $\quad Pv\, \dfrac{\overline{\underline{\omega}}}{r^m} = \dfrac{1}{m-1}\, \underline{\partial}\, \dfrac{1}{r^{m-1}}$

(iii) $\quad \underline{\partial} \left(Pv\, \dfrac{\overline{\underline{\omega}}}{r^m} \right) = -\underline{\partial}\, U_{-m} = -T_{-m-1} = -Fp\, \dfrac{1}{r^{m+1}}$

(iv) $\quad \Delta \left(Pv\, \dfrac{\overline{\underline{\omega}}}{r^m} \right) = \underline{\partial}(T_{-m-1}) = -(m+1)U_{-m-2} = (m+1)Fp\, \dfrac{\overline{\underline{\omega}}}{r^{m+2}}.$

As

$$\int_{S^{m-1}} \underline{\omega}\, dS(\underline{\omega}) = 0$$

(see section 3) the distribution $Pv\, \frac{\overline{\underline{\omega}}}{r^m}$ is a typical example of a convolution operator (see [9], [12]):

$$Pv\, \frac{\overline{\underline{\omega}}}{r^m} * \phi = \lim_{\substack{\varepsilon \to 0 \\ >}} \int_{\mathbb{R}^m \setminus B(0,\varepsilon)} \frac{\underline{y} - \underline{x}}{|\underline{y} - \underline{x}|}\, \frac{\phi(\underline{x})}{|\underline{y} - \underline{x}|^m}\, dV(\underline{x})$$

giving rise to the Hilbert transform of the test function ϕ (see [8] and [5]).

6.3.

Start with the observation that, according to Proposition I.5.2, we have for $\lambda = -m+1$:

$$\underline{\partial}\, U_{-m+1} = U_{-m+1}\, \underline{\partial} = -a_m\, \delta(\underline{x})$$

confirming that

$$-\frac{1}{a_m}\, U_{-m+1} = \frac{1}{a_m}\, \frac{\overline{\underline{\omega}}}{r^{m-1}}$$

is the fundamental solution of the Dirac operator in $\mathbb{R}^m$ (see Section 2).
Next, observe that, according to the Proposition 5.1, we have

$$\underline{\partial}^2 T_{-m+2} = \underline{\partial}((-m+2)U_{-m+1})) = (m-2)a_m\, \delta(\underline{x})$$

confirming that

$$-\frac{1}{a_m}\,\frac{1}{m-2}\, T_{-m+2} = -\frac{1}{a_m}\,\frac{1}{m-2}\,\frac{1}{r^{m-2}}$$

is the fundamental solution of the Laplace operator in $\mathbb{R}^m$.
Hence, find recursively the fundamental solutions of the natural powers of the Dirac operator, and at the same time of the Laplace operator, to be

$$\underline{\partial}^{2l}\left(\frac{1}{2l-2}\cdots\frac{1}{4}\,\frac{1}{2}\,\frac{1}{a_m}\,\frac{1}{m-2}\,\frac{1}{m-4}\cdots\frac{1}{m-2l}\, T_{-m+2l}\right) = \delta(\underline{x})$$

and

$$\underline{\partial}^{2l+1}\left(-\frac{1}{2l}\cdots\frac{1}{4}\,\frac{1}{2}\,\frac{1}{a_m}\,\frac{1}{m-2}\,\frac{1}{m-4}\cdots\frac{1}{m-2l}\, U_{-m+2l+1}\right) = \delta(\underline{x}).$$

Part II: New Clifford Distributions

7. Clifford analysis (II)

In the second part of this paper a fundamental rôle is played by the so-called inner and outer spherical monogenics.
Start with a homogeneous polynomial $P_p(\underline{x})$ of degree p which we take to be vector valued and left (and hence also right) monogenic. Recall from section 2 that a function f is left (right) monogenic in an open region Ω of $\mathbb{R}^m$ if $\underline{\partial} f = 0$ (respectively $f\underline{\partial} = 0$) in Ω.
These polynomials enjoy the following calculus rules:

$$\begin{aligned}
\underline{\partial} P_p(\underline{x}) &= P_p(\underline{x})\underline{\partial} &&= 0 \\
\underline{\partial}(\underline{x}\, P_p(\underline{x})) &= (P_p(\underline{x})\,\underline{x})\underline{\partial} &&= -(m+2p)P_p(\underline{x}) \\
\underline{\partial}(P_p(\underline{x})\,\underline{x}) &= (\underline{x}\, P_p(\underline{x}))\underline{\partial} &&= (m-2)P_p(\underline{x}), \qquad p \neq 0
\end{aligned}$$

$$\begin{aligned}
\underline{\partial}(\underline{x}\, P_p(\underline{x})\,\underline{x}) &= -(m+2p+2)\, P_p(\underline{x})\,\underline{x} - (m-2)\,\underline{x}\, P_p(\underline{x}), \quad p \neq 0 \\
(\underline{x}\, P_p(\underline{x})\,\underline{x})\underline{\partial} &= -(m+2p+2)\,\underline{x}\, P_p(\underline{x}) - (m-2)\, P_p(\underline{x})\,\underline{x}, \quad p \neq 0
\end{aligned}$$

and also

$$\Delta\, P_p(\underline{x}) = \Delta(\underline{x}\, P_p(\underline{x})) = \Delta(P_p(\underline{x})\, \underline{x}) = 0$$

$$\Delta(\underline{x}\, P_p(\underline{x})\, \underline{x}) = -2(m-2)P_p(\underline{x})$$

since the Dirac operator $\underline{\partial}$ factorizes the Laplace operator in $\mathbb{R}^m$: $-\underline{\partial}^2 = \Delta$.

For the proofs of some of these basic formulae we refer the reader to [6]. Note that these formulae are special cases of more general formulae, proved in [1] and [4], involving monogenic homogeneous polynomials which are multi-vector valued. However in order to making the paper self-contained we prove the most general of the formulae listed above.

Lemma 7.1. *If $P_p(\underline{x})$ is a vector valued, monogenic, homogeneous polynomial of degree p, then*

$$\underline{\partial}(\underline{x}\, P_p(\underline{x})\, \underline{x}) = -(m+2p+2)\, P_p(\underline{x})\, \underline{x} - (m-2)\, \underline{x}\, P_p(\underline{x}), \quad p \neq 0.$$

Proof. We have consecutively

$$\begin{aligned}
\underline{\partial}(\underline{x}\, P_p\, \underline{x}) &= -m\, P_p\, \underline{x} + \sum_j e_j\, \underline{x}\, \partial_{x_j}\, (P_p\, \underline{x}) \\
&= -m\, P_p\, \underline{x} - \sum_j \underline{x}\, e_j\, \partial_{x_j}(P_p\, \underline{x}) - 2\sum_j x_j\, \partial_{x_j}(P_p\, \underline{x}) \\
&= (-m-2p-2)\, P_p\, \underline{x} - \underline{x}\, \underline{\partial}(P_p\, \underline{x}) \\
&= -(m+2p+2)\, P_p\, \underline{x} - (m-2)\, \underline{x}\, P_p
\end{aligned}$$

since

$$\underline{\partial}(P_p\, \underline{x}) = \sum_j e_j\, P_p\, e_j = m\, P_p - 2\, P_p$$

□

These vector valued monogenic homogeneous polynomials $P_p(\underline{x})$ may be realized under the action of the Dirac operator upon real-valued harmonic homogeneous polynomials S_{p+1} of degree $(p+1)$:

$$P_p(\underline{x}) = \underline{\partial}\, S_{p+1}(\underline{x}).$$

We then have

$$\underline{x}\, P_p(\underline{x}) = \underline{x} \bullet \underline{\partial}\, S_{p+1}(\underline{x}) + \underline{x} \wedge \underline{\partial}\, S_{p+1}(\underline{x})$$

and

$$P_p(\underline{x})\, \underline{x} = \underline{x} \bullet \underline{\partial}\, S_{p+1}(\underline{x}) - \underline{x} \wedge \underline{\partial}\, S_{p+1}(\underline{x})$$

from which it follows that

$$\underline{x}\, P_p(\underline{x}) + P_p(\underline{x})\, \underline{x} = 2\underline{x} \bullet \underline{\partial}\, S_{p+1}(\underline{x}) = -2\langle\, \underline{x}, \underline{\partial}\,\rangle\, S_{p+1}(\underline{x}) = -2(p+1)S_{p+1}(\underline{x})$$

is scalar valued.

By taking restrictions of the polynomials $P_p(\underline{x})$ to the unit sphere S^{m-1} , we obtain so-called inner spherical monogenics $P_p(\underline{\omega})$, for which the following formulae hold:

$$\begin{aligned}
\underline{\partial}\, P_p(\underline{\omega}) &= -\frac{p}{r}\, \underline{\omega}\, P_p(\underline{\omega}) \\
P_p(\underline{\omega})\underline{\partial} &= -\frac{p}{r}\, P_p(\underline{\omega})\, \underline{\omega} \\
\partial_{\underline{\omega}}\, P_p(\underline{\omega}) &= -p\, \underline{\omega}\, P_p\, (\underline{\omega}) \\
P_p(\underline{\omega})\, \partial_{\underline{\omega}} &= -p\, P_p(\underline{\omega})\, \underline{\omega} \\
\underline{\omega}\, \partial_{\underline{\omega}}\, P_p(\underline{\omega}) &= P_p(\underline{\omega})\partial_{\underline{\omega}}\, \underline{\omega} = p\, P_p(\underline{\omega}) \\
\partial_{\underline{\omega}}(\underline{\omega}\, P_p(\underline{\omega})) &= (P_p(\underline{\omega})\underline{\omega})\partial_{\underline{\omega}} = -(m+p-1)P_p(\underline{\omega}) \\
\partial_{\underline{\omega}}(P_p(\underline{\omega})\, \underline{\omega}) &= (\underline{\omega}\, P_p(\underline{\omega}))\partial_{\underline{\omega}} = (m-2)P_p(\underline{\omega}) - (p+1)\underline{\omega}\, P_p(\underline{\omega})\, \underline{\omega}, \quad p \neq 0 \\
\partial_{\underline{\omega}}(\underline{\omega}\, P_p(\underline{\omega})\, \underline{\omega}) &= -(m+p)P_p(\underline{\omega})\, \underline{\omega} - (m-2)\underline{\omega}\, P_p(\underline{\omega}), \quad p \neq 0 \\
(\underline{\omega}\, P_p(\underline{\omega})\, \underline{\omega})\partial_{\underline{\omega}} &= -(m+p)\, \underline{\omega}\, P_p(\underline{\omega}) - (m-2)P_p(\underline{\omega})\, \underline{\omega}, \quad p \neq 0 \\
\partial^2_{\underline{\omega}}\, P_p(\underline{\omega}) &= P_p(\underline{\omega})\, \partial^2_{\underline{\omega}} = p(m+p-1)\, P_p(\underline{\omega}) \\
\Delta^*\, P_p(\underline{\omega}) &= (-p)(p+m-2)\, P_p(\underline{\omega})
\end{aligned}$$

Note again that these formulae are special cases of more general ones, proved in [1], involving multi-vector spherical monogenics; however as an illustration we prove the following

Lemma 7.2. *If $P_p(\underline{\omega})$ is a vector valued spherical monogenic of degree p then*

$$\partial_{\underline{\omega}}(P_p(\underline{\omega})\, \underline{\omega}) = -(p+1)\, \underline{\omega}\, P_p(\underline{\omega}) + (m-2)\, P_p(\underline{\omega}).$$

Proof. Using the expression of the Dirac operator in spherical co-ordinates, we have consecutively

$$\begin{aligned}
\partial_{\underline{\omega}}(P_p(\underline{\omega})\, \underline{\omega}) &= r\, \underline{\partial}(P_p(\underline{\omega})\, \underline{\omega}) = r\, \underline{\partial}(\frac{1}{r^{p+1}}\, P_p(\underline{x})\, \underline{x}) \\
&= r\, (\underline{\partial}\frac{1}{r^{p+1}})\, P_p(\underline{x})\, \underline{x} + \frac{1}{r^p}\, \underline{\partial}(P_p(\underline{x})\, \underline{x}) \\
&= \underline{\omega}\, (-p-1)\, \frac{1}{r^{p+1}}\, P_p(\underline{x})\, \underline{x} + \frac{1}{r^p}\, (m-2)\, P_p(\underline{x}) \\
&= -(p+1)\, \underline{\omega}\, P_p(\underline{\omega})\, \underline{\omega} + (m-2)\, P_p(\underline{\omega})
\end{aligned}$$

□

Given an inner spherical monogenic $P_p(\underline{\omega})$ then obviously

$$r^p\, P_p(\underline{\omega}) = P_p(\underline{x})$$

is a left and right monogenic homogeneous polynomial the restriction of which to the unit sphere is precisely $P_p(\underline{\omega})$.
At the same time the functions

$$\frac{1}{r^{m+p-1}}\, \underline{\omega}\, P_p(\underline{\omega}) = \frac{1}{r^{m+2p}}\, \underline{x}\, P_p(\underline{x}) = Q_p^{(l)}(\underline{x})$$

and

$$\frac{1}{r^{m+p-1}}\, P_p(\underline{\omega})\, \underline{\omega} = \frac{1}{r^{m+2p}}\, P_p(\underline{x})\, \underline{x} = Q_p^{(r)}(\underline{x})$$

are left, respectively right, monogenic homogeneous functions of order $-(m+p-1)$ in the complement of the origin. Their restrictions to the unit sphere S^{m-1}:

$$Q_p^{(l)} = \underline{\omega}\, P_p(\underline{\omega}) \quad \text{and} \quad Q_p^{(r)} = P_p(\underline{\omega})\, \underline{\omega}$$

are called outer spherical monogenics.
With the above notations we have

$$\underline{\omega}\, P_p(\underline{\omega}) + P_p(\underline{\omega})\, \underline{\omega} = -2(p+1)\, S_{p+1}(\underline{\omega}).$$

Also note that the inner spherical monogenics $P_p(\underline{\omega})$ and the outer spherical monogenics $\underline{\omega}\, P_p(\underline{\omega})$ and $P_p(\underline{\omega})\, \underline{\omega}$ are special cases of spherical harmonics.

8. Generalized spherical means (II)

In Section 3 we studied the generalized spherical means $\Sigma^{(0)}$ and $\Sigma^{(1)}$ used afterwards in the definition of the distributions T_λ and U_λ. In view of the definition of new distributions in Section 9 in which the spherical monogenics will play a rôle, we now introduce the necessary corresponding spherical means.

Let $\phi(\underline{x})$ be a scalar valued test function in $\mathbb{R}^m$, and let $P_p(\underline{x})$ be a vector valued, monogenic, homogeneous polynomial of degree $p \neq 0$ as introduced in the previous Section 7. The spherical mean $\Sigma^{(0)}$ of part I is now generalized, depending on the parity of p, as follows:

$$\Sigma_{2k}^{(0)}[\phi] = \Sigma^{(0)}[P_{2k}(\underline{\omega})\phi(\underline{x})] = \frac{1}{a_m}\int_{S^{m-1}} P_{2k}(\underline{\omega})\phi(\underline{x})\, dS(\underline{\omega}), \qquad k = 1, 2, \ldots$$

and

$$\Sigma_{2k+1}^{(0)}[\phi] = \Sigma^{(0)}[r\, P_{2k+1}(\underline{\omega})\phi(\underline{x})] = \frac{r}{a_m}\int_{S^{m-1}} P_{2k+1}(\underline{\omega})\phi(\underline{x})\, dS(\underline{\omega}), \; k = 0, 1, 2, \ldots$$

Note that if $p = 0$ and $P_0(\underline{x}) = 1$, then $\Sigma_0^{(0)}[\phi] = \Sigma^{(0)}[\phi]$. Note also the extra factor r in the definition of $\Sigma_{2k+1}^{(0)}$; in this way this spherical mean vanishes at the origin $r = 0$ (a property which will exploited in the sequel) and moreover the formulae established for $\Sigma_{2k+1}^{(0)}$ and $\Sigma_{2k}^{(0)}$ become symmetric.

We now examine the behaviour at the origin $r = 0$ of the derivatives of these vector valued spherical means.

Proposition 8.1. *If $\phi(\underline{x})$ is a scalar valued test function, then the spherical means $\Sigma_p^{(0)}[\phi]$ are even test functions on the real r-axis, with derivatives of odd order vanishing at the origin $r = 0$, while for the derivatives of even order one has:*

$$\left\{\partial_r^{2l}\, \Sigma_{2k}^{(0)}[\phi]\right\}_{r=0} = \frac{(2l)!}{(2k+2l)!}\, \frac{1}{C(k+l)}\, \left\{\Delta_m^{k+l}(\phi(\underline{x})P_{2k}(\underline{x}))\right\}_{\underline{x}=0}$$

and

$$\left\{\partial_r^{2l}\,\Sigma_{2k+1}^{(0)}[\phi]\right\}_{r=0} = \frac{(2l)!}{(2k+2l)!}\,\frac{1}{C(k+l)}\left\{\Delta_m^{k+l}(\phi(\underline{x})P_{2k+1}(\underline{x}))\right\}_{\underline{x}=0}.$$

Proof. First consider the case where $p = 2k$ and observe that

$$\Sigma^{(0)}[P_{2k}(\underline{x})\phi(\underline{x})] = r^{2k}\,\Sigma^{(0)}[P_{2k}(\underline{\omega})\phi(\underline{x})] = r^{2k}\,\Sigma_{2k}^{(0)}[\phi]$$

whence

$$\left\{\partial_r^{2k+j}\Sigma^{(0)}[P_{2k}(\underline{x})\phi(\underline{x})]\right\}_{r=0} = \frac{(2k+j)!}{j!}\left\{\partial_r^{j}\Sigma_{2k}^{(0)}[\phi]\right\}_{r=0}$$

If $j = 2l+1$, then

$$\left\{\partial_r^{2l+1}\,\Sigma_{2k}^{(0)}[\phi]\right\}_{r=0} = 0$$

while for $j = 2l$ we get

$$\begin{aligned}\left\{\partial_r^{2l}\Sigma_{2k}^{(0)}[\phi]\right\}_{r=0} &= \frac{(2l)!}{(2k+2l)!}\left\{\partial_r^{2k+2l}\Sigma^{(0)}[P_{2k}(\underline{x})\phi(\underline{x})]\right\}_{r=0}\\ &= (-1)^{k+l}\frac{(2l)!}{(2k+2l)!}\,\frac{1}{C(k+l)}\left\{\underline{\partial}^{2k+2l}(\phi(\underline{x})P_{2k}(\underline{x}))\right\}_{x=0}\end{aligned}$$

In the case where $p = 2k+1$, start with

$$\Sigma^{(0)}[P_{2k+1}(\underline{x})\phi(\underline{x})] = r^{2k}\Sigma^{(0)}[r\;P_{2k+1}(\underline{\omega})\phi(\underline{x})] = r^{2k}\,\Sigma_{2k+1}^{(0)}[\phi]$$

to obtain, in a similar way, the desired result. □

Remark 8.2. The above results for the values at the origin of the even order derivatives of the spherical means $\Sigma_p^{(0)}[\phi]$ may be rewritten in terms of distributions as:

$$\left\{\partial_r^{2l}\,\Sigma_{2k}^{(0)}[\phi]\right\}_{r=0} = \frac{(2l)!}{(2k+2l)!}\,\frac{1}{C(k+l)}\langle\, P_{2k}(\underline{x})\;\Delta_m^{k+l}\;\delta(\underline{x}),\phi(\underline{x})\,\rangle$$

and

$$\left\{\partial_r^{2l}\,\Sigma_{2k+1}^{(0)}[\phi]\right\}_{r=0} = \frac{(2l)!}{(2k+2l)!}\,\frac{1}{C(k+l)}\langle\, P_{2k+1}(\underline{x})\;\Delta_m^{k+l}\;\delta(\underline{x}),\phi(\underline{x})\,\rangle$$

where the obtained vector valued distributions may act from the left as well as from the right.

Remark 8.3. As $P_p(\underline{\omega})$ is a spherical harmonic, we have that

$$\int_{S^{m-1}} P_p(\underline{\omega})\;dS(\underline{\omega}) = 0, \qquad p = 1,2,3,\ldots,$$

while for $p = 0$

$$\int_{S^{m-1}} dS(\underline{\omega}) = a_m.$$

Hence

$$\left\{\Sigma_{2k}^{(0)}[\phi]\right\}_{r=0} = \frac{1}{(2k)!}\,\frac{1}{C(k)}\;\left\{\Delta^k(\phi(\underline{x})P_{2k}(\underline{x}))\right\}_{\underline{x}=0} = 0 \quad , \quad k \neq 0,$$

while

$$\left\{\Sigma_0^{(0)}[\phi]\right\}_{r=0} = \phi(0).$$

The spherical mean $\Sigma^{(1)}[\phi]$ of part I is now generalized as follows:

(i) $\Sigma_{2k}^{(1)}[\phi] = \Sigma^{(0)}[\underline{\omega}\, P_{2k}(\underline{\omega})\phi(\underline{x})] = \dfrac{1}{a_m}\displaystyle\int_{S^{m-1}} \underline{\omega}\, P_{2k}(\underline{\omega})\phi(\underline{x})\, dS(\underline{\omega})$

(ii) $\Sigma_{2k+1}^{(1)}[\phi] = \Sigma^{(0)}[r\, \underline{\omega}\, P_{2k+1}(\underline{\omega})\phi(\underline{x})] = \dfrac{r}{a_m}\displaystyle\int_{S^{m-1}} \underline{\omega}\, P_{2k+1}(\underline{\omega})\phi(\underline{x})\, dS(\underline{\omega})$

(iii) $\Sigma_{2k}^{(3)}[\phi] = \Sigma^{(0)}[P_{2k}(\underline{\omega})\, \underline{\omega}\, \phi(\underline{x})] = \dfrac{1}{a_m}\displaystyle\int_{S^{m-1}} P_{2k}(\underline{\omega})\, \underline{\omega}\, \phi(\underline{x})\, dS(\underline{\omega})$

(iv) $\Sigma_{2k+1}^{(3)}[\phi] = \Sigma^{(0)}[P_{2k+1}(\underline{\omega})\, r\, \underline{\omega}\, \phi(\underline{x})] = \dfrac{r}{a_m}\displaystyle\int_{S^{m-1}} P_{2k+1}(\underline{\omega})\, \underline{\omega}\, \phi(\underline{x})\, dS(\underline{\omega}).$

Note that if $p = 0$ and $P_0(\underline{x}) = 1$ then $\Sigma_0^{(1)}[\phi] = \Sigma_0^{(3)}[\phi] = \Sigma^{(1)}[\phi]$.

The next proposition summarizes the properties of these spherical means; its proof is similar to that of Proposition 8.1.

Proposition 8.4. *If $\phi(\underline{x})$ is a scalar valued test function, then the spherical means $\Sigma_p^{(1)}[\phi]$ and $\Sigma_p^{(3)}[\phi]$ are odd test functions on the real r-axis with derivatives of even order vanishing at $r = 0$, while for the derivatives of odd order one has:*

$$\left\{\partial_r^{2l+1}\Sigma_{2k}^{(1)}[\phi]\right\}_{r=0}$$
$$= (-1)^{k+l+1}\frac{(2l+1)!}{(2k+2l+1)!}\,\frac{1}{C(k+l+1)}\,\langle\, \left(\underline{\partial}^{2k+2l+1}\delta(\underline{x})\right) P_{2k}(\underline{x}), \phi(\underline{x})\,\rangle$$

$$\left\{\partial_r^{2l+1}\Sigma_{2k+1}^{(1)}[\phi]\right\}_{r=0}$$
$$= (-1)^{k+l+1}\frac{(2l+1)!}{(2k+2l+1)!}\,\frac{1}{C(k+l+1)}\,\langle\, \left(\underline{\partial}^{2k+2l+1}\delta(\underline{x})\right) P_{2k+1}(\underline{x}), \phi(\underline{x})\,\rangle$$

$$\left\{\partial_r^{2l+1}\Sigma_{2k}^{(3)}[\phi]\right\}_{r=0}$$
$$= (-1)^{k+l+1}\frac{(2l+1)!}{(2k+2l+1)!}\,\frac{1}{C(k+l+1)}\,\langle\, P_{2k}(\underline{x}) \left(\underline{\partial}^{2k+2l+1}\delta(\underline{x})\right), \phi(\underline{x})\,\rangle$$

$$\left\{\partial_r^{2l+1}\Sigma_{2k+1}^{(3)}[\phi]\right\}_{r=0}$$
$$= (-1)^{k+l+1}\frac{(2l+1)!}{(2k+2l+1)!}\,\frac{1}{C(k+l+1)}\,\langle\, P_{2k+1}(\underline{x}) \left(\underline{\partial}^{2k+2l+1}\delta(\underline{x})\right), \phi(\underline{x})\,\rangle.$$

Remark 8.5. The above obtained distributional expressions need an additional consideration w.r.t. left or right action upon the test function ϕ. We focus on one case :

$$\left\{\underline{\partial}^{2k+2l+1}(P_{2k}(\underline{x})\phi(\underline{x}))\right\}_{\underline{x}=0},$$

the other cases being treated similarly.
If left action on the test function is concerned, we have readily

$$\begin{aligned}\langle\ \delta(\underline{x}),\ \underline{\partial}^{2k+2l+1}(P_{2k}(\underline{x})\phi(\underline{x})\ \rangle &= -\langle\ \underline{\partial}^{2k+2l+1}\delta(\underline{x}),\ P_{2k}(\underline{x})\phi(\underline{x})\ \rangle \\ &= -\langle\ (\underline{\partial}^{2k+2l+1}\delta(\underline{x}))P_{2k}(\underline{x}),\phi(\underline{x})\ \rangle.\end{aligned}$$

In the case of right action on the test function, we have

$$\begin{aligned}&\langle\ \underline{\partial}^{2k+2l+1}(P_{2k}(\underline{x})\phi(\underline{x})),\delta(\underline{x})\ \rangle \\ &= \langle\ (-1)^{k+l}\underline{\partial}\ \Delta_m^{k+l}(P_{2k}(\underline{x})\phi(\underline{x})),\delta(\underline{x})\ \rangle \\ &= (-1)^{k+l}\langle\ \underline{\partial}(P_{2k}(\underline{x})\phi(\underline{x})),\Delta_m^{k+l}\delta(\underline{x})\ \rangle \\ &= (-1)^{k+l}\langle\ \underline{\partial}\ \phi(\underline{x})P_{2k}(\underline{x}),\Delta_m^{k+l}\delta(\underline{x})\ \rangle \\ &= (-1)^{k+l}\langle\ \underline{\partial}\ \phi(\underline{x}),\ P_{2k}(\underline{x})(\Delta_m^{k+l}\ \delta(\underline{x}))\ \rangle \\ &= (-1)^{k+l+1}\langle\ \phi(\underline{x}),\ \underline{\partial}(P_{2k}(\underline{x})(\Delta_m^{k+l}\ \delta(\underline{x})))\ \rangle \\ &= (-1)^{k+l+1}\langle\ \phi(\underline{x}),\ \underline{\partial}(\Delta_m^{k+l}\ \delta(\underline{x}))P_{2k}(\underline{x})\ \rangle \\ &= -\ \langle\ \phi(\underline{x}),(\underline{\partial}^{2k+2l+1}\delta(\underline{x}))P_{2k}(\underline{x})\ \rangle.\end{aligned}$$

This shows the result to be independent of a left or a right action.

Remark 8.6. As $\underline{\omega}\ P_p(\underline{\omega})$ and $P_p(\underline{\omega})\ \underline{\omega}$ are spherical harmonics, we have that

$$\int_{S^{m-1}}\underline{\omega}\ P_p(\underline{\omega})\ dS(\underline{\omega}) = \int_{S^{m-1}}P_p(\underline{\omega})\ \underline{\omega}\ dS(\underline{\omega}) = 0\quad,\qquad p=0,1,2,\ldots.$$

Hence

$$\left\{\partial_r\ \Sigma_{2k+1}^{(1)}[\phi]\right\}_{r=0} = \frac{(-1)^{k+1}}{(2k+1)!\ C(k+1)}\langle\ \underline{\partial}^{2k+1}\delta(\underline{x})\ P_{2k+1}(\underline{x}),\phi(\underline{x})\ \rangle = 0$$

and

$$\left\{\partial_r\ \Sigma_{2k+1}^{(3)}[\phi]\right\}_{r=0} = \frac{(-1)^{k+1}}{(2k+1)!\ C(k+1)}\langle\ P_{2k+1}(\underline{x})\ \underline{\partial}^{2k+1}\delta(\underline{x})\ ,\phi(\underline{x})\ \rangle = 0.$$

The spherical means $\Sigma_p^{(0)}, \Sigma_p^{(1)}$ and $\Sigma_p^{(3)}$ are interrelated by the action of the Dirac operator as shown in the next proposition.

Proposition 8.7. *For a scalar valued test function ϕ one has:*

(i) $\Sigma_{2k}^{(0)}[\phi\ \underline{\partial}] = \left(\partial_r + \dfrac{m+2k-1}{r}\right)\Sigma_{2k}^{(1)}[\phi]$

(ii) $\Sigma_{2k+1}^{(0)}[\phi\ \underline{\partial}] = \left(\partial_r + \dfrac{m+2k-1}{r}\right)\Sigma_{2k+1}^{(1)}[\phi]$

(iii) $\Sigma_{2k}^{(0)}[\underline{\partial}\ \phi] = \left(\partial_r + \dfrac{m+2k-1}{r}\right)\Sigma_{2k}^{(3)}[\phi]$

(iv) $\Sigma^{(0)}_{2k+1}[\underline{\partial}\ \phi] = \left(\partial_r + \frac{m+2k-1}{r}\right)\Sigma^{(3)}_{2k+1}[\phi]$

where $\Sigma^{(0)}_p[\phi\ \underline{\partial}]$ *stands for* $\Sigma^{(0)}[\underline{\partial}\ \phi\ P_p(\underline{\omega})]$ *and* $\Sigma^{(0)}_p[\underline{\partial}\ \phi]$ *stands for* $\Sigma^{(0)}[P_p(\underline{\omega})\underline{\partial}\ \phi]$.

Proof. Let f be an $\mathbb{R}_{0,m}$-valued function, then it may be proved in a similar way as in Proposition 3.1, that

$$\Sigma^{(0)}[\underline{\partial}\ f] = \left(\partial_r + \frac{m-1}{r}\right)\Sigma^{(1)}[f].$$

For $f = P_p(\underline{x})\phi(\underline{x})$, ϕ being a scalar valued test function, we get

$$\Sigma^{(0)}[\underline{\partial}\ \phi\ P_p(\underline{x})] = \left(\partial_r + \frac{m-1}{r}\right)\Sigma^{(1)}[P_p(\underline{x})\phi(\underline{x})].$$

In the case where $p = 2k$, this yields

$$r^{2k}\Sigma^{(0)}[\underline{\partial}\ \phi\ P_{2k}(\underline{\omega})] = (2k+m-1)r^{2k-1}\Sigma^{(1)}[P_{2k}(\underline{\omega})\phi(\underline{x})] + r^{2k}\ \partial_r\ \Sigma^{(1)}[P_{2k}(\underline{\omega})\phi(\underline{x})]$$

or

$$\Sigma^{(0)}_{2k}[\phi\ \underline{\partial}] = \left(\partial_r + \frac{m+2k-1}{r}\right)\Sigma^{(1)}_{2k}[\phi].$$

In the case where $p = 2k+1$, we get

$$\begin{aligned} &r^{2k+1}\Sigma^{(0)}[\underline{\partial}\ \phi\ P_{2k+1}(\underline{\omega})] \\ &\quad = (2k+m-1)r^{2k-1}\Sigma^{(1)}[r\ P_{2k+1}(\underline{\omega})\phi(\underline{x})] + r^{2k}\ \partial_r\ \Sigma^{(1)}[r\ P_{2k+1}(\underline{\omega})\phi(\underline{x})] \end{aligned}$$

or

$$\Sigma^{(0)}_{2k+1}[\phi\ \underline{\partial}] = \left(\partial_r + \frac{m+2k-1}{r}\right)\Sigma^{(1)}_{2k+1}[\phi].$$

Next, we also have

$$\Sigma^{(0)}[f\ \underline{\partial}] = \left(\partial_r + \frac{m-1}{r}\right)\Sigma^{(0)}[f\ \underline{\omega}]$$

which, for $f = P_p(\underline{x})\phi(\underline{x})$, turns into

$$\Sigma^{(0)}[P_p(\underline{x})\ \underline{\partial}\ \phi] = \left(\partial_r + \frac{m-1}{r}\right)\Sigma^{(0)}[P_p(\underline{x})\ \underline{\omega}\ \phi(\underline{x})].$$

In the case where $p = 2k$, this yields

$$\begin{aligned} &r^{2k}\ \Sigma^{(0)}[P_{2k}(\underline{\omega})\ \underline{\partial}\phi] \\ &\quad = (2k+m-1)r^{2k-1}\ \Sigma^{(0)}[P_{2k}(\underline{\omega})\ \underline{\omega}\ \phi(\underline{x})] + r^{2k}\ \partial_r\ \Sigma^{(0)}[P_{2k}(\underline{\omega})\ \underline{\omega}\ \phi(\underline{x})] \end{aligned}$$

or

$$\Sigma^{(0)}_{2k}[\underline{\partial}\ \phi] = \left(\partial_r + \frac{m+2k-1}{r}\right)\Sigma^{(3)}_{2k}[\phi].$$

In the case where $p = 2k+1$, we get

$$\begin{aligned} r^{2k+1}\Sigma^{(0)}[P_{2k+1}(\underline{\omega})\ \underline{\partial}\ \phi] &= (2k+m-1)r^{2k-1}\ \Sigma^{(0)}[r\ P_{2k+1}(\underline{\omega})\ \underline{\omega}\ \phi(\underline{x})] \\ &\quad + r^{2k}\ \partial_r\ \Sigma^{(0)}[r\ P_{2k+1}(\underline{\omega})\ \underline{\omega}\ \phi(\underline{x})] \end{aligned}$$

or

$$\Sigma_{2k+1}^{(0)}[\underline{\partial}\,\phi] = \left(\partial_r + \frac{m+2k-1}{r}\right)\Sigma_{2k+1}^{(3)}[\phi]. \qquad \square$$

Proposition 8.8. *For a scalar valued test function ϕ one has:*

$$\text{(i)}\ r\,\Sigma_{2k}^{(1)}[\phi\,\underline{\partial}] = \frac{r}{a_m}\int_{S^{m-1}} \underline{\partial}\,\phi\,\underline{\omega}\,P_{2k}(\underline{\omega})\,dS(\underline{\omega})$$
$$= (-r\,\partial_r + 2k)\Sigma_{2k}^{(0)}[\phi]$$

$$\text{(ii)}\ r\,\Sigma_{2k+1}^{(1)}[\phi\,\underline{\partial}] = \frac{r^2}{a_m}\int_{S^{m-1}} \underline{\partial}\,\phi\,\underline{\omega}\,P_{2k+1}(\underline{\omega})\,dS(\underline{\omega})$$
$$= (-r\,\partial_r + 2k+2)\Sigma_{2k+1}^{(0)}[\phi]$$

$$\text{(iii)}\ r\,\Sigma_{2k}^{(3)}[\underline{\partial}\,\phi] = \frac{r}{a_m}\int_{S^{m-1}} P_{2k}(\underline{\omega})\,\underline{\omega}\,\underline{\partial}\,\phi\ dS(\underline{\omega})$$
$$= (-r\,\partial_r + 2k)\Sigma_{2k}^{(0)}[\phi]$$

$$\text{(iv)}\ r\,\Sigma_{2k+1}^{(3)}[\underline{\partial}\,\phi] = \frac{r^2}{a_m}\int_{S^{m-1}} P_{2k+1}(\underline{\omega})\,\underline{\omega}\,\underline{\partial}\,\phi\ dS(\underline{\omega})$$
$$= (-r\,\partial_r + 2k+2)\Sigma_{2k+1}^{(0)}[\phi]$$

The proof is similar to that of Proposition 8.7.

9. New Clifford distributions

The technique used in this section for introducing new Clifford distributions is essentially the same as in the first part of this paper; we use spherical co-ordinates, the fundamental distribution $F_p r_+^{\mu}$ and the new spherical means of the foregoing section 8.

Let λ be a complex parameter and let ϕ be a scalar valued test funtion. We define the vector valued distributions $T_{\lambda,p}$ by:

$$\langle\ T_{\lambda,2k}, \phi\ \rangle = a_m\ \langle\ Fp\ r_+^{\mu+2k}, \Sigma_{2k}^{(0)}[\phi]\ \rangle$$

and

$$\langle\ T_{\lambda,2k+1}, \phi\ \rangle = a_m\ \langle\ Fp\ r_+^{\mu+2k}, \Sigma_{2k+1}^{(0)}[\phi]\ \rangle$$

where we have put

$$\mu = \lambda + m - 1$$

In the same order of ideas as in part I, we also introduce the distributions $U_{\lambda,p}$ and $V_{\lambda,p}$ by:

$$\langle\ U_{\lambda,2k}, \phi\ \rangle = a_m \langle\ Fp\ r_+^{\mu+2k}, \Sigma_{2k}^{(1)}[\phi]\ \rangle$$
$$\langle\ U_{\lambda,2k+1}, \phi\ \rangle = a_m \langle\ Fp\ r_+^{\mu+2k}, \Sigma_{2k+1}^{(1)}[\phi]\ \rangle$$
$$\langle\ V_{\lambda,2k}, \phi\ \rangle = a_m \langle\ Fp\ r_+^{\mu+2k}, \Sigma_{2k}^{(3)}[\phi]\ \rangle$$
$$\langle\ V_{\lambda,2k+1}, \phi\ \rangle = a_m \langle\ Fp\ r_+^{\mu+2k}, \Sigma_{2k+1}^{(3)}[\phi]\ \rangle$$

Example 9.1. For $\lambda = -m-p$, we have for $p = 2k$:

$$\begin{aligned}\langle T_{-m-2k,2k}, \phi \rangle &= a_m \langle Fp\, r_+^{-1}, \Sigma_{2k}^{(0)}[\phi] \rangle \\ &= \int_0^{+\infty} \frac{1}{r}\, dr \int_{S^{m-1}} P_{2k}(\underline{\omega})\, \phi(\underline{x})\, dS(\underline{\omega}) \\ &= \int_{\mathbb{R}^m} \frac{1}{r^m} P_{2k}(\underline{\omega})[\phi(\underline{x}) - \phi(\underline{0})]\, dV(\underline{x}) \\ &= \lim_{\substack{\varepsilon \to 0 \\ >}} \int_{\mathbb{R}^m \setminus B(0,\varepsilon)} \frac{1}{r^m} P_{2k}(\underline{\omega})\, \phi(\underline{x})\, dV(\underline{x})\end{aligned}$$

where we have used the fact (see Remark 8.3) that

$$\int_{S^{m-1}} P_{2k}(\underline{\omega})\, dS(\underline{\omega}) = 0$$

Similarly we obtain for $p = 2k+1$ that

$$\langle T_{-m-2k-1,2k+1}, \phi \rangle = \lim_{\substack{\varepsilon \to \\ >}} \int_{\mathbb{R}^m \setminus B(0,\varepsilon)} \frac{1}{r^m} P_{2k+1}(\underline{\omega})\, \phi(\underline{x})\, dV(\underline{x})$$

The distribution $T_{-m-p,p}$ thus turns out to be a so-called *principal value distribution* (see [9]):

$$T_{-m-p,p} = Pv\, \frac{P_p(\underline{\omega})}{r^m}$$

which yields another example of a convolution operator (see also 6.2).

In the same order of ideas we find

$$U_{-m-p,p} = Pv\, \frac{\underline{\omega}\, P_p(\underline{\omega})}{r^m}$$

and

$$V_{-m-p,p} = Pv\, \frac{P_p(\underline{\omega})\, \underline{\omega}}{r^m}$$

to be examples of such kind of distributions. Moreover we have

$$Pv\, \frac{\underline{\omega}\, P_p(\underline{\omega})}{r^m} + Pv\, \frac{P_p(\underline{\omega})\, \underline{\omega}}{r^m} = -2(p+1)\, Pv\, \frac{S_{p+1}(\underline{\omega})}{r^m}$$

showing that the $U_{-m-p,p}$ and $V_{-m-p,p}$ distributions may be used for decomposing the scalar valued principal value distribution $Pv\, \frac{1}{r^m}\, S_{p+1}(\underline{\omega})$ of [9] and [12].

Other examples of the $T_{\lambda,p}$, $U_{\lambda,p}$ and $V_{\lambda,p}$ distributions are given in the sections 10 and 11.

10. The action of the Dirac operator on the distributions $T_{\lambda,p}$

We expect the distributions $T_{\lambda,p}$, $U_{\lambda,p}$ and $V_{\lambda,p}$ to be interrelated by the action of the Dirac operator. As $T_{\lambda,p}$ is vector valued, it makes sense to distinguish between an action of $\underline{\partial}$ from the left or from the right.

As it is clear that the parity of the degree p of the spherical monogenic P_p influences the calculations and the results, we introduce the notation p_e ("even part of p") by

$$\begin{cases} p_e = 2k & \text{if} \quad p = 2k \\ p_e = 2k & \text{if} \quad p = 2k+1 \end{cases}$$

In a series of propositions we summarize the results of the calculations of $\underline{\partial}\, T_{\lambda,p}$ and $T_{\lambda,p}\, \underline{\partial}$ in the several cases to be distinguished; their proofs are similar to that of Proposition 5.1. Note that, as the odd order derivatives of $\Sigma_p^{(0)}[\phi]$ vanish at $r = 0$, we have to expect a singular term involving odd order derivatives of the delta distribution, only if $\lambda + m + p_e$ is even.

Proposition 10.1. *For $\lambda \in \mathbb{C}$ and $p \in \mathbb{N}$ such that $\lambda + m - 1 + p_e \neq 0, -1, -2, -3, \ldots$ one has*

$$\underline{\partial}\, T_{\lambda,p} = \lambda\, U_{\lambda-1,p}$$

and

$$T_{\lambda,p}\, \underline{\partial} = \lambda\, V_{\lambda-1,p}.$$

Example 10.2. Take $\lambda + 2k = 0$; then $\mu + p_e = m - 1$ and

$$\begin{aligned} \langle\, T_{-2k,2k}, \phi \,\rangle &= a_m \langle\, Fp\, r_+^{m-1},\ \Sigma_{2k}^{(0)}[\phi] \,\rangle \\ &= \int_0^{+\infty} r^{m-1}\, dr \int_{S^{m-1}} P_{2k}(\underline{\omega}) \phi(\underline{x})\, dS(\underline{\omega}) \\ &= \int_{\mathbb{R}^m} P_{2k}(\underline{\omega}) \phi(\underline{x})\, dV(\underline{x}) \end{aligned}$$

whence

$$T_{-2k,2k}(\underline{x}) = P_{2k}(\underline{\omega}).$$

If $\lambda + 2k + 1 = 0$ then $\mu + p_e = m - 2$ and

$$\begin{aligned} \langle\, T_{-2k-1,2k+1}, \phi \,\rangle &= a_m \langle\, Fp\, r_+^{m-2},\ \Sigma_{2k+1}^{(0)}[\phi] \,\rangle \\ &= \int_0^{+\infty} r^{m-2}\, dr \int_{S^{m-1}} P_{2k+1}(\underline{\omega}) \phi(\underline{x})\, dS(\underline{\omega}) \\ &= \int_{\mathbb{R}^m} P_{2k+1}(\underline{\omega}) \phi(\underline{x})\, dV(\underline{x}) \end{aligned}$$

whence

$$T_{-2k-1,2k+1}(\underline{x}) = P_{2k+1}(\underline{\omega}).$$

In a similar way it is shown that

$$U_{-p-1,p}(\underline{x}) = \frac{1}{r}\, \underline{\omega}\, P_p(\underline{\omega})$$

and

$$V_{-p-1,p}(\underline{x}) = \frac{1}{r}\, P_p(\underline{\omega})\underline{\omega}.$$

By Proposition 10.1 we get:

$$\underline{\partial}\, T_{-p,p}\,(\underline{x}) = (-p)\, U_{-p-1,p}(\underline{x})$$

and

$$T_{-p,p}\,\underline{\partial} = (-p)\, V_{-p-1,p}(\underline{x})$$

or thus

$$\underline{\partial}\, P_p(\underline{\omega}) = (-p)\, \frac{1}{r}\, \underline{\omega}\, P_p(\underline{\omega})$$

and

$$P_p(\underline{\omega})\, \underline{\partial} = (-p)\, \frac{1}{r}\, P_p(\underline{\omega})\, \underline{\omega}$$

confirming known formulae (see Section 7).

Proposition 10.3. *For $\lambda \in \mathbb{C}$ and $p \in \mathbb{N}$ such that $\lambda + m - 1 + p_e = 0$ one has*

$$\underline{\partial}\, T_{-m+1-p_e,p} = -(m + p_e - 1) U_{-m-p_e,p}$$

and

$$T_{-m+1-p_e,p}\, \underline{\partial} = -(m + p_e - 1) V_{-m-p_e,p}.$$

Remark 10.4. The distribution considered in the above proposition is, by definition:

$$\langle\, T_{-m-2k+1,2k}(\underline{x}), \phi(\underline{x})\, \rangle = a_m \langle\, Fp\, r_+^0, \Sigma_{2k}^{(0)}[\phi]\, \rangle$$

$$= \int_0^{+\infty} dr \int_{S^{m-1}} P_{2k}(\underline{\omega})\phi(r\,\underline{\omega})\, dS(\underline{\omega}) = \int_{\mathbb{R}^m} \frac{1}{r^{m-1}}\, P_{2k}(\underline{\omega})\, \phi(\underline{x})\, dV(\underline{x})$$

whence

$$T_{-m-2k+1,2k}(\underline{x}) = \frac{1}{r^{m-1}}\, P_{2k}(\underline{\omega})$$

and similarly

$$T_{-m-2k+1,2k+1}(\underline{x}) = \frac{1}{r^{m-2}}\, P_{2k+1}(\underline{\omega})$$

which are clearly locally integrable functions in $\mathbb{R}^m$.
We also have

$$\langle\, \phi, U_{-m-2k,2k}(\underline{x})\, \rangle = a_m \langle\, Fp\, r_+^{-1}, \Sigma_{2k}^{(1)}[\phi]\, \rangle$$

and hence

$$U_{-m-2k,2k}(\underline{x}) = Fp\, \frac{1}{r^m}\, \underline{\omega}\, P_{2k}(\underline{\omega}) = Pv\, \frac{\underline{\omega}\, P_{2k}(\underline{\omega})}{r^m}$$

and similarly

$$U_{-m-2k,2k+1}(\underline{x}) = \frac{1}{r^{m-1}}\, \underline{\omega}\, P_{2k+1}(\underline{\omega}) = a_m\, E_m(\underline{x})\, P_{2k+1}(\underline{\omega}).$$

The formulae of Proposition 10.3 may thus be rewritten as:

$$\underline{\partial}\left(\frac{1}{r^{m-1}}\, P_{2k}(\underline{\omega})\right) = (m + 2k - 1)\, Pv\, \frac{\overline{\underline{\omega}}}{r^m}\, P_{2k}(\underline{\omega})$$

and

$$\underline{\partial}\left(\frac{1}{r^{m-2}}\, P_{2k+1}(\underline{\omega})\right) = -(m+2k-1)\; a_m\; E_m(\underline{x})\; P_{2k+1}(\underline{\omega}).$$

Proposition 10.5. *For $\lambda \in \mathbb{C}$ and $p \in \mathbb{N}$ such that $\lambda+m-1+p_e = -s$, $s = 1,2,3,\ldots$ one has*

(i) $\underline{\partial}\, T_{-m+1-2k-2l,2k} = -(m+2k+2l-1)\; U_{-m-2k-2l,2k}$, $l = 1,2,3,\ldots$

(ii) $$\underline{\partial}\, T_{-m-2k-2l,2k} = -(m+2k+2l)\; U_{-m-2k-2l-1,2k} + a_m(-1)^{k+l+1}\frac{(\underline{\partial}^{2k+2l+1}\delta(\underline{x}))P_{2k}(\underline{x})}{(2k+2l+1)!C(k+l+1)},$$
$l = 0,1,2,3,\ldots$

(iii) $\underline{\partial}\, T_{-m+1-2k-2l,2k+1} = -(m+2k+2l-1)\; U_{-m-2k-2l,2k+1}$, $l = 1,2,3,\ldots$

(iv) $$\underline{\partial}\, T_{-m-2k-2l,2k+1} = -(m+2k+2l)\; U_{-m-2k-2l-1,2k+1} + a_m(-1)^{k+l+1}\frac{(\underline{\partial}^{2k+2l+1}\delta(\underline{x}))P_{2k+1}(\underline{x})}{(2k+2l+1)!C(k+l+1)},$$
$l = 0,1,2,3,\ldots$

and similar formulae for $T_{\lambda,p}\,\underline{\partial}$.

11. The action of the Dirac operator on the distributions $U_{\lambda,p}$ and $V_{\lambda,p}$

In this section we study the action of the Dirac operator from the left on $U_{\lambda,p}$ considered as a left distribution, and the action from the right on $V_{\lambda,p}$ considered as a right distribution. For the action of the Dirac operator from the right on $U_{\lambda,p}$ and from the left on $V_{\lambda,p}$, we refer the reader to [3] which is a continuation of the present paper.

It turns out that the mentioned action of the Dirac operator reproduces – in the general case – the distributions $T_{\lambda-1,p}$ .The proofs of the following propositions are similar to that of Proposition 5.2. Note that, as the even order derivatives of $\Sigma_p^{(1)}[\phi]$ and $\Sigma_p^{(3)}[\phi]$ vanish at $r = 0$, we have to expect a singular term involving even order derivatives of the delta distribution, only if $\lambda + m + p_e$ is odd.

Proposition 11.1. *For $\lambda \in \mathbb{C}$ and $p \in \mathbb{N}$ such that $\lambda+m-1+p_e \neq 0,-1,-2,-3,\ldots$ one has:*

$$\underline{\partial}\, U_{\lambda,2k} = V_{\lambda,2k}\,\underline{\partial} = -(\lambda+m-1+4k)\; T_{\lambda-1,2k}$$

and

$$\underline{\partial}\, U_{\lambda,2k+1} = V_{\lambda,2k+1}\,\underline{\partial} = -(\lambda+m+1+4k)\; T_{\lambda-1,2k+1}.$$

Example 11.2. We verify the formulae of Proposition 11.1 in the specific case where, for $\lambda = 1$, we have

$$U_{1,p} = r^{p+1}\; \underline{\omega}\; P_p(\underline{\omega})$$

The left hand side of the formula then takes the form

$$\underline{\partial}\, U_{1,p} = -(m+2p)\, r^p\, P_p(\underline{\omega})$$

while on the right hand side we have for $p = 2k$

$$-(\lambda + m - 1 + 4k)\, T_{\lambda-1,2k} = -(m+4k)\, T_{0,2k} = -(m+4k)\, r^{2k}\, P_{2k}(\underline{\omega})$$

and for $p = 2k+1$:

$$-(\lambda+m+1+4k)\, T_{\lambda-1,2k+1} = -(m+4k+2)\, T_{0,2k+1} = -(m+4k+2)\, r^{2k+1}\, P_{2k+1}(\underline{\omega}).$$

Example 11.3. Take $\lambda = -2k$ and $p = 2k$; then $\mu + p_e = m-1$ and

$$\begin{aligned}\langle\, U_{-2k,2k}, \phi\, \rangle &= a_m \langle\, Fp\, r_+^{m-1}, \Sigma_{2k}^{(1)}[\phi]\, \rangle \\ &= \int_0^{+\infty} r^{m-1}\, dr \int_{S^{m-1}} \underline{\omega}\, P_{2k}(\underline{\omega})\, \phi(\underline{x})\, dS(\underline{\omega})\end{aligned}$$

whence

$$U_{-2k,2k}(\underline{x}) = \underline{\omega}\, P_{2k}(\underline{\omega})$$

If $\lambda = -2k-1$ and $p = 2k+1$, then $\mu + p_e = m-2$ and

$$\begin{aligned}\langle\, U_{-2k-1,2k+1}, \phi\, \rangle &= a_m \langle\, Fp\, r_+^{m-2}, \Sigma_{2k+1}^{(1)}[\phi]\, \rangle \\ &= \int_0^{+\infty} r^{m-2}\, dr \int_{S^{m-1}} r\, \underline{\omega}\, P_{2k+1}(\underline{\omega})\, \phi(\underline{x})\, dS(\underline{\omega})\end{aligned}$$

whence

$$U_{-2k-1,2k+1} = \underline{\omega}\, P_{2k+1}(\underline{\omega}).$$

We thus obtain in this specific case the left and right outer spherical monogenics (see Section 7):

$$U_{-p,p} = \underline{\omega}\, P_p(\underline{\omega}) = Q_p^{(l)}(\underline{\omega})$$

and

$$V_{-p,p} = P_p(\underline{\omega})\, \underline{\omega} = Q_p^{(r)}\, (\underline{\omega}).$$

The formulae of Proposition 11.1 yield:

$$\underline{\partial}\, U_{-2k,2k} = V_{-k,2k}\, \underline{\partial} = -(m+2k-1)\, T_{-2k-1,2k}$$

and

$$\underline{\partial}\, U_{-2k-1,2k+1} = V_{-2k-1,2k+1}\, \underline{\partial} = -(m+2k)\, T_{-2k-2,2k+1},$$

or, in terms of the outer spherical monogenics,

$$\underline{\partial}(\underline{\omega}\, P_{2k}(\underline{\omega})) = (P_{2k}(\underline{\omega})\, \underline{\omega})\underline{\partial} = -(m+2k-1)\, \frac{1}{r}\, P_{2k}(\underline{\omega})$$

and

$$\underline{\partial}(\underline{\omega}\, P_{2k+1}(\underline{\omega})) = (P_{2k+1}(\underline{\omega})\, \underline{\omega})\underline{\partial} = -(m+2k)\, \frac{1}{r}\, P_{2k+1}(\underline{\omega})$$

in accordance with the formulae of Section 7.

Proposition 11.4. *For $\lambda \in \mathbb{C}$ and $p \in \mathbb{N}$ such that $\lambda + m - 1 + p_e = 0$ one has*

$$\begin{aligned} \underline{\partial}\, U_{-m-2k+1,2k} &= V_{-m-2k+1,2k}\, \underline{\partial} \\ &= -2k\, T_{-m-2k,2k} - a_m \frac{1}{(2k)!} \frac{1}{C(k)} P_{2k}(\underline{x})\, \Delta_m^k\, \delta(\underline{x}) \end{aligned}$$

and

$$\begin{aligned} &\underline{\partial}\, U_{-m-2k+1,2k+1} = V_{-m-2k+1,2k+1}\, \underline{\partial} \\ &= -(2k+2)\, T_{-m-2k,2k+1} - a_m \frac{1}{(2k)!} \frac{1}{C(k)} P_{2k+1}(\underline{x})\, \Delta_m^k\, \delta(\underline{x}) \end{aligned}$$

Proposition 11.5. *For $\lambda \in \mathbb{C}$ and $p \in \mathbb{N}$ such that $\lambda + m - 1 + p_e = -s$, $s = 1, 2, 3, \ldots$ one has:*

(i) $$\begin{aligned} \underline{\partial}\, U_{-m-2k-2l+1,2k} &= (2l-2k)\, T_{-m-2k-2l,2k} \\ &\quad - a_m \frac{1}{(2k+2l)!} \frac{1}{C(k+l)} P_{2k}(\underline{x})\, \Delta_m^{k+l}\, \delta(\underline{x}), \end{aligned}$$
$l = 1, 2, 3, \ldots$

(ii) $\underline{\partial}\, U_{-m-2k-2l,2k} = (2l-2k+1)\, T_{-m-2k-2l-1,2k}, \quad l = 0, 1, 2, \ldots$

(iii) $$\begin{aligned} \underline{\partial}\, U_{-m-2k-2l+1,2k+1} &= (2l-2k-2)\, T_{-m-2k-2l,2k+1} \\ &\quad - a_m \frac{1}{(2k+2l)!} \frac{1}{C(k+l)} P_{2k+1}(\underline{x})\, \Delta_m^{k+l}\, \delta(\underline{x}), \end{aligned}$$
$l = 1, 2, 3, \ldots$

(iv) $\underline{\partial}\, U_{-m-2k-2l,2k+1} = (2l-2k-1)\, T_{-m-2k-2l-1,2k+1}, \quad l = 0, 1, 2, \ldots$

References

[1] F. Brackx, B. De Knock and H. De Schepper, *Multi-vector Spherical Monogenics, Spherical Means and Distributions in Clifford Analuysis* (to appear)

[2] F. Brackx, R. Delanghe and F. Sommen, *Clifford Analysis*, Pitman Publ., 1982

[3] F. Brackx, R. Delanghe and F. Sommen, *Spherical Means, Distributions and Convolution Operators in Clifford Analysis*, Chinese Annals of Mathematics, **24B**:2 (2003), 133–146

[4] F. Brackx, R. Delanghe and F. Sommen, *Differential Forms and / or Multi-vector Functions*, to be published in the Bulletin of the Belgian Mathematical Society – Simon Stevin

[5] R. Delanghe, *Some Remarks on the Principal Value Kernel in* $\mathbb{R}^m$, Complex Variables: Theory and Application, **47** (2002), 653–662

[6] R. Delanghe, F. Sommen and V. Souček, *Clifford Algebra and Spinor-Valued Functions*, Kluwer Academic Publ., 1992

[7] I.M. Gelfand and G.E. Shilov, *Generalized Functions*, vol 1, Academic Press, 1964

[8] J. Gilbert and M. Murray, *Clifford Algebra and Dirac Operators in Harmonic Analysis*, Cambridge Univ. Press, 1991

[9] J. Horváth: *Singular Integral Operators and Spherical Harmonics*, Trans. Amer. Math. Soc. **82** (1950) 52–63

[10] F. John, *Plane Waves and Spherical Means*, Springer, 1955

[11] L. Schwartz: *Théorie des Distributions*, Hermann, 1966

[12] E. Stein and G. Weiß, *Introduction to Fourier Analysis on Euclidean spaces*, Princeton Univ. Press, 1971

[13] F. Sommen, *Spin Groups and Spherical Means.* In: *Clifford Algebras and Their Applications in Mathematical Physics*, J.S.R. Chisholm and A.K. Common (eds.), Nato ASI Series, D. Reidel Publ. Co., 1985

Fred Brackx, Richard Delanghe, and Frank Sommen
Ghent University
Department of Mathematical Analysis
B-9000 Gent
Belgium
e-mail: `fb@cage.rug.ac.be`
`rd@cage.rug.ac.be`
`fs@cage.rug.ac.be`

Trends in Mathematics: *Advances in Analysis and Geometry*, 97–112

Hypermonogenic Functions and their Cauchy-Type Theorems

Sirkka-Liisa Eriksson and Heinz Leutwiler

Abstract. Let $C\ell_n$ be the (universal) Clifford algebra generated by $e_1, \ldots, e_n$ satisfying $e_ie_j + e_je_i = -2\delta_{ij}$, $i, j = 1, \ldots, n$. The Dirac operator in $C\ell_n$ is defined by $D = \sum_{i=0}^{n} e_i \frac{\partial}{\partial x_i}$, where $e_0 = 1$. The modified Dirac operator is introduced for $k \in \mathbb{R}$ by $M_k f = Df + k\frac{Q'f}{x_n}$, where $'$ is the main involution and Qf is given by the decomposition $f(x) = Pf(x) + Qf(x)\, e_n$ with $Pf(x), Qf(x) \in C\ell_{n-1}$. A $k+1$-times continuously differentiable function $f : \Omega \to C\ell_n$ is called k-hypermonogenic in an open subset Ω of $\mathbb{R}^{n+1}$, if $M_k f(x) = 0$ outside the hyperplane $x_n = 0$. Note that 0-hypermonogenic functions are monogenic and $n-1$-hypermonogenic functions are hypermonogenic as defined by the authors in [15]. The power function x^m is hypermonogenic. The set of k-hypermonogenic functions is a right $C\ell_{n-1}$-module. We state a Cauchy type theorem for k-hypermonogenic functions. We also prove an integral formula for the P-part of an hypermogenic function.

Mathematics Subject Classification (2000). Primary 30G35; Secondary 30A05, 30F45.

Keywords. Monogenic, hypermonogenic, Dirac operator, hyperbolic metric.

1. Introduction

We consider an extension of classical complex analysis to higher dimensions. The higher dimensional algebra extending the field of complex numbers is the universal Clifford algebra $C\ell_n$ generated by the elements $e_1, \ldots, e_n$ satisfying the relation $e_ie_j + e_je_i = -2\delta_{ij}$, where δ_{ij} is the usual Kronecker delta. The elements $x = x_0 + x_1e_1 + \ldots + x_ne_n$ for $x_0, \ldots, x_n \in \mathbb{R}$ are called *paravectors*. The coordinate x_0 of a paravector x is called the real part of x. The set $\mathbb{R}^{n+1}$ is identified with the set of paravectors.

This work was supported by the Academy of Finland.

In the complex plane, at least locally, any holomorphic function f is a conjugate gradient of a harmonic function. This means that a holomorphic function f can be written locally in the form $f = \frac{\partial H}{\partial x} - i\frac{\partial H}{\partial y}$ for some harmonic function H. The starting point of modified Clifford analysis, noticed by the second author in [22] and [23], is that, for any $m \in \mathbb{N}$, the power function

$$f(x) = x^m$$

is paravector-valued for any paravector x (see eg. [7]) and satisfies the property

$$f(x) = \frac{\partial h}{\partial x_0} - \sum_{i=1}^{n} e_i \frac{\partial h}{\partial x_i}, \tag{1.1}$$

where

$$h(x) = \frac{1}{m+1} \operatorname{Re} x^{m+1}$$

(see e.g. [14, p. 231]). The function h is not harmonic in the usual sense. However, it is harmonic with respect to the hyperbolic metric

$$ds^2 = \frac{dx_0^2 + \ldots + dx_n^2}{x_n^2}. \tag{1.2}$$

Indeed, as noticed in [22], the function h satisfies the Laplace-Beltrami equation

$$x_n \triangle h - (n-1)\frac{\partial h}{\partial x_n} = 0, \tag{1.3}$$

associated to the Riemannian metric (1.2). The second author in [22] and [23] initiated the research on those functions f, called H-solutions, admitting locally the representation (1.1) for some function h satisfying the equation (1.3). Another characterization of H-solutions is that they are paravector-valued solutions $f = u_0 + u_1 e_1 + \ldots + u_n e_n$ of the following generalized Cauchy-Riemann system

$$\begin{aligned} &x_n\left(\tfrac{\partial u_0}{\partial x_0} - \textstyle\sum_{i=1}^{n} \tfrac{\partial u_i}{\partial xi}\right) + (n-1)\,u_n = 0, \\ &\tfrac{\partial u_i}{\partial x_k} = \tfrac{\partial u_k}{\partial x_i}, \quad i,k = 1,\ldots,n, \\ &\tfrac{\partial u_0}{\partial x_k} = -\tfrac{\partial u_k}{\partial x_0}, \quad k = 1,\ldots,n. \end{aligned} \tag{H}$$

(see [23]). Using differential forms, the system (H) holds if and only if the 1-form

$$\sigma = \operatorname{Re}(f(x)\,dx) = x_0 dx_0 - x_1 dx_1 - \ldots - x_n dx_n$$

is harmonic in the sense of W.V.D. Hodge (see e.g. [33]). The H-solutions are notably studied in [18], [19], [20], [3], [4], [14], [6], [10], [8], [5], [22], [23], [24] and [25].

The image space of H-solutions consists just of paravectors of the Clifford algebra. In order to extend these functions to the total Clifford algebra valued functions we first review some important concepts. The main involution $' : C\ell_n \to C\ell_n$ is the algebra isomorphism defined by $e_0' = 1$ and $e_i' = -e_i$ for $i = 1,\ldots,n$. If $\nu \subset \{1,\ldots,n\}$, we denote $e_\nu = e_{j_1} \ldots e_{j_k}$, where $j_i \in \nu$ and $j_1 < j_2 < \ldots < j_k$.

The involution $\widehat{\ }: C\ell_n \to C\ell_n$ is defined by $\widehat{e_n} = -e_n$, $\widehat{e_i} = e_i$ for $i = 0, \ldots, n-1$ and $\widehat{ab} = \widehat{a}\widehat{b}$. It is easy to calculate that for arbitrary $a \in C\ell_n$

$$a' e_n = e_n \widehat{a} \qquad \text{and} \qquad e_n a' = \widehat{a} e_n. \tag{1.4}$$

The anti-automorphism $* : C\ell_n \to C\ell_n$, called *reversion*, is defined by $e_i^* = e_i$ for $i = 0, \ldots, n$ and $(ab)^* = b^* a^*$. The conjugation $\overline{a}$ is given by $\overline{a} = (a')^*$.

Any element $a \in C\ell_n$ may uniquely be decomposed as $a = b + c e_n$ for $b, c \in C\ell_{n-1}$(the Clifford algebra generated by $e_1, \ldots, e_{n-1}$; further $C\ell_0 = \mathbb{R}$). Using this decomposition we define the mappings $P : C\ell_n \to C\ell_{n-1}$ and $Q : C\ell_n \to C\ell_{n-1}$ by $Pa = b$ and $Qa = c$. Note that if $w \in Cl_n$ then

$$\begin{aligned} Qw &= \frac{e_n w' - w e_n}{2} = \frac{\widehat{w} - w}{2} e_n, \qquad (1.5)\\ Pw &= \frac{w - e_n w' e_n}{2} = \frac{w + \widehat{w}}{2}. \end{aligned}$$

The following calculation rules are proved in ([15, Lemma 2] and [16, Lemma 1])

$$P(ab) = (Pa) Pb + (Qa) Q(b'), \tag{1.6}$$

$$Q(ab) = (Pa) Qb + (Qa) P(b'), \tag{1.7}$$

$$Q(ab) = a Qb + (Qa) b'. \tag{1.8}$$

We use the abbreviated notation $P'a = (Pa)'$ and $Q'a = (Qa)'$.

Note that in the case $n = 1$, we have $C\ell_1 = \mathbb{C}$, $Pa = \operatorname{Re} a$ and $Qa = \operatorname{Im} a$.

2. Hypermonogenic functions

Let Ω be an open subset of $\mathbb{R}^{n+1}$. We denote $e_0 = 1$. The left Dirac operator in $C\ell_n$ is defined by $D_l f = \sum_{i=0}^n e_i \frac{\partial f}{\partial x_i}$ and the right Dirac operator by $D_r f = \sum_{i=0}^n \frac{\partial f}{\partial x_i} e_i$, for a mapping $f : \Omega \to C\ell_n$, whose components are continuously differentiable. The operators $\overline{D_l}$ and $\overline{D_r}$ are defined by $\overline{D_l} f = \sum_{i=0}^n \overline{e_i} \frac{\partial f}{\partial x_i}$ and $\overline{D_r} f = \sum_{i=0}^n \frac{\partial f}{\partial x_i} \overline{e_i}$. We abbreviate $D_l f = Df$ if there is no confusion.

Let Ω be an open subset of $\mathbb{R}^{n+1}$ and k be a non-negative integer. We introduce the modified Dirac operators M_k^l , $\overline{M}_k^l$, M_k^r and $\overline{M}_k^r$ by

$$M_k^l f(x) = D_l f(x) + k \frac{Q' f}{x_n}$$

$$M_k^r f(x) = D_r f(x) + k \frac{Q f}{x_n}$$

and

$$\overline{M}_k^l f(x) = \overline{D}_l f(x) - k \frac{Q' f}{x_n},$$

$$\overline{M}_k^r f(x) = \overline{D}_r f(x) - k \frac{Q f}{x_n},$$

where $f \in C^1(\Omega, C\ell_n)$. The operator M_k^l is also denoted by M_k and M_{n-1}^l by M.

Definition 2.1. Let Ω be an open subset of $\mathbb{R}^{n+1}$. A mapping $f:\Omega\to C\ell_n$ is called left *k-hypermonogenic*, if $f\in C^{k+1}(\Omega)$ and $M_k^l f(x)=0$ for any $x\in\Omega$ with $x_n\neq 0$. The $n-1$-hypermonogenic functions are briefly called hypermonogenic and the 0-hypermonogenic functions are called monogenic. Right hypermonogenic functions are defined similarly.

Hypermonogenic functions were introduced in [15] and further developed in [16], [11], [12]. In $\mathbb{R}^3$ under the name hyperholomorphic functions they were investigated by W. Hengartner and the second author ([21]) and in $\mathbb{R}^4$ by the authors [17]. Notice that paravector-valued hypermonogenic functions are H-solutions. For the general reference to the properties of monogenic functions we refer to [2] or in the quaternionic case to [32].

We review some basic properties of hypermonogenic functions. The generalization of Cauchy-Riemann equations is stated next.

Theorem 2.2 ([15, Proposition 3]). *Let Ω be an open subset of $\mathbb{R}^{n+1}$ and $f:\Omega\to C\ell_n$ be a mapping with continuous partial derivatives. The equation $M_k^l f=0$ is equivalent with the following system of equations*

$$\begin{aligned} &x_n\left(D_{n-1}(Pf)-\tfrac{\partial(Q'f)}{\partial x_n}\right)+kQ'f=0,\\ &D_{n-1}(Qf)+\tfrac{\partial(P'f)}{\partial x_n}=0, \end{aligned}\tag{2.1}$$

where $D_{n-1}=\sum_{i=0}^{n-1}e_i\frac{\partial}{\partial x_i}$.

Note that if $k=0$ and $n=1$, the system (2.1) is just the usual complex Cauchy-Riemann system.

Theorem 2.3. *Let $f:\Omega\to C\ell_n$ be twice continuously differentiable. Then*

$$\begin{aligned} M_k\overline{M}_kf=\overline{M}_kM_kf&=\Delta Pf-\frac{k}{x_n}\frac{\partial Pf}{\partial x_n}+\left(\Delta Qf-\frac{k}{x_n}\frac{\partial Qf}{\partial x_n}+k\frac{Qf}{x_n^2}\right)e_n\\ &=\Delta f-\frac{k}{x_n}\frac{\partial f}{\partial x_n}+k\frac{Qf}{x_n^2}e_n\\ &=\Delta f-\frac{k}{x_n}\frac{\partial f}{\partial x_n}+k\frac{e_nf'e_n+f}{2x_n^2}. \end{aligned}$$

Using the preceding lemma we obtain.

Corollary 2.4 ([11]). *Let $f:\Omega\to C\ell_n$ be twice continuously differentiable on an open subset Ω of $\mathbb{R}^{n+1}$. If $M_kf=0$ in $\Omega\backslash\{x\,|\,x_n=0\}$, then the components of Pf satisfy the equation*

$$x_n\Delta u-k\frac{\partial u}{\partial x_n}=0\tag{2.2}$$

and the components of Qf satisfy the equation

$$x_n^2\Delta u-kx_n\frac{\partial u}{\partial x_n}+ku=0.\tag{2.3}$$

The differential equation (2.2), called Weinstein equation, is the Laplace-Beltrami equation for the Riemannian space defined by the metric

$$ds^2 = \frac{\sum_{i=0}^{n} dx_i^2}{x_n^{\frac{2k}{n-1}}}.$$

The most important example of a hypermonogenic function is the power function given next.

Proposition 2.5 ([15]). *For any $m \in \mathbb{Z}$ the mapping x^m, with $x = \sum_{i=0}^{n} x_i e_i$, is hypermonogenic in $\mathbb{R}^{n+1}$.*

Hypermonogenic functions form also a module.

Proposition 2.6 ([11]). *The set of k-hypermonogenic functions in an open subset Ω of $\mathbb{R}^{n+1}$ is a right $C\ell_{n-1}$-module.*

There is a way to extend holomorphic functions to hypermonogenic functions, the so-called Fueter construction.

Example ([22, Corollary 2.3]). Let $f = u + iv$ be holomorphic on an open set $\Omega \subset \mathbb{R}_+^2$. Then the function $\widetilde{f}$ defined by

$$\widetilde{f}\,(x_0, \ldots, x_n) = u\Big(x_0, \sqrt{x_1^2 + \ldots + x_n^2}\Big) + \frac{x_1 e_1 + \ldots + x_n e_n}{\sqrt{x_1^2 + \ldots + x_n^2}}\, v\Big(x_0, \sqrt{x_1^2 + \ldots + x_n^2}\Big)$$

is hypermonogenic on the set $\Big\{(x_0, \ldots, x_n) \mid \Big(x_0, \sqrt{x_1^2 + \ldots + x_n^2}\Big) \in \Omega\Big\}$.

Using the preceding example we see that the exponential and the logarithmic functions have also extensions to hypermonogenic functions, as stated in [22].

Theorem 2.7 ([11]). *Let $f : \Omega \to Cl_n$ be k-hypermonogenic. Then the functions $\frac{\partial f}{\partial x_i}$, for $i = 0, 1, \ldots, n-1$, and*

$$g = \frac{\partial f}{\partial x_n} e_n + k \frac{Qf}{x_n} = \frac{\partial f}{\partial x_n} e_n - \overline{D} f'$$

are k-hypermonogenic in Ω.

Classical holomorphic functions in the complex plane are (locally) conjugate gradients of harmonic functions. A similar result holds also for k-hypermonogenic functions with respect to hyperbolic harmonic functions introduced next.

Definition 2.8. A twice continuously differentiable function $f : \Omega \to Cl_n$ is called *k-hyperbolic harmonic*, if $M_k \overline{M}_k f = 0$. In the case $k = n-1$ we call $n-1$-hyperbolic harmonic functions just hyperbolic harmonic.

For real-valued functions the definition of $n-1$-hyperbolic harmonic functions coincides with the usual definition of hyperbolic harmonic (or hyperharmonic) functions in the hyperbolic model of the upper half space (see [1]).

Theorem 2.9 ([11]). *Let Ω be an open subset of $\mathbb{R}^{n+1}$ and $f : \Omega \to C\ell_n$ be $k+1$ times continuously differentiable. Then f is k-hypermonogenic if and only if there exists locally a k-hyperbolic harmonic mapping H with values in $C\ell_{n-1}$ satisfying $\overline{D}H = f$.*

The preceding theorem directly implies the following result.

Corollary 2.10. *A mapping f is k-hypermonogenic in $\Omega \subset \mathbb{R}^{n+1}$ if and only if there exist locally paravector-valued k-hypermonogenic functions g_α such that*

$$f = \sum_{\alpha \subset \{1,\dots,n-1\}} g_\alpha e_\alpha .$$

In complex analysis, a function f is holomorphic if and only if f and zf are harmonic. A similar result holds also for hypermonogenic functions.

Theorem 2.11 ([11]). *Let $f : \Omega \to Cl_n$ be $k+1$ times continuously differentiable. Then f is k-hypermonogenic if and only if f and xf are k-hyperbolic harmonic functions.*

There is the following correspondence between k- and $k+2s$-hypermonogenic functions.

Theorem 2.12 ([11]). *Let $f : \Omega \to Cl_n$ be k-hypermonogenic. Then there exists locally a $k+2s$-hypermonogenic function $g : \Omega \to Cl_n$ such that $f = \triangle^s g$. Conversely, if a mapping $g : \Omega \to Cl_n$ is $k+2s$-hypermonogenic, then $f = \triangle^s g$ is k-hypermonogenic.*

Hypermonogenic functions are also related to $\triangle^m$-monogenic functions, that is functions satisfying $D \triangle^m f = 0$, considered in complex Clifford algebras by J. Ryan in [30] and in real Clifford algebras under the name holomorphic Cliffordian functions by G. Laville and I. Ramadanoff in [27] . They have also been investigated by M. Sce in [31] and in the quaternionic case by L. Pernas [28].

Theorem 2.13 ([11]). *If a function $f : \Omega \to Cl_n$ satisfies $D \triangle^m f = 0$, then there exist, locally, $2i$-hypermonogenic functions $g_{ij} : \Omega \to Cl_n$ $(j = 1, \dots, 4)$ such that*

$$\sum_{i=1}^{m} \left(g_{i1} + g_{i2} e_n + \frac{\partial}{\partial x_n} \left(g_{i3} + g_{i4} e_n \right) \right) = f.$$

The next theorem states how Möbius transformations act on k-hypermonogenic functions.

Theorem 2.14 ([12]). *Let Ω be an open set contained in $\mathbb{R}^{n+1}$ and $T : \Omega \to \mathbb{R}^{n+1}$ be the Möbius transformation mapping the upper half space onto itself induced by $\begin{pmatrix} a & b \\ c & d \end{pmatrix}$, where $-c^{-1}d \notin \Omega$. If f is (left) k-hypermonogenic on $T(\Omega)$, then the function F defined by*

$$F(x) = \frac{(cx+d)^{-1}}{|cx+d|^{n-k-1}} f(T(x))$$

is also (left)k-hypermonogenic on Ω.

3. Integral formulas

So far there was no integral formula for hypermonogenic functions in the general case. Since in the case n odd hypermonogenic functions are holomorphic Cliffordian (that is, they satisfy $D \triangle^{\frac{n-1}{2}} f = 0$), the following formula, proved for example in [11], holds also for hypermonogenic functions.

Theorem 3.1. *Let $\Omega \subset \mathbb{R}^{n+1}$ be open, $n = 2m+1$ odd and $f : \Omega \to Cl_n$ an n times continuously differentiable function. Let C be an $n+1$-chain in Ω. If $f : \Omega \to Cl_n$ satisfies $D \triangle^m f = 0$, then*

$$\omega_{n+1} a_m f(x) = \sum_{i=0}^{m} \int_{\partial C} \triangle^{m-i} (y-x)^{-1} \, d\sigma(y) \triangle^i f(y)$$
$$- \sum_{i=0}^{m-1} \int_{\partial C} D \triangle^{m-i-1} (y-x)^{-1} \, d\bar{\sigma}(y) D \triangle^i f(y),$$

where ω_{n+1} is the surface measure of the $n+1$-dimensional unit ball and a_k is recursively defined by

$$a_k = -2(n-2k+1) k a_{k-1}, \qquad a_0 = 1.$$

We will verify for the P-part a simpler integral formula, depending only on f and not on its derivatives. It is based on the relations of Dirac operator and the modified Dirac operator stated next.

Lemma 3.2. *Let f be twice continuously differentiable. Then*

$$D_l\left(\frac{f}{x_n^k}\right) = \frac{M_k^l f}{x_n^k} - k \frac{P' f}{x_n^{k+1}} e_n, \tag{3.1}$$

$$D_r\left(\frac{f}{x_n^k}\right) = \frac{M_k^r f}{x_n^k} - k \frac{P f}{x_n^{k+1}} e_n, \tag{3.2}$$

$$P\left(M_k^l f\right) = x_n^k P\left(D_l\left(\frac{f}{x_n^k}\right)\right), \tag{3.3}$$

$$Q\left(M_k^l f\right) = Q\left(D_l f\right), \tag{3.4}$$

$$D_l\left(\frac{Q f e_n}{x_n^k}\right) = \frac{M_k^l\left(Q f e_n\right)}{x_n^k}. \tag{3.5}$$

Proof. We just compute

$$D_l\left(\frac{f}{x_n^k}\right) = \frac{D_l f}{x_n^k} - k\frac{e_n f}{x_n^{k+1}} = \frac{D_l f}{x_n^k} + k\frac{Q' f}{x_n^{k+1}} - k\frac{P' f e_n}{x_n^{k+1}}$$
$$= \frac{M_k^l f}{x_n^k} - k\frac{P' f e_n}{x_n^{k+1}},$$

which implies the first equality. If we take P from both sides of this equality we obtain the third equality. The fourth equality follows directly from the definition of M_k^l. The last equality follows from the first one if we replace f by $(Qf)e_n$. □

Let Ω be an open subset of $\mathbb{R}^{n+1} \setminus \{x_n = 0\}$ and K an $n+1$-chain satisfying $\overline{K} \subset \Omega$. Define an n-form by

$$d\sigma_k = \frac{1}{x_n^k} \sum_{i=0}^{n} (-1)^i e_i dx_0 \wedge \ldots \wedge dx_{i-1} \wedge dx_{i+1} \wedge \ldots dx_n$$

and an $n+1$-form by

$$dm_k = \frac{1}{x_n^k} dx_0 \wedge \ldots \wedge dx_n.$$

Lemma 3.3. *If $f \in C^1(\Omega, C\ell_n)$, then*

$$\begin{aligned}
\int_{\partial K} \left((d\sigma_k + d\sigma_0) f + (\widehat{d\sigma_k} - \widehat{d\sigma_0}) \widehat{f} \right) &= 2 \int_{\partial K} \left(P(d\sigma_k f) + Q(d\sigma_0 f) e_n \right) \\
&= 2 \int_K \left(P(M_k f) dm_k + Q(M_k f) e_n dm_0 \right) \\
&= \int_K \left(M_k f + \widehat{M_k f} \right) dm_k \\
&\quad + \int_K \left(M_k f - \widehat{M_k f} \right) dm_0.
\end{aligned}$$

Proof. Recall first that

$$\int_{\partial K} g d\sigma_0 f = \int_K \left((D_r g) f + g D_l f \right) dm_0, \tag{3.6}$$

see [2, 9.2 Proposition, p. 52]. Using this and the previous Lemma we obtain

$$\begin{aligned}
\int_{\partial K} d\sigma_k f &= \int_K D_l \left(\frac{f}{x_n^k} \right) dm_0 = \int_K \frac{M_k^l f}{x_n^k} - k \frac{P' f e_n}{x_n^{k+1}} dm_0 \\
&= \int_K M_k^l f - k \frac{P' f e_n}{x_n} dm_k \\
\int_{\partial K} d\sigma_0 f &= \int_K D_l f dm_0.
\end{aligned}$$

Hence we have

$$\begin{aligned}
\int_{\partial K} P(d\sigma_k f) &= \int_K P\left(M_k^l f\right) dm_k \\
\int_{\partial K} Q(d\sigma_0 f) &= \int_K Q\left(M_k^l f\right) dm_0.
\end{aligned}$$

Hence

$$\int_{\partial K} P(d\sigma_k f) + \int_{\partial K} Q(d\sigma_0 f) e_n = \int_K P\left(M_k^l f\right) dm_k + \int_K Q\left(M_k^l f\right) e_n dm_0.$$

Using equalities (1.5) we obtain the last equality. □

Theorem 3.4. *A function $f \in C^1(\Omega, C\ell_n)$ is k-hypermonogenic in an open subset $\Omega \subset \mathbb{R}^{n+1} \backslash \{x_n = 0\}$ if and only if the property*

$$\int_{\partial K} P(d\sigma_k f) + \int_{\partial K} Q(d\sigma_0 f) e_n = 0$$

or equivalently

$$\int_{\partial K} \left((d\sigma_k + d\sigma_0) f + (\widehat{d\sigma_k} - \widehat{d\sigma_0}) \widehat{f} \right) = 0$$

holds for any $n+1$-chain K satisfying $\bar{K} \subset \Omega$.

Theorem 3.5. *Let Ω be an open subset of $\mathbb{R}^{n+1} \backslash \{x_n = 0\}$ and K an $n+1$-chain satisfying $\overline{K} \subset \Omega$. If $f, g \in C^1(\Omega, C\ell_n)$and $D_r g \in Cl_{n-1}$, then*

$$\int_{\partial K} P(g d\sigma_k f) = \int_K P\left(M_k^r g P f + g M_k^l f\right) dm_k.$$

Proof. Using (3.6) and Lemma 3.2 we obtain

$$\begin{aligned}
\int_{\partial K} g d\sigma_k f &= \int_{\partial K} \frac{g}{x_n^k} d\sigma_0 P f + \int_{\partial K} g d\sigma_0 \frac{Q f e_n}{x_n^k} \\
&= \int_K \left(D_r \left(\frac{g}{x_n^k} \right) P f + \frac{g}{x_n^k} D_l P f + g \frac{M_k^l (Q f e_n)}{x_n^k} + \frac{D_r g Q f e_n}{x_n^k} \right) dm_0 \\
&= \int_K \left(M_k^r g P f + g M_k^l f - \frac{k P g P' f e_n}{x_n} + D_r g Q f e_n \right) dm_k.
\end{aligned}$$

Hereby we used that $D_l(Pf) = M_k^l(Pf)$. Since $D_r g \in Cl_{n-1}$, by assumption, the result follows. □

Lemma 3.6. *The function*

$$\begin{aligned}
g(x) &= \frac{x_n^{n-1}}{2y_n} \left(\frac{(x-y)^{-1}}{|x-y|^{n-1} |x-\widehat{y}|^{n-1}} - \frac{(x-\widehat{y})^{-1}}{|x-y|^{n-1} |x-\widehat{y}|^{n-1}} \right) \\
&= x_n^{n-1} \frac{(x-y)^{-1}}{|x-y|^{n-1}} e_n \frac{(x-\widehat{y})^{-1}}{|x-\widehat{y}|^{n-1}} = \frac{1}{2^{2n-1} y_n^n} \overline{D}^x \left(\int_{\frac{|x-y|}{|x-\widehat{y}|}}^1 \frac{(1-s^2)^{n-1}}{s^n} ds \right)
\end{aligned}$$

is left and right hypermonogenic on $\mathbb{R}^{n+1} \backslash \{y, \widehat{y}\}$ for each y with $y_n \neq 0$.

Proof. Note first that $\frac{(x-y)^{-1}}{|x-y|^{n-1}}$ and $\frac{(x-\widehat{y})^{-1}}{|x-\widehat{y}|^{n-1}}$ are monogenic. Hence

$$\begin{aligned}
2y_n D g = x_n^{n-1} &\left(\frac{(1-n)(x-\widehat{y})(x-y)^{-1}}{|x-y|^{n-1} |x-\widehat{y}|^{n+1}} + \frac{(n-1)(x-y)(x-\widehat{y})^{-1}}{|x-y|^{n+1} |x-\widehat{y}|^{n-1}} \right) \\
&+ 2(n-1) y_n x_n^{n-2} e_n \frac{(x-y)^{-1}}{|x-y|^{n-1}} e_n \frac{(x-\widehat{y})^{-1}}{|x-\widehat{y}|^{n-1}}.
\end{aligned}$$

Since $x-\widehat{y}=x-y+2y_ne_n$ and $x-y=x-\widehat{y}-2y_ne_n$ we infer

$$\begin{aligned}2y_nDg &= \frac{(1-n)\,x_n^{n-1}}{|x-y|^{n-1}\,|x-\widehat{y}|^{n+1}}+\frac{(1-n)\,x_n^{n-1}\,(2y_ne_n\ (\overline{x}-\overline{y}))}{|x-y|^{n+1}\,|x-\widehat{y}|^{n+1}}\\ &+\frac{(n-1)\,x_n^{n-1}}{|x-y|^{n+1}\,|x-\widehat{y}|^{n-1}}+\frac{x_n^{n-1}\,(n-1)\left(-2y_ne_n\ \left(\overline{x}-\overline{\widehat{y}}\right)\right)}{|x-y|^{n+1}\,|x-\widehat{y}|^{n+1}}\\ &+2\,(n-1)\,y_nx_n^{n-2}e_n\frac{(x-y)^{-1}}{|x-y|^{n-1}}e_n\frac{(x-\widehat{y})^{-1}}{|x-\widehat{y}|^{n-1}}.\end{aligned}$$

Since

$$\frac{1}{|x-y|^2}-\frac{1}{|x-\widehat{y}|^2}=\frac{4x_ny_n}{|x-y|^2\,|x-\widehat{y}|^2}$$

and $(\overline{x-y})\,e_n=e_n\,(\widehat{x}-\widehat{y})=e_n\,(x-\widehat{y}-2x_ne_n)$ we obtain

$$\begin{aligned}2y_nDg &= \frac{4\,(n-1)\,x_n^ny_n+(1-n)\,2x_n^{n-1}y_ne_n\,(\overline{x}-\overline{y})-2\,(n-1)\,x_n^{n-1}y_ne_n\left(\overline{x}-\overline{\widehat{y}}\right)}{|x-y|^{n+1}\,|x-\widehat{y}|^{n+1}}\\ &+\frac{(n-1)\,4x_n^{n-1}y_ne_n\left(\overline{x}-\overline{\widehat{y}}\right)}{|x-y|^{n+1}\,|x-\widehat{y}|^{n+1}}-\frac{2\,(n-1)\,x_n^{n-2}y_n}{|x-y|^{n+1}\,|x-\widehat{y}|^{n-1}}\\ &=\frac{4\,(n-1)\,x_n^ny_n+4\,(n-1)\,x_n^{n-1}y_n^2}{|x-y|^{n+1}\,|x-\widehat{y}|^{n+1}}-\frac{2\,(n-1)\,x_n^{n-2}y_n}{|x-y|^{n+1}\,|x-\widehat{y}|^{n-1}}.\end{aligned}$$

Hence we have

$$\begin{aligned}&\frac{2y_n}{n-1}\left(Dg+(n-1)\,\frac{Q'g}{x_n}\right)\\ &=\frac{4x_n^ny_n+4x_n^{n-1}y_n^2}{|x-y|^{n+1}\,|x-\widehat{y}|^{n+1}}-\frac{2x_n^{n-2}y_n}{|x-y|^{n+1}\,|x-\widehat{y}|^{n-1}}\\ &\quad+\frac{x_n^{n-2}\,(y_n-x_n)}{|x-y|^{n+1}\,|x-\widehat{y}|^{n-1}}+\frac{x_n^{n-2}\,(y_n+x_n)}{|x-y|^{n-1}\,|x-\widehat{y}|^{n+1}}\\ &=\frac{4x_n^ny_n+4x_n^{n-1}y_n^2}{|x-y|^{n+1}\,|x-\widehat{y}|^{n+1}}+\frac{x_n^{n-2}\,(y_n+x_n)\left(-\frac{1}{|x-y|^2}+\frac{1}{|x-\widehat{y}|^2}\right)}{|x-y|^{n-1}\,|x-\widehat{y}|^{n-1}}\\ &=\frac{4x_n^ny_n+4x_n^{n-1}y_n^2-4x_n^{n-1}\,(y_n+x_n)\,y_n}{|x-y|^{n+1}\,|x-\widehat{y}|^{n+1}}=0.\end{aligned}$$

□

The kernel g was calculated by the second author in [22]. Note that the point-pair-invariance of the hyperbolic distance yields

$$\frac{|x-y|}{|x-\widehat{y}|}=\frac{|\varphi(x)-\varphi(y)|}{\left|\varphi(x)-\widehat{\varphi(y)}\right|}$$

for each Möbius transformation mapping the upper half space onto itself.

Theorem 3.7. *Let Ω be an open subset of $\mathbb{R}^{n+1}_+$ (or $\mathbb{R}^{n+1}_-$) and K an $n+1$-chain satisfying $\overline{K} \subset \Omega$. If f is hypermonogenic in Ω and $y \in K$, then*

$$\begin{aligned}
Pf(y) &= \frac{(2y_n)^n}{\omega_{n+1}} \int_{\partial K} P\left(x_n^{n-1} \frac{(x-y)^{-1}}{|x-y|^{n-1}} e_n \frac{(x-\widehat{y})^{-1}}{|x-\widehat{y}|^{n-1}} d\sigma_{n-1}(x) f(x) \right) \\
&= \frac{(2y_n)^n}{\omega_{n+1}} \int_{\partial K} P\left(\frac{(x-y)^{-1}}{|x-y|^{n-1}} e_n \frac{(x-\widehat{y})^{-1}}{|x-\widehat{y}|^{n-1}} d\sigma_0(x) f(x) \right) \\
&= \frac{(2y_n)^n}{2\omega_{n+1}} \int_{\partial K} \frac{(x-y)^{-1}}{|x-y|^{n-1}} e_n \frac{(x-\widehat{y})^{-1}}{|x-\widehat{y}|^{n-1}} d\sigma_0(x) f(x) \\
&\quad - \frac{(2y_n)^n}{2\omega_{n+1}} \int_{\partial K} \frac{(\widehat{x}-\widehat{y})^{-1}}{|x-y|^{n-1}} e_n \frac{(\widehat{x}-y)^{-1}}{|\widehat{x}-y|^{n-1}} \widehat{d\sigma_0}(x) \widehat{f}(x)
\end{aligned}$$

where ω_{n+1} is the surface measure of the unit ball in $\mathbb{R}^{n+1}$.

Proof. Using Theorem 3.5 we obtain

$$\begin{aligned}
&\int_{\partial K} P\left(x_n^{n-1} \frac{(x-y)^{-1}}{|x-y|^{n-1}} e_n \frac{(x-\widehat{y})^{-1}}{|x-\widehat{y}|^{n-1}} d\sigma_{n-1} f \right) \\
&= \int_{\partial(K \setminus B_r(y))} P(g(x)\, d\sigma_{n-1} f) + \int_{\partial B_r(y)} P(g(x)\, d\sigma_{n-1} f) \\
&= \int_{\partial B_r(y)} P(g(x)\, d\sigma_{n-1} f) = P\left(\int_{\partial B_r(y)} \frac{(x-y)^{-1}}{|x-y|^{n-1}} e_n \frac{(x-\widehat{y})^{-1}}{|x-\widehat{y}|^{n-1}} d\sigma_0 f \right).
\end{aligned}$$

The preceding Lemma implies that the function $g(x) = x_n^{n-1} \frac{(x-y)^{-1}}{|x-y|^{n-1}} e_n \frac{(x-\widehat{y})^{-1}}{|x-\widehat{y}|^{n-1}}$ is paravector-valued. Thus we have

$$g(x) = g(x)^* = x_n^{n-1} \frac{(x-\widehat{y})^{-1}}{|x-\widehat{y}|^{n-1}} e_n \frac{(x-y)^{-1}}{|x-y|^{n-1}}.$$

Hence

$$\begin{aligned}
&\int_{\partial K} P\left(x_n^{n-1} \frac{(x-y)^{-1}}{|x-y|^{n-1}} e_n \frac{(x-\widehat{y})^{-1}}{|x-\widehat{y}|^{n-1}} d\sigma_{n-1}(x) f(x) \right) \\
&= P\left(\int_{\partial B_r(y)} \frac{(x-\widehat{y})^{-1}}{|x-\widehat{y}|^{n-1}} e_n \frac{(x-y)^{-1}}{|x-y|^{n-1}} d\sigma_0(x) f(x) \right) \\
&= P\left(\int_{\partial B_r(y)} \frac{(x-\widehat{y})^{-1}}{|x-\widehat{y}|^{n-1}} e_n \frac{(x-y)^{-1}}{r^{n-1}} \frac{(x-y)}{r} f(x)\, dS(x) \right),
\end{aligned}$$

where S is the usual surface measure of the ball $B_r(y)$. When $r \to 0$, we obtain the result. □

The Q-part of the hypermonogenic functions satisfies the following result.

Theorem 3.8. *Let Ω be an open subset of $\mathbb{R}^{n+1}_+$ and K an $n+1$-chain satisfying $\overline{K} \subset \Omega$. If f is hypermonogenic in Ω and $y \in K$, then*

$$\frac{\partial\left(\frac{Qf(y)}{y_n^{n-1}}\right)}{\partial y_n}$$
$$= -\frac{2y_n}{\omega_{n+1}} \int_{\partial K} P\left(\overline{D}^y_{n-1}\left(x_n^{n-1}\frac{(x-y)}{|x-y|^{n+1}}e_n\frac{(x-\widehat{y})}{|x-\widehat{y}|^{n+1}}\right) d\overline{\sigma_{n-1}}(x)\, f'(x)\right)$$

where ω_{n+1} is the surface measure of the unit ball in $\mathbb{R}^{n+1}$.

Proof. Assume that f is hypermonogenic in Ω. Using Lemma 2.2 we obtain

$$\frac{\partial \frac{Qf(y)}{y_n^{n-1}}}{\partial y_n} = \frac{\overline{D}^y_{n-1}(P'f)}{y_n^{n-1}}.$$

Hence by the previous theorem we conclude the result. □

Lastly we consider the question how uniquely the P- or Q-parts determine a k-hypermonogenic function.

Theorem 3.9. *If f and g are k-hypermonogenic in Ω and $Pf = Pg$, then there exists a monogenic function w depending only on $x_0, \ldots, x_{n-1}$ such that $f = g + wx_n^k e_n$.*

Proof. Assume that f and g are k-hypermonogenic in Ω and $Pf = Pg$. Then applying Theorem 2.2 we deduce

$$\frac{\partial \frac{Q(f-g)}{x_n^k}}{\partial x_n} = 0$$
$$D_{n-1}\left(\frac{Q(f-g)}{x_n^k}\right) = 0.$$

Hence there exists a monogenic function $w(x_0, \ldots, x_{n-1})$ such that $\frac{Q(f-g)}{x_n^k} = w(x_0, \ldots, x_{n-1})$. This means that

$$f = g + w(x_0, \ldots, x_{n-1})\, x_n^k e_n.$$ □

Theorem 3.10. *Let h be a $C\ell_{n-1}$-valued k-hyperbolic harmonic function in an open subset Ω of $\mathbb{R}^{n+1}_+$. Then there exists locally a k-hypermonogenic function f such that $Pf = h$.*

Proof. Let a ball $B_r(a)$ satisfy $\overline{B_r}(a) \subset \Omega$. Let $x = (x_0, \ldots, x_n) = (\tilde{x}, x_n)$ for any $x \in \mathbb{R}^{n+1}$. Then there exists a mapping s from $(B(a,r) \cap \{x \mid x_n = a_n\})^{\sim}$ into $C\ell_{n-1}$ satisfying the equation

$$D_{n-1}s(\tilde{x}) = \frac{1}{a_n^k}\frac{\partial h'}{\partial x_n}(\tilde{x}, a_n).$$

Define a function f by

$$\begin{aligned} Qf(x) &= x_n^k \left(\int_{a_n}^{x_n} \frac{\overline{D_{n-1} h'(\tilde{x},t)}}{t^k} dt - s(\tilde{x}) \right), \\ Pf(x) &= h(x). \end{aligned} \tag{3.7}$$

We prove that f satisfies the system of Theorem 2.2. The first equation follows directly from differentiating $Q'f$ with respect to x_n. For the second equation we infer

$$D_{n-1} Qf(x) = x_n^k \left(\int_{a_n}^{x_n} \frac{\triangle_{n-1} h'(\tilde{x},t)}{t^k} dt - \frac{1}{a_n^k} \frac{\partial h'}{\partial x_n}(\tilde{x}, a_n) \right). \tag{3.8}$$

Since h is k-hyperbolic harmonic

$$\begin{aligned} \frac{\triangle_{n-1} h'(\tilde{x},t)}{t^k} &= -\frac{1}{t^k} \frac{\partial^2 h'(\tilde{x},t)}{\partial t^2} + k \frac{1}{t^{k+1}} \frac{\partial h'(\tilde{x},t)}{\partial t} \\ &= -\frac{\partial \left(\frac{1}{t^k} \frac{\partial h'}{\partial x_n}(\tilde{x},t) \right)}{\partial t}. \end{aligned}$$

Substituting this into (3.8) we obtain the second equation. □

Lemma 3.11. *If f and g are k-hypermonogenic in Ω and $Qf = Qg$, then there exists a monogenic function w depending only on $x_0, \ldots, x_{n-1}$ such that $f = g + w$.*

Proof. Assume that f and g are k-hypermonogenic in Ω and $Qf = Qg$. Then applying Theorem 2.2 we deduce

$$\begin{aligned} \frac{\partial P(f-g)}{\partial x_n} &= 0 \\ D_{n-1}(P(f-g)) &= 0. \end{aligned}$$

Hence there exists a monogenic function $w(x_0, \ldots, x_{n-1})$ such that $P(f-g) = w(x_0, \ldots, x_{n-1})$. This means that

$$f = g + w(x_0, \ldots, x_{n-1}).$$

□

Addendum. The results of this paper were presented at the conference on Clifford Algebras and Applications in Macao in August 2002. In June 2003 the first author found the analogous integral formula for the Q-part, presented at the ICIAM 2003 in Sydney. We just include one of the results from [13].

Theorem 3.12. *Let Ω be an open subset of $\mathbb{R}_{+}^{n+1}$ (or $\mathbb{R}_{-}^{n+1}$) and K an $n+1$-chain satisfying $\overline{K} \subset \Omega$. If f is hypermonogenic in Ω and $y \in K$, then*

$$
\begin{aligned}
Qf(y) &= \frac{2^n y_n^{n-1}}{\omega_{n+1}} \int_{\partial K} Q\left(q(x,y)\, d\sigma_0(x)\, f(x)\right) \\
&= \frac{2^{n-1} y_n^{n-1}}{\omega_{n+1}} \int_{\partial K} \widehat{q(x,y)} d\widehat{\sigma}_0(x)\, \widehat{f}(x)\, e_n - \int_{\partial K} q(x,y)\, d\sigma_0(x)\, f(x)\, e_n
\end{aligned}
$$

where

$$
q(x,y) = \frac{(x-\widehat{y})^{-1}}{|x-\widehat{y}|^{n-1}} (x - Py) \frac{(x-y)^{-1}}{|x-y|^{n-1}}
$$

and ω_{n+1} is the surface measure of the unit ball in $\mathbb{R}^{n+1}$.

Combining Theorem 3.7 with Theorem 3.12 we thus obtain a Cauchy-type integral representation for hypermonogenic functions.

References

[1] Ahlfors, L.V., *Möbius transformations in several dimensions.* Ordway Lectures in Math., University of Minnesota, 1981.

[2] Brackx, R. , Delanghe and F. Sommen, *Clifford Analysis*, Pitman, Boston, London, Melbourne, 1982.

[3] Cerejeiras, P., *Decomposition of analytic hyperbolically harmonic functions.* Proceedings of Clifford Algebras and their Applications Aachen, Kluwer, Dordrecht (1998), 45–51.

[4] Cnops, J., *Hurwitz pairs and applications of Möbius transformation*, Habilitation thesis, Univ. Gent, 1994.

[5] Eriksson-Bique, S.-L., *Comparison of quaternionic analysis and its modification,* Harlow: Longman. Pitman Res. Notes Math. Ser. 394, 109–121 (1998).

[6] Eriksson-Bique, S.-L, On modified Clifford analysis, *Complex Variables* **45** (2001), 11–32.

[7] Eriksson-Bique, S.-L., The binomial theorem for hypercomplex numbers, *Ann. Acad. Sci. Fenn.* **24** (1999), 225–229.

[8] Eriksson-Bique, S.-L., *Real analyticity in modified Clifford analysis,* Submitted for publication, 10 pages.

[9] Eriksson-Bique, S.-L., *Hyperholomorphic functions in $\mathbb{R}^4$*, In: Proceedings of the 2nd conference on quaternionic structures in mathematics and Physics, World scientific, London, Singapore (2001), 125–135.

[10] Eriksson-Bique, S.-L., *A correspondence of hyperholomorphic and monogenic functions in $\mathbb{R}^4$*. In: Proceedings of Clifford analysis and its Applications in Praha, NATO Science Series II. Mathematics 25 (2001), 71–80.

[11] Eriksson-Bique, S.-L., *$k-$hypermonogenic functions,* Submitted for publication, 15 pages.

[12] Eriksson-Bique, S.-L.: *Möbius transformations and k-hypermonogenic functions*, Submitted for publication, 18 pages.

[13] Eriksson, S.-L., *Integral formulas for hypermonogenic functions*, Submitted for publication.

[14] Eriksson-Bique, S.-L., Leutwiler, H., On modified quaternionic analysis in $\mathbb{R}^3$. *Arch. Math.* **70** (1998), 228–234.

[15] Eriksson-Bique, S.-L., and H. Leutwiler, *Hypermonogenic functions*, In: Clifford Algebras and their Applications in Mathematical Physics, Vol. 2: Clifford Analysis, Ryan, J. and Sprössig, W., eds., Birkhäuser, Boston, 2000, 287–302.

[16] Eriksson-Bique, S.-L., Leutwiler, H.: Hypermonogenic functions and Möbius transformations. *Adv. Appl. Clifford Algebras* **11 (S2)** (2001), 67–76.

[17] Eriksson-Bique, S.-L., Leutwiler, H.: Hyperholomorphic functions, *Computional Methods and Function Theory* **1** (2002), 1–15.

[18] Hempfling, Th., *Beiträge zur Modifizierten Clifford-Analysis,* Dissertation, Univ. Erlangen-Nürnberg, August 1997.

[19] Hempfling, Th., Multinomials in modified Clifford analysis, *C. R. Math. Rep. Acad. Sci. Canada* **18** (2,3) (1996), 99–102.

[20] Hempfling, Th., H. Leutwiler, *Modified quaternionic analysis in* $\mathbb{R}^4$. In: Proceedings of Clifford Algebras and their Application in Mathematical Physics, Aachen 1996, Fundamental theories of physics, 94, Kluwer, Dordrecht (1998), 227–237.

[21] Hengartner, W., H. Leutwiler, Hyperholomorphic functions in $\mathbb{R}^3$. *Adv. Appl. Clifford Algebras* **11 (S1)** (2001), 247–259.

[22] Leutwiler, H., Modified Clifford analysis, *Complex Variables* **17** (1992), 153–171.

[23] Leutwiler, H., Modified quaternionic analysis in $\mathbb{R}^3$. *Complex Variables* **20** (1992), 19–51.

[24] Leutwiler, H., More on modified quaternionic analysis in $\mathbb{R}^3$, *Forum Math.* **7** (1995), 279–305.

[25] Leutwiler, H., Rudiments of a function theory in $\mathbb{R}^3$, *Expo. Math* **14** (1996), 097-123.

[26] Leutwiler, H., *Quaternionic analysis in* $\mathbb{R}^3$ *versus its hyperbolic modification,* In: Proceedings of Clifford analysis and its Applications in Praha, NATO Science Series II. Mathematics 25 (2001), 193–211.

[27] Laville, G.; Ramadanoff, I., Holomorphic Cliffordian functions. *Adv. Appl. Clifford Algebr.* **8** No.2 (1998), 323-340.

[28] Pernas, L., Holomorphie quaternionienne. *Adv. Appl. Clifford Algebr.* **8** No. 2 (1998), 283-298.

[29] Qian, T., Generalization of Fueter's result to $\mathbb{R}^{n+1}$, *Atti Accad. Naz. Lincei. Rend. Cl. Sci. Fis. Mat. Nat* **(9) 8** (1997), 111–117.

[30] Ryan, J. Iterated Dirac operators in C_n. *Z. Anal. Anwendungen* **9** (1990), no. 5, 385–401

[31] Sce, M., Osservazioni sulle serie di potenze nei moduli quadratici. (Italian) *Atti Accad. Naz. Lincei. Rend. Cl. Sci. Fis. Mat. Nat.* **(8) 23** (1957), 220–225.

[32] Sudbery, A., Quaternionic analysis, *Math. Proc. Cambridge Philos. Soc.* **85** (1979), 199–225.

[33] Warner, F.W., *Foundations of differentiable manifolds and Lie groups.* Scott, Foresman and Company, Glenview (Illinois), London, 1971.

Sirkka-Liisa Eriksson
Department of Mathematics
University of Joensuu
P.O. Box 111
Fin-80101-Joensuu
Finland
e-mail: Sirkka-Liisa.Eriksson.Bique@joensuu.fi

Heinz Leutwiler
University of Erlangen-Nürnberg
Department of Mathematics
Bismarckstrasse 1 1/2
D-91054 Erlangen
Germany
e-mail: leutwil@mi.uni-erlangen.de

Trends in Mathematics: *Advances in Analysis and Geometry*, 113–129

On Series Expansions of Hyperholomorphic $\mathbf{B}^{\mathbf{q}}$ Functions

K. Gürlebeck and A. El-Sayed Ahmed

Abstract. In the previous article [11], $\mathbf{B}^{\mathbf{q}}$ spaces of hyperholomorphic functions were studied and it was shown that these spaces form a scale of subspaces, all included in the hyperholomorphic Bloch space. Here, we study the problem if these inclusions within the scale and with respect to the Bloch space are strict. Main tool is the characterization of $\mathbf{B}^{\mathbf{q}}$ functions by the coefficients of certain series expansions.

Mathematics Subject Classification (2000). Primary 30G35; Secondary 46E15.

Keywords. Quaternionic Analysis, $\mathbf{B}^{\mathbf{q}}$-spaces.

1. Introduction

$\mathbf{Q_p}$ spaces of holomorphic functions for the first time appeared in the paper by Aulaskari and Lappan [3]. Since then a great number of articles and monographs were devoted to their investigation (see e.g., [2], [4], [8], [9], [23] and [24]). On the other hand $\mathbf{B}^{\mathbf{q}}$ spaces of holomorphic functions were introduced by Stroethoff [21]. The studies of these spaces in Clifford analysis began with the papers [12] and [7], followed by [10], [11], [12] and [13]. In the present paper our aim is to extend some results obtained by Miao [19] to the case of hyperholomorphic functions.

First, we describe the spaces we are interested in. Let $\Delta = \{z : |z| < 1\}$ be the unit disk in the complex plane $\mathbb{C}$. Then the well known Bloch space is:

$$\mathcal{B} = \{f : f \text{ analytic in } \Delta \text{ and } \sup_{z\in\Delta}(1 - |z|^2)|f'(z)| < \infty\}.$$

We mention [1] as a general reference for the Bloch space in the complex plane $\mathbb{C}$.

Let $0 < q < \infty$. Then the Besov-type spaces

$$\mathbf{B^q} = \{f : f \text{ analytic in } \Delta \text{ and} \sup_{a\in\Delta} \int_{\Delta} |f'(z)|^q (1-|z|^2)^{q-2} (1-|\varphi_a(z)|^2)^2 d\sigma_z < \infty\}$$

are introduced and studied intensively (see [21]). Here, φ_a always stands for the Möbius transformation $\varphi_a(z) = \frac{a-z}{1-\bar{a}z}$.

To introduce the meaning of hyperholomorphic functions let $\mathbb{H}$ be the set of real quaternions. This means, elements of $\mathbb{H}$ are of the form

$$a = \sum_{k=0}^{3} a_k e_k, \quad \text{where} \quad \{a_k | k \in \mathbb{N}_3^0 := \mathbb{N}_3 \cup \{0\}; \mathbb{N}_3 := \{1,2,3\}\} \subset \mathbb{R};$$

$e_0 = 1$ the unit; e_1, e_2, e_3 are called imaginary units, and they define arithmetic rules in $\mathbb{H}$; by definition $e_k^2 = -e_0$, $k \in \mathbb{N}_3$; $e_1e_2 = -e_2e_1 = e_3$; $e_2e_3 = -e_3e_2 = e_1$; $e_3e_1 = -e_1e_3 = e_2$. Sometimes we use $\mathbf{i}$, $\mathbf{j}$ and $\mathbf{k}$ instead of e_1, e_2 and e_3.

Natural operations of addition and multiplication in $\mathbb{H}$ turn it into a skew-field. The main involution in $\mathbb{H}$, the quaternionic conjugation, is defined by

$$\overline{e_0} := e_0; \quad \overline{e_k} := -e_k; \quad \textit{for} \quad k \in \mathbb{N}_3,$$

and it extends onto $\mathbb{H}$ by $\mathbb{R}$-linearity, i.e., for $a \in \mathbb{H}$

$$\bar{a} := \sum_{k=0}^{3} \overline{a_k e_k} = \sum_{k=0}^{3} a_k \bar{e_k} = a_0 - \sum_{k=1}^{3} a_k e_k.$$

Note that

$$\bar{a}a = a\bar{a} = \sum_{k=0}^{3} a_k^2 = |a|^2_{\mathbb{R}^4} =: |a|^2_{\mathbb{H}}$$

Therefore, for $a \in \mathbb{H} \setminus \{0\}$ the quaternion

$$a^{-1} := \frac{1}{|a|^2}\bar{a}$$

is an inverse to a. Whereas the above mentioned properties are analogous to the complex one-dimensional case we have for the quaternionic conjugation that for any $a, b \in \mathbb{H}$

$$\overline{ab} = \bar{b}\,\bar{a}.$$

Let Ω be a domain in $\mathbb{R}^3$, then we shall consider $\mathbb{H}$-valued functions defined in Ω (depending on $x = (x_0, x_1, x_2)$):

$$f : \Omega \longrightarrow \mathbb{H}.$$

The notation $C^p(\Omega; \mathbb{H}), p \in \mathbb{N} \cup \{0\}$, has the usual component-wise meaning. On $C^1(\Omega; \mathbb{H})$ we define a generalized Cauchy-Riemann operator D by

$$D(f) := \sum_{k=0}^{2} e_k \frac{\partial f}{\partial x_k} =: \sum_{k=0}^{2} e_k\, \partial_k f.$$

D is a right-linear operator with respect to scalars from $\mathbb{H}$. The operator $\overline{D}$

$$\overline{D}(f) := \sum_{k=0}^{2} \bar{e}_k \frac{\partial f}{\partial x_k} =: \sum_{k=0}^{2} \bar{e}_k \, \partial_k f$$

is the adjoint Cauchy-Riemann operator. The solutions of $Df = 0,\ x \in \Omega$, are called (left) hyperholomorphic (or monogenic) functions and generalize the class of holomorphic functions from the one-dimensional complex function theory. Let $\triangle$ be the three-dimensional Laplace operator $\triangle := \sum\limits_{k=0}^{2} \partial_k^2$. Then on $C^2(\Omega; \mathbb{H})$ analogously to the complex case the following equalities hold:

$$\triangle = D\overline{D} = \overline{D}D.$$

Let $f : B_1(0) \mapsto \mathbb{H}$ be a hyperholomorphic function. According to [12], we have the definition of the quaternionic Bloch semi-norm as follows:

$$\mathcal{B}(f) = \sup_{x \in B_1(0)} (1 - |x|^2)^{3/2} |\overline{D} f(x)|,$$

where $B_1(0)$ stands for the unit ball in $\mathbb{R}^3$. This definition leads to the following

Definition 1.1. The spatial (or hyperholomorphic) Bloch space $\mathcal{B}$ is the right $\mathbb{H}$-module of all hyperholomorphic functions $f : B_1(0) \mapsto \mathbb{H}$ with $\mathcal{B}(f) < \infty$.

For more information about the study of quaternionic Bloch spaces, we refer to [7], [11], [12], and [13].

By using the adjoint generalized Cauchy-Riemann operator $\overline{D}$ instead of the derivative $f'(z)$, the quaternionic Möbius transformation $\varphi_a(x) = (a-x)(1-\bar{a}x)^{-1}$, $\mathbf{B^q}$ spaces of quaternion-valued functions were defined in [10] by

$$\mathbf{B^q} = \left\{ f \in \ker D : \sup_{a \in B_1(0)} \int_{B_1(0)} |\overline{D} f(x)|^q \left(1 - |x|^2\right)^{\frac{3}{2}q - 3} \left(1 - |\varphi_a(x)|^2\right)^3 dB_x < \infty \right\},$$

where $0 < q < \infty$ and the exponent three is related to the real space dimension. From [10] it is known that the $\mathbf{B^q}$ spaces describe the Bloch space $\mathcal{B}$ equivalently by the integral norms of $\mathbf{B^q}$, $\forall q > 0$.

Also, in [11] we studied $\mathbf{B^q}$ spaces of hyperholomorphic functions by employing the weight function $\left(1 - |\varphi_a(z)|^2\right)^2$ instead of $\left(1 - |\varphi_a(z)|^2\right)^3$ as follows:

$$\mathbf{B^q} = \left\{ f \in \ker D : \sup_{a \in B_1(0)} \int_{B_1(0)} |\overline{D} f(x)|^q \left(1 - |x|^2\right)^{\frac{3}{2}q - 3} \left(1 - |\varphi_a(x)|^2\right)^2 dB_x < \infty \right\}.$$

The most important results are that in this way one can describe a "wider" scale of $\mathbf{B^q}$ spaces of hyperholomorphic functions. It was shown that this exponent $2 = 3 - 1 (= n - 1)$ generates a new scale of spaces, not equivalent to the hyperholomorphic Bloch space for the whole range of q. Also, it was proved in [11] for $1 \leq q < p < \infty$ that

$$\mathbf{B^q} \subset \mathbf{B^p}.$$

It remained open in [11] if these inclusions are strict and to describe a rich class of $\mathbf{B^q}$-functions explicitly. These problems will be solved in the present paper. We will restrict us for simplicity to $\mathbb{R}^3$ and quaternion-valued functions as a model case. For more details about quaternionic analysis and general Clifford analysis, we refer to [5], [6], [14], [15], [16], [20] and [22].

We will need the following lemmas in the sequel:

Lemma 1.2. [13] *Let* $|a| < 1$. *Then*

$$\int_{\partial B_1(0)} \frac{1}{|1-\bar{a}ry|^4} d\Gamma_y = \frac{4\pi}{\left(1-|a|^2\right)^2}.$$

Lemma 1.3. [18] *Let* $\alpha > 0$, $p > 0$, $n \geq 0$, $a_n \geq 0$, $I_n = \{k : 2^n \leq k < 2^{n+1}, k \in \mathbb{N}\}$, $t_n = \sum_{k\in I_n} a_k$ *and* $f(r) = \sum_{n=1}^{\infty} a_n r^n$. *Then there exists a constant* K *depending only on* p *and* α *such that*

$$\frac{1}{K}\sum_{n=0}^{\infty} 2^{-n\alpha} t_n^p \leq \int_0^1 (1-r)^{\alpha-1} f(r)^p \, dr \leq K \sum_{n=0}^{\infty} 2^{-n\alpha} t_n^p.$$

2. Power series structure of hyperholomorphic functions

The major difference to power series in the complex case consists of the absence of regularity of the basic variable $x = x_0 + x_1\mathbf{i} + x_2\mathbf{j}$ and of all its natural powers x^n, $n = 2, \ldots$. This means that we should expect other types of terms which could be designated as generalized powers. Indeed, following [17] we use a pair $\underline{y} = (y_1, y_2)$ of two regular variables (c.f. [6], [14]) given by

$$y_1 = x_1 - \mathbf{i}x_0 \quad \text{and} \quad y_2 = x_2 - \mathbf{j}x_0$$

and a multi-index $\nu = (\nu_1, \nu_2)$, $|\nu| = (\nu_1 + \nu_2)$ to define the ν-power of $\underline{y}$ by a $|\nu|$-ary product.

Definition 2.1. Let ν_1 elements of the set $a_1, \ldots, a_{|\nu|}$ be equal to y_1 and ν_2 elements be equal to y_2. Then the ν-power of $\underline{y}$ is defined by

$$\underline{y}^{\nu} := \frac{1}{|\nu|!} \sum_{(i_1,\ldots,i_{|\nu|})\in\pi(1,\ldots|\nu|)} a_{i_1} a_{i_2} \cdots a_{i_{|\nu|}} \tag{2.1}$$

where the sum runs over *all* permutations of $(1, \ldots, |\nu|)$.

Remark 2.2. Evidently for a fixed value of $|\nu| = d$ there exist exactly $(d+1)$ different ν-powers of $\underline{y}$. To distinguish among them we sometimes also use the notation $\underline{y}^{\nu} = {y_1}^{\nu_1} \times {y_2}^{\nu_2} = {y_2}^{\nu_2} \times {y_1}^{\nu_1}$ but the meaning of the last expressions is slightly different with the usual one in commutative rings and should be understood in the sense of formula (2.1). We will set parentheses if the separated powers of y_1 or y_2 have to be understood in the ordinary way. Notice that the algebraic fundamentals for such a definition of generalized powers lie in the application of the symmetric product of d elements of a non-commutative ring like discussed in [17]. In this sense the variables $y_k,\ k = 1, 2$, themselves are symmetric products of $x = x_0 + x_1\mathbf{i} + x_2\mathbf{j}$ with $(-\mathbf{i})$ resp. $(-\mathbf{j})$ in the form

$$y_1 = x_1 - \mathbf{i}x_0 = -\frac{1}{2}(\mathbf{i}x + x\mathbf{i}) \quad \text{and} \quad y_2 = x_2 - \mathbf{j}x_0 = -\frac{1}{2}(\mathbf{j}x + x\mathbf{j}).$$

With this the definition of the ν-power of $\underline{y}$, Theorem 2 in [17], implies that all polynomials in $y_k,\ k = 1, 2$, homogeneous of degree $|\nu|$ and of the form

$$f_\nu(y_1, y_2) = \underline{y}^\nu$$

with $\nu = (\nu_1, \nu_2)$ an arbitrary multi-index, are both left and right hyperholomorphic and $\mathbb{H}$-linearly independent. Therefore they can serve as basis for generalized power series. In particular, we are interested in *left* power series with center at the origin and ordered by such homogeneous polynomials. It was shown in [17], that the general form of the Taylor series of left hyperholomorphic functions around the origin is given by

$$P(x) = \sum_{n=0}^{\infty}\Big(\sum_{|\nu|=n} \underline{y}^\nu c_\nu\Big), \quad \text{with} \quad c_\nu \in \mathbb{H}. \tag{2.2}$$

In Section 3 we need the following results.

Theorem 2.3. *Let $g(x)$ be left hyperholomorphic in a neighborhood of the origin with the Taylor series given in the form* (2.2). *Then it holds*

$$\Big|\frac{1}{2}\overline{D}g(x)\Big| \;\leq\; \sum_{n=1}^{\infty} n\Big(\sum_{|\nu|=n} |c_\nu|\Big)|x|^{n-1}. \tag{2.3}$$

For the proof of this theorem we refer to [13].

In order to formulate the next theorem we introduce the abbreviated notation $H_n(x) := \sum_{|\nu|=n} \underline{y}^\nu c_\nu$ for such a homogeneous hyperholomorphic polynomial of degree n and consider hyperholomorphic functions composed by $H_n(x)$ in the following form:

$$f(x) = \sum_{n=0}^{\infty} H_n(x) b_n, \quad b_n \in \mathbb{H}.$$

Taking into account formula (2.3), we see that

$$\Big|\frac{1}{2}\overline{D}f(x)\Big| \leq \sum_{n=1}^{\infty} n\Big(\sum_{|\nu|=n} |c_\nu|\Big)|b_n||x|^{n-1}.$$

This is the motivation for another shorthand notation,

$$a_n := \Big(\sum_{|\nu|=n} |c_\nu|\Big)|b_n|, \quad (a_n \geq 0),$$

and we get finally

$$|\frac{1}{2}\overline{D}f(x)| \leq \sum_{n=1}^{\infty} na_n|x|^{n-1}.$$

3. Fourier coefficients of hyperholomorphic $\mathbf{B^q}$ functions

From the study of $\mathbf{Q_p}$ spaces of quaternion-valued functions in the three dimensional case it is known that a certain class of hyperholomorphic functions belonging to $\mathbf{Q_p}$ spaces can be characterized by their Taylor or Fourier coefficients (see [13]). This makes it natural to look for adequate properties for $\mathbf{B^q}$ spaces of quaternion-valued functions. The results presented in this paper extend some results due to Miao (see [19]) to quaternionic analysis.

Theorem 3.1. *Let* $0 < q < \infty$, $I_n = \{k : 2^n \leq k < 2^{n+1}; k \in \mathbb{N}\}$, *let* $f(x) = \sum_{n=0}^{\infty} H_n(x)b_n$, $b_n \in \mathbb{H}$ *be homogeneous hyperholomorphic polynomials as defined before, and let* a_n *defined as above. Then*

$$\sum_{n=0}^{\infty} 2^{q(1-\frac{n}{2})}\left(\sum_{k\in I_n}|a_k|\right)^q < \infty \Longrightarrow f \in \mathbf{B^q}.$$

Proof. Suppose that $\sum\limits_{n=0}^{\infty} 2^{q(1-\frac{n}{2})}\left(\sum_{k\in I_n}|a_k|\right)^q < \infty$. Then, using Lemma 1.2 and the equality

$$(1-|\varphi_a(x)|^2) = \frac{(1-|a|^2)(1-|x|^2)}{|1-\bar{a}x|^2}$$

we obtain that

$$\begin{aligned}
&\int_{B_1(0)} |\frac{1}{2}\overline{D}f(x)|^q (1-|x|^2)^{\frac{3}{2}q-3}(1-|\varphi_a(x)|^2)^2\, dB_x \\
= &\int_{B_1(0)} \left|\frac{1}{2}\overline{D}\left(\sum_{n=0}^{\infty} H_n(x)\, b_n\right)\right|^q \frac{(1-|x|^2)^{\frac{3}{2}q-1}(1-|a|^2)^2}{|1-\bar{a}x|^4} dB_x \\
\leq &\int_{B_1(0)} \left(\sum_{n=1}^{\infty} n\, a_n|x|^{n-1}\right)^q \frac{(1-|x|^2)^{\frac{3}{2}q-1}(1-|a|^2)^2}{|1-\bar{a}x|^4}\, dB_x \\
= &\int_0^1 \left(\sum_{n=1}^{\infty} n\, a_n\, r^{n-1}\right)^q (1-r^2)^{\frac{3}{2}q-1}(1-|a|^2)^2 \int_{\partial B_1(0)} \frac{1}{|1-\bar{a}ry|^4}\, d\Gamma_y r^2\, dr \\
\leq &\ 2^{\frac{3}{2}q-1}\int_0^1 \left(\sum_{n=1}^{\infty} n\, a_n\, r^{n-1}\right)^q (1-r)^{\frac{3}{2}q-1}(1-|a|^2)^2 . \frac{4\pi}{(1-|a|^2)^2}\, dr \\
\leq &\ \eta \int_0^1 \left(\sum_{n=1}^{\infty} n\, a_n\, r^{n-1}\right)^q (1-r)^{\frac{3}{2}q-1}\, dr,
\end{aligned}$$

where $\eta = \pi 2^{\frac{3}{2}q+1}$. Using Lemma 1.3 in the last inequality, we get that

$$\int_{B_1(0)} \left|\frac{1}{2}\overline{D}f(x)\right|^q \left(1-|x|^2\right)^{\frac{3}{2}q-3}\left(1-|\varphi_a(x)|^2\right)^2 dB_x \leq \eta K \sum_{n=0}^{\infty} 2^{-\frac{3}{2}nq}\Big(\sum_{k\in I_n} k\, a_k\Big)^q.$$

Since

$$t_n = \sum_{k\in I_n} k\, a_k \; < 2^{n+1} \sum_{k\in I_n} a_k,$$

we have

$$\int_{B_1(0)} \left|\frac{1}{2}\overline{D}f(x)\right|^q \left(1-|x|^2\right)^{\frac{3}{2}q-3}\left(1-|\varphi_a(x)|^2\right)^2 dB_x \leq \eta K \sum_{n=0}^{\infty} 2^{q(1-\frac{n}{2})}\Big(\sum_{k\in I_n} a_k\Big)^q.$$

Therefore,

$$\|f\|_{B^q} \leq \lambda \sum_{n=0}^{\infty} 2^{q(1-\frac{n}{2})} \left(\sum_{k\in I_n} |a_k|\right)^q < \infty,$$

where λ is a constant. Hereafter, λ stands for absolute constants, which may be different constants from one occurrence to the next.

The last inequality implies that $f \in \mathbf{B^q}$ and the proof of our theorem is completed. □

In the following theorem, our aim is to consider the converse direction of Theorem 3.1. We will restrict ourselves to hyperholomorphic homogeneous polynomials of the form

$$H_{n,\alpha}(x) = \left(y_1\alpha_1 + y_2\alpha_2\right)^n = \sum_{k=0}^{n} \binom{n}{k} {y_1}^{n-k} \times {y_2}^k {\alpha_1}^{n-k} {\alpha_2}^k,$$

where $\alpha_i \in \mathbb{R}$, $i = 1, 2$. The hypercomplex derivative is given by

$$\left(-\frac{1}{2}\overline{D}\right) H_{n,\alpha}(x) = n H_{n-1,\alpha}(x)\left(\alpha_1 i + \alpha_2 j\right). \tag{3.1}$$

Proposition 3.2. *Let $\alpha = (\alpha_1, \alpha_2)$, $\alpha_i \in \mathbb{R}$, $i = 1, 2$ be the vector of real coefficients defining the hyperholomorphic homogeneous polynomial $H_{n,\alpha}(x) = \left(y_1\alpha_1 + y_2\alpha_2\right)^n$. Suppose that $|\alpha|^2 = {\alpha_1}^2 + {\alpha_2}^2 \neq 0$. Then,*

$$\|H_{n,\alpha}\|^q_{L_q(\partial B_1)} = 2\pi\sqrt{\pi}\, |\alpha|^{nq}\, \frac{\Gamma(\frac{n}{2}q+1)}{\Gamma(\frac{n}{2}q+\frac{3}{2})}, \qquad \textit{where } 0 < q < \infty.$$

Proof. Since

$$
\begin{aligned}
&\|H_{n,\alpha}\|^q_{L_q(\partial B_1)} \qquad (3.2)\\
&= \int_0^{2\pi}\int_0^{\pi} \left[\left(\sin^2\phi_1(\alpha_1\cos\phi_2+\alpha_2\sin\phi_1)^2+({\alpha_1}^2+{\alpha_2}^2)\cos^2\phi_1\right)^n\right]^{\frac{q}{2}} \sin\phi_1 d\phi_1 d\phi_2\\
&= \int_0^{2\pi}\int_0^{\pi} \left[\left(|\alpha|^2+|\alpha|^2\sin^2\phi_1\left[\sin^2(\phi_2+\omega)-1\right)^n\right]^{\frac{q}{2}} \sin\phi_1 d\phi_1 d\phi_2\\
&= \int_0^{2\pi}\int_0^{\pi} \left[\left(|\alpha|^2-|\alpha|^2\sin^2\phi_1\cos^2(\phi_2+\omega)\right]\right)^n\right]^{\frac{q}{2}} \sin\phi_1 d\phi_1 d\phi_2\\
&= |\alpha|^{nq}\int_0^{2\pi}\int_0^{\pi} \left[\left(1-\sin^2\phi_1\cos^2(\phi_2+\omega)\right]\right)^n\right]^{\frac{q}{2}} \sin\phi_1 d\phi_1 d\phi_2,
\end{aligned}
$$

where ω is defined by

$$\sin\omega := \frac{\alpha_1}{\sqrt{{\alpha_1}^2+{\alpha_2}^2}} \quad \text{and} \quad \cos\omega := \frac{\alpha_2}{\sqrt{{\alpha_1}^2+{\alpha_2}^2}},$$

equation (3.2) will reduce to

$$
\begin{aligned}
&\|H_{n,\alpha}\|^q_{L_q(\partial B_1)}\\
&= |\alpha|^{nq}\sum_{k=0}^{\infty}(-1)^k\binom{\frac{nq}{2}}{k}\left(\int_0^{2\pi}[\cos(\phi_2+\omega)]^{2k}d\phi_2\right)\left(\int_0^{\pi}[\sin(\phi_1)]^{2k+1}d\phi_1\right).
\end{aligned}
$$

Using integration by parts, it follows that

$$I_k := \int_0^{2\pi}[\cos(\phi_2+\omega)]^{2k}d\phi_2 = \pi\frac{(2k-1)!!}{2^{k-1}(k)!}.$$

Also,

$$I_k^* := \int_0^{\pi}[\sin(\phi_1)]^{2k+1}d\phi_1 = \frac{2^{k+1}(k)!}{(2k+1)!!}.$$

Therefore, we obtain that

$$
\begin{aligned}
\|H_{n,\alpha}\|^q_{L_q(\partial B_1)} &= \pi|\alpha|^{nq}\sum_{k=0}^{\infty}(-1)^k\binom{\frac{nq}{2}}{k}\left(\frac{(2k-1)!!}{2^{k-1}(k)!}\,\frac{2^{k+1}(k)!}{(2k+1)!!}\right)\\
&= 4\pi|\alpha|^{nq}\sum_{k=0}^{\infty}(-1)^k\binom{\frac{nq}{2}}{k}\frac{1}{2k+1}.
\end{aligned}
$$

Now, we calculate the sum of the series

$$\sum_{k=0}^{\infty}(-1)^k\binom{\frac{nq}{2}}{k}\frac{1}{2k+1}.$$

Let

$$F(t_1)=\sum_{k=0}^{\infty}(-1)^k\binom{\frac{nq}{2}}{k}\frac{1}{2k+1}t_1^{2k+1}.$$

Then,

$$\frac{dF(t_1)}{dt_1}=F'(t_1)=\sum_{k=0}^{\infty}(-1)^k\binom{\frac{nq}{2}}{k}t_1^{2k}=(1-t_1^2)^{\frac{nq}{2}}$$

and

$$\begin{aligned} F(1) &= \sum_{k=0}^{\infty}(-1)^k\binom{\frac{nq}{2}}{k}\frac{1}{2k+1}=\int_0^1(1-t_1^2)^{\frac{nq}{2}}\,dt_1=\int_0^1 t_2{}^{-\frac{1}{2}}(1-t_2)^{\frac{n}{2}q}\,dt_2 \\ &= B(\frac{1}{2},\frac{n}{2}q+1)=\frac{1}{2}\sqrt{\pi}\frac{\Gamma(\frac{n}{2}q+1)}{\Gamma(\frac{n}{2}q+\frac{3}{2})}. \end{aligned}$$

We obtain

$$\|H_{n,\alpha}\|^q_{L_q(\partial B_1)}=2\pi\sqrt{\pi}\,|\alpha|^{nq}\,\frac{\Gamma(\frac{n}{2}q+1)}{\Gamma(\frac{n}{2}q+\frac{3}{2})},$$

and our proposition is proved. □

Now, using formula (3.1), we obtain

$$\frac{\|-\frac{1}{2}\overline{D}H_{n,\alpha}\|^q{}_{L_q(\partial B_1)}}{\|H_{n,\alpha}\|^q{}_{L_q(\partial B_1)}}=n^q\,\frac{B(\frac{1}{2},\frac{n-1}{2}q+1)}{B(\frac{1}{2},\frac{n}{2}q+1)}\geq\lambda\,n^q$$

where $\frac{B(\frac{1}{2},\frac{n-1}{2}q+1)}{B(\frac{1}{2},\frac{n}{2}q+1)}>0,\ \forall\, n$, and

$$\lim_{n\to\infty}\frac{B(\frac{1}{2},\frac{n-1}{2}q+1)}{B(\frac{1}{2},\frac{n}{2}q+1)}=1.$$

Corollary 3.3. *We have*

$$\frac{\|-\frac{1}{2}\overline{D}H_{n,\alpha}\|_{L_q(\partial B_1)}}{\|H_{n,\alpha}\|_{L_q(\partial B_1)}}\geq\lambda n,\quad\forall q,\ 0<q<\infty.$$

Corollary 3.4. *Suppose that $q\geq 2$. Then,*

$$\frac{\|-\frac{1}{2}\overline{D}H_{n,\alpha}\|^2_{L_2(\partial B_1)}}{\|H_{n,\alpha}\|^2_{L_q(\partial B_1)}}\geq\lambda n^{\frac{2+3q}{2q}}.$$

Proof. We consider the following equalities:

$$\frac{\|-\frac{1}{2}\overline{D}H_{n,\alpha}\|^2_{L_2(\partial B_1)}}{\|H_{n,\alpha}\|^2_{L_q(\partial B_1)}} = \frac{\|-\frac{1}{2}\overline{D}H_{n,\alpha}\|^2_{L_2(\partial B_1)}}{\|-\frac{1}{2}\overline{D}H_{n,\alpha}\|^2_{L_q(\partial B_1)}} \frac{\|-\frac{1}{2}\overline{D}H_{n,\alpha}\|^2_{L_q(\partial B_1)}}{\|H_{n,\alpha}\|^2_{L_q(\partial B_1)}}.$$

Then, from (3.1) and Proposition 3.2, we obtain

$$\frac{\|-\frac{1}{2}\overline{D}H_{n,\alpha}\|^2_{L_2(\partial B_1)}}{\|-\frac{1}{2}\overline{D}H_{n,\alpha}\|^2_{L_q(\partial B_1)}} = \lambda \frac{\Gamma(n)}{\Gamma(n+\frac{1}{2})}\left(\frac{\Gamma\left(\frac{(n-1)}{2}q+\frac{3}{2}\right)}{\Gamma\left(\frac{(n-1)}{2}q+1\right)}\right)^{\frac{2}{q}}.$$

Using $\frac{\Gamma(n)}{\Gamma(n+\frac{1}{2})}n^{\frac{1}{2}} \to 1$ as $n \to \infty$, we conclude that

$$\frac{\Gamma(n)}{\Gamma(n+\frac{1}{2})}n^{\frac{1}{2}}\left(\frac{\Gamma\left(\frac{(n-1)}{2}q+\frac{3}{2}\right)}{\Gamma\left(\frac{(n-1)}{2}q+1\right)}n^{-\frac{1}{2}}\right)^{\frac{2}{q}} \to 1$$

and, applying Corollary 3.3 we proved that

$$\frac{\|-\frac{1}{2}\overline{D}H_{n,\alpha}\|^2_{L_2(\partial B_1)}}{\|H_{n,\alpha}\|^2_{L_q(\partial B_1)}} \geq \lambda n^{\frac{2+3q}{2q}},$$

where λ is a constant not depending on n. □

Theorem 3.5. *Let* $2 \leq q < \infty$ *and let*

$$f(x) = \left(\sum_{n=0}^{\infty} \frac{H_{n,\alpha}}{\|H_{n,\alpha}\|_{L_q(\partial B_1)}} a_n\right) \in \mathbf{B^q}.$$

Then,

$$\sum_{n=0}^{\infty} 2^{q\left(1-\frac{n}{2}\left(\frac{3q-2}{2q}+1\right)\right)}\left(\sum_{k\in I_n}|a_k|\right)^q < \infty.$$

Proof. From the definition of $\mathbf{B^q}$, we have

$$\begin{aligned}\|f\|^q_{B^q} &\geq \int_{B_1(0)} \left|-\frac{1}{2}\overline{D}f(x)\right|^q (1-|x|^2)^{\frac{3}{2}q-1} dB_x \\ &= \int_{B_1(0)} \left|\sum_{n=0}^{\infty}\left[\frac{(-\frac{1}{2}\overline{D}H_{n,\alpha})}{\|H_{n,\alpha}\|_{L_q(\partial B_1)}}\right]a_n\right|^q (1-|x|^2)^{\frac{3}{2}q-1} dB_x = J. \quad (3.3)\end{aligned}$$

$\left[\frac{(-\frac{1}{2}\overline{D}H_{n,\alpha})}{\|H_{n,\alpha}\|_{L_q(\partial B_1)}}\right]$ is a homogeneous hyperholomorphic polynomial of degree $n-1$ and it can be written in the form

$$\left[\frac{(-\frac{1}{2}\overline{D}H_{n,\alpha})}{\|H_{n,\alpha}\|_{L_q(\partial B_1)}}\right] = r^{(n-1)}\,\Phi_n(\phi_1,\phi_2), \quad (3.4)$$

where

$$\Phi_n(\phi_1,\phi_2) := \left(\left[\frac{(-\frac{1}{2}\overline{D}H_{n,\alpha})}{\|H_{n,\alpha}\|_{L_q(\partial B_1)}}\right]\right)_{\partial B_1}. \tag{3.5}$$

Now, using the quaternion-valued inner product

$$\langle f,g\rangle_{\partial B_1(0)} = \int_{\partial B_1(0)} \overline{f(x)}g(x)\,d\Gamma_x,$$

the orthogonality of the spherical hyperholomorphic polynomials $\Phi_n(\phi_1,\phi_2)$ (see [6]) in $L_2(\partial B_1(0))$, these hyperholomorphic polynomials are called spherical monogenics therein), and substitute from (3.4) and (3.5) to (3.3), we obtain

$$\begin{aligned} J &= \int_0^1 \int_{\partial B_1(0)} \left(\left|\sum_{n=0}^{\infty} r^{(n-1)}\,\Phi_n(\phi_1,\phi_2)a_n\right|^2\right)^{\frac{q}{2}} r^2(1-r^2)^{\frac{3}{2}q-1} d\Gamma_x dr \\ &= \int_0^1 \int_{\partial B_1(0)} \left(\sum_{n=0}^{\infty}\sum_{j=0}^{\infty} \overline{a}_n r^{2(n-1)}\,\overline{\Phi}_n(\phi_1,\phi_2)\Phi_j(\phi_1,\phi_2)a_j\right)^{\frac{q}{2}} r^2(1-r^2)^{\frac{3}{2}q-1} d\Gamma_x dr \end{aligned}$$

Using Hölder's inequality, for $1 \le q < \infty$, we have

$$\int_{\partial B_1(0)} |f(x)|^q\, d\Gamma_x \ge (4\pi)^{1-q} \left|\int_{\partial B_1(0)} f(x)\, d\Gamma_x\right|^q.$$

From the last inequality, we obtain for $2 \le q < \infty$ that

$$\begin{aligned} J &\ge (4\pi)^{1-\frac{q}{2}} \int_0^1 \left(\sum_{n=0}^{\infty} |a_n|^2 r^{2(n-1)}\,\|\Phi_n(\phi_1,\phi_2)\|^2_{L_2(\partial B_1)}\right)^{\frac{q}{2}} r^2(1-r^2)^{\frac{3}{2}q-1}\,dr \\ &\ge (4\pi)^{1-\frac{q}{2}} \int_0^1 \left(\sum_{n=0}^{\infty} |a_n|^2 r^{2(n-1)}\,\|\Phi_n(\phi_1,\phi_2)\|^2_{L_2(\partial B_1)}\right)^{\frac{q}{2}} r^3(1-r^2)^{\frac{3}{2}q-1}\,dr. \end{aligned} \tag{3.6}$$

Using Corollary 3.4 we obtain

$$\|\Phi_n(\phi_1,\phi_2)\|^2_{L_2(\partial B_1)} = \frac{\|-\frac{1}{2}\overline{D}H_{n,\alpha}\|^2_{L_2(\partial B_1)}}{\|H_{n,\alpha}\|^2_{L_q(\partial B_1)}} \ge \lambda n^{\frac{2+3q}{2q}}.$$

Then (3.6) will reduce to

$$\begin{aligned}
J &\geq (4\pi)^{1-\frac{q}{2}}\lambda_1 \int_0^1 \left(\sum_{n=0}^{\infty} n^{\frac{2+3q}{2q}} |a_n|^2 r^{2(n-1)}\right)^{\frac{q}{2}} r^3(1-r^2)^{\frac{3}{2}q-1}\, dr \\
&= \lambda_2 \int_0^1 \left(\sum_{n=0}^{\infty} n^{\frac{2+3q}{2q}} |a_n|^2 r^{2(n-1)}\right)^{\frac{q}{2}} r^3(1-r^2)^{\frac{3}{2}q-1}\, dr \\
&= \frac{\lambda_2}{2} \int_0^1 \left(\sum_{n=0}^{\infty} n^{\frac{2+3q}{2q}} |a_n|^2 r_1^{(n-1)}\right)^{\frac{q}{2}} r_1(1-r_1)^{\frac{3}{2}q-1}\, dr_1 \\
&\geq \lambda_3 \int_0^1 \left(\sum_{n=0}^{\infty} n^{\frac{2+3q}{2q}} |a_n|^2 r_1^{n}\right)^{\frac{q}{2}} (1-r_1)^{\frac{3}{2}q-1}\, dr_1 \qquad (3.7)
\end{aligned}$$

where λ_1, λ_2, and λ_3 are constants not depending on n. Applying Lemma 1.3 in (3.7), we obtain that

$$\|f\|_{B^q}^q \geq J \geq \frac{\lambda_3}{K} \sum_{n=0}^{\infty} 2^{-\frac{3}{2}qn} \Big(\sum_{k\in I_n} k^{\frac{2+3q}{2q}} |a_k|^2\Big)^{\frac{q}{2}}.$$

Since

$$\sum_{k\in I_n} k^{\frac{2+3q}{2q}} |a_k|^2 > (2^n)^{\frac{2+3q}{2q}} \sum_{k\in I_n} |a_k|^2.$$

Then,

$$\|f\|_{B^q}^q \geq J \geq C \sum_{n=0}^{\infty} 2^{-\frac{nq}{2}(\frac{3q-2}{2q})} \left(\sum_{k\in I_n} |a_k|^2\right)^{\frac{q}{2}},$$

where C be a constant not depending on n. Using Hölder's inequality, we obtain

$$\sum_{k\in I_n} |a_k|^2 \geq \frac{1}{2^n} \left(\sum_{k\in I_n} |a_k|\right)^2.$$

Therefore,

$$\begin{aligned}
\|f\|_{B^q}^q &\geq J \geq C_1 \sum_{n=0}^{\infty} 2^{q\left(1-\frac{n}{2}(\frac{3q-2}{2q})\right)} \frac{1}{2^n} \left(\sum_{k\in I_n} |a_k|\right)^q, \\
&\geq \sum_{n=0}^{\infty} 2^{q\left(1-\frac{n}{2}(\frac{3q-2}{2q})\right)} \frac{1}{2^{n\frac{q}{2}}} \left(\sum_{k\in I_n} |a_k|\right)^q \\
&= \sum_{n=0}^{\infty} 2^{q\left(1-\frac{n}{2}(\frac{3q-2}{2q}+1)\right)} \left(\sum_{k\in I_n} |a_k|\right)^q,
\end{aligned}$$

where C_1 be a constant not depending on n. Hence we deduce that,

$$\sum_{n=0}^{\infty} 2^{q\left(1-\frac{n}{2}\left(\frac{3q-2}{2q}+1\right)\right)} \left(\sum_{k\in I_n} |a_k|\right)^q < \infty. \qquad \square$$

Corollary 3.6. *Let f be a hyperholomorphic function in $B_1(0)$. Then for $2 \leq q < \infty$ and $1 < |\alpha| < \infty$, we have that*

$$f(x) = \left(\sum_{n=0}^{\infty} \frac{H_{n,\alpha}}{\|H_{n,\alpha}\|_{L_q(\partial B_1)}} a_n\right) \in \mathbf{B^q} \Longleftrightarrow \sum_{n=0}^{\infty} 2^{q(1-\frac{n}{2}(\frac{5}{2}-\frac{1}{q}))} \left(\sum_{k\in I_n} |a_k|\right)^q < \infty.$$

Proof. "$\Longrightarrow$": This direction can be proved directly from Theorem 3.5.
"$\Longleftarrow$": The proof of this direction follows as in the proof of Theorem 3.1 by employing the function $\frac{H_{n,\alpha}}{\|H_{n,\alpha}\|_{L_q(\partial B_1)}}$ (where $1 < |\alpha| < \infty$) instead of $H_n(x)$. So we obtain

$$\begin{aligned}
&\int_{B_1(0)} \left|\frac{1}{2}\overline{D}f(x)\right|^q (1-|x|^2)^{\frac{3}{2}q-3}(1-|\varphi_a(x)|^2)^2 dB_x \\
\leq\ & k_1 \int_0^1 \left(\sum_{n=1}^{\infty} \frac{n\, a_n\, r^{n-1}}{n^{\frac{1}{2}}|\alpha|^n}\right)^q (1-r^2)^{\frac{3}{2}q-1}(1-|a|^2)^2 \int_{\partial B_1(0)} \frac{1}{|1-\bar{a}ry|^4} d\Gamma_y\, r^2 dr \\
\leq\ & 2^{\frac{3}{2}q-1} k_1 \int_0^1 \left(\sum_{n=1}^{\infty} \frac{n}{|\alpha|^n} a_n\, r^{n-1}\right)^q (1-r)^{\frac{3}{2}q-1}(1-|a|^2)^2 . \frac{4\pi}{(1-|a|^2)^2} dr \\
=\ & k_2 \int_0^1 \left(\sum_{n=1}^{\infty} \frac{n}{|\alpha|^n} a_n\, r^{n-1}\right)^q (1-r)^{\frac{3}{2}q-1} dr \\
\leq\ & k_3(|\alpha|) \int_0^1 \left(\sum_{n=1}^{\infty} a_n r^{n-1}\right)^q (1-r)^{\frac{3}{2}q-1} dr
\end{aligned}$$

where $k_2 = \pi 2^{\frac{3}{2}q+1} k_1$, k_1 is a constant not depending on n, and $k_3(|\alpha|)$ is a constant not depending on n but on k_2 and $|\alpha|$. Using Lemma 1.3 in the last inequality, we get

$$\int_{B_1(0)} \left|\frac{1}{2}\overline{D}f(x)\right|^q (1-|x|^2)^{\frac{3}{2}q-3}(1-|\varphi_a(x)|^2)^2 dB_x \leq k_3(|\alpha|)\, K \sum_{n=0}^{\infty} 2^{-\frac{3}{2}nq} \left(\sum_{k\in I_n} a_k\right)^q$$

which implies that

$$\begin{aligned}
&\int_{B_1(0)} \left|\frac{1}{2}\overline{D}f(x)\right|^q (1-|x|^2)^{\frac{3}{2}q-3}(1-|\varphi_a(x)|^2)^2 dB_x \\
&\leq k_3(|\alpha|)\, K \sum_{n=0}^{\infty} 2^{q(1-\frac{n}{2}(\frac{5}{2}-\frac{1}{q}))} \left(\sum_{k\in I_n} a_k\right)^q
\end{aligned}$$

Therefore,

$$\|f\|_{B^q} \leq \lambda \sum_{n=0}^{\infty} 2^{q(1-\frac{n}{2}(\frac{5}{2}-\frac{1}{q}))} \left(\sum_{k\in I_n} |a_k|\right)^q < \infty. \qquad \square$$

Corollary 3.7. *Let f be a hyperholomorphic function in $B_1(0)$. Then for $q = 2$, we have that*

$$f(x) = \sum_{n=0}^{\infty} \frac{H_{2^n,\alpha}}{\|H_{2^n,\alpha}\|_{L_2(\partial B_1)}} a_{2^n} \in \mathbf{B^2} = \mathbf{Q_2} \Longleftrightarrow \sum_{n=0}^{\infty} 2^{2(1-\frac{n}{2})} |a_{2^n}|^2 < \infty,$$

where

$$\mathbf{Q_2} = \left\{ f \in \ker D : \sup_{a\in B_1(0)} \int_{B_1(0)} \left|\overline{D}f(x)\right|^2 \left(1-|\varphi_a(x)|^2\right)^2 dB_x < \infty \right\} \quad (see\ [12])$$

Proof. The proof of this corollary follows directly from Theorem 3.1 by using the same steps as in Theorem 3.5 with keeping in mind that we have only $|a_{2^n}|$ but not $\sum_{k\in I_n} |a_k|$. $\square$

4. Strict inclusions of hyperholomorphic $\mathbf{B}^q$ functions

In this section we study the equivalence between hyperholomorphic $\mathbf{B}^q$ functions and their coefficients by series expansions in homogeneous hyperholomorphic polynomials. Finally, we prove that the inclusions $B^{q_1} \subset B^q$, $2 \leq q_1 < q < \infty$ are strict.

Theorem 4.1. *Let $2 \leq q < \infty$ and let*

$$f(x) = \left(\sum_{n=0}^{\infty} \frac{n\, H_{n,\alpha}}{\|H_{n,\alpha}\|_{L_q(\partial B_1)}} a_n\right) \in \mathbf{B^q}.$$

Then,

$$\sum_{n=0}^{\infty} 2^{q(1-\frac{n}{2})} \left(\sum_{k\in I_n} |a_k|\right)^q < \infty.$$

Proof. This theorem can be proved by using the following inequality

$$\frac{\| -n(\frac{1}{2}\overline{D}H_{n,\alpha})\|^2_{L_2(\partial B_1)}}{\|H_{n,\alpha}\|^2_{L_q(\partial B_1)}} \geq \lambda n^3$$

and the same steps used in the proof of Theorem 3.5. $\square$

The rigorous statement of our idea is given by the next theorem.

Theorem 4.2. *Let f be a hyperholomorphic function in $B_1(0)$. Then for $2 \leq q < \infty$, we have that*

$$f(x) = \sum_{n=0}^{\infty} \frac{n\, H_{n,\alpha}}{\|H_{n,\alpha}\|_{L_q(\partial B_1)}} a_n \in \mathbf{B^q} \Longleftrightarrow \sum_{n=0}^{\infty} 2^{q(1-\frac{n}{2})} \left(\sum_{k\in I_n} |a_k|\right)^q < \infty.$$

The proof follows from Theorems 3.1 and 4.1.

Remark 4.3. It should be remarked here that our functions

$$f(x) = \sum_{n=0}^{\infty} \frac{n\, H_{n,\alpha}}{\|H_{n,\alpha}\|_{L_q(\partial B_1)}}\, a_n$$

are more general than the functions introduced in [13]. In [13] only powers $n = 2^k$ were allowed in the series expansion. Moreover, $\mathbf{B}^q$ functions can be characterized (where $2 \leq q < \infty$) by their coefficients of a series expansion with non-normalized functions as it was shown in Theorem 4.2 for general α $(0 < |\alpha| < \infty)$.

Corollary 4.4. *The inclusions $\mathbf{B^{q_1}} \subset \mathbf{B^q}$ are strict for all $2 \leq q_1 < q < \infty$.*

Proof. Let

$$f(x) = \sum_{n=0}^{\infty} \frac{n\, H_{n,\alpha}}{\|H_{n,\alpha}\|_{L_q(\partial B_1)}} a_n, \quad H_{n,\alpha}(x) = (y_1\alpha_1 + y_2\alpha_2)^n, \quad |\alpha|^2 = \alpha_1^2 + \alpha_2^2 \neq 0,$$

and $|a_n| = \frac{1}{2^{\frac{q_1}{q}(1-\frac{n}{2})}}$. Then,

$$\sum_{n=0}^{\infty} 2^{q(1-\frac{n}{2})} |a_n|^q = \sum_{n=0}^{\infty} \frac{1}{2^{(q-q_1)(\frac{n}{2}-1)}} < \infty, \qquad \forall q > q_1$$

and

$$\sum_{n=0}^{\infty} 2^{q(1-\frac{n}{2})} |a_n|^q = \sum_{n=0}^{\infty} 1 = \infty.$$

By Theorem 3.1 and Theorem 3.5 we have that $f \in \mathbf{B^q}$ but $f \notin \mathbf{B^{q_1}}$, so the inclusions are strict. □

Remark 4.5. We would like to emphasize that the motivation for this work lies in the definition of weighted $\mathbf{B^q}$ spaces as given in [11] not that used in [10]. The strict inclusions ensure that the $\mathbf{B^q}$-spaces form a scale consisting in spaces, all different from the Bloch space.

References

[1] J. Anderson, J. Clunie and Ch. Pommerenke: On Bloch functions and normal functions, *J. Reine Angew. Math.* **270** (1974), 12–37.

[2] R. Aulaskari, D. Girela and H. Wulan: Taylor coefficients and mean growth of the derivative of $\mathbf{Q_p}$ functions, *J. Math. Anal. Appl.* **258**, No.2 (2001), 415–428.

[3] R. Aulaskari and P. Lappan: *Criteria for an analytic function to be Bloch and a harmonic or meromorphic function to be normal*, Complex Analysis and its Applications (eds Y. Chung-Chun et al.), Pitman Research Notes in Mathematics 305, Longman (1994), 136–146.

[4] R. Aulaskari, J. Xiao and R. Zhao: On subspaces and subsets of BMOA and UBC, *Analysis* **15** (1995), 101–121.

[5] F. Brackx, J.S.R. Chisholm and V. Soucek (eds.): *Clifford analysis and its applications.* Proceedings of the NATO advanced research workshop, Prague, Czech Republic, October 30–November 3, 2000. NATO Science Series II: Mathematics, Physics and Chemistry. 25. Dordrecht: Kluwer Academic Publishers. (2001).

[6] F. Brackx, R. Delanghe and F. Sommen: *Clifford Analysis*, Pitman Research Notes in Math., Boston, London, Melbourne, 1982.

[7] J. Cnops and R. Delanghe: Möbius invariant spaces in the unit ball, *Applicable Analysis* (1-2) **73** (2000), 45–64.

[8] M. Essen, S. Janson, L. Peng and J. Xiao: $\mathbf{Q}$ spaces of several real variables, *Indiana Univ. Math. J.* **49**, No.2 (2000), 575–615.

[9] M. Essen and J. Xiao: $\mathbf{Q_p}$ spaces-a survey, Complex function spaces, Proceedings of the Summer School, Mekrijärvi, Finland, August 30 – September 3, 1999, *Rep. Ser. Univ. Joensuu.* **4** (2001), 11–60.

[10] K. Gürlebeck and A. El-Sayed Ahmed: *Integral norms for hyperholomorphic Bloch-functions in the unit ball of* $\mathbf{R}^3$, to appear in: *Progress in Analysis*, Proceedings of the 3rd International ISAAC Congress, Volume I, Editors H. Begehr, R. Gilbert, M.W. Wong, World Scientific, New Jersey, London, Singapore, Hong Kong, 2003

[11] K. Gürlebeck and A. El-Sayed Ahmed: *On* $\mathbf{B}^q$ *spaces of hyperholomorphic functions and the Bloch space in* $\mathbf{R}^3$, to appear in the proceedings of the 9th ICFI DCAA held in Hanoi, 2001 (Kluwer Academic Publishers).

[12] K. Gürlebeck, U. Kähler, M. Shapiro and L.M. Tovar: On $\mathbf{Q_p}$-spaces of quaternion-valued functions, *Complex Variables Theory Appl.* **39** (1999), 115–135.

[13] K. Gürlebeck and H.R. Malonek: On strict inclusions of weighted Dirichlet spaces of monogenic functions, *Bull. Austral. Math. Soc.* **64** (2001), 33–50.

[14] K. Gürlebeck and W. Sprößig: *Quaternionic Analysis and elliptic boundary value problems*, Int. Ser. Num. Math. 89, Birkhäuser Verlag Basel, 1990.

[15] Gürlebeck, K. and W. Sprößig: Quaternionic and Clifford calculus for Engineers and Physicists, John Wiley &. Sons, Chichester, 1997.

[16] V.V. Kravchenko and M.V. Shapiro: *Integral representations for spatial models of mathematical physics*, Pitman Research Notes in Mathematics Series 351 Harlow: Longman, 1996.

[17] M. Malonek: Power series representation for monogenic functions in $\mathbf{R}^{m+1}$ based on a permutational product, *Complex Variables Theory Appl.* **15** (1990), 181–191.

[18] M. Mateljevic, M. Pavlovic: L_p-behavior of power series with positive coefficients and Hardy spaces, *Proc. Amer. Math. Soc.* (2) **87** (1983), 309–316.

[19] J. Miao: A property of analytic functions with Hadamard gaps, *Bull. Austral. Math. Soc.* **45** (1992), 105–112.

[20] J. Ryan and W. Sprößig (eds.): *Clifford algebras and their applications in mathematical physics.* Papers of the 5th international conference, Ixtapa-Zihuatanejo, Mexico, June 27–July 4, 1999. Volume 2: Clifford analysis. Progress in Physics 19. Boston, MA: Birkhäuser. (2000).

[21] K. Stroethoff: Besov-type characterisations for the Bloch space, *Bull. Austral. Math. Soc.* **39** (1989), 405–420.

[22] A. Sudbery: Quaternionic Analysis, *Math. Proc. Cambr. Phil. Soc.* **85** (1979), 199–225.

[23] J. Xiao: *Holomorphic* **Q** *classes*, Lecture Notes in Mathematics, Berlin, Springer, 2001.

[24] R. Zhao: A general family of function spaces, *Annales Academiae Scientiarum Fennicae. Series A I. Mathematica. Dissertationes.* **105**. Helsinki: Suomalainen Tiedeakatemia, (1996), 1–56.

K. Gürlebeck
Bauhaus-Universität Weimar
Institut für Mathematik und Physik
Coudray-Str. 13 B
D-99423 Weimar
Germany
e-mail: guerlebe@fossi.uni-weimar.de

A. El-Sayed Ahmed
South Valley University
Faculty of Science
Department of Mathematics
Sohag
Egypt
e-mail: ahsayed80@hotmail.com

Trends in Mathematics: *Advances in Analysis and Geometry*, 131–147

Pointwise Convergence of Fourier Series on the Unit Sphere of $\mathbf{R}^4$ with the Quaternionic Setting

Shuang Liu and Tao Qian

Abstract. We offer a new approach to convergence of Fourier series on the unit sphere of the four-dimensional Euclidean space. The approach is via the quaternionic analysis setting with a crucial application of Fueter's theorem. Analogs to the Riemann-Lebesgue theorem, localization principle and a Dini's type pointwise convergence theorem are proved.

Mathematics Subject Classification (2000). Primary: 42B05, 30G35; Secondary: 42C10.

Keywords. Quaternion space, Fourier series, unit sphere, Dini's type pointwise convergence theorems, Fueter's theorem.

1. Introduction

There have been many investigations on convergence and summability of Fourier-Laplace series on unit spheres of higher dimensional Euclidean spaces (see [WL] and its references). Except for the very lowest dimensional cases, pointwise convergence could be said to be relatively ignored. The case $n = 2$ is simply Fourier series on the circle (see, for instance, [Zy]). Dirichlet ([Di]) gave the first detailed study for the case $n = 3$ on the so called Laplace series. Koschmieder studied the case $n = 4$. In 1976 E.L. Roetman considered the general cases, and, under certain conditions, reduced the convergence problem for $n = 2k + 2$ to $n = 2$; and for $n = 2k + 3$ to $n = 3$ ([Ro]). C. Meaney studied the problem, as well as some related ones, including the cases where functions are in L^p spaces (see [Me] and its references).

In this note we offer a different approach using Clifford analysis. The expectation is to use the complex structure, viz. the Cauchy type methods. On one hand, in the lowest dimensional case $n = 2$ it involves complex analysis and operator theory and has obtained delicate results; while on the other hand in higher dimensional cases results obtained are less and weaker. With Clifford analysis one is able to develop an analogous function theory in relation to operator theory on higher dimensional spheres. Indeed, one can define a Hilbert transform and an algebra of singular integral operators on spheres ([Q1], [Q3]). We restrict ourselves to the quaternionic space which is equivalent to the four-dimensional Euclidean space.

The study in [Ro], and those of the others mentioned above, heavily depends on properties of spherical harmonics, and especially on properties of Legendre polynomials or other types of special functions. Our approach, however, depends on Fueter's theorem on inducing quaternionic regular functions from holomorphic functions of one complex variable. In particular, the modified Dirichlet kernels in the quaternionic case are induced from the Dirichlet kernels in one complex variable using the theorem. The study is much along the same line as in [Q1], that is, we stay away from special functions, and we are based on the Clifford analysis setting. The convergence results presented are not included in the above mentioned literature. In the present paper we wish to stress on the principles and do not make deliberate efforts to make the results to be best possible under the type of conditions assumed.

The convenience of quaternions is that they form a non-commutative associative division algebra. A higher-dimensional Euclidean space is only a linear algebra. The approach developed in the quaternionic case, however, can be adapted to higher dimensional cases. In the latter the simplification role of our approach to the theory is clearer. It requires somehow more abstract and different treatments and worths a separate paper.

In §2 we introduce Fueter's theorem and monomial functions in the quaternionic space as well as convolution integral expressions of Laurent series by monomial functions. In §3 formulas for the monomial functions and the quaternionic Dirichlet kernels are deduced. In §4 we state and prove a Riemann-Lebesgue theorem, a localization theorem, and a Dini's type convergence theorem in the context.

2. Fueter's Theorem and its Applications to Laurent Series Expansion

Quaternionic analysis introduces a complex structure in the four-dimensional Euclidean space which is in great similarity to the complex structure of the two-dimensional complex plane. The Hamilton quaternionic space is the universal algebra generated by $\mathbf{e}_1, \mathbf{e}_2$ with the properties $\mathbf{e}_1^2 = \mathbf{e}_2^2 = -1, \mathbf{e}_1\mathbf{e}_2 = -\mathbf{e}_2\mathbf{e}_1$. We set $\mathbf{i}_0 = 1, \mathbf{i}_1 = \mathbf{e}_1, \mathbf{i}_2 = \mathbf{e}_2, \mathbf{i}_3 = \mathbf{e}_1\mathbf{e}_2$ as the usual canonical basis. There follows from the properties of $\mathbf{e}_1$ and $\mathbf{e}_2$ that

$$\mathbf{i}_1\mathbf{i}_2 = -\mathbf{i}_2\mathbf{i}_1 = \mathbf{i}_3, \quad \mathbf{i}_2\mathbf{i}_3 = -\mathbf{i}_3\mathbf{i}_2 = \mathbf{i}_1, \quad \mathbf{i}_3\mathbf{i}_1 = -\mathbf{i}_1\mathbf{i}_3 = \mathbf{i}_2,$$

and

$$\mathbf{i}_0^2 = 1, \quad \mathbf{i}_1^2 = \mathbf{i}_2^2 = \mathbf{i}_3^2 = -1.$$

Let $\mathbf{H}$ and $\mathbf{H}^{\mathbf{C}}$ denote the algebras of the real and complex Hamilton's quaternions, respectively.

A general quaternion is of the form $q = \sum_{l=0}^{3} q_l \mathbf{i}_l = q_0 + \underline{q}$, where $q_l \in \mathbf{R}$ or $q_l \in \mathbf{C}$, depending on $q \in \mathbf{H}$ or $q \in \mathbf{H}^{\mathbf{C}}$, q_0 and $\underline{q} = q_1\mathbf{i}_1 + q_2\mathbf{i}_2 + q_3\mathbf{i}_3$ are the real and imaginary part of q, respectively. The quaternionic conjugate of q, denoted by $\bar{q}$, is defined to be $\bar{q} = q_0 - \underline{q}$. For any non-zero element $q \in \mathbf{H}$, there exists an inverse $q^{-1} \in \mathbf{H} : q^{-1} = \frac{\bar{q}}{|q|^2}$ such that $q^{-1}q = qq^{-1} = 1$. The natural inner product between q and q' in $\mathbf{H}^{\mathbf{C}}$, denoted by $< q, q' >$, is the number $\sum_{l=0}^{3} q_l \bar{q}'_l$, and the norm of q associated with this inner product is $|q| = (\sum_{l=0}^{3} |q_l|^2)^{\frac{1}{2}}$, and we have $|qq'| = |q||q'|$. The angle between q and q' in $\mathbf{H}$, denoted by $\arg(q, q')$, is defined to be $\arccos \frac{<q,q'>}{|q||q'|}$, where the inverse function arccos takes values in $[0, \pi]$. Denote, in similarity to the one complex variable case,

$$D = \frac{1}{2} \sum_{l=0}^{3} \frac{\partial}{\partial q_l} \mathbf{i}_l$$

as the *Cauchy-Riemann operator*. Functions to be concerned will be $\mathbf{H}^{\mathbf{C}}$-valued, but defined in subsets of $\mathbf{H}$. We will assume the existence of partial derivatives of any order of the functions under study, whenever these are involved. The operator D can be applied to a function $f = f_0\mathbf{i}_0 + f_1\mathbf{i}_1 + f_2\mathbf{i}_2 + f_3\mathbf{i}_3$ from the left- and the right-hand side in the manners

$$Df(q) = \frac{1}{2} \sum_l \sum_k \frac{\partial f_k}{\partial q_l} \mathbf{i}_l \mathbf{i}_k, \quad fD(q) = \frac{1}{2} \sum_l \sum_k \frac{\partial f_k}{\partial q_l} \mathbf{i}_k \mathbf{i}_l,$$

respectively.

If $Df = 0$ or $fD = 0$, then f is said to be *left-regular* or *right-regular*, respectively. A function which is both left- and right-regular is said to be *regular*. The *conjugate Cauchy-Riemann operator* will also be used:

$$\bar{D} = \frac{1}{2}\left(\frac{\partial}{\partial q_0} - \sum_{l=1}^{3} \frac{\partial}{\partial q_l} \mathbf{i}_l\right).$$

We denote

$$\underline{D} = \frac{1}{2} \sum_{l=1}^{3} \frac{\partial}{\partial q_l} \mathbf{i}_l \quad \textit{and} \quad \triangle = 4D\bar{D} = \sum_{l=0}^{3} \frac{\partial^2}{\partial q_l^2}.$$

Note that for any left-regular function f we have

$$\bar{D}f = \frac{\partial}{\partial q_0} f = -\underline{D}f. \tag{1}$$

Assume that Ω is a bounded, open and connected set in $\mathbf{R}^4$ that is considered to be identical with $\mathbf{H}$ under the mapping: $q \to (q_0, q_1, q_2, q_3)$. Assume that Ω has a Lipschitz boundary, and f, g are respectively left- and right-regular functions defined in a neighborhood of the closure of Ω. Then there holds the *Cauchy-Fueter theorem*

$$\int_{\partial\Omega} g(q)n(q)f(q)d\sigma(q) = 0,$$

where $d\sigma$ is the surface area measure and $n(q)$ is the outward pointing unit normal to $\partial\Omega$ at $q \in \partial\Omega$. Under the conditions, if $q \in \Omega$, then

$$f(q) = \frac{1}{2\pi^2}\int_{\partial\Omega} E(q'-q)n(q')f(q')d\sigma(q'),$$

known as *Cauchy-Fueter formula*, where $E(q) = \frac{\bar{q}}{|q|^4}$ is the *Cauchy-Fueter kernel.* Denote by I the *Kelvin inversion* defined by

$$I(f)(q) = E(q)f(q^{-1}).$$

Define, for $k \in \mathbf{Z}_+$, where $\mathbf{Z}_+$ is the set of positive integers,

$$P^{(-k)}(q) = \frac{(-1)^{(k-1)}}{(k-1)!}\bar{D}^{(k-1)}E(q), \quad \text{and} \quad P^{(k-1)} = I(P^{(-k)}).$$

Since E is regular, owing to the relations in (1), $P^{(-k)}$ has alternative expressions:

$$P^{(-k)}(q) = \frac{(-1)^{k-1}}{(k-1)!}(\frac{\partial}{\partial q_0})^{k-1}E(q) = \frac{1}{(k-1)!}\underline{D}^{k-1}E(q).$$

Functions $P^{(k)}, k \in \mathbf{Z}$, where $\mathbf{Z}$ is the set of all integers, are called *monomial functions.*

It can be easily shown:

Lemma 1. *For $k \in \mathbf{Z}_+$, $P^{(-k)}$ is regular outside a neighborhood of the origin and homogeneous of degree $-2-k$; $P^{(k-1)}$ is a polynomial of $q_l, l = 0,1,2,3$, regular and homogeneous of degree $k-1$. Moreover,*

$$P^{(k-1)}(q) = \bar{q}|q|^{2k}P^{(-k)}(\bar{q}).$$

Remark. In the quaternionic space, we consider the vector $q = q_0 + \underline{q}$ where $\underline{q} = q_1\mathbf{i}_1 + q_2\mathbf{i}_2 + q_3\mathbf{i}_3$ can be written in the form $\underline{q} = \mathbf{e}|\underline{q}|$, where $\mathbf{e} = \frac{\underline{q}}{|\underline{q}|}$ and $\mathbf{e}^2 = -1$. This shows that the variable $q_0 + \mathbf{e}|\underline{q}|$ behaves like the complex variable $z = x+iy$, under the correspondence $x \to q_0, y \to |\underline{q}|, i \to \mathbf{e}$.

Fueter's theorem provides a method to induce regular functions of one quaternionic variable from holomorphic functions of one complex variable. Now assume that f^0 is holomorphic, defined in a relatively open set O in the upper half complex plane, $f^0(z) = u(x,y) + iv(x,y), z = x+iy$, where u and v are real-valued. Then Fueter's theorem says that $\triangle\vec{f^0}$ is a quaternionic regular function in the relatively open set $\vec{O} = \{q = q_0 + \underline{q} \in \mathbf{H} : (q_0, |\underline{q}|) \in O\}$, where

$$\vec{f^0} = u(q_0, |\underline{q}|) + \mathbf{e}_{\underline{q}}v(q_0, |\underline{q}|),$$

and $\mathbf{e}_{\underline{q}} = \frac{\underline{q}}{|\underline{q}|}$. We will call $\vec{f^0}$ the *induced function* from f^0 and $\vec{O}$ the *induced set* from O. The following relations are useful:

$$\bar{D}\vec{f^0} = \overrightarrow{(\frac{df^0}{dz})}, \quad \bar{D}\triangle\vec{f^0} = \triangle\overrightarrow{(\frac{df^0}{dz})},$$

where

$$\frac{d}{dz} = \frac{1}{2}(\frac{\partial}{\partial x} - \frac{\partial}{\partial y}),$$

and

$$\triangle\vec{f^0}(q) = \frac{2}{|\underline{q}|}\frac{\partial u}{\partial y}(q_0, |\underline{q}|) + 2\mathbf{e}_{\underline{q}}(\frac{1}{|\underline{q}|}\frac{\partial v}{\partial y}(q_0, |\underline{q}|) - \frac{1}{|\underline{q}|^2}v(q_0, |\underline{q}|)). \tag{2}$$

We refer the reader to [De] or [Su] for proofs of these equalities and the two-sided regularity of $\triangle\vec{f^0}$.

Denote by τ the mapping

$$\tau : f^0 \to -\frac{1}{4}\triangle\vec{f^0}.$$

Note that τ is linear with respect to addition and multiplication by real scalars. As is shown in [Su],

$$E(q) = \tau((\cdot)^{-1})(q),$$

which is

$$P^{(-1)}(q) = \tau((\cdot)^{-1})(q)$$

in our notation. In general, we have

Lemma 2 ([Q1]).

$$P^{(-k)} = \tau((\cdot)^{-k}), k \in \mathbf{Z}_+.$$

The following relation is anticipated which can be used to alternatively define $P^{(k)}$ for $k \in \{0\} \cup \mathbf{Z}_+$.

Lemma 3 ([Q2]).

$$I(\tau((\cdot)^{-k})) = \tau((\cdot)^{k+1}),$$

and thus

$$P^{(k-1)} = \tau((\cdot)^{k+1}), k \in \mathbf{Z}_+.$$

Lemma 2 and Lemma 3 establish a corresponding relationship between the sequence

$$\{\ldots, z^{-3}, z^{-2}, z^{-1}, z^2, z^3, \ldots\}$$

and the sequence

$$\{\ldots, P^{(-3)}, P^{(-2)}, P^{(-1)}, P^{(0)}, P^{(1)}, P^{(2)}, \ldots\},$$

and therefore a possible relation between Laurent series in the complex plane and those in the quaternionic space.

As example, the kernel for Hilbert transform on the unit sphere $\mathbf{S}^3$ can be obtained through the one on the unit circle. Indeed, in the holomorphic extension sense, there follows

$$\begin{aligned}
\sum_{k=1}^{\infty} P^{(k)}(q) + \sum_{-\infty}^{-1} -P^{(k)}(q) &= \tau(\sum_{k=1}^{\infty} z^k + \sum_{-\infty}^{-1} -z^k) \\
&= \tau(\frac{1+z}{1-z}) \\
&= \tau(-1 + \frac{2}{1-z}) \\
&= \tau(\frac{2}{1-z}) \\
&= 2E(1-q),
\end{aligned}$$

a function regularly defined everywhere except at $q = 1$ ([Q1]).

The unit circle, or the one dimensional unit sphere, in the complex plane $\mathbf{C}$ is denoted by $\mathbf{S}^1$. Denote by $A(\mathbf{S}^1)$ the class of the functions analytic in some annulus containing $\mathbf{S}^1$. It is well known that the restrictions on $\mathbf{S}^1$ of the functions in $A(\mathbf{S}^1)$ form a dense subclass of $L^2(\mathbf{S}^1)$. By the theory of Laurent series, we have that for any $f \in A(\mathbf{S}^1)$,

$$f(z) = \sum_{k=-\infty}^{\infty} c_k z^k,$$

where

$$c_k = \frac{1}{2\pi i} \int_{\mathbf{S}^1} \xi^{-k} f(\xi) \frac{d\xi}{\xi}, \quad k \in \mathbf{Z}.$$

Note that the term $c_k z^k$ is the projection of the function f onto the space of holomorphic functions which have homogeneity degree k. The projection operator, denoted by P_k, is a convolution operator on the unit circle using the function $p^{(k)}(z) = z^n$ as kernel:

$$P_k(f)(z) = c_k z^k = \frac{1}{2\pi i} \int_{\mathbf{S}^1} p^{(k)}(\xi^{-1} z) f(\xi) \frac{d\xi}{\xi}, \quad k \in \mathbf{Z}.$$

Thus, the Laurent series of a function $f^0 \in A(\mathbf{S}^1)$ has the expression

$$f^0(z) = \sum_{k=-\infty}^{\infty} \frac{1}{2\pi i} \int_{\mathbf{S}^1} p^{(k)}(\xi^{-1} z) f^0(\xi) \frac{d\xi}{\xi}, \quad k \in \mathbf{Z}.$$

In the quaternionic space we consider the class of functions

$$A(\mathbf{S}^3) = \{f(q) : f(q) \quad \text{is left-regular in} \quad 1 - s < |q| < 1 + s \quad \text{for some} \quad s > 0\}.$$

It was proved in [Q1] that the restrictions on $\mathbf{S}^3$ of the functions in the class $A(\mathbf{S}^3)$ form a dense subclass of $L^2(\mathbf{S}^3)$.

Lemma 4 ([Q1]). *If $f \in \mathcal{A}$, then*

$$f(q) = \sum_{k=-\infty}^{\infty} \frac{1}{2\pi^2} \int_{\mathbf{S}^3} P^{(k)}(p^{-1}q)E(p)n(p)f(p)d\sigma(p), \ \ 1 - s < |q| < 1 + s.$$

Comparing the above lemma with Laurent series expansions in one complex variable, we can see that the functions $P^{(k)}$ and $p^{(k)}$ play the same role in their respective spaces, and, the striking fact is: $\tau(p^{(k)}) = P^{(k)}, k < 0$; and $\tau(p^{(k+2)}) = P^{(k)}, k \geq 0$.

We note that for any k,

$$P^{(k)}(p^{-1}q)E(p)$$

is regular in both p and q ([Q1]). Since $E(p)n(p) = 1$ on the sphere, owing to the orthogonality of spherical regular functions of different homogeneities ([BDS]), we have

$$\int_{\mathbf{S}^3} P^{(k)}(p^{-1}q)d\sigma(p) = 0, \quad |q| = 1, \ \ k \neq 0; \tag{3}$$

and, since $P^{(0)} = I(P^{(-1)}) = I(E) = 1$, we have

$$\frac{1}{2\pi^2} \int_{\mathbf{S}^3} P^{(0)}(p^{-1}q)d\sigma(p) = 1, \quad |q| = 1. \tag{4}$$

These results will be used in §4.

3. Dirichlet Kernel and Its Estimates

The Fourier series of a square integrable function f has the form

$$\sum_{n=-\infty}^{\infty} c_n z^n,$$

where $z^n = e^{inx}$ and $c_n = \frac{1}{2\pi}\int_0^{2\pi} f(x)e^{-inx}dx$. If N is a nonnegative integer, then the Nth symmetric partial sum of the Fourier series of f is

$$s_N f(x) = \sum_{|n|\leq N} c_n e^{inx}.$$

Inserting the integral expression for c_n we have that

$$\begin{aligned} s_N f(x) &= \sum_{|n|\leq N} \frac{1}{2\pi} \int_0^{2\pi} f(y)e^{in(x-y)}dy \\ &= \frac{1}{2\pi} \int_0^{2\pi} f(y)\widetilde{d}_N(x-y)dy \\ &= \widetilde{d}_N * f(x), \end{aligned}$$

where for $x \not\equiv 0 \pmod{2\pi}$

$$\widetilde{d}_N(x) = \sum_{|n|\le N} e^{inx} = \frac{\sin(N+\frac{1}{2})x}{\sin\frac{1}{2}x}. \tag{5}$$

If x is congruent to $0 \pmod{2\pi}$, then $\widetilde{d}_N(x)$ has the value $2N+1$, which is obtained by continuous extension of the last expression. The function $\widetilde{d}_N$ is the classical *Dirichlet kernel* on the circle.

Note that $\widetilde{d}_N$ is a trigonometric polynomial of degree N which is even in x and satisfies

$$\frac{1}{2\pi}\int_{-\pi}^{\pi}\widetilde{d}_N(x)dx = \frac{1}{\pi}\int_0^{\pi}\widetilde{d}_N(x)dx = 1.$$

It can be shown (see, for instance, [Zy]) that

$$\|\widetilde{d}_N\|_1 = \frac{4}{\pi^2}\log N + O(1).$$

The Dirichlet kernel in the complex plane is

$$d_N(z) = \sum_{|n|\le N} z^n, \tag{6}$$

where z is any complex number. We note that $d_N(e^{ix}) = \widetilde{d}_N(x)$.

Parallel to the discussion in the complex plane, and in accordance with Lemma 4, for any function f integrable on the sphere $\mathbf{S}^3$, we can associate it with a Fourier series

$$\frac{1}{2\pi^2}\sum_{k=-\infty}^{\infty}\int_{\mathbf{S}} P^{(k)}(p^{-1}q)E(p)n(p)f(p)d\sigma(p),$$

where $\mathbf{S}$ is the abbreviation of $\mathbf{S}^3$ that will be kept the same in the sequel. The above series is also written

$$\sum_{k=-\infty}^{\infty} P_k(f)(q),$$

where

$$P_k(f)(q) = \frac{1}{2\pi^2}\int_{\mathbf{S}} P^{(k)}(p^{-1}q)E(p)n(p)f(p)d\sigma(p)$$

is the projection of f onto the space of left-regular functions of k-homogeneity.

The partial sum $S_N(f)$ and the associated *Dirichlet kernel* are defined to be

$$S_N(f)(q) = \sum_{|k|\le N} P_k(f)(q) \quad\text{and}\quad D_N(q) = \sum_{|k|\le N} P^{(k)}(q).$$

The study in the previous section shows that the mapping τ does not map $d_N(z)$ to $D_N(q)$, but

$$\tau(d_N)(q) = \sum_{k=-N}^{N-2} P^{(k)}(q).$$

Define

$$D'_N(q) = \tau(d_N)(q).$$

The latter is called the *modified Dirichlet kernel.* The partial sum corresponding to the modified Dirichlet kernel is

$$\sum_{k=-N}^{N-2} P_k(f)(q).$$

Owing to the Riemann-Lebesgue theorem proved in the next section, under the set conditions, there holds

$$\lim_{N\to\infty} S_N(f)(q) - S'_N(q) = 0.$$

So, in relation to pointwise convergence under the conditions, to consider $S_N(f)(q)$ and to consider $S'_N(f)(q)$ are the same.

Below we first apply τ to the terms z^k, and get the expressions of $P^{(k)}(q)$ on the unit sphere $\mathbf{S}$.

Lemma 5. *For* $k \in \mathbf{Z}$,

$$P^{(k)}(q) = -\frac{\cos\theta}{2\sin\theta}\frac{\partial u}{\partial\theta} - \frac{1}{2}\frac{\partial v}{\partial\theta} - \mathbf{e}_{\underline{q}}\left(\frac{\cos\theta}{2\sin\theta}\frac{\partial v}{\partial\theta} - \frac{1}{2}\frac{\partial u}{\partial\theta} - \frac{v(\theta)}{2\sin^2\theta}\right), \tag{7}$$

where $\mathbf{e}_{\underline{q}} = \frac{\underline{q}}{|\underline{q}|}$, $z^k = u_k(r,\theta) + iv_k(r,\theta) = r^k\cos k\theta + ir^k\sin k\theta$,
for $k < 0$, $u = u_k$, $v = v_k$;
and for $k \geq 0$, $u = u_{k+2}$, $v = v_{k+2}$.

Proof. From Lemma 2 and 3, we have $\tau(z^k) = P^{(k)}(q)$ for $k < 0$; and $\tau(z^{k+2}) = P^{(k)}(q)$ for $k \geq 0$. By using the relation (2), we have

$$P^{(k)}(q) = -\frac{1}{2|\underline{q}|}\frac{\partial u}{\partial y}(q_0,|\underline{q}|) - \mathbf{e}_{\underline{q}}\frac{1}{2}\left(\frac{1}{|\underline{q}|}\frac{\partial v}{\partial y}(q_0,|\underline{q}|) - \frac{1}{|\underline{q}|^2}v(q_0,|\underline{q}|)\right). \tag{8}$$

For $z = x + iy, x = r\cos\theta, y = r\sin\theta$, we have, on the unit sphere ($r = 1$),

$$\frac{\partial u}{\partial y} = \frac{\partial u}{\partial\theta}\frac{\partial\theta}{\partial y} + \frac{\partial u}{\partial r}\frac{\partial r}{\partial y} = \frac{\partial u}{\partial\theta}\frac{\cos\theta}{r} + \frac{\partial u}{\partial r}\sin\theta = \frac{\partial u}{\partial\theta}\cos\theta + \frac{\partial u}{\partial r}\sin\theta,$$

$$\frac{\partial v}{\partial y} = \frac{\partial v}{\partial\theta}\frac{\partial\theta}{\partial y} + \frac{\partial v}{\partial r}\frac{\partial r}{\partial y} = \frac{\partial v}{\partial\theta}\frac{\cos\theta}{r} + \frac{\partial v}{\partial r}\sin\theta = \frac{\partial v}{\partial\theta}\cos\theta + \frac{\partial v}{\partial r}\sin\theta.$$

Substituting these into the formula (8) while taking into account the Cauchy-Riemann equations in polar coordinates,

$$\begin{aligned}\frac{\partial u}{\partial r} &= \frac{1}{r}\frac{\partial v}{\partial\theta},\\ \frac{\partial v}{\partial r} &= -\frac{1}{r}\frac{\partial u}{\partial\theta},\end{aligned}$$

we have

$$\begin{aligned}
P^{(k)}(q) &= -\frac{1}{2\sin\theta}(\frac{\partial u}{\partial\theta}\cos\theta + \frac{\partial v}{\partial\theta}\sin\theta) \\
&\quad -\mathbf{e}_{\underline{q}}\frac{1}{2}(\frac{1}{\sin\theta}(\frac{\partial v}{\partial\theta}\cos\theta - \frac{\partial u}{\partial\theta}\sin\theta) - \frac{v(\theta)}{\sin^2\theta}) \\
&= -\frac{\cos\theta}{2\sin\theta}\frac{\partial u}{\partial\theta} - \frac{1}{2}\frac{\partial v}{\partial\theta} + \mathbf{e}_{\underline{q}}(-\frac{\cos\theta}{2\sin\theta}\frac{\partial v}{\partial\theta} + \frac{1}{2}\frac{\partial u}{\partial\theta} + \frac{v(\theta)}{2\sin^2\theta}).
\end{aligned}$$

Since

$$\frac{\partial v}{\partial\theta} = k\cos k\theta, \qquad \frac{\partial u}{\partial\theta} = -k\sin k\theta,$$

we further obtain

$$P^{(k)}(q) = \frac{k\sin(k-1)\theta}{2\sin\theta} + \mathbf{e}_{\underline{q}}(\frac{\sin k\theta}{2\sin^2\theta} - \frac{k\cos(k-1)\theta}{2\sin\theta}), \quad k<0;$$

and

$$P^{(k)}(q) = \frac{(k+2)\sin(k+1)\theta}{2\sin\theta} + \mathbf{e}_{\underline{q}}(\frac{\sin(k+2)\theta}{2\sin^2\theta} - \frac{(k+2)\cos(k+1)\theta}{2\sin\theta}), \quad k\geq 0.$$

□

Next we compute the modified Dirichlet kernels.

Lemma 6. *If $d_N(z) = u + iv$, then*

$$\begin{aligned}
D_N'(q) &= -\frac{\cos\theta}{2\sin\theta}\frac{\partial u}{\partial\theta} + \mathbf{e}_{\underline{q}}\frac{1}{2}\frac{\partial u}{\partial\theta} \\
&= \frac{\cos\theta[(N+1)\sin N\theta - N\sin(N+1)\theta]}{2\sin\theta(1-\cos\theta)} \\
&\quad +\mathbf{e}_{\underline{q}}\frac{N\sin(N+1)\theta - (N+1)\sin N\theta}{2(1-\cos\theta)},
\end{aligned} \tag{9}$$

where $\theta = \arccos q_0 \in [0,\pi]$. Moreover, we have

$$D_N(q) = O(N^3) \ \ as \ \ \theta\to 0.$$

Proof. For $d_N(z) = u + iv$, the relation (2) gives

$$D_N'(q) = -\frac{1}{2|\underline{q}|}\frac{\partial u}{\partial y}(q_0,|\underline{q}|) - \mathbf{e}_{\underline{q}}\frac{1}{2}(\frac{1}{|\underline{q}|}\frac{\partial v}{\partial y}(q_0,|\underline{q}|) - \frac{1}{|\underline{q}|^2}v(q_0,|\underline{q}|)). \tag{10}$$

We first work out u and v in polar coordinates. We have

$$\begin{aligned} d_N(z) &= \sum_{|n|\le N} z^n \\ &= \frac{1}{z^N(1-z)} - \frac{z^{N+1}}{1-z} \\ &= \frac{(1+r^{2N+2})\cos N\theta - (r+r^{2N+1})\cos(N+1)\theta}{r^N(r^2-2r\cos\theta+1)} \\ &\quad +i\frac{(r-r^{2N+1})\sin(N+1)\theta + (r^{2N+2}-1)\sin N\theta}{r^N(r^2-2r\cos\theta+1)} \\ &= u(r,\theta)+iv(r,\theta). \end{aligned}$$

Accordingly,

$$\frac{\partial u}{\partial y} = \frac{\partial u}{\partial \theta}\frac{\partial \theta}{\partial y} + \frac{\partial u}{\partial r}\frac{\partial r}{\partial y} = \frac{\partial u}{\partial \theta}\frac{\cos\theta}{r} + \frac{\partial u}{\partial r}\sin\theta,$$

$$\frac{\partial v}{\partial y} = \frac{\partial v}{\partial \theta}\frac{\partial \theta}{\partial y} + \frac{\partial v}{\partial r}\frac{\partial r}{\partial y} = \frac{\partial v}{\partial \theta}\frac{\cos\theta}{r} + \frac{\partial v}{\partial r}\sin\theta.$$

Invoking the Cauchy-Riemann equations in polar coordinates, restricted on the sphere, we have

$$\begin{aligned} \frac{\partial u}{\partial \theta} &= \frac{N\sin(N+1)\theta - (N+1)\sin N\theta}{1-\cos\theta}, \\ \frac{\partial u}{\partial r} &= \frac{\partial v}{\partial \theta} = 0, \\ \frac{\partial v}{\partial r} &= -\frac{\partial u}{\partial \theta}. \end{aligned}$$

Note that when $r=1$ the value of function $v(r,\theta)$ is zero, and the partial derivatives of u and v only depend on the angle θ.

Substituting these results into (9), we obtain

$$\begin{aligned} D'_N(q) &= -\frac{1}{2|\underline{q}|}\frac{\partial u}{\partial y}(q_0,|\underline{q}|) - \mathbf{e}_{\underline{q}}\frac{1}{2}\left(\frac{1}{|\underline{q}|}\frac{\partial v}{\partial y}(q_0,|\underline{q}|) - \frac{1}{|\underline{q}|^2}v(q_0,|\underline{q}|)\right) \\ &= -\frac{\cos\theta}{2\sin\theta}\frac{\partial u}{\partial \theta} - \mathbf{e}_{\underline{q}}\frac{1}{2}\frac{\partial v}{\partial r} \\ &= -\frac{\cos\theta}{2\sin\theta}\frac{\partial u}{\partial \theta} + \mathbf{e}_{\underline{q}}\frac{1}{2}\frac{\partial u}{\partial \theta} \\ &= \frac{\cos\theta[(N+1)\sin N\theta - N\sin(N+1)\theta]}{2\sin\theta(1-\cos\theta)} \\ &\quad +\mathbf{e}_{\underline{q}}\frac{N\sin(N+1)\theta - (N+1)\sin N\theta}{2(1-\cos\theta)}. \end{aligned}$$

The kernel $D'_N(q)$ may be written

$$D'_N(q) = R(\theta) + \mathbf{e}_q I(\theta).$$

Using Taylor series expansions of sine and cosine functions at the origin, we obtain the estimates:

$$\begin{aligned}\lim_{\theta\to 0} R(\theta) &= -\frac{1}{6}N(N+1)(2N+1);\\ \lim_{\theta\to 0} I(\theta) &= 0.\end{aligned}$$

□

4. Convergence Results

The classical Riemann-Lebesgue theorem for Fourier series asserts that for any function in $L^1([0,2\pi])$ its Fourier series coefficients c_k enjoy $|c_k| \to 0$, as $|k| \to \infty$. The Riemann-Lebesgue theorem for Fourier transforms says that for any function $f \in L^1(\mathbf{R}^n)$, there follows $\hat{f}(\xi) \to 0$, as $|\xi| \to \infty$. On higher-dimensional spheres there do not exist the concept "Fourier coefficients" (not like in the classical case where $c_k z^k$ has two separated parts c_k and z^k), for the projection $P_k(f)$ corresponds to a vector-valued function in a multi-dimensional space: There are several coefficients depending on the base that is chosen to express general vectors in the space. Therefore, we should consider the whole projection, and the assertion $P_k(f)(q) \to 0$ is a generalization of the classical Riemann-Lebesgue theorem (see Theorem 1 below).

Averages of spherical functions on two-dimensional spheres are involved. Except the $\mathbf{e}_{\underline{q}}$ part, the terms in $P^{(k)}(q)$ and the modified Dirichlet kernels $D'_N(q)$ on $\mathbf{S}$ depend only on the angle $\theta = \arccos q_0, \theta \in [0,\pi]$. This corresponds to represent a quaternionic number q as $q = q_0\mathbf{i}_0 + q_1\mathbf{i}_1 + q_2\mathbf{i}_2 + q_3\mathbf{i}_3 = q_0\mathbf{i}_0 + \underline{q} = \cos\theta\mathbf{i}_0 + \mathbf{e}_{\underline{q}}\sin\theta$, where $\mathbf{e}_{\underline{q}} = \frac{\underline{q}}{|\underline{q}|} = \mathbf{i}_1\cos\alpha + \mathbf{i}_2\sin\alpha\cos\beta + \mathbf{i}_3\sin\alpha\sin\beta$. Now let θ be fixed and take average of a function $f \in L^1(\mathbf{S})$ over the two-dimensional sphere on which all the points q satisfy $q_0 = \cos\theta$. This average is denoted by

$$G_{\mathbf{i}_0}(f)(\theta) = \int_{\mathbf{S}^2} f(\cos\theta + \mathbf{e}_{\underline{\xi}}\sin\theta)d\sigma_2(\underline{\xi}),$$

where $\underline{\xi}$ is the spherical variable on the unit sphere $\mathbf{S}^2$ that may be easily written out in terms of α and β as we did for $\mathbf{e}_{\underline{q}}$, and $d\sigma_2(\underline{\xi})$ is the normalized surface area measure on the sphere $\mathbf{S}^2$. We call $G_{\mathbf{i}_0}(f)(\theta)$ *the average of* f *about* $\mathbf{i}_0$ *in angle* θ.

Now let q be a fixed point on $\mathbf{S}$. We can similarly formulate $G_q(f)(\theta)$ to be the average of a function f over the two dimensional sphere whose points p satisfy $\arg(p,q) = \theta$, and phrased as *the average of* f *around* q *in angle* θ.

Note that in considering convergence problems it suffices to assume f to be scalar-valued. Without repeating, we will always assume this for the rest of the paper. The Riemann-Lebesgue theorem may be formulated as follows.

Theorem 1 (Riemann-Lebesgue theorem). *Assume that $f \in L^1(\mathbf{S})$. Let q be a fixed point on $\mathbf{S}$ and $G_q(f)(\theta)$ the average of f around q in angle θ. Denote $G_q(\theta) = G_q(f)(\theta)$ and $G_q^{\mathbf{e}}(\theta) = G_q(\mathbf{e}_{\underline{(\cdot)^{-1}q}} f(\cdot))(\theta)$. Assume that both $G_q(f)$ and $G_q^{\mathbf{e}}$ are absolutely continuous in $[0,\pi]$, and, with their derivatives denoted by F_q and $F_q^{\mathbf{e}}$, respectively, the functions $\sin(\cdot)F_q(\cdot)$ and $\sin(\cdot)F_q^{\mathbf{e}}(\cdot)$ are in $L^1([0,\pi])$. Then*

$$\lim_{|k|\to\infty} P_k(f)(q) = 0.$$

Proof. Since on the unit sphere $E(p)n(p) = 1$, the integration formula of $P_k(f)$ is abbreviated as

$$P_k(f)(q) = \int_{\mathbf{S}} P^{(k)}(p^{-1}q)f(p)d\sigma(p),$$

where $p^{-1} = \bar{p} = p_0 - p_1\mathbf{i}_1 - p_2\mathbf{i}_2 - p_3\mathbf{i}_3$ and $p^{-1}q = (p_0q_0 + p_1q_1 + p_2q_2 + p_3q_3) + \mathrm{Im}(p^{-1}q) = \cos\theta + \mathbf{e}_{\underline{p^{-1}q}}\sin\theta$.

Substituting the expression (7) for $P^{(k)}(p^{-1}q)$ into the integral expression and, based on Fubini's theorem, decomposing the integral on the sphere $\mathbf{S}$ into the iterated integral composed by one in angle θ with respect to the direction of q and the other in the 2-dimensional sphere orthogonal with the q direction, we have

$$\begin{aligned} P_k(f)(q) &= \omega_2\int_0^\pi \sin^2\theta[-\frac{\cos\theta}{2\sin\theta}\frac{\partial u}{\partial\theta} - \frac{1}{2}\frac{\partial v}{\partial\theta}]G_q(f)(\theta) \\ &\quad -\omega_2\int_0^\pi \sin^2\theta[\frac{\cos\theta}{2\sin\theta}\frac{\partial v}{\partial\theta} - \frac{1}{2}\frac{\partial u}{\partial\theta} - \frac{v(\theta)}{2\sin^2\theta})]G_q(\mathbf{e}_{\underline{(\cdot)^{-1}q}}f(\cdot))(\theta)d\theta \\ &= \omega_2\int_0^\pi (-\frac{1}{4}\sin 2\theta\frac{\partial u}{\partial\theta} - \frac{1}{2}\sin^2\theta\frac{\partial v}{\partial\theta})G_q(\theta)d\theta \\ &\quad -\omega_2\int_0^\pi (\frac{1}{4}\sin 2\theta\frac{\partial v}{\partial\theta} - \frac{1}{2}\sin^2\theta\frac{\partial u}{\partial\theta} - \frac{v(\theta)}{2})G_q^{\mathbf{e}}(\theta)d\theta, \end{aligned}$$

where ω_2 is the surface area of the two dimensional unit sphere. Since G_q and $G_q^{\mathbf{e}}$ are absolutely continuous, that validates taking integration by parts, we have

$$\begin{aligned} P_k(f)(q) &= \omega_2\int_0^\pi u(1,\theta)(\frac{\cos 2\theta}{2}G_q(\theta) + \frac{\sin 2\theta}{4}F_q(\theta))d\theta \\ &\quad +\omega_2\int_0^\pi v(1,\theta)(\frac{\sin 2\theta}{2}G_q(\theta) + \frac{\sin^2\theta}{2}F_q(\theta))d\theta \\ &\quad +\omega_2\int_0^\pi v(1,\theta)(\frac{\cos 2\theta}{2}G_q^{\mathbf{e}}(\theta) + \frac{\sin 2\theta}{4}F_q^{\mathbf{e}}(\theta))d\theta \\ &\quad -\omega_2\int_0^\pi u(1,\theta)(\frac{\sin 2\theta}{2}G_q^{\mathbf{e}}(\theta) + \frac{\sin^2\theta}{2}F_q^{\mathbf{e}}(\theta))d\theta \\ &\quad +\omega_2\int_0^\pi \frac{v(1,\theta)}{2}G_q^{\mathbf{e}}(\theta)d\theta. \end{aligned}$$

Recall that

$$u(1,\theta) = \cos k'\theta, \quad v(1,\theta) = \sin k'\theta,$$

where $k' = k$ for $k < 0$; and $k' = k+2$ for $k \geq 0$. The assumptions on $G_q, G_q^{\mathbf{e}}, F_q, F^{\mathbf{e}}$ allow us to use the Riemann-Lebesgue theorem on $[0, 2\pi]$, and we obtain

$$\lim_{|k|\to\infty} P_k(f)(q) = 0. \qquad \square$$

Corollary 1. *Let $f \in L^1(\mathbf{S})$ and q be a fixed point on $\mathbf{S}$. Assume that G_q is absolutely continuous and its derivative F_q satisfies $\sin(\cdot)F_q(\cdot) \in L^1([0, \pi]$. Then for the Fourier series of f in spherical harmonics:*

$$f(q) \sim \sum_{k=0}^{\infty} f_k(q),$$

we have

$$\lim_{k\to\infty} f_k(q) = 0.$$

Proof. It is a general result that f_k, as a k-spherical harmonic on the three-dimensional unit sphere, has a unique decomposition $f_k = g_k + g_{-k-2}$, where g_k and g_{-k-2} are spherical regular functions ([BDS]) of, respectively, homogeneity degree k and $-k-2$. It turns out that, in the notation of the proof of Theorem 1, $g_k = P_k(f), g_{-k-2} = P_{-k-2}(f)$. In fact, the partial sum $\sum_{k=0}^{N} f_k(q)$ is identical with the modified partial sum $S'_{N+2}(f)(q)$ induced from the modified Dirichlet kernel $D'_{N+2}(q)$. Since f_k is scalar-valued, we have

$$f_k = \mathrm{Re}(P_k(f)) + \mathrm{Re}(P_{-k-2}(f)).$$

As consequence, in the proof of Theorem 1 the average $G_q^{\mathbf{e}}$ and the related $F_q^{\mathbf{e}}$ are cancelled out, and thus are irrelevant. Thus, without the assumptions on $G_q^{\mathbf{e}}$ and $F_q^{\mathbf{e}}$ the result still holds. $\square$

The corollary may be regarded as Riemann-Lebesgue theorem on the sphere with the traditional setting. We will call the series in the corollary a *scalar Fourier series.* Due to the fact that all entries of the series, as well as the function itself, are scalar-valued, it is clear in the proof that only assuming the conditions on G and F is sufficient. Thanks to this corollary we can ignore any finite but limited number of entries in the end of a partial sum of a scalar Fourier series, or any finite but limited pairs $P_k(f)$ and $P_{-k-2}(f)$ in the two ends of a modified partial sum when dealing with pointwise convergence under the set conditions on G_q and F_q. In the sequel we will concentrate in the modified partial sums as they are equivalent with the partial sums of scalar Fourier series.

Localization Principle in the context is as follows.

Theorem 2. *Let $f \in L^1(\mathbf{S})$ and q be a fixed pint on $\mathbf{S}$. If f vanishes in some neighborhood of q, G_q is absolutely continuous and $F_q = G'_q$ is integrable, then for the modified partial sums we have*

$$\lim_{N\to\infty} S'_N(f)(q) = 0.$$

Proof. In view of the proof of Corollary 1, the modified partial sums are real-valued and we can restrict ourselves to the scalar-part of the integrals under study. Using the formula for D'_N, and performing integration by parts, we have

$$\begin{aligned} S'_N(f)(q) &= \frac{\omega_2}{2\pi^2}\int_0^\pi \sin^2\theta\left(-\frac{\cos\theta}{2\sin\theta}\frac{\partial u}{\partial\theta}\right)G_q(\theta)d\theta \\ &= \frac{\omega_2}{2\pi^2}\int_0^\pi -\frac{\sin 2\theta}{4}\frac{\partial u}{\partial\theta}G_q(\theta)d\theta \\ &= \frac{\omega_2}{2\pi^2}\int_0^\pi u(1,\theta)\left(\frac{\cos 2\theta}{2}G_q(\theta)+\frac{\sin 2\theta}{4}F_q(\theta)\right)d\theta. \end{aligned}$$

Since f vanishes in a neighborhood of q, there exists $\delta > 0$ such that $f(p) = 0$, $0 \le \theta \le \delta$, $\theta = \arg(p,q)$. Recalling

$$u(1,\theta) = \frac{\sin(N+\frac{1}{2})\theta}{\sin\frac{\theta}{2}},$$

we have

$$\begin{aligned} S'_N(f)(q) &= \frac{\omega_2}{2\pi^2}\int_\delta^\pi \frac{\sin(N+\frac{1}{2})\theta\cos 2\theta}{2\sin\frac{\theta}{2}}G_q(\theta)d\theta \\ &\quad+\frac{\omega_2}{2\pi^2}\int_\delta^\pi \frac{\sin(N+\frac{1}{2})\theta\sin 2\theta}{4\sin\frac{\theta}{2}}F_q(\theta)d\theta. \end{aligned}$$

Since $F_q(\theta), G_q(\theta)$ are integrable in the interval (δ,π), on letting N go to infinity and applying the classical Riemann-Lebesgue theorem, we conclude

$$\lim_{N\to\infty} S'_N(f)(q) = 0. \qquad \square$$

Denote

$$G_q(0) = \lim_{\theta\to 0} G_q(\theta),$$

if the limit exists.

A Dini's type convergence theorem is as follows.

Theorem 3. *Let $f \in L^1(\mathbf{S})$ and q be a fixed point on $\mathbf{S}$. If, in addition to absolute continuity of G_q in $[0,\pi]$ and integrability of F_q in $(0,\delta)$, for some $\delta > 0$, we assume that $G(0)$ exists and $\frac{G(\theta)-G(0)}{\theta}$ is integrable in $(0,\delta)$, then*

$$\lim_{N\to\infty} S'_N(f)(q) = G_q(0).$$

If, in particular, f is continuous at q, then

$$\lim_{N\to\infty} S'_N(f)(q) = f(q).$$

Proof. It suffices to show

$$\lim_{N\to\infty}(S'_N(f)(q) - G_q(0)) = 0.$$

From the relations (3) and (4) we have

$$\frac{1}{2\pi^2}\int_{\mathbf{S}} D'_N(p^{-1}q)d\sigma(p) = 1.$$

We are reduced to show

$$\lim_{N\to\infty}\frac{1}{2\pi^2}\int_{\mathbf{S}} D'_N(p^{-1}q)(f(p)-G_q(0))d\sigma(p) = 0.$$

The last integral is scalar-valued and thus is induced from the scalar part of the modified Dirichlet kernel, viz.

$$\frac{1}{2\pi^2}\int_{\mathbf{S}} \mathrm{Re}(D'_N(p^{-1}q))(f(p)-G_q(0))d\sigma(p).$$

Substituting the expression of $D'_N(p^{-1}q)$ and writing the integral into an iterated integral, the above becomes

$$\frac{1}{2\pi^2}\int_0^{\pi}\sin^2\theta(G_q(\theta)-G_q(0))(-\frac{\cos\theta}{2\sin\theta}\frac{\partial u}{\partial\theta})d\theta.$$

Replacing u by $\frac{\sin(N+\frac{1}{2})\theta}{\sin\frac{\theta}{2}}$ and integrating by parts, we have

$$\begin{aligned}&\lim_{N\to\infty}(S'_N(f)(q)-G_q(0))\\ &= \lim_{N\to\infty}\{\frac{\omega_2}{2\pi^2}\int_0^{\pi}\sin(N+\frac{1}{2})\theta\cos 2\theta\frac{(G_q(\theta)-G_q(0))}{2\sin\frac{\theta}{2}}d\theta\\ &+\frac{\omega_2}{2\pi^2}\int_0^{\pi}\frac{\sin(N+\frac{1}{2})\theta\sin 2\theta}{4\sin\frac{\theta}{2}}\frac{d}{d\theta}(G_q(\theta)-G_q(0))d\theta\}.\end{aligned}$$

Owing to the assumptions on G_q and F_q we can apply the classical Riemann-Lebesgue theorem to the last two integrals, and obtain

$$\lim_{N\to\infty} S'_N(f)(q) = G_q(0). \qquad \square$$

References

[BDS] F. Brackx, R. Delanghe and F. Sommen, *Clifford Analysis,* Research Notes in Mathematics, Vol. **76**, Pitman Advanced Publishing Company, Boston, London, Melbourne (1982).

[De] C.A. Deavours, *The quaternion calculus,* Amer. Math. Monthly **80** (1973), 995–1008.

[Di] P.G.L. Dirichlet , *Sur les séries dont le terme général dépend de deux angle, et qui servent á exprimer des fonctions arbitraires entre des limites données,* J. Reine Angew. Math. **17** (1873), 35–56.

[Fu] R. Fueter, *Die Funktionentheorie der Differentialgleichungen* $\Delta u = 0$ *und* $\Delta\Delta u = 0$ *mit vier reellen Variablen,* Comm. Math. Helv. **7**(1935), 307–330.

[Ko] L. Koschmieder, *Unmittelbarer Beweis der Konvergenz einiger Reihen, die von mehreren Veränderlichen abhängen,* Monatsh Math. Phys. **41** (1934) 58–63.

[Me] C. Meaney, *Divergence Jacobi Polynomial Series,* Proceedings of the American Mathematical Society, vol. **87**, Number 3, March 1983, 459–462.

[Q1] T. Qian, *Singular integrals on star-shaped Lipschitz surfaces in the quaternionic space,* Math. Ann., **310** (4) (April, 1998), 601-630.

[Q2] T. Qian, *Generalization of Fueter's result to* $\mathbf{R}^{n+1}$, Rend. Mat. Acc. Lincei, s.9, v.8 (1997), 111–117.

[Q3] T. Qian, *Fourier analysis on star-shaped Lipschitz surfaces,* J. of Func. Anal, Volume **188**, Issue 2, February, 2002.

[Ro] E.L. Roetman, *Pointwise convergence for expansios in surface harmonics of arbitrary dimension,* J. Reine. Angew. Math. **282** (1976), 1–10.

[St] E.M. Stein, *Singular integrals and differentiability properties of functions,* Princeton University Press, 1970.

[WL] K-Y, Wang and L-Q. Li, *Harmonic analysis and approximation on te unit sphere,* Science Press, Beijing/New York, 2000.

[Zy] A. Zygmund, *Trigonometric Series (Second Edition),* 2 vols., Cambridge, Eng. 1959.

Shuang Liu and Tao Qian
Faculty of Science and Technology
University of Macau
P.O. Box 3001
Macau
e-mail: `shnliu@ucdavis.edu`
`fsttq@umac.mo`

Trends in Mathematics: *Advances in Analysis and Geometry*, 149–160

Cauchy Kernels for some Conformally Flat Manifolds

John Ryan

Abstract. Here we will consider examples of conformally flat manifolds that are conformally equivalent to open subsets of the sphere S^n. For such manifolds we shall introduce a Cauchy kernel, Cauchy integral formula for sections taking values in a spinor bundle and annihilated by a Dirac operator, or generalized Cauchy-Riemann operator. Basic properties of this kernel are examined. We also introduce a Green's kernel and a Green's formula for harmonic sections in this context.

Mathematics Subject Classification (2000). 30G35, 42B30, 53C27, 58J32.

Keywords. Conformally flat manifolds, surgery, Cauchy kernels, Dirac operator.

1. Introduction

In recent years Clifford analysis has proved to be a powerful tool in analyzing problems in n real variables using techniques akin to those employed in the complex plane. In euclidean space the Dirac operator takes the place of the Cauchy-Riemann operator, and one uses analogues of Cauchy kernels to study properties of solutions to the Dirac equation. In turn this analysis reveals a wide array of extremely useful tools including singular Cauchy transforms, Hardy spaces, Kerzmann-Stein kernels, Szegö kernels and Bergman kernels. All of these, and other operators, prove to be very useful in solving boundary value problems and other problems related to classical harmonic analysis. See for instance [18, 15, 14]. In contrast Dirac operators have proved to play important and even crucial roles in mathematical physics and geometry. See for instance [2, 13, 20]. Also in recent times there have been several attempts to reconcile these two approaches to applications of Dirac operators. See for instance [5, 7, 19]. In these works, amongst other things, Cauchy integral formulas for solutions to general Dirac operators are introduced and applied in the setting of general Riemannian manifolds. However, so far in this wider context there is a general lack of an ability to perform explicit computations. This is in sharp contrast to the euclidean setting. However, if one restricts attention to a particular class of manifolds one can obtain, at least for a number of examples, explicit representations of the Cauchy and Green's kernels. This allows one to be able to compute more directly in such settings. The setting we are referring to is the class of conformally flat manifolds. Conformally flat manifolds have been studied extensively in various contexts aside from Clifford analysis. See for instance [4, 6, 25]. They are manifolds that posses an atlas whose transition functions are Möbius transformations. As such they include the n dimensional analogues of the compact Riemann surfaces, namely the sphere, the torus and the k-handled sphere. Other examples of conformally flat manifolds, that do not arise as analogues of Riemann surfaces, are given in [10].

In a number of papers the author and his collaborators, [11, 12, 16, 17, 23, 24] use conformal transformations to develop aspects of Clifford analysis, function theory, potential theory and classical harmonic analysis over certain examples of conformally flat manifolds. The construction that enables one to carry out the calculations and obtain explicit formulas is due to Ahlfors and Vahlen. See for instance [1, 22, 27]. In those works it is shown that any Möbius transformation over the one point compactification of R^n can be expressed as $(ax+b)(cx+d)^{-1}$ where a, b, c and d are members of a Clifford algebra generated from R^n and satisfy certain constraints. See[1] and elsewhere for details.

In [23, 24] we use Cayley transformations to develop basic properties of Clifford analysis and its applications to harmonic analysis on spheres and hyperbolas. Similar results using a different approach were developed by Van Lancker, [28] for the sphere. In fact in [28] a broader function theory is developed as the author considers a family of Dirac type operators. As the hyperbola and the sphere are both real submanifolds of the complex sphere $\{(z_1, \ldots, z_n \in \mathbf{C}^n : z_1^2 + \ldots z_n^2 = 1\}$ many of the results obtained in [28] automatically extend from the context of the sphere to the context of the hyperbola via holomorphic continuation arguments. This is also true for results obtained in [16, 17] where the authors, amongst other results, construct a Green's formula for functions that are harmonic on a domain in S^n.

One way to explicitly construct a conformally flat manifold is to factor out a simply connected subdomain of the sphere or R^n by a Kleinian subgroup of the Möbius group that acts discontinuously on the subdomain. In [11] we consider the cases where the domain is R^n and the Kleinian group is an integer lattice. This enabled the authors in [11] to develop large aspects of Clifford analysis and its applications to harmonic analysis in the context of cylinders and tori. In [12] we extend this technique to establish similar results in the context of amongst other manifolds, including real projective space and the Hopf manifold. The Hopf manifold is $S^1 \times S^{n-1}$ and is obtained by factoring the domain $R^n \backslash \{0\}$ by the discrete subgroup $\{2^k : k \in \mathbf{Z}\}$.

Using Cayley transformations from R^n into S^n one may see that the sphere S^n is a conformally flat manifold. Also by factoring out a simply connected subdomain of either R^n or S^n by a Kleinian group that acts on the domain discontinuously one may construct other examples of conformally flat manifolds. These examples include cylinders, tori, real projective spaces and the Hopf manifold $S^1 \times S^{n-1}$. Cauchy kernels for Dirac operators and Green's kernels for Laplacians are explicitly constructed for spinor bundles over all these manifolds in [23, 24, 16, 17, 11, 12, 28].

While factoring out domains by Kleinian groups is one useful way of constructing conformally flat manifolds here we will look at another technique for constructing some examples of conformally flat manifolds where again we can obtain explicit formulas for the Cauchy kernel and other related kernels. These manifolds are obtained using Möbius transformations to glue together n-spheres or copies of R^n or pairs of hyperbolas. If we glue together a total of k such spheres to an n-sphere we obtain a manifold that can be seen as a sphere with k bumps, or warts, on it. In turn we might also put a finite number of bumps or warts on each wart via the same process. We can continue to place finite numbers of warts on already existing warts. Again this is done via the same surgery or gluing process. If we do not glue any of the warts to more than one sphere or wart we obtain a manifold that is homeomorphic to S^n. As such this type of manifold is simply connected. However its conformal structure is more complicated than that of the sphere. In particular the construction of the spinor bundle structures on these manifolds is more complicated than the construction for the sphere embedded in R^n.

The purpose in this paper is to give an explicit construction of the Cauchy kernels for such manifolds. This construction makes use of the unique continuation property for solutions to the Dirac equation considered here. We also show that a number of known properties of the Cauchy kernel in euclidean space readily carry over to the context considered here. This includes a Hardy space decomposition of L^p spaces for strongly Lipschitz hypersurfaces lying in the manifold. This Hardy space decomposition is expressed in terms of solutions to a Dirac equation defined on the complementary domains to the hypersurface and with appropriate continuation properties to the boundary. The techniques used here mimic existing techniques from classical harmonic analysis in Euclidean space, see for instance [26].

Acknowledgement. The author is grateful to David Calderbank and Michael Eastwood for drawing his attention to the type of conformally flat manifolds introduced here.

2. Preliminaries

We will consider R^n as embedded in the real 2^n dimensional Clifford algebra Cl_n such that under Clifford algebra multiplication $x^2 = -\|x\|^2$ for each $x \in R^n$. So each non-zero vector $x \in R^n$ has a multiplicative inverse given by the Kelvin inverse of x, namely $x^{-1} = \frac{-x}{\|x\|^2}$. If $e_1, \ldots, e_n$ is an orthonormal basis for R^n then $1, e_1, \ldots, e_n, \ldots, e_{j_1} \ldots e_{j_r}, \ldots, e_1 \ldots e_n$ is a basis for Cl_n. Here $1 \leq r \leq n$ and $j_1 < \ldots j_r$. Regarding this basis as an orthonormal basis for Cl_n then the norm, $\|A\|$ of a vector $A \in Cl_n$ can be introduced in the usual way. We will also need the anti-automorphism $\sim: Cl_n \to Cl_n :\sim (e_{j_1} \ldots e_{j_r}) = e_{j_r} \ldots e_{j_1}$. As usual we write $\tilde{A}$ for $\sim A$.

As shown in [1] and elsewhere any Möbius transformation $y = \psi(x)$ over the one point compactification of R^n can be expressed as $(ax+b)(cx+d)^{-1}$ where a, b, c and d are elements of Cl_n and satisfy certain constraints. Those constraints are detailed in [1, 12, 22] and elsewhere. If we regard R^n as embedded in R^{n+1} then Cl_n is a subalgebra of Cl_{n+1}. Further we may regard R^{n+1} as having as orthonormal basis $e_1, \ldots, e_n, e_{n+1}$. In this case we have the Cayley transformation $c(x) = (e_{n+1}x + 1)(x + e_{n+1})^{-1}$ which maps R^n homeomorphically to $S^n \backslash \{e_{n+1}\}$. The Dirac operator in euclidean space is the differential operator $D = \Sigma_{j=1}^n e_j \frac{\partial}{\partial x_j}$. Any Cl_n valued, differentiable function f defined on a domain U in R^n is called a left Clifford holomorphic function or left monogenic function if $Df(x) = 0$ on U. If g is also defined on U and takes values in Cl_n then if g is differentiable then g is called a right Clifford holomorphic function, or right monogenic function if $g(x)D = 0$. Here $g(x)D = \Sigma_{j=1}^n \frac{\partial g(x)}{\partial x_j} e_j$. Basic properties of left and right monogenic functions are described in [3] and elsewhere. The function $G(x) = \frac{x}{\|x\|^n}$ is an example of a function that is both left and right Clifford holomorphic. It is shown in [21] and elsewhere that if $f(y)$ is left Clifford holomorphic in the variable y then $J_1(\psi, x) f(\psi(x))$ is left Clifford holomorphic in the variable x where $J_1(\psi, x) = \frac{\widetilde{cx+d}}{\|cx+d\|^n}$. Moreover, [21], $G(\psi(x) - \psi(y)) = \tilde{J}_1(\psi, y)^{-1} G(x-y) J_1(\psi, x)^{-1}$. Similarly, [21], if $f(y)$ is harmonic then $J_2(\psi, x) f(\psi(x))$ is harmonic in x where $J_2(\psi, x) = \frac{1}{(n-2)\|cx+d\|^{n-2}}$.

In [23] the author notes that if $y = c^{-1}(x)$ where C^{-1} is the inverse of the Cayley transformation then any left Clifford holomorphic function $f(y)$ defined on a domain U in R^n is transformed to a function $J(c^{-1}, x) f(c^{-1}(x))$ defined on a domain $c^{-1}(U)$ lying on S^n. In [23] it is shown that such a function is annihilated by a Dirac operator D_s acting over S^n. An explicit expression for D_s is determined in [8], see also [16, 24]. For each $x \in c^{-1}(U)$ the expression $J(c^{-1}, x) f(c^{-1}(x))$ takes its values in a 2^n dimensional subspace of Cl_{n+1}. What we are describing is a bundle structure over $S^n \backslash \{e_{n+1}\}$. In [23] it is shown via the Cayley transformation that the Cauchy kernel for S^n is $\frac{x-y}{\|x-y\|^n} = \frac{x-y}{2^n(1 - <x,y>)^n}$ where $x, y \in S^{n-1}$. As noted in [23, 24] there is also a Cayley transformation from the open unit disc in R^n to the hyperbola $R^{n,1}$ =span$< e_1, \ldots, e_n, ie_{n+1} >$. Specifically the hyperbola H_n is the manifold $\{\underline{x} \in R^{n,1} : \underline{x} = x_1 e_1 + \ldots + x_n e_n + x_{n+1} i e_{n+1}$ where $x_{n+1} > 0$ and $x_{n+1}^2 - x_1^2 - \ldots - x_n^2 = 1\}$. In this case, again by this Cayley transformation, one sees [23, 24] that the Cauchy kernel is $\frac{x-y}{((x-y)^2)^{\frac{n}{2}}}$. In fact by noting that $iH_n \subset S_{\mathbf{C}}^n = \{z_1 e_1 + \ldots z_{n+1} e_{n+1} \in \mathbf{C} : z_1^2 + \ldots + z_{n+1}^2 = 1\}$ as is S^n one may see that the Cauchy kernels for the sphere and iH_n are restrictions to these real manifolds of the kernel $\frac{(z-w)}{((z-w)^2)^{\frac{n}{2}}}$ where z and $w \in S^n_{\mathbf{C}}$. In turn the Dirac operator D_{iH_n} for iH_n is the restriction to iH_n of the holomorphic continuation to $S^n_{\mathbf{C}}$ of the Dirac operator D_s associated to S^n.

It should be noted that $D^2 = -\triangle$, where $\triangle$ is the Laplacian for euclidean n-space. The fundamental solution to Laplace's equation is $\frac{1}{(n-2)\|x-y\|^{\frac{n-2}{2}}}$. Via the Cayley transformation this kernel transforms to

$$G_{2,s}(u,v) = \frac{1}{(n-2)\|u-v\|^{\frac{n-2}{2}}}$$

for the sphere. Here $u, v \in S^n$. In [16, 17] we explicitly construct a Laplacian $\triangle_s$ over the sphere. The kernel $G_{2,s}(u,v)$ is the Green's kernel for this spherical Laplacian. See [16] for details.

It is now time to introduce conformally flat manifolds.

Definition 1. An n-dimensional manifold M is said to be *conformally flat* if there is an atlas $\mathcal{A}$ of M such that for any pair of chart maps ϕ_1, U_1 and ϕ_2, U_2 the transition function $\psi_{12} = \phi_2\phi_1^{-1}$ is a Möbius transformation wherever this function is defined.

Using Cayley transformations it is clear that S^n and H_n are examples of conformally flat manifolds.

Definition 2. Two n dimensional conformally flat manifolds M_1 and M_2 are said to be *conformally equivalent* if there is a diffeomeorphism $\Psi : M_1 \to M_2$ which with respect to the atlases $\mathcal{A}_1$ and $\mathcal{A}_2$ of M_1 and M_2 is locally a Möbius transformation.

The manifolds we shall consider here are conformally equivalent to connected, open subsets of the sphere, particularly spheres with a finite number of points removed.

We shall now introduce appropriate spinor bundles over conformally flat manifolds. Given a conformally flat manifold M then following [12] one can identify a pair of points $(y, Y) \in (\phi_2(U_2), Cl_n)$ with either $(x, J_1(\psi_{12}, x)Y)$ or $(x, -J_1(\psi_{12}, x)Y)$ where $y = \psi_{12}(x) = (ax+b)(cx+d)^{-1} = (-ax-b)(-cx-d)^{-1}$. It is the two distinct ways of representing the same Möbius transformations in the previous sentence that gives rise to the choice of signs in the construction of these fibers. If a choice of signs in the previous construction can be made for each pair of chart maps in $\mathcal{A}$ that is compatible then we have constructed a spinor bundle $S(M)$ over M and M is regarded as a conformally flat spin manifold. The sphere S^n is an example of such a manifold. In constructing such a bundle we have used the conformal weight functions that preserve Clifford holomorphy. It follows that for any subdomain U of a conformally flat spin manifold we can introduce a section $f : U \to S(M)$ such that locally f pulls back to a left Clifford holomorphic function. Such a section is called a left Clifford holomorphic section. It also follows that we can introduce a Dirac operator D_M that locally pulls back to the euclidean Dirac operator. Moreover $D_M f = 0$ for any left Clifford holomorphic section. It is in this way that conformally flat spin manifolds may be seen as natural generalizations of Riemann surfaces.

Similarly for our conformally flat manifold M one may also follow [12] and identify a pair of points $(y, Y) \in (\phi_2(U_2), Cl_n)$ with $(x, J_2(\psi_{12}, x)Y)$ where, as before, $y = \psi_{12}(x)$ and $J_2(\psi_{12}, x) = \frac{1}{(n-2)\|cx+d\|^{n-2}}$. Note here that as the conformal weight function $J_2(\psi_{12}, x)$ is scalar valued, unlike the conformal weight function $J_1(\psi_{12}, x)$ which contains the term $cx + d$, there is no choice of signs here. It follows that the previous construction can be made for each pair of chart maps in $\mathcal{A}$ and we obtain a bundle $H(M)$ over M. It follows that for any subdomain U of a conformally flat spin manifold we can introduce a section $h : U \to H(M)$ such that f locally pulls back to a harmonic function. Such a section is called a harmonic section. It also follows that we can introduce a Laplacian $\triangle_M$ that locally pulls back to the euclidean Laplacian. It should be noted that the two bundles E and F over M are different bundles.

3. Construction of the Manifolds

We begin by considering two copies, A_1 and A_2 of the annulus $A(0, \frac{1}{r}, r) = \{x \in R^n : \frac{1}{r} < \|x\| < r\}$. We identify each point $x \in A_1$ with the point $-x^{-1} \in A_2$. This is done via a Möbius transformation $\psi' : A_1 \to A_2$. We now consider two copies, S_1 and S_2, of the unit sphere S^n lying in R^{n+1}. Let R_1 be a copy of R^n containing the annulus A_1 and R_2 be a copy of R^n containing A_2. We have Cayley transformations $c_1 : R_1 \to S_1$ and $c_2 : R_2 \to S_2$. Let $C_1 = c_1(A_1)$ and $C_2 = c_2(A_2)$. In fact both C_1 and C_2 are annuli on the spheres S_1 and S_2 respectively. Let $\overline{B}_1$ be the copy in R_1 of the closure of the ball in R^n centered at the origin and of radius $\frac{1}{r}$. We may similarly define $\overline{B}_2$ in R_2. For $i = 1, 2$ let $S_i' = S_i \backslash c_i(\overline{B}_i)$. We adjoin S_1' and S_2' by identifying points in C_1 with points in C_2 via the Möbius transformation $\psi : C_1 \to C_2$ where $\psi = c_2 \psi' c_1^{-1}$. In this way we have used Möbius transformations to "glue" together the two spheres S_1 and S_2. As we have only used Möbius transformations in this "gluing" process the resulting manifold M is conformally flat. We denote this manifold by $S_1 \wedge S_2(r)$. The reason for the r appearing here is that our construction of the manifold depends on our choice of r in the outer radii of the annuli A_1 and A_2. So in fact we have constructed a whole family of conformally flat manifolds. In fact using dilation and inversion it may fairly easily be seen that each of these manifolds is conformally equivalent to the sphere, S^n embedded in R^{n+1} even though $S_1 \wedge S_2(r)$ is not embedded in R^{n+1}. In this way $S_1 \wedge S_2(r)$ is a different representation of S^n than the one usually used in Clifford analysis, see for instance [23, 24, 16, 17, 27].

As mentioned in our introduction the conformally flat manifold $S_1 \wedge S_2(r)$ can be regarded as an n-sphere with a bump or wart. Furthermore we can in turn attach another sphere S_3 to either S_1 or S_2 by the same techniques that we used here to "glue" S_1 and S_2. We would denote such a manifold as $S_1 \wedge S_2 \wedge S_3(r)$. Here we will not consider attaching S_3 to both S_1 and S_2. This particular construction leads to a manifold which is a sphere with a handle. The process we are describing is an elementary form of surgery.

In general we may choose a finite number of points $x_1, \ldots, x_k$ lying in the sphere S^n. We now excise k nonintersecting closed caps $C(x_j)$ from S^n. Each cap $C(x_j)$ is centered at x_j, where $j = 1, \ldots, k$. By the process just described using Cayley transforms we now may attach k spheres $S^1, \ldots, S^k$ to this open subset of S^n. Again we obtain a conformally flat manifold that is conformally equivalent to the sphere S^n.

Similarly we may attach copies of R^n to S^n. To do this consider $A(0, \frac{1}{r}, \infty) = \{x \in R^n : \|x\| > \frac{1}{r}\}$. Now using a Cayley transform c we identify $A(0, \frac{1}{r}, r) \subset A(0, \frac{1}{r}, \infty)$ with the spherical annulus $c(A(0, \frac{1}{r}, r) \subset S^n \backslash c(\overline{B}(0\frac{1}{r})$. In this way we "glued" one copy of R^n to S^n. We will denote this manifold by $R_1 \wedge S_2$. This manifold is conformally equivalent to the sphere with one point removed, or equivalently R^n. However, it is another representation of those manifolds.

By picking finitely many points $x_1, \ldots, x_k$ on S^n we may adapt the process just outlined, and attach finitely many copies of R^n to S^n. Using Kelvin inversion and other Möbius transformations one may see that such a manifold is conformally equivalent to $S^n \backslash \{x_1, \ldots, x_k\}$.

Of course one may repeat this process and "glue" copies of either R^n or S^n to the spheres or copies of R^n that have already been attached to S^n, and so on. If we avoid multiply gluing a sphere or euclidean space to several parts of another sphere or euclidean space we end up with a manifold that is conformally equivalent to either S^n or $S^n \backslash \{x_1, \ldots, x_k\}$.

We can similarly remove a closed disc from one copy of R^n and also from another copy of R^n. Using Kelvin inversion we may now identify two annuli on the remaining open sets Here the annuli have same centers and inner radii as the removed discs. The manifold we obtain is conformally equivalent to $S^n \backslash \{\pm e_n\}$ or to $R^n \backslash \{0\}$.

Similarly we may consider the image, $c_2(B(0,r)$, of a ball $B(0,r)$, $\subset B(0,1)$, under the Cayley transformation c_2. This gives rise to a cap $C(ie_{n+1}, r)$ on the hyperbola H_n centered at ie_{n+1}. As noted in [9, 24] and elsewhere the conformal group acting on the unit ball in R^n is equivalent to the Lorentz group $SO(n,1)$ acting on H_n. So under any action T of the Lorentz group $T(C(f_{n+1}, r)$ is conformally equivalent to $B(0,r)$. Now it is straightforward to introduce, via c_2 and Lorentz transformations, analogues of annuli on hyperbolas and to use the same type of surgery techniques used earlier in this section, namely Kelvin inversion, dilation and translation, to identify together finitely many hyperbolas. If we identify two hyperbolas in this way we get a manifold that is conformally equivalent to an annulus in R^n. It is diffeomorphic to $(0,1) \times S^{n-1}$, $\mathbf{R} \times S^{n-1}$ and $S^n \backslash \{\pm e_n\}$. If we identify three such hyperbolas in this way we get a manifold that is diffeomorphic to a ball with three infinite cylinders attached. The conformally flat manifold here is conformally equivalent to the n-sphere with three non-intersecting closed balls removed, which in turn is conformally equivalent to an open ball in R^n with two non-intersecting closed balls removed.

For all these types of conformally flat manifolds we are able to construct a Cauchy kernel.

4. Construction of the Cauchy Kernel and some Consequences

We will start by working simply with the case $M = S_1 \wedge S_2$. All the other cases are relatively straightforward extensions of this one case. We will denote the kernel by $C_M(x,y)$. This will be a Cl_{n+1} valued function defined on $M \times M \backslash \{(x,x) : x \in M\}$. When x and y both belong to S'_j, for $j = 1,2$, then $C_M(x,y) = \frac{x-y}{\|x-y\|^n}$. Here we are regarding each S'_j as being embedded in a copy of R^{n+1}. Let us denote the part of S'_1 that is identified with part of S'_2 by S_{12}, and the part of S'_2 that is identified with part of S'_1 with S'_2 by S_{21}. Recall that S_{12} and S_{21} are annuli on spheres that are identified with each other via a Möbius transformation $\psi_{12} : S_{12} \to S_{21}$. Let $M_1 = S'_1 \backslash S_{12}$ and $M_2 = S'_2 \backslash S_{21}$. Suppose now that $x \in S'_1$ and y belongs to the part of $S_1 \wedge S_2$ where S_{12} and S_{21} are identified. Then in this case y can be identified with $y_1 \in S_{12}$ or $y_2 \in S_{21}$ where $\psi_{12}(y_1) = y_2$. In this case the kernel $C_M(x,y)$ is again $\frac{x-y_1}{\|x-y_1\|^n}$, which in turn may be identified with $\frac{x-\psi_{12}^{-1}(y_2)}{\|x-\psi_{12}^{-1}(y_2)\|^n} J(\psi_{12}^{-1}, y_2)$. Furthermore if x also belongs to S_{12} then $x_2 = \psi_{12}(x)$ and $C_M(x,y)$ can be identified with $J(\psi_{12}^{-1}, x_2) \frac{\psi_{12}^{-1}(x_2)-\psi_{12}^{-1}(y_2)}{\|\psi_{12}^{-1}(x_2)-\psi_{12}^{-1}(y_2)\|^n} J(\psi_{12}^{-1}, y_2) = \frac{x_2-y_2}{\|x_2-y_2\|^n}$.

Last of all we turn to the case where $x \in S'_1$ and $y \in S'_2$. In this case we note that the Möbius transformation ψ_{12} has a unique continuation to a Möbius transformation $\Psi_{12} : N_1 \to N_2$ where $N_j = S_j \backslash M_j$. In this case $y \in N_2$ and $y = \Psi_{12}(y_1)$ for some $y_1 \in N_1$. In this case $C_M(x,y) = \frac{x-\Psi_{12}^{-1}(y)}{\|x-\Psi_{12}^{-1}(y)\|^n} J(\Psi_{12}^{-1}, y)$.

While we have not covered all possibilities of locations in x and y on M in this construction all remaining possibilities can be constructed easily by adapting the existing constructions presented so far here. However our construction yields a kernel $K(x,y)$ defined for all $(x,y) \in (S_1 \wedge S_2(r) \times S_1 \wedge S_2(r))$ and $x \neq y$.

As stated at the beginning of this section the construction of the Cauchy kernel just given for the conformally flat manifold $S_1 \wedge S_2(r)$ is readily adapted to all the other examples of representations of conformally flat manifolds constructed using surgery in the previous section. So in fact if we let M denote one of those examples we now have a Cauchy kernel $K_M(x,y)$ where $(x,y) \in M \times M$ and $x \neq y$.

From now on let M denote one of the examples of a conformally flat manifold constructed in the previous section. In our construction of the Cauchy kernel $K(x,y)$ and the observed automatic generalization to the other conformally flat manifolds it should be noted that we have constructed a spinor bundle $S(M)$ over M. As pointed out earlier here we may now introduce a Dirac operator D_M for $S(M)$ and for each open subset U we may consider left Clifford holomorphic sections $f : U \to S(M)$ where $D_M f = 0$.

Of course one may also similarly set up right Clifford holomorphic sections. In this case the sections take values in a spinor bundle analogous to $E(M)$ but constructed with multiplication of the conformal weight functions appearing on the right instead of on the left.

Definition 3. A $(n-1)$-dimensional surface, or hypersurface, S, lying in M is called a *Lipschitz hypersurface* if locally S is the image under Möbius transformations of Lipschitz surfaces lying in R^n. If the Lipschitz constants for these hypersurfaces in R^n are bounded then S is said to be *strongly Lipschitz.*

We now have the following version of Cauchy's integral formula for the class of manifolds considered here.

Theorem 1. *Suppose that ∂U is the boundary of a domain U lying in a domain $V \subset M$. Suppose that the closure of V is compact and ∂U is strongly Lipschitz. Furthermore suppose $f : V \to S(M)$ and that f is left Clifford holomorphic. Then for each $y \in U$*

$$f(y) = \frac{1}{\omega_n} \int_{\partial U} K_M(x, y) n(x) f(x) d\sigma_x$$

where $n(x)$ is the unit vector in the tangent space TM_x that is orthogonal to the tangent space $T\partial U_x$, and points outward from U. Further σ is the Lebesgue measure on ∂U.

Having obtained this Cauchy kernel and Cauchy integral formula standard arguments developed in [19] and elsewhere give us the following decomposition result.

Theorem 2. *Suppose that S is a Lipschitz hypersurface lying in M and the complement of S has two components, $S^{\pm}$. Then for $1 < p < \infty$ we have that*

$$L^p(S) = H^p(S^+) \oplus H^p(S^-)$$

where $L^p(S)$ is the space of $S(M)$ valued sections on S that are L^p integrable, and $H^p(S^{\pm})$ is the Hardy p-space of left Clifford holomorphic sections defined on $S^{\pm}$ with L^p non-tangential extension to S.

As seen earlier the construction of the spinor bundle $S(M)$ over M using Möbius transformations can readily be adapted using the conformal weight function J_2 to construct a bundle $H(M)$ over M. For this bundle we have an associated Laplacian $\triangle_M$ and for U an open subset of M a section $h : U \to H(M)$ is said to be a harmonic section if $\triangle_M h = 0$. Furthermore by similar arguments to those used to construct the kernel $K_M(x, y)$ we can replace the weight function J_1 by the harmonic weight function J_2 to produce the Green's kernel $G_M(x, y)$ of M.

Adapting arguments in [21] we have:

Theorem 3. *Suppose U is an open, connected subset of M and $u_1 : U \to S(M)$ is a C^1 section with compact support. Then for each $y \in U$*

$$u_1(y) = \frac{1}{\omega_n} \int_U K_M(x, y)(D_M u(x)) dU(x) \tag{1}$$

where $dU(x)$ is a volume element of U.

Proof (Outline). If $w = \psi(x)$ and $v = \psi(y)$ we have previously noted that $G(v - w) = J_1(\psi, y)^{-1}G(x - y)J_1(\psi, x)^{-1}$. Furthermore the Dirac operator D is intertwined by $J_1(\psi, x)$ and $J_{-1}(\psi, x)$ where $J_{-1}(\psi, x) = \frac{\widetilde{cx+d}}{\|cx+d\|^{n+2}}$. See [21] for details. Also the Jacobian for a change of variable under the Möbius transformation ψ over a domain on M is $\frac{1}{\|cx+d\|^{2n}}$. It follows that the right hand side of Equation 1 is locally invariant under Möbius transformations. As the spinor bundle is constructed from equivalence classes of fibers obtained in local co-ordinates the right hand side of Equation 1 becomes globally invariant under local changes of co-ordinates under Möbius transformations. As Equation 1 is locally valid the result now follows. □

Similarly one may readily adapt arguments of [21] and the previous out line proof of Theorem 3 to deduce the following

Theorem 4. *Suppose U and M are as in the previous theorem and $u_2 : U \to H(M)$ is a C^2 section with compact support. Then for each $y \in U$*

$$u_2(y) = \frac{1}{\omega_n}\int_U G_M(x, y)(\triangle_M u_2(x))dU(x).$$

Of course the sections u_1 and u_2 both have trivial extensions to all of M by simply defining them to be zero on the remainder of M.

Theorem 5. *Suppose U and V are as in Theorem 1 and that the boundary of V is piecewise smooth. Suppose also that $h : V \to H(M)$ is a harmonic section then for each $y \in V$*

$$h(y) = \frac{1}{\omega_n}\int_{\partial U}(K_M(x, y)n(x)h(x) - (G_M(x, y)n(x)(D_M h(x)))d\sigma(x).$$

Proof. Let us first multiply h by a C^∞ partition of unity function $p(x)$ where $p(x) = 1$ on U and is zero in a neighborhood of the boundary of V. Let us denote $p(x)h(x)$ by $h_1(x)$. Then by the previous theorem

$$h_1(y) = \frac{1}{\omega_n}\int_V G_M(x, y)(\triangle_M h_1(x))dV(x)$$

for each $y \in U$.

Let us now subdivide the region $V\backslash U$ via a grid. So The closure of $V\backslash U$ is a finite union of n dimensional manifolds M_j, where $j = 1, \ldots K$, with boundaries ∂M_j. Moreover $M_i \cap M_j$ for $i \neq j$ is either empty or subset of the boundaries. Moreover, each M_j is a subset of an open subset of M that is via a Möbius transformation homeomorphic to an open subset of R^n. It follows from Stokes' Theorem that

$$\int_{M_j} G_M(x, y)(\triangle_M h_1(x)dM_j(x) = \int_{\partial M_j}(K_M(x, y)n(x)h_1(x)$$
$$-G_M(x, y)n(x)(D_M h_1(x)))d\sigma(x).$$

Summing these integrals over j and noting that $\triangle_M h_1 = 0$ on U we have from Stokes' Theorem

$$h(y) = \frac{1}{\omega_n} \int_{\partial U \cup \Theta_1} (K_M(x,y)n(x)h_1(x)$$
$$-G_M(x,y)n(x)(D_M h_1(x)))d\sigma(x)$$

where Θ_1 is a union of $(n-1)$-dimensional manifolds lying in M_j for $j = 1, \ldots, K$.

As we can vary the support of the partition function p we may also vary the set Θ_1. In this way we can obtain a sequence of such sets Θ_k, where $k = 1, \ldots$. Such that $\Theta_{k+1} \subset \Theta_k$ and $\cap_{k=1}^{\infty} \Theta_k$ is empty. Furthermore h_1 is replaced by a sequence of functions h_k such that $h_k(x) = h(x)$ on U and the support of h_{k+1} is contained in an open subset of the support of h_k. It follows that

$$\lim_{k \to \infty} \int_{\Theta_k} (K_M(x,y)n(x)h_k(x) - G_M(x,y)n(x)(D_M h_k(x)))d\sigma(x) = 0.$$

The result follows. □

It now follows that many standard results in Clifford analysis more or less automatically extend to the context described here, including applications to harmonic analysis and boundary value problems.

References

[1] L. Ahlfors, *Möbius transformations expressed through* 2×2 *matrices of Clifford numbers*, Complex Variables **5** (1986), 215–224.

[2] N. Berline, E. Getzler and M. Vergne, *Heat Kernels and Dirac Operators*, Springer-Verlag, Heidelberg, 1992.

[3] F. Brackx, R. Delanghe and F. Sommen, *Clifford Analysis*, Pitman, London, 1982.

[4] T. Branson, G. Olafsson and P. Gilkey, *Invariants of conformally flat manifolds*, Transactions of the A. M. S. **347** (1995), 939–954.

[5] D. Calderbank, *Dirac operators and Clifford analysis on manifolds with boundary*, Max Plank Institute of Mathematics, Bonn, preprint no 96-131, 1996.

[6] A. Chang, J. Qing, P. Yang, *Compactification of a class of conformally flat manifolds*, Invent. Math. **142** (2000), 65–93.

[7] J. Cnops, *An Introduction to Dirac Operators on Manifolds*, Progress in Mathematical Physics, Birkhaüser, Boston, 2002.

[8] J. Cnops and H. Malonek, *An introduction to Clifford analysis* Textos de Matematica, Serie B, 7, Universidade de Coimbra, Departmento de Matematica, Coimbra, 1995.

[9] J. Elstrodt, F. Grunewald and J. Mennicke, *Vahlen's group of Clifford matrices*, Math. Z. **196** (1987), 369–390.

[10] W. Goldman and Y. Kamishima, *Conformal automorphism and conformally flat manifolds*, Transactions of the AMS **323** (1991), 797–810.

[11] R. S. Krausshar and J. Ryan, *Clifford and harmonic analysis on spheres and tori*, to appear in Revista Matematica Iberoamericana.

[12] R. S. Krausshar and J. Ryan, *Some conformally flat spin manifolds, Dirac operators and automorphic forms*, to appear.

[13] H. B. Lawson and M-L. Michelson, *Spin Geometry*, Princeton University Press, Princeton, 1989.

[14] C. Li, A. McIntosh and T. Qian, *Clifford algebras, Fourier transformations, and singular convolution operators on Lipschitz surfaces*, Revista Matematica Iberoamericana **10** (1994), 665–721.

[15] C. Li, A. McIntosh and S. Semmes, *Convolution singular integrals on Lipschitz surfaces*, J. of the AMS. **5** (1992), 455–481.

[16] H. Liu and J. Ryan, *Clifford analysis techniques for spherical pde*, J. Fourier Anal. Appl. **8** (2002), 535–563.

[17] H. Liu and J. Ryan, *The conformal Laplacian on spheres and hyperbolas via Clifford analysis*, Clifford Analysis and its Applications, F. Brackx et al, editors, Kluwer, Dordrecht, 2001, 255–266.

[18] A. McIntosh, *Clifford algebras, Fourier theory, singular integrals and harmonic functions on Lipschitz domains*, Clifford Algebras in Analysis and Related Topics, edited by J. Ryan, CRC Press, Boca Raton, 1996, 33–87.

[19] M. Mitrea, *Generalized Dirac operators on non-smooth manifolds and Maxwell's equations*, J. Fourier Anal. Appl. **7** (2001), 207–256.

[20] J. D. Moore, *Lectures on Seiberg-Witten Invariants*, Lecture Notes in Mathematics, no 1629, Springer Verlag, Heidelberg, 2001.

[21] J. Peetre and T. Qian, *Möbius covariance of iterated Dirac operators*, J. Australian Math. Soc., Ser. A **56** (1994), 403–414.

[22] I. Porteous, *Clifford algebras and the Classical Groups*, Cambridge University Press, Cambridge, 1995.

[23] J. Ryan, *Dirac operators on spheres and hyperbolae*, Bolletin de la Sociedad Matematica a Mexicana **3** (1996), 255–270.

[24] J. Ryan, *Clifford analysis on spheres and hyperbolae*, Mathematical Methods in the Applied Sciences **20** (1997), 1617–1624.

[25] R. Schoen and S-T. Yau, *Conformally flat manifolds, Kleinian groups and scalar curvature*, Inventiones Mathematicae **92** (1988), 47–71.

[26] E. Stein, *Singular Integrals and Differentiability Properties of Functions*, Princeton University Press, Princeton, 1970.

[27] K. Th. Vahlen, *Über Bewegungen und Complexe Zahlen*, Math. Ann. **55** (1902), 585–593.

[28] P. Van Lancker, *Clifford analysis on the sphere*, Clifford Algebras and their Applications in Mathematical Physics, Aachen 1996, V. Dietrich et al, editors, Kluwer, Dordrecht, 1998, 201–215.

John Ryan
Department of Mathematics, University of Arkansas
Fayetteville, AR 72701
USA
e-mail: jryan@uark.edu

Trends in Mathematics: *Advances in Analysis and Geometry*, 161–185

Clifford Analysis on the Space of Vectors, Bivectors and ℓ-vectors

Irene Sabadini and Frank Sommen

Abstract. In this paper we propose several extensions of Clifford analysis. First, we recall the most important definitions and properties for the usual Dirac operator, i.e. the operator defined for functions of a vector variable, and we discuss the importance of using differential forms. Next we discuss Clifford analysis in several vector variables and in particular so called "symplicial monogenics", that are related to the representation theory of Spin(m). Then we study several canonically defined Clifford algebras valued systems of differential equations defined on the space of bivectors, for which we provide a rather complete theory, and more in general on the space of ℓ-vectors.

Acknowledgments. The second author has been sponsored for this research by: FWO-Krediet aan Navorsers: 1.5.106.02.

Mathematics Subject Classification (2000). 30G35.

1. Introduction

Clifford analysis mostly involves the theory of nullsolutions to the Dirac operator $\partial_{\underline{x}} = \sum e_j \partial_{x_j}$, where e_j, $j = 1, \ldots, m$ are the generators of the standard Clifford algebra. A first extension of this setting involves the use of differential forms on the space of vectors of the type $F = \sum F_A dx_{a_1} \ldots dx_{a_k}$, whose components are Clifford algebra valued. If the components are monogenic, the forms are said to be monogenic. Differential forms allow the definition of Dirac operators on embedded surfaces. In section 2 we recall the most important definitions and properties for Clifford algebras and we discuss the problem of characterizing when an ℓ-vector X is a ℓ-blade, i.e. a pure wedge product $X = \underline{x}_1 \wedge \ldots \wedge \underline{x}_\ell$ of vectors. For a bivector to be a blade it is necessary and sufficient that X^2 is scalar; for ℓ-vectors the characterization is much more complicated. In section 3 we recall the basic definition and elementary properties of monogenic functions, i.e. solutions of the Dirac operator. For the current state of the art on monogenic function theory and its applications, the reader is referred to the survey paper [7] and to the other references in the main text. In that section we also discuss the use of differential form calculus in Clifford analysis. In section 4 we present an overview on the theory of functions $f(\underline{x}_1, \ldots, \underline{x}_\ell)$ of vector variables in the kernel of several Dirac operators. Of special importance are functions of a wedge product $f(\underline{x}_1, \ldots, \underline{x}_\ell) = F(\underline{x}_1 \wedge \ldots \wedge \underline{x}_\ell)$, F being defined on the space $\mathbb{R}_{m;\ell}$ of ℓ-vectors, and more in general functions of the form $f(\underline{x}_1, \ldots, \underline{x}_\ell) = F(\underline{x}_1, \underline{x}_1 \wedge \underline{x}_2, \ldots, \underline{x}_1 \wedge \ldots \wedge \underline{x}_\ell)$. Monogenic functions of this form are called "symplicial monogenics": they form a refinement of the harmonic functions of a matrix-variable discussed in [12] and play a crucial role in the representation theory of Spin(m). This discussion naturally leads to the consideration of functions defined on the space $\mathbb{R}_{m;\ell}$ of ℓ-vectors, and to the study of canonically defined systems of partial differential equations defined there, with values in a Clifford algebra. These are systems on $\mathbb{R}_{m;\ell}$ which correspond to the embedding of symplicial monogenics. This extended form of Clifford analysis is at its very beginning and still contains many unanswered problems. In this section we discuss two open questions that will be treated in this paper and we show how the Fischer decomposition gives an answer to the first of them. In section 5 we consider the case of systems on the space $\mathbb{R}_{m;2}$ of bivectors and we are able to provide a rather complete theory also giving an answer to the second question we addressed. Finally, in section 6 we discuss the ℓ-vector case, discussing many open problems while in section 7 we shortly describe the case of monogenic systems on the Stiefel manifold.

2. The standard Clifford algebra

We start this section by recalling some standard notions about Clifford algebras and some basic results that will be useful in the sequel. For a complete treatment of these topic, we refer the reader to [8], [13], [22].

Definition 2.1. The standard *Clifford algebra* $\mathbb{R}_m$ is the free associative algebra generated by the set of symbolic vectors $\{e_1, \ldots, e_m\}$ together with the defining relations $e_j e_k + e_k e_j = -2\delta_{jk}$, δ_{jk} being the Kronecker delta.

Remark 2.2. A canonical basis for $\mathbb{R}_m$ is obtained as follows: let $A = \{a_1, \ldots, a_k\}$ be a subset of $M = \{1, \ldots, m\}$ ordered in the lexicographic way $a_1 < \ldots < a_k$ and let $|A|$ be its cardinality; if we put $e_A = e_{a_1} \ldots e_{a_k}$, we can write any element $a \in \mathbb{R}_m$ as

$$a = \sum_{A \subseteq M} a_A e_A = \sum_{k=0}^{m} [a]_k$$

where $[a]_k = \sum_{|A|=k} a_A e_A$ is called the k-vector part of a.

The following result is well known:

Proposition 2.3. *The set $\{e_A\}$ turns out to be a basis for $\mathbb{R}_m$, so that dim $\mathbb{R}_m = 2^m$.*

Denoting by $\mathbb{R}_{m;k}$ the subspace of all k-vectors in $\mathbb{R}_m$, i.e. the image of $[\cdot]_k$, we have a multivector decomposition, also called "multivector structure":

$$\mathbb{R}_m = \mathbb{R}_{m;0} \oplus \mathbb{R}_{m;1} \oplus \ldots \oplus \mathbb{R}_{m;m}.$$

Note that $\mathbb{R}_{m;0} = \mathbb{R}$ is the subspace of scalars and $[a] = [a]_0$ is the scalar part of a Clifford number a. We also have that $\mathbb{R}_{m;1} = \mathbb{R}^m$ is the Euclidean vector space generated by the units $e_1, \ldots, e_m$ also called the space of Clifford vectors. From now on, the elements in $\mathbb{R}_{m;1}$ with the exception of the units e_i, will be underlined. Because of the above multivector structure, general Clifford numbers are also called multivectors. The set $\mathbb{R}_{m;m}$ contains elements of the form se_M, where $s \in \mathbb{R}$ and the element $e_M = e_1 \ldots e_m$ is called the (unit) pseudoscalar.

Remark 2.4. Note that the multivector structure is not unique. Indeed, Clifford algebras may be introduced with respect to general quadratic forms g_{jk}, replacing the defining relations by

$$\varepsilon_j \varepsilon_k + \varepsilon_k \varepsilon_j = -2g_{jk};$$

by Sylvester's law one can always choose g_{jk} to be of the form $g_{jk} = -\delta_{jk}$ for $j = 1, \ldots, p$, $g_{jk} = \delta_{jk}$ for $j = p+1, \ldots, p+q$ and $g_{jk} = 0$ otherwise.

There are several isomorphisms between Clifford algebras of different signatures (see [8], [18], [22]), so that a Clifford algebra may admit several multivector structures. For analysis, it however suffices to consider the complexified Clifford algebras $\mathbb{C}_m = \mathbb{R}_m \otimes_{\mathbb{R}} \mathbb{C}$ since a general Clifford algebra $\mathbb{R}_{p,q,r}$ with generators $\{\epsilon_1, \ldots, \epsilon_p; e_1, \ldots, e_q; f_1, \ldots, f_r\}$ and relations $\epsilon_j^2 = 1$, $e_j^2 = -1$, $f_j^2 = 0$ may always be embedded in $\mathbb{C}_{p+q+2r}$ putting

$$\epsilon_j = ie_{j+q}, \qquad f_j = \frac{1}{2}(e_{2j+p+q-1} - ie_{2j+p+q}).$$

The invariant automorphisms on $\mathbb{R}_m$, i.e. the automorphisms leaving the multivector structure invariant, are:

(i) the main involution $\widetilde{ab} = \tilde{a}\tilde{b}$; $\tilde{e}_j = -ej$;
(ii) the conjugation $\overline{ab} = \bar{b}\bar{a}$; $\bar{e}_j = -e_j$;
(iii) the reversion $(ab)^* = b^* a^*$; $e_j^* = e_j$;

while on the complex Clifford algebra $\mathbb{C}_m$ we may also consider the

(iv) Hermitian conjugation $(a + ib)^+ = \bar{a} - i\bar{b}$; $a, b \in \mathbb{R}_m$.

Remark 2.5. The main involution has two eigenspaces $\mathbb{R}_{m,+}$ and $\mathbb{R}_{m,-}$ also referred to as the spaces of multivectors of even or odd type and $\mathbb{R}_{m,+}$ is the so-called even subalgebra which is isomorphic to $\mathbb{R}_{m-1}$ if we identify the bivectors $e_1 e_m = E_1$, ..., $e_{m-1} e_m = E_{m-1}$ with the new Clifford algebra generators $E_1, \ldots, E_{m-1}$. We also use the notation $a = [a]_+ + [a]_-$ with $[a]_\pm \in \mathbb{R}_{m,\pm}$.

Remark 2.6. The Hermitian conjugation leads to a Hermitian inner product on $\mathbb{C}_m$ given by $\{a, b\} = [a^+ b]_0$. In the odd case $m = 2n+1$, the pseudoscalar $e_M = e_1 \ldots e_m$ is central and for $m = 4l+1$ we have that $e_M^2 = -1$ while for

$m = 4l+3$, $e_M^2 = +1$. Hence in the complex Clifford algebra $\mathbb{C}_{2n+1}$, putting either $W = e_M$ or $W = ie_M$ there always exists a central element W with $W^2 = 1$; we may introduce the projectors $\frac{1}{2}(1 \pm W)$ giving rise to a direct sum decomposition into mutually annihilating subalgebras

$$\mathbb{C}_m = \frac{1}{2}(1+W)\mathbb{C}_{m;+} \oplus \frac{1}{2}(1-W)\mathbb{C}_{m;+},$$

where, by Remark 2.5, $\mathbb{C}_{m;+}$ is isomorphic to $\mathbb{C}_{2n}$ so that it suffices to determine $\mathbb{C}_{2n}$. This follows from the isomorphisms

$$\mathbb{C}_{2m} = \mathbb{C} \otimes \mathbb{R}_{m,m}, \quad \mathbb{R}_{m,m} = \mathrm{End}(\mathbb{R}_m) = \mathbb{R}(2^m)$$

so that $\mathbb{C}_{2m} = \mathbb{C}(2^m)$ is the full complex $2^m \times 2^m$ matrix algebra.

The isomorphism $\mathbb{R}_{m,m} = \mathrm{End}(\mathbb{R}_m)$ is basically equivalent to the definition of Clifford algebras in terms of spinor spaces (see [5]) and follows from the consideration of the basic endomorphisms

$$e_j : a \to e_j a; \qquad e_{j|} : a \to \tilde{a} e_j$$

which satisfy the defining relations of $\mathbb{R}_{m,m}$ (see also [12], [33]). The above leads to the introduction of the Witt basis

$$f_j = \frac{1}{2}(e_j - e_{j|}), \qquad f'_j = \frac{1}{2}(e_j + e_{j|})$$

with the following identities

$$f_j f_k = -f_k f_j, \qquad f'_j f'_k = -f'_k f'_j, \qquad f_j f'_k + f'_k f_j = -\delta_{jk}, \qquad f_k^+ = -f'_k.$$

Moreover, introducing the primitive idempotent $I = I_1 \dots I_m$, $I_j = f_j f_j^+$, the vector space $\mathbb{R}_m$ may also be identified with the minimal left ideal $S_m = \mathbb{R}_{m,m} I = \mathbb{R}_m I$ and the action of $a \in \mathrm{End}(\mathbb{R}_m)$ on $b \in \mathbb{R}_m$ (denoted by $a[b]$) is obtained from the left multiplication

$$a[b]I = abI.$$

From the geometrical point of view, not only the Clifford product but also the multivector structure plays an essential role; it is determined by the identification $\mathbb{R}^m = \mathbb{R}_{m;1}$ and the definition of a Grassmann product (also called wedge product) in terms of the Clifford product by

$$\underline{v}_1 \wedge \dots \wedge \underline{v}_k = \frac{1}{k!} \sum \mathrm{sgn}\pi \; \underline{v}_{\pi(1)} \cdots \underline{v}_{\pi(k)} = [\underline{v}_1 \cdots \underline{v}_k]_k$$
$$= \frac{1}{2}(\underline{v}_1 \wedge \dots \wedge \underline{v}_{k-1} \underline{v}_k + (-1)^{k+1} \underline{v}_k \underline{v}_1 \wedge \dots \wedge \underline{v}_{k-1}).$$

Definition 2.7. The elements of the pure wedge product form $v = \underline{v}_1 \wedge \dots \wedge \underline{v}_k$ are called *pure k-vectors* or *k-blades.*

If $m \leq 3$, it can be shown that all k-vectors are k-blades, but this is no more true when $m > 3$.

Although the k-blades span all of $\mathbb{R}_{m;k}$, they form a conic surface inside $\mathbb{R}_{m;k}$ consisting of rays λv, $\lambda > 0$ based on the compact submanifold $\mathcal{G}(m,k)$ of unit k-blades i.e. satisfying the condition $v^2 = +1$ or $v^2 = -1$, depending on k.

Remark 2.8. Note that $\mathcal{G}(m,k)$ corresponds to the manifold of all oriented k-subspaces of $\mathbb{R}^m$, which is of crucial importance in projective geometry; it is a double covering of the classical Grassmann manifold $G(m,k)$ consisting of k-dimensional subspaces of $\mathbb{R}^m$.

Let us now introduce the following:

Definition 2.9. The *spin group* is the set given by

$$\mathrm{Spin}(m) = \{s = \underline{w}_1 \dots \underline{w}_{2\ell} \quad : \underline{w}_j \in \mathbb{R}_{m;1},\ \underline{w}_j^2 = -1\}.$$

It is well known that there is a double covering $\phi : \mathrm{Spin}(m) \to SO(m)$ given by $\phi(s)\underline{x} = s\underline{x}\bar{s}$ which extends to an element of $\mathrm{End}(\mathbb{R}_m)$ using the same formula $\phi(s)a = sa\bar{s}$. Note that for $k < [m/2]$ the above representation leaves $\mathbb{R}_{m;k}$ invariant and the restrictions of ϕ to $\mathbb{R}_{m;k}$ are irreducible representations of $\mathrm{Spin}(m)$. For $k = m/2$, m even, the space $\mathbb{R}_{m;k}$ splits further into two inequivalent irreducible representations. One may also consider the representation of $\mathrm{Spin}(m)$ obtained by left multiplication $l(s)a = sa$, which becomes irreducible on minimal left ideals (spinor spaces). The above representation constitutes the complete set of fundamental irreducible representations of $\mathrm{Spin}(m)$ from which all other may be obtained using tensor products (Cartan composition method).
The Lie algebra of $\mathrm{Spin}(m)$ is described in the following:

Proposition 2.10. *The Lie algebra of Spin(m) is the set of Clifford bivectors $b \in \mathbb{R}_{m;2}$.*

For the proof of this result the reader may consult see [12], [19], or [22]. The set of bivectors admits a decomposition in terms of 2-blades (see [19]) as follows

Proposition 2.11. *Bivectors b admit an eigenvalue decomposition of the form*

$$b = r_1 B_1 + \dots + r_n B_n, \qquad n = [m/2],$$

where $r_i \in \mathbb{R}$, $r_1 \geq r_2 \geq \dots \geq r_n \geq 0$ and $B_j = \underline{v}_j \wedge \underline{w}_j$ are commuting unit 2-blades, i.e. $B_j B_k = B_k B_j$ and $B_j^2 = -1$.

It is still an open problem whether similar special decompositions may be obtained for the general k-vectors. In any case, the above decomposition leads to the following important lemma:

Lemma 2.12. *An element $b \in \mathbb{R}_{m;2}$ is a 2-blade if and only if b^2 is scalar.*

Proof. If $b \in \mathbb{R}_{m;2}$ is a 2-blade, $b = \underline{v} \wedge \underline{w}$ then a simple computation shows that b^2 is a scalar. Let us suppose that the bivector b is such that b^2 is scalar. Then, using the previous decomposition, we have

$$b^2 = (r_1 B_1 + \dots + r_n B_n)^2 = -\sum_{i=1}^{n} r_i^2 + \sum_{1 \leq i < j \leq n} r_i r_j B_i B_j$$

that is scalar if and only if $r_i r_j = 0$ for $1 \leq i < j \leq n$. Let us fix the index $1 < j \leq n$: then either $r_i = 0$ for all i and, choosing $j = n$ we get the statement, or $r_j = 0$; in this second case, we can iterate the argument and we get $r_j = 0$ for all $1 < j \leq n$. The statement follows. □

Remark 2.13. One may wonder whether the above property generalizes to k-vectors X i.e. whether X^2 is scalar implies that X is a k-blade (the converse being obvious), but this is not the case. The first case for which the space of trivectors has a dimension greater than that of the space of 3-blades appears for $m = 6$, i.e. $\mathbb{R}_{6;3}$. In fact the trivector $e_{123} + e_{456}$ is not a blade and has a scalar square. It may easily be verified that any trivector for which the square X^2 is scalar and the equation $X\underline{v} - \underline{v}X$ has a nonzero vector solution $\underline{v}$, must be a 3-blade.

In the case of k-vectors, we mention the following result (see [10]):

Proposition 2.14. *Let $k \leq \frac{m}{2}$, then a k-vector $X \in \mathbb{R}_{m;k}$ is a k-blade if and only if for any $(k-1)$-vector Y the wedge product $[YX]_1 \wedge X$ vanishes, where $[YX]_1$ denotes the vector part of YX.*

Remark 2.15. Now clearly it suffices to consider the case $Y = e_{a_1 \ldots a_{k-1}}$ and one gets a system of quadratic polynomial equations for the cones of k-blades.

3. Clifford analysis on the space of vectors

Clifford analysis started with the attempt to generalize the Cauchy–Riemann system $(\partial_x + i\partial_y)f(x,y) = 0$ to the higher dimensional setting. Among the most successful higher dimensional generalization there is the theory of holomorphic functions in several complex variables and the Hamilton–Fueter system

$$(\partial_x + i\partial_y + j\partial_z + k\partial_u)f(x,y,z,u) = 0.$$

This system may be incorporated into Clifford analysis by putting $i = e_1$, $j = e_2$, $k = e_1e_2$ or by putting $i = e_2e_3$, $j = e_3e_1$, $k = e_1e_2$: it is an example of generalized Cauchy–Riemann operators which is not vector valued. In the higher dimensional setting $\mathbb{R}^m$ the most commonly used operator is the Dirac operator

$$\partial_{\underline{x}} = \sum_{j=1}^{m} e_j \partial_{x_j},$$

acting on $\mathbb{C}_m$-valued functions defined on $\mathbb{R}^m$, and the Weyl operator given by

$$\partial_{x_0} + \partial_{\underline{x}} = \partial_{x_0} + \sum_{j=1}^{m} e_j \partial_{x_j},$$

which may be interpreted as the modified Dirac operator

$$-e_{m+1}(e_1\partial_{x_1} + \ldots + e_{m+1}\partial_{x_{m+1}})$$

by putting $x_{m+1} = x_0$ and using the fact that the bivectors $E_j = -e_{m+1}e_j$ play the role of new vectors generating the Clifford algebra $\mathbb{R}_m$. For the basic function theoretic properties of monogenic functions, which are nullsolutions of generalized Cauchy–Riemann systems, we refer (without claiming completeness) to the books

[4], [8], [12] and [18]. The three most fundamental theorems which give rise to special solutions, and which one would try to generalize to general systems of partial differential equations in Clifford analysis, are

(i) the *Cauchy-Kowalewski formula* $f(x_0, \underline{x}) = \exp(-x_0\partial_{\underline{x}})f(\underline{x})$, solving the Weyl equation;

(ii) the *Fischer decomposition* stating that every homogeneous polynomial $R_k(\underline{x})$ of degree k in $\mathbb{R}^m$ admits a unique decomposition of the form: $R_k(\underline{x}) = \sum_{s=0}^{k} \underline{x}^s P_{k-s}(\underline{x})$, where $P_l(\underline{x})$, $l = 0, \ldots, k$ is spherical monogenic of degree l (i.e. a homogeneous monogenic polynomial of degree l);

(iii) the *Cauchy-Pompeiu formula* which is a direct consequence of Stokes' formula together with the identity

$$d(fd\sigma g) = \big((f\partial_{\underline{x}})g + f(\partial_{\underline{x}}g)\big)\, dx_1 \ldots dx_m,$$

where $d\sigma$ is the Clifford $m-1$ form given by

$$d\sigma = \sum(-1)^{j+1}e_j\widehat{dx_j}, \qquad \widehat{dx_j} = dx_1 \ldots dx_{j-1}dx_{j+1} \ldots dx_m$$

In distributional form, in the case $\partial_{\underline{x}}g = 0$ and $f(\underline{x})$ is equal to the fundamental solution $E(\underline{x} - \underline{y})$, it gives rise to the Cauchy integral formula.

A collection of interesting first order systems was presented in our paper [26], which includes also the complexes for certain special examples. We will list here a few examples that follow directly from the Dirac operator itself.

Example 3.1. In the even dimensional case $m = 2n$ we may put $e_{n+j} = ie_{j|}$, where $\{e_j, e_{j|} \ : \ j = 1, \ldots, n\}$ are the generators of $\mathrm{End}(\mathbb{R}_m)$ which may act on spinor valued function of the form fI, I being the primitive idempotent. Choosing f to take values in $\mathbb{R}_{n,\pm}$ then we get an equation of the form

$$\partial_{\underline{x}}f \pm if\partial_{\underline{y}} = 0$$

with $\partial_{\underline{x}} = \sum_{j=1}^{n} e_j\partial_{x_j}$, $\partial_{\underline{y}} = \sum_{j=1}^{n} e_j\partial_{y_j}$. In this way one essentially obtains the Dirac equation but with less components. Note also that the Dirac equation in the phase space $\mathbb{R}^{m,m}$ is given by the ultra-hyperbolic system

$$\partial_{\underline{x}}f = \pm f\partial_{\underline{y}}.$$

Example 3.2. In terms of the Witt basis we have that $e_j = f_j + f_j'$; $e_{j+n} = i(f_j - f_j')$ and so, putting $y_j = x_{j+n}$, the Dirac operator may be rewritten as:

$$\partial_{\underline{x}} = \sum f_j'(\partial_{x_j} - i\partial_{y_j}) + \sum f_j(\partial_{x_j} + i\partial_{y_j}) = 2(\partial_{Z^+} - \partial_Z),$$

$$\partial_Z = \sum f_j^+\partial_{z_j}, \qquad \partial_{Z^+} = \sum f_j\partial_{\bar{z}_j}.$$

Example 3.3. Also the classical Hodge system (see [14]) may be reobtained from Clifford analysis. On one hand, taking $f : \mathbb{R}^m \to \mathbb{R}_{m;k}$ a k-vector field, then the condition $\partial_{\underline{x}}f = 0$ leads to the Hodge system

$$\partial_{\underline{x}} \wedge f = \frac{1}{2}(\partial_{\underline{x}}f + (-1)^k f\partial_{\underline{x}}) = 0, \qquad \partial_{\underline{x}} \cdot f = \frac{1}{2}(\partial_{\underline{x}}f - (-1)^k f\partial_{\underline{x}}) = 0,$$

that is equivalent to $\partial_{\underline{x}} f = f\partial_{\underline{x}} = 0$. Conversely, the relations $\partial_{\underline{x}} f = f\partial_{\underline{x}} = 0$ together also imply that the k-vector part $[f]_k$ of f is still monogenic, i.e. $\partial_{\underline{x}}[f]_k = 0$, leading to the Hodge system.

A different generalization of the classical Hodge system is the so-called "modified Clifford analysis" which uses the Poincaré metric instead of the Euclidean metric to derive the governing system of differential equations, and Hodge's definition of harmonicity via the Hodge-star-operator (see [11], [16], [17]).

We finish this section with an argument in favor of combining the use of differential forms with Clifford numbers, in spite of the fact that (see example 3.3) k-vectors behave much in the same way like k-forms. But from a geometric and analytic point of view, they are different tools, each of them leading to a much wider class of interesting Clifford analysis systems on the space of vectors. This point of view has lead to interesting results (see [7], [9], [20], as well as [29], [32]). The importance of special Clifford differential forms (from now on called Cliffforms, for short) is already clear from the use of the special $(m-1)$-form $d\sigma$ leading to the Cauchy–Pompeiu formula. For a good reference on difforms we refer to [37].

Definition 3.4. A general *k-Cliffform* is an expression of the type

$$F = \sum F_A dx_A,$$

where the F_A are Clifford algebra valued functions and $dx_A = dx_{a_1} \dots dx_{a_k}$, where $A = \{a_1, \dots, a_k\} \subseteq \{1, \dots, m\}$, $a_1 < \dots < a_k$.

If one defines the exterior derivative by $d = \sum dx_j \partial_{x_j}$, one has the fundamental Stokes' formula

$$\int_U dF = \int_{\partial U} F.$$

Let us define some endomorphisms on Cliffforms, called basic contraction operators and denoted by $\partial_{x_j}\rfloor$, as

$$\partial_{x_j}\rfloor(dx_k F) = \delta_{jk} F - dx_k \partial_{x_j}\rfloor F,$$

so that they behave like differential operators. The Lie derivative of an operator $v = \sum v_j(x)\partial_{x_j}$ is then defined in terms of the operators d and the contraction $v\rfloor = \sum v_j(x)\partial_{x_j}\rfloor$ and acts on Cliffforms as $\mathcal{L}_v F = v\rfloor dF + dv\rfloor F$. This definition is classical but the new thing about Cliffforms is that one may choose the components $v_j(x)$ to be Clifford algebra valued, leading to many new special operators on Cliffforms; the most important is the Dirac operator which corresponds to the case $v_j(x) = e_j$. It is given by the "first basic identity" (see [29])

$$\mathcal{L}_{\partial_{\underline{x}}} F = \partial_{\underline{x}} F = d\partial_{\underline{x}}\rfloor F + \partial_{\underline{x}}\rfloor dF$$

and, since v_j are constants, $\partial_{\underline{x}}$ acts componentwise. In particular we have that $d\sigma_{\underline{x}} = \partial_{\underline{x}}\rfloor dV_{\underline{x}}$ with $dV_{\underline{x}} = dx_1 \dots dx_m$ so that the above Lie derivative definition readily leads to the basic formula

$$d(d\sigma_{\underline{x}} f) = (d\partial_{\underline{x}}\rfloor + \partial_{\underline{x}}\rfloor d) dV_{\underline{x}} f = dV_{\underline{x}} \partial_{\underline{x}} f$$

which hence becomes a special case. The Cauchy–Pompeiu form $f d\sigma g$ was generalized in [29] to the case of general k-Clifforms as follows: let F be a k-Clifform and G be an ℓ-Clifform; then the monogenic coupling of F and G is defined by

$$\{F, G\} = (F\partial_{\underline{x}}\rfloor)G - F(\partial_{\underline{x}}\rfloor G)$$

and we have the second basic identity

$$d(\{F, G\}) = \{dF, G\} + (-1)^{k-1}\{F, dG\} + (-1)^{k-1}[(F\partial_{\underline{x}})G + F(\partial_{\underline{x}}G)]$$

which, in the case $\ell = k - m$, simplifies to the "duality principle"

$$d(\{F, G\}) = (-1)^{k-1}[(F\partial_{\underline{x}})G + F(\partial_{\underline{x}}G)].$$

Using Stokes'theorem, one obtains Cauchy–Pompeiu formulae for Clifforms. Moreover the analogue of the Cauchy kernel is obtained as follows: introduce the form

$$dV_k(\underline{x}, \underline{y}) = \sum_{|A|=k} \mathrm{sgn}A \; dx_A dy_{M\setminus A}$$

where $A = \{a_1, \ldots, a_k\} \subseteq M = \{1, \ldots, m\}$ with $a_1 < \ldots < a_k$ and where $\mathrm{sgn}A$ is the signature of the permutation $(a_1, \ldots, a_k, b_1, \ldots, b_{m-k})$ of $(1, \ldots, m)$, with $M \setminus A = \{b_1, \ldots, b_{m-k}\}$, $b_1 < \ldots < b_{m-k}$. Then the indicatrix of an $(m-k)$-chain S is defined by

$$I(S)(\underline{x}) = \int_{\underline{y}\in S} E(\underline{x} - \underline{y}) dV_k(\underline{x}, \underline{y}),$$

where

$$E(\underline{x}) = \frac{1}{A_m} \frac{\underline{x}}{|\underline{x}|^m}$$

is the Cauchy kernel. It is not hard to prove that $dI(S)(\underline{x}) = \pm I(\partial S)$ so that in case of an $(m-k)$-chain S, we have both the closedness $dI(S) = 0$ as well as the monogenicity $\partial_{\underline{x}} I(S) = 0$ of the k-Clifform $I(S)$. From the first basic identity it then also follows that the contraction $\partial_{\underline{x}}\rfloor I(S)(\underline{x})$ is a closed $(k-1)$-form so that for any $(k-1)$-cycle S' in $\mathbb{R}^m \setminus S$, the number

$$I(S', S) = \int_{\underline{x}\in S'} \int_{\underline{y}\in S} \partial_{\underline{x}}\rfloor dV_k(\underline{x}, \underline{y}) E(\underline{x} - \underline{y})$$

is invariant under deformations of S or S' within the same homology; it is in fact the winding or linking number of S' and S.

Remark 3.5. In a problem session (see [24]) S. Semmes described Clifford analysis as a codimension one theory due to the fact that, in most papers, only surface integrals over codimension one hypersurfaces are considered. The above treatment of Clifforms clearly extends Clifford analysis on the space of vectors to the higher codimension case. Yet, it is also clear to us that, without the simultaneous use of both Clifford numbers and differential forms, such generalization would be hard to obtain.

Also the Dirac operator on embedded surfaces can be obtained from the first basic identity. Indeed, let U be an oriented k-surface in $\mathbb{R}^m$ and $C \subseteq U$ be a compact set with smooth boundary; then, using Stokes' theorem, we obtain the general relation for k-Cliffforms

$$\int_{\partial C} \partial_{\underline{x}} \rfloor F = \int_C (\partial_{\underline{x}} - \partial_{\underline{x}} \rfloor d) F$$

Next, consider the basic one-form $d\underline{x} = \sum e_j dx_j$ (see also [32]); then it is not hard to see that

$$d\underline{x}^m = e_M dV_{\underline{x}}, \qquad d\underline{x}^{m-1} = \pm e_M d\sigma_{\underline{x}}, \qquad d\underline{x}^k = \sum_{|A|=k} e_A dx_A$$

are the oriented surface elements and one also has that $\partial_{\underline{x}} \rfloor d\underline{x}^k = \gamma d\underline{x}^{k-1}$, γ being a constant. For any function $f(\underline{x})$ in space one may consider the k-form $F = d\underline{x}^k f(\underline{x})$ and wonder what the closedness equations $dF = 0$ mean. Except for $k = m - 1$ it gives rise to a strongly overdetermined system with trivial solutions. But the restricted closedness condition

$$d\partial_{\underline{x}} \rfloor F_{|U} = d d\underline{x}^{k-1} f(\underline{x})_{|U} = 0$$

corresponds to the tangential Dirac equation for the restriction $f_{|U}$ and one also has the Cauchy theorem

$$\int_{\partial C} \partial_{\underline{x}} \rfloor F = 0.$$

4. Clifford analysis in several vector variables

If the Dirac operator $\partial_{\underline{x}}$ is considered as a generalization of the Cauchy–Riemann operator $\partial_x + i\partial_y$, it is natural to think of the function theory for several Dirac operators $\partial_{\underline{x}_1}, \ldots, \partial_{\underline{x}_\ell}$, with $\partial_{\underline{x}_j} = \sum e_k \partial_{x_{jk}}$, as a generalization of the theory of several complex variables. In particular, one may investigate functions of several vector variables $\underline{x}_1, \ldots, \underline{x}_\ell$, $\underline{x}_j = \sum e_k x_{jk}$ satisfying the monogenicity equations $\partial_{\underline{x}_j} f(\underline{x}_1, \ldots, \underline{x}_\ell) = 0$, $j = 1, \ldots, \ell$ and one may try to generalize properties valid for monogenic functions in one vector variable to the several vector variable case. This includes generalizations of the Cauchy–Kowalewski theorem (see [6], [15], [30]) and the Fischer decomposition discussed in [6] and [34] in the two variable case and [31] in the several variable case. Moreover there is also a generalization of the Dolbeault complex studied in [1], [3], [25] for the case of several quaternion variables and in [27], [28] for several (abstract) vector variables. In our papers [30] and [36] we also defined and studied so called "symplicial monogenics". These are monogenic functions $f(\underline{x}_1, \ldots, \underline{x}_\ell)$ of the symplicial form $f(\underline{x}_1, \ldots, \underline{x}_\ell) = F(\underline{x}_1 \wedge \ldots \wedge \underline{x}_\ell)$, F being defined on an open subset of the set of ℓ-blades in $\mathbb{R}_{m;\ell}$. In most applications we will be working with functions $F(X)$, $X \in \mathbb{R}_{m;\ell}$, which are defined in an open conical subset U of $\mathbb{R}_{m;\ell}$ (i.e. $U = rU$, $r > 0$) and which satisfy a homogeneity condition of the form $F(rX) = r^a F(X)$. Of special interest is the case in which F is a homogeneous polynomial of degree k on $\mathbb{R}_{m;\ell}$. This leads to the following

Definition 4.1. *A spherical monogenic of degree k on the Grassmannian* $\mathcal{G}(m,\ell)$ is a function $f(\underline{x}_1,\ldots,\underline{x}_\ell)$ of several vector variables $\underline{x}_1,\ldots,\underline{x}_\ell$ satisfying the monogenicity conditions $\partial_{\underline{x}_j} f(\underline{x}_1,\ldots,\underline{x}_\ell) = 0$, $j = 1,\ldots,\ell$, which is of the special form $f(\underline{x}_1,\ldots,\underline{x}_\ell) = F(\underline{x}_1 \wedge \ldots \wedge \underline{x}_\ell)$ where $F(X)$, $X \in \mathbb{R}_{m;\ell}$ is a homogeneous polynomial of degree k. For simplicity, we speak of "Grassmann monogenics of type ℓ and degree k".

Remark 4.2. The above notion of spherical monogenics on $\mathcal{G}(m,\ell)$ generalizes the notion of harmonic functions (in particular spherical harmonics) of a matrix variable, which are functions of the form $f(\underline{x}_1,\ldots,\underline{x}_\ell) = F(\underline{x}_1 \wedge \ldots \wedge \underline{x}_\ell)$ satisfying the harmonicity equations

$$\langle \partial_{\underline{x}_i}, \partial_{\underline{x}_j} \rangle f(\underline{x}_1,\ldots,\underline{x}_\ell) = \sum \partial_{x_{ik}} \partial_{x_{jk}} f(x_{1k},\ldots,x_{\ell k}) = 0$$

(see also [12]). Note that non trivial spherical monogenics on $\mathcal{G}(m,\ell)$ only exist for $\ell \leq [m/2]$.

More in general, one may consider the set of ℓ-tuples of the form

$$(\underline{x}_1, \underline{x}_1 \wedge \underline{x}_2, \ldots, \underline{x}_1 \wedge \ldots \wedge \underline{x}_\ell) \in \mathbb{R}_{m;1} \times \ldots \times \mathbb{R}_{m;\ell}$$

and consider monogenic functions $f(\underline{x}_1,\ldots,\underline{x}_\ell)$ of several vector variables of the special form $f(\underline{x}_1,\ldots,\underline{x}_\ell) = F(\underline{x}_1, \underline{x}_1 \wedge \underline{x}_2, \ldots, \underline{x}_1 \wedge \ldots \wedge \underline{x}_\ell)$ where $F(X_1, X_2, \ldots, X_\ell)$, $X_j \in \mathbb{R}_{m;j}$ is defined in an open subset of $\mathbb{R}_{m;1} \times \ldots \times \mathbb{R}_{m;\ell}$ and satisfies a homogeneity condition of the form $F(r_1 X_1, \ldots, r_\ell X_\ell) = r_1^{a_1} \ldots r_\ell^{a_\ell} F(X_1, \ldots, X_\ell)$. Functions of this type are called symplicial monogenics. Moreover we may consider the so called Stiefel manifold $St(m,\ell)$, which is the manifold of orthonormal ℓ-frames $(\underline{w}_1, \ldots \underline{w}_\ell)$ in $\mathbb{R}^m$. It can be identified with the manifold of ℓ-tuples of blades of the form $(\underline{w}_1, \underline{w}_1 \wedge \underline{w}_2, \ldots, \underline{w}_1 \wedge \ldots \wedge \underline{w}_\ell)$, with $(\underline{w}_1, \ldots, \underline{w}_\ell)$ orthonormal. Then we introduce the following

Definition 4.3. *A spherical monogenic of degree* $(k_1,\ldots,k_\ell)$ *on the Stiefel manifold* $St(m,\ell)$ is a monogenic function of several vector variables $f(\underline{x}_1 \wedge \ldots \wedge \underline{x}_\ell)$ of the symplicial form $f(\underline{x}_1,\ldots,\underline{x}_\ell) = F(\underline{x}_1, \underline{x}_1 \wedge \underline{x}_2, \ldots, \underline{x}_1 \wedge \ldots \wedge \underline{x}_\ell)$, where the function $F(X_1,\ldots,X_\ell)$ is a polynomial on $\mathbb{R}_{m;1} \times \ldots \times \mathbb{R}_{m;\ell}$ that is homogeneous of degree k_j in the j-vector variable X_j. We will call those functions symplicial monogenics.

Non trivial spherical monogenics on Stiefel manifolds only exist for $\ell \leq [m/2]$ and, without disturbing the generality we may in fact assume that $\ell = n = [m/2]$. One may also define spherical monogenics on various kinds of Flag-manifolds, but they are in fact special cases of symplicial monogenics on $St(m,[m/2])$. Note that, using the above notations, the function $f(\underline{x}_1,\ldots,\underline{x}_n)$ is also a homogeneous polynomial in each of the vector variables $\underline{x}_1,\ldots,\underline{x}_n$, in fact it is homogeneous of degree $k_1 + \ldots + k_n$ in $\underline{x}_1$, of degree $k_2 + \ldots + k_n$ in $\underline{x}_2$, etc., and of degree k_n in $\underline{x}_n$.

Moreover, when considering the representation of Spin(m) defined for functions of several vector variables by

$$L(s)f(\underline{x}_1,\ldots,\underline{x}_n)=sf(\bar{s}\underline{x}_1s,\ldots,\bar{s}\underline{x}_ns),$$

we proved in [36] that the space of symplicial spherical monogenics of degree $(k_1,\ldots,k_n)$ on the Stiefel manifold, forms a model for an irreducible representation of weight $(k_1+\ldots+k_n+\frac{1}{2},\ldots,k_n+\frac{1}{2})$. For the integer weight case one may use harmonic functions of a matrix variable as has been exposed in [12]. For more details about Spin(m) representations we also refer to the book [12], to the famous [35] and also to our paper [30]. As a consequence, all finite-dimensional representations of Spin(m) can be modelled by using symplicial monogenics on which one may consider either the above representation L for half-integer weight, and the representation H given by

$$H(s)f(\underline{x}_1,\ldots,\underline{x}_n)=sf(\bar{s}\underline{x}_1s,\ldots,\bar{s}\underline{x}_ns)\bar{s};$$

for more details, we refer the reader to [30]. These representations L and H may also act on the functions $F(X_1,\ldots,X_n)$, $X_j\in\mathbb{R}_{m;j}$, defined on the multivector space by direct extension of the $SO(m)$-representation $\underline{x}\to s\underline{x}\bar{s}$, $\underline{x}\in\mathbb{R}^m=\mathbb{R}_{m;1}$ to $a\to sa\bar{s}$, $a\in\mathbb{R}_m$, i.e.

$$L(s)F(X_1,\ldots,X_n)=sF(\bar{s}X_1s,\ldots,\bar{s}X_ns),\qquad H(s)F=L(s)F\bar{s}$$

and in the case $f(\underline{x}_1,\ldots,\underline{x}_n)=F(\underline{x}_1,\underline{x}_1\wedge\underline{x}_2,\ldots,\underline{x}_1\wedge\ldots\wedge\underline{x}_n)$, both definitions are equivalent. The proof of the main result in [36] is based on the fact that the space of all symplicial monogenics of degree $(k_1,\ldots,k_n)$ are generated by plane-wave type polynomials of the form

$$\langle\underline{x}_1,T_1\rangle^{k_1}\langle\underline{x}_1\wedge\underline{x}_2,T_1\wedge T_2\rangle^{k_2}\ldots\langle\underline{x}_1\wedge\ldots\wedge\underline{x}_n,T_1\wedge\ldots\wedge T_n\rangle^{k_n}T_1\wedge\ldots\wedge T_n$$

where $\langle a,b\rangle=[\bar{a}b]_0$ is the standard bilinear form on $\mathbb{C}_m$, and where the set of vectors $T_1,\ldots,T_n\in\mathbb{C}_{m;1}$ satisfy the nullrelations $T_j^2=0$ together with the anti-commutativity $T_jT_k=-T_kT_j$. In other words, $(T_1,\ldots,T_n)$ spans an n-dimensional complex subspace of the nullcone. Putting $n=[m/2]$, we get an example of such a frame i.e. the Witt basis $T_j=f_j=\frac{1}{2}(e_{2j-1}-ie_{2j})$. This example also illustrates why one cannot have any monogenic plane waves of variables $\underline{x}_1\wedge\ldots\wedge\underline{x}_\ell$ with $\ell>n=[m/2]$. When proven the fact that such monogenic plane waves generate the spaces of all symplicial monogenics, it is easy to derive from it models for irreducible representations of Spin(m) for monogenic plane waves: they correspond to maximal vectors of irreducible Spin(m)-representation spaces, which also generate these spaces as Spin(m)-modules. For more information about this nomenclature, we again refer to the extended treatment in [12] and our papers [30] and [36]. To conclude this section, note that in the above reasoning, we made use of functions F depending on multivector variables of the form $F(X_1,\ldots,X_n)$, $X_\ell\in\mathbb{R}_{m;\ell}$ and in particular of polynomials in these variables, but we point out that we only need the restriction of these functions to the special submanifolds of ℓ-blades, i.e. to ℓ-vectors of the form $X_\ell=\underline{x}_1\wedge\ldots\wedge\underline{x}_\ell$. Of course, it is clear that a polynomial $P(X_\ell)$, $X_\ell\in\mathbb{R}_{m;\ell}$ is by no means determined by its restriction to the manifold

$X_\ell = \underline{x}_1 \wedge \ldots \wedge \underline{x}_\ell$, except the case where $P(X_\ell)$ is a linear function, which is considered in [13]. This book is to our knowledge the first in which the multivector variables

$$X = \sum X_\ell, X_\ell = \sum_{|A|=\ell} X_A e_A,$$

and derivatives

$$\partial_X = \sum \partial_{X_\ell}, \qquad \partial_{X_\ell} = \sum_{|A|=\ell} \partial_{X_A} e_A$$

were studied in detail, although, from a historical point of view, the multivector derivative was already considered in [23]. Also in our paper [31] we used the ℓ-vector derivatives ∂_{X_ℓ} in connection with the general theory of spherical means, during a time when [13] was not yet published. An important result (see [13]) is the fact that, for scalar linear functions of a ℓ-vector variable (outer morphisms), one has the relation

$$\partial_{X_\ell} P(X)_{|X_\ell = \underline{x}_1 \wedge \ldots \wedge \underline{x}_\ell} = C_\ell \partial_{\underline{x}_1} \wedge \ldots \wedge \partial_{\underline{x}_\ell} P(\underline{x}_1 \wedge \ldots \wedge \underline{x}_\ell), \tag{1}$$

i.e. the ℓ-vector derivative is expressible in terms of the "symplicial derivative" $\partial_{\underline{x}_1} \wedge \ldots \wedge \partial_{\underline{x}_\ell}$. This property clearly does not hold for polynomials of a ℓ-vector variable which are homogeneous of a degree higher than one, in fact one may easily find such nonzero polynomials $P(X_\ell)$ for which $P(\underline{x}_1 \wedge \ldots \wedge \underline{x}_\ell) = 0$. Hence we are faced with at least two fundamental problems for scalar-valued polynomials on the ℓ-vector space $\mathbb{R}_{m;\ell}$:

Problem 1. Does there exist a space $\mathcal{P}(k, \ell)$ of real valued homogeneous polynomials of degree k on $\mathbb{R}_{m;\ell}$ such that any $P(X_\ell) \in \mathcal{P}(k, \ell)$ is canonically determined by its restriction to the manifold of ℓ-blades $X_\ell = \underline{x}_1 \wedge \ldots \wedge \underline{x}_\ell$ and such that for any real valued homogeneous scalar polynomial $R(X_\ell)$ of degree k on $\mathbb{R}_{m;\ell}$ there exists a unique polynomial $P(X_\ell) \in \mathcal{P}(k, \ell)$ for which

$$R(\underline{x}_1 \wedge \ldots \wedge \underline{x}_\ell) = P(\underline{x}_1 \wedge \ldots \wedge \underline{x}_\ell)?$$

Problem 2. Does the identity (1) established in [13] for linear functions (i.e. expressing the partial derivatives of a linear function in terms of symplicial derivatives) generalize to the polynomials of the special type described in Problem 1?

Needless to say that the above problems can be stated for Clifford algebra valued polynomials by considering their (scalar-valued) components. In the following sections, we will illustrate how Fischer's decomposition gives an answer to Problem 1. We also expect that Problem 2 has an affirmative answer since we were able to verify the relation in some special cases. Moreover, a complete description will be obtained for the canonical extensions of symplicial monogenics in all cases: for bivectors, for ℓ-vectors and for functions of the general Stiefel-type $F(\underline{x}_1, \underline{x}_1 \wedge \underline{x}_2, \ldots, \underline{x}_1 \wedge \ldots \wedge \underline{x}_n)$ by simply using the chain rule; this leads to canonically defined over-determined elliptic systems on the space of multivectors.

5. Clifford analysis on the space of bivectors

We first consider the case $\mathbb{R}_4$, i.e. the Clifford algebra with generators e_1, e_2, e_3, e_4. This algebra is very important for applied mathematics, indeed, the 3 dimensional projective space $\mathbb{RP}^3$ corresponds to the manifold of rays

$$\mathrm{Ray}(\underline{w}) = \{\lambda \underline{w} : \quad \lambda \neq 0, \quad \underline{w} = \sum_{j=1}^{4} w_j e_j \in \mathbb{R}^4, \underline{w} \neq 0\}.$$

The case $w_4 \neq 0$ corresponds to the 3 dimensional affine space and w_4 may then be chosen such that $w_4 = 1$. Next, let $\underline{u}, \underline{v} \in \mathbb{R}^4$ represent two points in $\mathbb{RP}^3$; then the line through these points is in fact the manifold of rays $\mathrm{Ray}(\underline{w})$, $\underline{w} = \lambda \underline{u} + \mu \underline{v} \in \mathrm{span}\{\underline{u}, \underline{v}\}$ where $\mathrm{span}\{\underline{u}, \underline{v}\}$ is represented by the "homogeneous 2-blade" $\lambda \underline{u} \wedge \underline{v}$. So, a function on the manifold of lines in $\mathbb{RP}^3$ may be thought of as a function on the manifold of homogeneous 2-blades which is representable as an even (or odd) function on the Grassmannian $\mathcal{G}(4,2)$. The 3D-X-ray-transform assigns to each function its integrals over lines in the 3-space, hence it may be represented as a function on the variety of 2-blades and Clifford analysis may be applied to it. Note that the canonical basis for $\mathbb{R}_{4;2}$ is the set

$$e_{14},\ e_{24},\ e_{34},\ e_{23},\ e_{31} = -e_{13},\ e_{12}$$

and every element $X \in \mathbb{R}_{4;2}$ may be written into the form

$$X = \sum_{i<j} x_{ij} e_{ij}, \qquad x_{ji} = -x_{ij},$$

from which it follows that

$$X^2 = -\sum_{i<j} x_{ij}^2 + e_{1234} K(X),$$

where

$$K(X) = x_{14}x_{23} + x_{24}x_{31} + x_{34}x_{12}$$

is the so called Plücker polynomial. By Lemma 2.12, it is clear that the equation $K(X) = 0$ is necessary and sufficient for X to be a 2-blade; the cone $K(X) = 0$ of bivectors corresponds to the classical Klein quadric of projective geometry, which describes the manifold of lines in $\mathbb{RP}^3$ or $\mathbb{CP}^3$. It is not hard to see that for polynomials $P(X)$ on $\mathbb{C}_{4;2}$ the following holds:

Proposition 5.1. *Let $P(X)$ be a polynomial on $\mathbb{C}_{4;2}$. Then*

$$P(X)_{|X=\underline{x}\wedge\underline{y}} = 0 \quad \text{if and only if} \quad P(X) = K(X)Q(X)$$

for some polynomial $Q(X)$.

Proof. The irreducible polynomial $K(X)$ generates a radical ideal so by the Hilbert Nullstellensatz, a polynomial vanishing on the 2-blades belongs to that ideal, i.e. it is divisible by $K(X)$. □

On the space of polynomials on $\mathbb{R}_{4;2}$, we may consider the Fischer-inner product

$$(R(X), S(X)) = [R(\partial_X)^+ S(X)]_{0|X=0}$$

which is positive definite on the space of these polynomials and where "+" denotes the Hermitian conjugate, while $R(\partial_X)$ is obtained replacing the variables x_{ij} by the corresponding partial derivatives $\partial_{x_{ij}}$. Then we have the following

Theorem 5.2 (Fischer decomposition). *Every polynomial $R(X)$ on $\mathbb{R}_{4;2}$ admits a unique decomposition of the form*

$$R(X) = P(X) + K(X)Q(X),$$

where $K(\partial_X)P(X) = 0$ and $Q(X)$ is a suitable polynomial.

Proof. The above decomposition is in fact orthogonal with respect to the Fischer inner product. □

The answer to Problem 1 is obtained by considering the polynomial solutions (real or Clifford algebra valued) to the scalar quadratic equation (Plücker PDE)

$$K(\partial_X)P(X) = (\partial_{x_{14}}\partial_{x_{23}} + \partial_{x_{24}}\partial_{x_{31}} + \partial_{x_{34}}\partial_{x_{12}})P(X) = 0.$$

Indeed, from the Fischer decomposition one immediately gets the following:

Corollary 5.3. *The restriction operator*

$$P(X) \to P(\underline{x} \wedge \underline{y})$$

of polynomials to the variety of 2-blades is injective on the space of solutions of Plücker PDE.

We also have the following result, the proof of which does not depend on the dimension $m = 4$ considered:

Theorem 5.4 (Higher order Hestenes–Sobczyk for plane waves). *The operator ∂_X coincides, up to a constant, with $\partial_{\underline{x}} \wedge \partial_{\underline{y}}$ when acting on the plane wave type polynomials $\langle \underline{x} \wedge \underline{y}, \underline{u} \wedge \underline{v}\rangle^k$.*

Proof. Let us consider a plane wave polynomial $\langle X, \underline{u} \wedge \underline{v}\rangle^k$ and the operator ∂_X. Easy computations show that

$$\partial_X \langle X, \underline{u} \wedge \underline{v}\rangle^k = \sum_{i,j} e_{ij}\partial_{x_{ij}} \langle X, \underline{u} \wedge \underline{v}\rangle^k = k\langle X, \underline{u} \wedge \underline{v}\rangle^{k-1}\underline{u} \wedge \underline{v}.$$

Restricting to the variety of 2-blades we get

$$\partial_X \langle \underline{x} \wedge \underline{y}, \underline{u} \wedge \underline{v}\rangle^k = k\langle \underline{x} \wedge \underline{y}, \underline{u} \wedge \underline{v}\rangle^{k-1}\underline{u} \wedge \underline{v}.$$

Let us consider now the operator $\partial_{\underline{x}} \wedge \partial_{\underline{y}}$ and its action on $\langle \underline{x} \wedge \underline{y}, \underline{u} \wedge \underline{v}\rangle^k$. Since $\partial_{\underline{x}} \wedge \partial_{\underline{y}} = \frac{1}{2}(\partial_{\underline{x}}\partial_{\underline{y}} - \partial_{\underline{y}}\partial_{\underline{x}})$ we will consider separately the action of $\partial_{\underline{x}}\partial_{\underline{y}}$ and $\partial_{\underline{y}}\partial_{\underline{x}}$. We have

$$\partial_{\underline{y}}\langle \underline{x} \wedge \underline{y}, \underline{u} \wedge \underline{v}\rangle^k = k\langle \underline{x} \wedge \underline{y}, \underline{u} \wedge \underline{v}\rangle^{k-1} \sum_{j<k} (x_j e_k - x_k e_j)(u_j v_k - u_k v_j)$$

and

$$\partial_{\underline{x}}\partial_{\underline{y}}\langle \underline{x}\wedge\underline{y},\underline{u}\wedge\underline{v}\rangle^k = k\langle \underline{x}\wedge\underline{y},\underline{u}\wedge\underline{v}\rangle^{k-1}\sum_{j<k} e_{jk}(u_jv_k - u_kv_j)$$

$$+k(k-1)\langle \underline{x}\wedge\underline{y},\underline{u}\wedge\underline{v}\rangle^{k-2}\left(\sum_{j,k} e_jy_k(u_jv_k-u_kv_j)\right)\left(\sum_{jk} x_je_k(u_jv_k-u_kv_j)\right).$$

Direct computations show that

$$\langle \underline{x}\wedge\underline{y},\underline{u}\wedge\underline{v}\rangle = \langle \underline{x},\underline{u}\rangle\langle \underline{y},\underline{v}\rangle - \langle \underline{x},\underline{v}\rangle\langle \underline{y},\underline{u}\rangle,$$

so that

$$\partial_{\underline{x}}\partial_{\underline{y}}\langle \underline{x}\wedge\underline{y},\underline{u}\wedge\underline{v}\rangle^k = k(\underline{uv}-\underline{vu})\langle \underline{x}\wedge\underline{y},\underline{u}\wedge\underline{v}\rangle^{k-1}$$

$$+k(k-1)\langle \underline{x}\wedge\underline{y},\underline{u}\wedge\underline{v}\rangle^{k-2}\left(\langle \underline{y},\underline{v}\rangle\underline{u} - \langle \underline{y},\underline{u}\rangle\underline{v}\right)\left(\langle \underline{x},\underline{u}\rangle\underline{v} - \langle \underline{x},\underline{v}\rangle\underline{u}\right).$$

Analogously, we have

$$\partial_{\underline{y}}\partial_{\underline{x}}\langle \underline{x}\wedge\underline{y},\underline{u}\wedge\underline{v}\rangle^k = k(\underline{vu}-\underline{uv})\langle \underline{x}\wedge\underline{y},\underline{u}\wedge\underline{v}\rangle^{k-1}$$

$$+k(k-1)\langle \underline{x}\wedge\underline{y},\underline{u}\wedge\underline{v}\rangle^{k-2}\left(\langle \underline{x},\underline{u}\rangle\underline{v} - \langle \underline{x},\underline{v}\rangle\underline{u}\right)\left(\langle \underline{y},\underline{v}\rangle\underline{u} - \langle \underline{y},\underline{u}\rangle\underline{v}\right).$$

Putting together, we get

$$\partial_{\underline{x}}\wedge\partial_{\underline{y}}\langle \underline{x}\wedge\underline{y},\underline{u}\wedge\underline{v}\rangle^k = k\langle \underline{x}\wedge\underline{y},\underline{u}\wedge\underline{v}\rangle^{k-1}\underline{u}\wedge\underline{v}$$

$$-\frac{k}{2}(k-1)\langle \underline{x}\wedge\underline{y},\underline{u}\wedge\underline{v}\rangle^{k-2}\left(\langle \underline{x},\underline{u}\rangle\underline{v} - \langle \underline{x},\underline{v}\rangle\underline{u}\right)\wedge\left(\langle \underline{y},\underline{v}\rangle\underline{u} - \langle \underline{y},\underline{u}\rangle\underline{v}\right) =$$

$$k\langle \underline{x}\wedge\underline{y},\underline{u}\wedge\underline{v}\rangle^{k-1}\underline{u}\wedge\underline{v} - \frac{k}{2}(k-1)\langle \underline{x}\wedge\underline{y},\underline{u}\wedge\underline{v}\rangle^{k-2}(\langle \underline{x},\underline{v}\rangle\langle \underline{y},\underline{u}\rangle - \langle \underline{x},\underline{u}\rangle\langle \underline{y},\underline{v}\rangle)\underline{u}\wedge\underline{v} =$$

$$k\langle \underline{x}\wedge\underline{y},\underline{u}\wedge\underline{v}\rangle^{k-1}\underline{u}\wedge\underline{v} + \frac{k}{2}(k-1)\langle \underline{x}\wedge\underline{y},\underline{u}\wedge\underline{v}\rangle^{k-1}\underline{u}\wedge\underline{v},$$

thus the statement is proved. □

Remark 5.5. Note that, as the Plücker PDE is linear with constant coefficients, partial derivatives of sufficiently regular solutions are again solutions and so identities of the Hestenes–Sobczyk type are also available for the higher order derivatives.

Next, let us study some canonical systems of Clifford-PDE on the space of bivectors $\mathbb{R}_{4;2}$.

(i) the bivector derivative $\partial_X = \sum e_{ij}\partial_{x_{ij}}$. This operator is not elliptic and also it is not a Dirac operator in the classical sense of leading to a factorization of the Laplace operator on $\mathbb{R}^6 = \mathbb{R}_{4;2}$. In fact, note that

$$\partial_X = e_{14}\partial_{x_{14}} + e_{24}\partial_{x_{24}} + e_{34}\partial_{x_{34}} - e_{1234}(e_{14}\partial_{x_{23}} + e_{24}\partial_{x_{31}} + e_{34}\partial_{x_{12}})$$

$$= (\frac{1}{2}(1-e_{1234})\partial_{\underline{x}} + \frac{1}{2}(1+e_{1234})\partial_{\underline{y}})e_4$$

where $\partial_{\underline{x}} = e_1\partial_{x_1} + e_2\partial_{x_2} + e_3\partial_{x_3}$ and $\partial_{\underline{y}} = e_1\partial_{y_1} + e_2\partial_{y_2} + e_3\partial_{y_3}$ are 3 dimensional Dirac operators.

Hence noting that e_{1234} commutes with bivectors, it suffices to consider solutions of the form $\frac{1}{2}(1-e_{1234})g(\underline{x},\underline{y}) + \frac{1}{2}(1+e_{1234})h(\underline{x},\underline{y})$, with $\partial_{\underline{x}}g(\underline{x},\underline{y}) =$

$\partial_{\underline{y}} h(\underline{x}, \underline{y}) = 0$. Needless to say that this system has many non-real analytic solutions so the system is certainly not elliptic. However we have the following result that allows us to say that ∂_X behaves like a square root of the Laplacian (i.e. a Dirac operator) on the subspace of solutions to Plücker PDE.

Proposition 5.6. *The operator ∂_X becomes elliptic when we restrict to solutions $f(X)$ which satisfy the Plücker PDE $K(\partial_X)f(X) = 0$.*

Proof. In fact we have $\partial_X^2 = -\Delta_X + K(\partial_X)e_{1234}$ so that solutions of $\partial_X f(X) = 0$ which also satisfy the Plücker PDE are harmonic, where defined. □

(ii) The above Dirac like system: $\partial_X f(X) = K(\partial_X)f(X) = 0$ is not the best possible choice to define a Dirac system on $\mathbb{R}_{4;2}$. Indeed, for $j = 1, 2, 3, 4$, the operators

$$[e_j \partial_X]_1 = \frac{1}{2}(e_j \partial_X + \partial_X e_j)$$

behave like standard Dirac operators on suitable subspaces of $\mathbb{R}_{4;2}$ and the sum of their squares is proportional to the 6-dimensional Laplacian while $\sum e_j [e_j \partial_X]_1$ is proportional to ∂_X itself. Hence the solutions to the overdetermined system

$$[e_j \partial_X]_1 f(X) = 0, \qquad j = 1, 2, 3, 4,$$

also satisfy the 6-dimensional Laplacian as well as the bivector derivative $\partial_X f(X) = 0$, so that they also satisfy the Plücker PDE. Then, the solutions to this system, which is also Spin(m)-invariant under the representation $s \to L(s) : f(X) \to s f(\bar{s} X s)$, really have all the good properties and deserve to be called "monogenic functions on $\mathbb{R}_{4;2}$". At this point, the following questions arise naturally: do there exist solutions to the above system? and what about their restrictions to the variety of 2-blades? The answer will be investigated for general spaces of bivectors $\mathbb{R}_{m;2}$.

Let us come back to the problem of the characterization of 2-blades: for general m, we have the bivector variable $X \in \mathbb{R}_{m;2}$ given by

$$X = \sum_{i<j} e_{ij} x_{ij}$$

having square

$$X^2 = -\sum_{i<j} x_{ij}^2 + \sum_{i<j<k<l} e_{ijkl} K_{ijkl}(X)$$

where $K_{ijkl}(X)$ are the special quadratic polynomials defined on $\mathbb{R}_{m;2}$ which correspond to the higher dimensional Plücker polynomials and which are related to the Grassmannian $G(m, 2)$ (see [2]): $K_{ijkl}(X) = x_{ij}x_{k\ell} - x_{ik}x_{j\ell} + x_{i\ell}x_{jk}$. In the sequel, let us call Plücker variety the variety of 2-blades, i.e. the variety defined by the vanishing of $K_{ijkl}(X) = 0$. We have the following

Theorem 5.7. *A polynomial $P(X)$ on $\mathbb{C}_{m;2}$ vanishes on the Plücker variety given by the equations $K_{ijkl}(X) = 0$, if and only if $P(X)$ is of the form $\sum K_{ijkl}(X)Q_{ijkl}(X)$ for suitable polynomials $Q_{ijkl}(X)$.*

Proof. The ideal I generated by the polynomials $K_{ijkl}(X)$ is radical, see [2] or [21], so any function vanishing on the Plücker variety must belong to I. □

Once again, it is natural to introduce the Fischer inner product by the above formula (similar to the case $m = 4$) and we have the

Theorem 5.8 (Fischer decomposition). *Every polynomial $R(X)$ admits a unique decomposition of the form*

$$R(X) = P(X) + \sum K_{ijkl}(X)Q_{ijkl}(X),$$

where $P(X)$ satisfies the Plücker PDE: $K_{ijkl}(\partial_X)P(X) = 0$.

Again the proof follows by Fischer orthogonality.

Remark 5.9. It is also clear that the generalization of Corollary 5.3 holds and the restriction operator to the variety of 2-blades: $P(X) \to P(\underline{x} \wedge \underline{y})$ is injective on the space of solutions of the Plücker PDE, and provides a solution to the Problem 1. Moreover, the proof given in the case $m = 4$ about the higher order Hestenes–Sobczyk identification for plane waves

$$\partial_X P(X) \to C_k \partial_{\underline{x}} \wedge \partial_{\underline{y}} f(\underline{x} \wedge \underline{y}),$$

has general validity since does not depend on the dimension m considered.

Now, again, we may have a look for canonically defined operators on the general space of bivectors and first of all we mention:

(i) The bivector derivative $\partial_X f(X) = \sum e_{ij}\partial_{x_{ij}} f(X) = 0$, which is a rather complicated non-elliptic system on $\mathbb{R}_{m;2}$.
(ii) The combined system: $\partial_X f(X) = 0$, $K_{ijkl}(\partial_X)f(X) = 0$, i.e. solutions of the Plücker PDE which also satisfy the bivector derivative. Solutions to this system satisfy the Laplace equation on $\mathbb{R}_{m;2}$, i.e. they are harmonic and so the above system is elliptic and could be called a generalized Dirac system.
(iii) One could also consider simultaneous solutions of the systems

$$\partial_X f(X) = 0 \qquad [\partial_X^2]_4 f(X) = \sum e_{ijkl}K_{ijkl}(\partial_X)f(X) = 0,$$

that are harmonic though they are not solutions of the Plücker PDE.
(iv) Like in the $m = 4$ case we may consider the "monogenicity conditions"

$$[e_j\partial_X]_1 f(X) = \frac{1}{2}(e_j\partial_X - \partial_X e_j)f(X) = 0,$$

which also imply both equations $\Delta_X f(X) = 0$ and $\partial_X f(X) = 0$; from this fact one readily obtains $[\partial_X^2]_4 f(X) = 0$.

We have the following:

Theorem 5.10 (Chain rule). *Let $f(X)$ be a scalar valued function on part of $\mathbb{R}_{m;2}$. Then for the restriction of f to the manifold of 2-blades we have that*

$$\partial_{\underline{x}} f(\underline{x} \wedge \underline{y}) = \frac{1}{2}(\underline{y}\partial_X - \partial_X \underline{y}) f(X)_{|X=\underline{x}\wedge\underline{y}}$$

Proof. Direct application of the chain rule leads to

$$\begin{aligned} \partial_{x_i} f\left(\sum e_{jk}(x_j y_k - x_k y_j)\right) &= \sum \partial_{x_i}(x_j y_k - x_k y_j)\partial_{X_{jk}} f(X) \\ &= \sum y_k \partial_{X_{ik}} f(X) - \sum y_j \partial_{X_{ji}} f(X) \end{aligned}$$

where we restrict to $X = \underline{x} \wedge \underline{y}$. □

The above also holds for Clifford algebra-valued functions and so we have

Corollary 5.11. *Let $f(X)$ satisfy the monogenicity conditions $[e_j \partial_X]_1 f(X) = 0$. Then the restriction $f(\underline{x} \wedge \underline{y})$ to the manifold of 2-blades is symplicial monogenic, i.e. $\partial_{\underline{x}} f(\underline{x} \wedge \underline{y}) = 0$.*

We still do not know from this whether non trivial solutions to the monogenicity conditions do exist; after all in the above result we have only considered these equations on the manifold of 2-blades. We have the following:

Theorem 5.12. *Let $f(\underline{x}\wedge\underline{y})$ be polynomial symplicial monogenic, i.e. $\partial_{\underline{x}} f(\underline{x}\wedge\underline{y}) = 0$. Then f admits a unique polynomial extension $f(X)$, $X \in \mathbb{R}_{m;2}$ satisfying both the Plücker PDE and the monogenicity conditions on $\mathbb{R}_{m;2}$.*

Proof. We use the fact here that the space of homogeneous polynomial symplicial monogenics is spanned by the plane wave type functions $\langle \underline{x} \wedge \underline{y}, T_1 \wedge T_2\rangle^k T_1 \wedge T_2$, (T_1, T_2) being a null 2-frame which admit the canonical extensions to $\mathbb{R}_{m;2}$: $\langle X, T_1 \wedge T_2\rangle^k T_1 \wedge T_2$. Now for any holomorphic function we have that

$$[e_j \partial_X]_1 h(\langle X, T_1 \wedge T_2\rangle) = [e_j(T_1 \wedge T_2)]_1 h'(\langle X, T_1 \wedge T_2\rangle)$$

while also

$$[e_j(T_1 \wedge T_2)]_1 (T_1 \wedge T_2) = 0.$$

The monogenicity conditions easily follow from this, while, having to do with functions of the plane wave variable $\langle X, T_1 \wedge T_2\rangle$, one also has the Plücker PDE. □

Remark 5.13. One may state that the monogenicity combined with the Plücker PDE leads to equations for symplicial monogenics on the bivector space. We only gave the proof for polynomials but the above property is expected to be valid in general. We also expect the Plücker PDE to follow from monogenicity.

6. Clifford analysis on the space of ℓ-vectors

Also in the case of an ℓ-vector variable, $\ell \leq m/2$, $X = X_\ell = \sum e_{a_1 \ldots a_\ell} X_{a_1 \ldots a_\ell}$ one can find quadratic polynomial equations for the manifold of ℓ-blades $X = \underline{x}_1 \wedge \ldots \wedge \underline{x}_\ell$, given by the Plücker polynomials. An elegant description of this manifold in simple Clifford algebra terms is still to be found. But in any case one can do the following.

Let I be the ideal of polynomials $T(X)$ for which $T(\underline{x}_1 \wedge \ldots \wedge \underline{x}_\ell) = 0$ and introduce the Fischer inner product by

$$(S(X), R(X)) = [S(\partial_X)^+ R(X)]_{0|(X=0)},$$

then we have the following:

Theorem 6.1 (Fischer decomposition). *Every polynomial $R(X)$ admits a unique orthogonal decomposition of the form $R(X) = P(X) + S(X)$, whereby $S(\underline{x}_1 \wedge \ldots \wedge \underline{x}_\ell) = 0$, i.e. $S \in I$ and $P(X)$ satisfies the Plücker PDE*

$$T(\partial_X)^+ P(X) = 0$$

for any Plücker polynomial T.

In any case, a polynomial satisfying the Plücker PDE is fully determined by its restriction to the manifold of blades and any restriction to this manifold of a homogeneous polynomial of degree k is also the restriction of a unique polynomial satisfying the Plücker PDE. We conjecture that a higher order Hestenes–Sobczyk for plane waves holds also in this case. More in general, we formulate the following conjecture that generalizes the result to homogenous polynomials solutions of the Plücker PDE:

Conjecture 6.2 (Higher order Hestenes–Sobczyk). *There exists a constant C (depending on ℓ and k) such that for any homogeneous polynomial solution of degree k of the Plücker PDE, the restriction of $\partial_X P(X)$ to the manifold of blades is given by the symplicial derivative $C\partial_{\underline{x}_1} \wedge \ldots \wedge \partial_{\underline{x}_\ell} P(\underline{x}_1 \wedge \ldots \wedge \underline{x}_\ell)$.*

Remark 6.3. Like in the case for bivectors, it may be verified by direct computation that a generalized Hestenes–Sobczyk relation is valid for plane wave functions $\langle X, u_1 \wedge \ldots \wedge u_\ell \rangle^\ell$. The verification is more complicated but requires no new ideas so it is left to the reader.

Next, let us consider special PDE defined for functions of a ℓ-vector variable.

(i) The fundamental operator is the ℓ-vector derivative $\partial_X = \sum e_{a_1 \ldots a_\ell} \partial_{X_{a_1 \ldots a_l}}$; it is in general not elliptic.

(ii) Let ∂_X be the ℓ-vector derivative and let Y be a fixed s-vector; then one may consider the Clifford product $Y\partial_X$ which admits a multivector decomposition of the form

$$Y\partial_X = [Y\partial_X]_{\ell+s} + [Y\partial_X]_{\ell+s-2} + \ldots + [Y\partial_X]_{|\ell-s|}.$$

Each of these terms is an interesting operator and one may consider systems consisting of such operators where the s-vector varies over the space of s-vectors or over a submanifold. In particular, when Y varies over the space of all s-vectors, it suffices to consider $Y \in \{e_A \ : \ A \subset M, \ |A| = s\}$.

(iii) The so called "monogenic system" corresponds to the special case $s = \ell - 1$, i.e. it consists of the collection of operators $[e_{b_1 \dots b_{\ell-1}} \sum e_{a_1 \dots a_\ell} \partial_{X_{a_1 \dots a_\ell}}]_1 = \partial_{B;X}$ and the equations $\partial_{B;X} f = 0$ are called "monogenicity conditions" as before.

We also have the following:

Theorem 6.4 (Chain rule). *Let f be a scalar function defined on an open subset of the space of ℓ-vectors $\mathbb{R}_{m;l}$. Then for the restriction to the manifold of ℓ-blades we have the formula*

$$\partial_{\underline{x}} f(\underline{x} \wedge \underline{u}_1 \wedge \dots \wedge \underline{u}_{\ell-1}) = \pm[\underline{u}_1 \wedge \dots \wedge \underline{u}_{\ell-1} \partial_X]_1 f(X)_{|X = \underline{x} \wedge \underline{u}_1 \wedge \dots \wedge \underline{u}_{\ell-1}}.$$

Proof. Applying the chain rule we have that

$$\frac{\partial}{\partial x_j} f(\underline{x} \wedge \underline{u}_1 \wedge \dots \wedge \underline{u}_{\ell-1}) = \sum \frac{\partial X_A}{\partial x_j} \partial_{X_A} f(X)_{|X = \underline{x} \wedge \underline{u}_1 \wedge \dots \wedge \underline{u}_{\ell-1}}$$

$$= \sum \langle e_A, \frac{\partial(\underline{x} \wedge \underline{u}_1 \wedge \dots \wedge \underline{u}_{\ell-1})}{\partial x_j} \rangle \partial_{X_A} f(X)_{|X = \underline{x} \wedge \underline{u}_1 \wedge \dots \wedge \underline{u}_\ell}$$

$$= \langle \partial_X, e_j \wedge \underline{u}_1 \wedge \dots \wedge \underline{u}_{\ell-1} \rangle f(X)_{|X = \underline{x} \wedge \underline{u}_1 \wedge \dots \wedge \underline{u}_\ell}$$

so that for scalar functions:

$$\partial_{\underline{x}} f(\underline{x} \wedge \underline{u}_1 \wedge \dots \wedge \underline{u}_\ell) = \sum e_j \langle ej \wedge \underline{u}_1 \wedge \dots \wedge \underline{u}_{\ell-1}, \partial_X \rangle f(X)_{|X}$$

and for any ℓ-vector Y one has that

$$\langle e_j \wedge \underline{u}_1 \wedge \dots \wedge \underline{u}_{\ell-1}, Y \rangle = \pm[e_j \underline{u}_1 \cdots \underline{u}_{\ell-1} Y]_0 = \pm \langle e_j, [\underline{u}_1 \cdots \underline{u}_{\ell-1} Y]_1 \rangle$$

from which one obtains that

$$\sum e_j \langle e_j \wedge \underline{u}_1 \wedge \dots \wedge \underline{u}_{\ell-1}, \partial_X \rangle = \pm[\underline{u}_1 \cdots \underline{u}_{\ell-1} \partial_X] = \pm[\underline{u}_1 \wedge \dots \wedge \underline{u}_{\ell-1} \partial_X]$$

leading to the required identity. □

The above identity of course also remains true for Clifford algebra valued functions and in particular we have the following:

Corollary 6.5. *Suppose that $f(X)$ satisfies the monogenicity conditions; then the restriction $f(\underline{x}_1 \wedge \dots \wedge \underline{x}_\ell)$ of f to the manifold of blades is symplicial monogenic.*

Conversely, at least for homogeneous polynomials we have:

Theorem 6.6. *Let $P(\underline{x}_1 \wedge \dots \wedge \underline{x}_\ell)$ be symplicial monogenic of degree k; then P admits a (of course unique) extension $P(X)$ to the space of ℓ-vectors satisfying the Plücker PDE together with the monogenicity conditions.*

Proof. The proof is again based on the main result in [36] which states that the space of symplicial monogenics of the above type is spanned by the maximal vectors i.e. by the plane waves of the form $\langle \underline{x}_1 \wedge \ldots \wedge \underline{x}_\ell, T_1 \wedge \ldots \wedge T_\ell \rangle^k T_1 \wedge \ldots \wedge T_\ell$, where $(T_1, \ldots, T_\ell)$ is an ℓ-nullframe, i.e. $T_i T_j = -T_j T_i$ for all $i, j = 1, \ldots, \ell$ examples of which are the Witt basis vectors $2T_j = e_{2j-1} - ie_{2j}$. For such solutions to exist we must have that $2\ell \leq m$. The extension of this to the space of ℓ-vectors given by $\langle X, T_1 \wedge \ldots \wedge T_\ell \rangle^k T_1 \wedge \ldots \wedge T_\ell$ clearly satisfies the Plücker PDE whereas for the monogenicity conditions we find that

$$[e_B \partial_X]_1 \langle X, T_1 \wedge \ldots \wedge T_\ell \rangle^k T_1 \wedge \ldots \wedge T_\ell$$
$$= k[e_B T_1 \wedge \ldots \wedge T_\ell]_1 \langle X, T_1 \wedge \ldots \wedge T_\ell \rangle^{k-1} T_1 \wedge \ldots \wedge T_\ell$$

and straightforward verification shows that for any subset B of M with cardinality $\ell - 1$,

$$[e_B T_1 \ldots T_\ell]_1 T_1 \ldots T_\ell = 0.$$

□

It remains to be verified also in this case that the monogenicity conditions are strong enough to imply the Plücker PDE. In spite of many problems that still remain to be considered it is clear from the above that the most important system of equations is obtained simply by the application of the chain rule along the manifold of ℓ-blades. The space of ℓ-vectors allows the consideration of many more canonical submanifolds and differential operators defined for them. For example, let X be an ℓ-vector variable and let Y be an s-vector variable; then the manifold of $(\ell + s)$-vectors of the simple form $X \wedge Y$ is a special submanifold of the space of all $(\ell + s)$-vectors and one can investigate in which cases one has higher order Hestenes–Sobczyk: $\partial_W f(W) \to C\partial_X \wedge \partial_Y f(X \wedge Y)$ for $W \to X \wedge Y$. In particular one may consider chain rules of the form

$$\partial_X f(X \wedge Y) = C[Y \partial_W]_\ell f(W)_{|W = X \wedge Y}$$

and consider several interesting operators and problems.

7. The case of Stiefel manifolds

More general than considering functions on the ℓ-vector space $\mathbb{R}_{m;\ell}$, one may consider functions $f(X_1, X_2, \ldots, X_\ell)$, $2\ell \leq m$ of multivector variables $X_j \in \mathbb{R}_{m;j}$ and investigate necessary and sufficient conditions (in case f is a polynomial) for this function f to be determined by the restriction to the "homogeneous Stiefel manifold" given by the restriction to the case

$$X_1 = \underline{x}_1, \quad X_2 = \underline{x}_1 \wedge \underline{x}_2, \ldots, \quad X_\ell = \underline{x}_1 \wedge \ldots \wedge \underline{x}_\ell.$$

Let us point out, though we are not going to discuss this matter in this paper, that using the Fischer inner product, this again leads to Plücker PDE. It is also clear that the good candidate for the monogenicity system consists of the totality of operators of the form

$$[e_B\partial_{X_j}]_1, \qquad B\subset M, \qquad |B| = j-1 \text{ and } j = 1,\ldots,\ell.$$

Moreover, we also have the chain rule (which follows from the previous ones) valid for all scalar functions

$$\begin{aligned}&\partial_{\underline{x}} f(\underline{x}, \underline{x}\wedge\underline{u}_1,\ldots,\underline{x}\wedge\underline{u}_1\wedge\ldots\wedge\underline{u}_{\ell-1})\\ &\qquad = \sum_{j=1}^{\ell} \pm[\underline{u}_1\wedge\ldots\wedge\underline{u}_{j-1}\partial_{X_j}]_1 f(X_1,\ldots,X_\ell)_{|X_j=\underline{x}\wedge\underline{u}_1\wedge\ldots\wedge\underline{u}_{j-1}}.\end{aligned}$$

Again the above also holds for Clifford algebra-valued material and the validity of the monogenicity equations for a functions $f(X_1,\ldots,X_\ell)$ readily entails that the restriction to the Stiefel manifold $f(\underline{x},\underline{x}\wedge\underline{u}_1,\ldots,\underline{x}\wedge\underline{u}_1\wedge\ldots\wedge\underline{u}_{\ell-1})$ is symplicial monogenic. Conversely, in the case of homogeneous symplicial monogenics of degree $(k_1,\ldots,k_\ell)$ we may extend the function $f(\underline{x},\underline{x}\wedge\underline{u}_1,\ldots,\underline{x}\wedge\underline{u}_1\wedge\ldots\wedge\underline{u}_{\ell-1})$ in a unique way to a function $f(X_1,\ldots,X_\ell)$ satisfying all the Plücker PDE together with all monogenicity conditions. In the case of plane wave solutions

$$\langle\underline{x},T_1\rangle^{k_1}\langle\underline{x}\wedge\underline{u}_1,T_1\wedge T_2\rangle^{k_2}\ldots\langle\underline{x}\wedge\underline{u}_1\wedge\ldots\wedge\underline{u}_{\ell-1},T_1\wedge\ldots\wedge T_\ell\rangle^{k_\ell}T_1\wedge\ldots\wedge T_\ell$$

$(T_1,\ldots,T_\ell)$ being an ℓ-nullframe, this extension is given by

$$\langle X_1,T_1\rangle^{k_1}\langle X_2,T_1\wedge T_2\rangle^{k_2}\ldots\langle X_\ell,T_1\wedge\ldots\wedge T_\ell\rangle^{k_\ell}T_1\wedge\ldots\wedge T_\ell,$$

and these plane wave functions span the space of all symplicial monogenics (see [36]).

References

[1] W.W. Adams, C.A. Berenstein, P. Loustaunau, I. Sabadini, D.C. Struppa, *Regular functions of several quaternionic variables and the Cauchy–Fueter complex*, J. Geom. Anal. **9** (1999), 1–16.

[2] E. Arbarello, M. Cornalba, P.A. Griffiths, J. Harris, *Geometry of Algebraic Curves*, Springer Verlag, New-York, 1985.

[3] W.W. Adams, P. Loustaunau, V. Palamodov, D.C. Struppa, *Hartogs' phenomenon for polyregular functions and projective dimension of related modules over a polynomial ring*, Ann. Inst. Fourier, **47** (1997), 623–640.

[4] F. Brackx, R. Delanghe, F. Sommen, *Clifford Analysis*, Pitman Res. Notes in Math. **76**, 1982.

[5] C. Chevalley, *The algebraic theory of spinors*, Columbia Univ. Press, New York, 1954.

[6] D. Constales, *The relative position of L^2-domains in complex and Clifford analysis*, Ph.D. thesis, Univ. Ghent, 1989–1990.

[7] R. Delanghe, *Clifford analysis: history and perspective*, Comp. Meth. Funct. Theor. **1** (2001), 107–153.

[8] R. Delanghe, F. Sommen, V. Soucek, *Clifford algebra and spinor-valued functions: a function theory for the Dirac operator*, Math. and Its Appl. **53**, Kluwer Acad. Publ., Dordrecht, 1992.

[9] A. Dimakis, F. Mueller-Hoisen, *Clifform calculations with applications to classical field theories*, Class. Quantum Gravity, **8** (1991), 2093–2132.

[10] M.G. Eastwood, P.W. Michor, *Some remarks on the Plücker relations*, Rend. Circ. Mat. Palermo Suppl., **63** (2000), 83–85.

[11] S.L. Eriksson-Bique, H. Leutwiler, *Hypermonogenic Functions*, in Clifford algebras and their applications in mathematical physics. Papers of the 5th international conference, Ixtapa-Zihuatanejo, Mexico, 1999. Clifford analysis. Boston, MA: Birkhuser. Prog. Phys. **19**, 287–302 (2000).

[12] J. Gilbert, M. Murray, *Clifford algebras and Dirac operators in harmonic analysis*, Cambridge Univ. Press **26**, Cambridge, 1990.

[13] D. Hestenes, G.Sobczyk, *Clifford algebras to geometric calculus*, D.Reidel, Dordrecht, 1985.

[14] W. Hodge, *The theory and applications of harmonic integrals*, Cambridge Univ. Press, Cambridge, 1959.

[15] G. Laville, *Une famille de solutions de l'equation de Dirac avec champ electromagnetique quelconque*, C. R. Acad. Sc. Paris, **296** (1983), 1029–1032.

[16] H. Leutwiler, *Modified Clifford Analysis*, Compl. Var., **17** (1992), 153–171.

[17] H. Leutwiler, *Modified Quaternionic Analysis in* R^3, Compl. Var., **20** (1992), 19–51

[18] P. Lounesto, *Spinor valued regular functions in hypercomplex analysis*, Ph.D. thesis, Helsinki, 1979.

[19] P. Lounesto, *Clifford algebras and spinors*, Cambridge Univ. Press, 1997.

[20] H. Malonek, *Contributions to a geometric function theory in higher dimensions by Clifford analysis methods: monogenic functions and M-conformal mappings*, F. Brackx et al. (eds.), NATO Science Series, II Math., Phys., Chem., **25** (2001), 213–222.

[21] L. Manivel, *Symmetric functions, Schubert polynomials and degeneracy loci*, Translated from the 1998 French, SMF/AMS Texts and Monographs, **6** Cours Spécialisés American Mathematical Society, Providence, RI; Société Mathématique de France, Paris, 2001.

[22] I.R. Porteous, *Topological Geometry*, Cambridge Univ. Press, Cambridge, 1981.

[23] G.B. Rizza, *Funzioni regolari nelle Algebre di Clifford*, Rend. Mat., XV (1956), 1–27.

[24] J. Ryan ed., *Clifford Algebras in Analysis and Related Topics*, Studies in Advanced Mathematics, CRC Press, 1996.

[25] I. Sabadini, M. Shapiro, D.C. Struppa, *Algebraic analysis of the Moisil–Theodorescu system*, Compl. Var. **40**, (2000) 333–357.

[26] I. Sabadini, F. Sommen, *Special first order systems in Clifford analysis and resolutions*, Zeit. Anal. Anw., **21**, n. 1 ,(2002) 27–55.

[27] I. Sabadini, F. Sommen, D.C. Struppa, *The Dirac complex on abstract vector variables: megaforms*, Exp. Math. **12** (2003), 351–364.

[28] I. Sabadini, F. Sommen, D.C. Struppa, P. Van Lancker, *Complexes of Dirac operators in Clifford algebras*, Math. Z. **239** (2002), 293–320.

[29] F. Sommen, *Monogenic Differential Calculus*, Trans. Amer. Math. Soc., **326** n. 2 (1991), 613–632.

[30] F. Sommen, *Clifford analysis in two and several vector variables*, Appl. An., **73** (1999), 225–253.

[31] F. Sommen, *Clifford analysis on the level of abstract vector variables*, NATO Science Series, II Math., Phys., Chem., **25** (2001), 303–322.

[32] F. Sommen, V. Soucek, *Monogenic differential forms*, Compl. Var.:Theor. Appl., **19** (1992), 81–90.

[33] F. Sommen, N. Van Acker, *$SO(m)$-invariant differential operators on Clifford algebra-valued functions*, Found. Phys. **23**, n. 11 (1993), 1491–1519.

[34] F. Sommen, N. Van Acker, *Functions of two vector variables*, Adv. Appl. Cliff. Alg. **4**, n.1 (1994) 65–72.

[35] E.M. Stein, G. Weiss, *Generalization of the Cauchy–Riemann equations and representations of the rotation group*, Amer. J. Math. **90** (1968), 163–196.

[36] P. Van Lancker, F. Sommen, D. Constales, *Models for irreducible representations of Spin(m)*, Adv. Appl. Cliff. Alg. **11** (2001), 271–289.

[37] C. von Westenholz, *Differential forms in mathematical physics*, Stud. Math. Appl. **3**, North Holland, Amsterdam, 1978.

Irene Sabadini
Dipartimento di Matematica
Politecnico di Milano
Via Bonardi, 9
20133 Milano
Italy
e-mail: `sabadini@mate.polimi.it`

Frank Sommen
Department of Mathematical Analysis
State University of Ghent
Galglaan,2
9000 Gent
Belgium
e-mail: `fs@cage.rug.ac.be`

Part B
Global Analysis and Differential Geometry

Trends in Mathematics: *Advances in Analysis and Geometry*, 189–208

Universal Bochner-Weitzenböck Formulas for Hyper-Kählerian Gradients

Yasushi Homma

Abstract. Hyper-Kählerian gradients on hyper-Kähler manifolds are first-order differential operators naturally defined by hyper-Kähler structure. We show that the principal symbols of hyper-Kählerian gradients are related to the enveloping algebra and Casimir elements of the symplectic group. In particular, we give universal Bochner-Weitzenböck formulas which are certain relations in the enveloping algebra. From the formulas, we construct Bochner-Weitzenböck formulas for hyper-Kählerian gradients.

Mathematics Subject Classification (2000). 58J60, 17B35, 53C26.

Keywords. Gradients, Bochner-Weitzenböck formulas, $Sp(n)$-modules, Casimir elements.

1. Introduction

Let (M, g, I, J, K) be a real $4n$-dimensional hyper-Kähler manifold. The hyper-Kähler structure gives a reduction of the frame bundle with the Levi-Civita connection. We denote the principal $Sp(n)$-bundle on M by $\mathbf{Sp}(M)$. Considering an irreducible unitary representation (π_ρ, V_ρ) of $Sp(n)$ with highest weight ρ, we have a Hermitian vector bundle $\mathbf{S}_\rho := \mathbf{Sp}(M) \times_\rho V_\rho$. The Levi-Civita connection induces a covariant derivative ∇ on $\mathbf{S}_\rho$,

$$\nabla : \Gamma(\mathbf{S}_\rho) \xrightarrow{\nabla} \Gamma(\mathbf{S}_\rho \otimes (\Lambda^{1,0}(M) \oplus \Lambda^{0,1}(M))).$$

Here, $\Lambda^{1,0}(M)$ and $\Lambda^{0,1}(M)$ are the $(1,0)$- and the $(0,1)$-cotangent bundles with respect to the complex structure I. Since both $\mathbf{S}_\rho$ and $\Lambda^{1,0}(M)$ are associated bundles to $\mathbf{Sp}(M)$, we decompose the tensor bundle $\mathbf{S}_\rho \otimes \Lambda^{1,0}(M)$ into the sum of irreducible bundles, $\mathbf{S}_\rho \otimes \Lambda^{1,0}(M) = \oplus_\nu \mathbf{S}_\nu$. Then we have a first-order differential operator as a component of the $(1,0)$-covariant derivative $\nabla^{1,0}$,

$$D_\nu : \Gamma(\mathbf{S}_\rho) \xrightarrow{\nabla^{1,0}} \Gamma(\mathbf{S}_\rho \otimes \Lambda^{1,0}(M)) \xrightarrow{\Pi_\nu} \Gamma(\mathbf{S}_\nu),$$

where Π_ν is the orthogonal projection from $\mathbf{S}_\rho \otimes \Lambda^{1,0}(M)$ onto $\mathbf{S}_\nu$. Since there is a real or quaternionic structure $\mathfrak{J}$ on each associated vector bundle, we twist D_ν by $\mathfrak{J}$ and have another first-order differential operator $\mathfrak{J}D_\nu\mathfrak{J}^{-1} : \Gamma(\mathbf{S}_\rho) \to \Gamma(\mathbf{S}_\nu)$ (see Section 4). In fact, we know that $\mathfrak{J}D_\nu\mathfrak{J}^{-1}$ is a component of $\nabla^{0,1}$. We call these operators $\{D_\nu, \mathfrak{J}D_\nu\mathfrak{J}^{-1}\}_\nu$ *hyper-Kählerian gradients* on M.

A usual gradient is a first-order differential operator as a component of covariant derivative on a Riemannian or spin manifold. In recent research for gradients by Thomas Branson et al., we know that gradients have importance and many applications in geometry and analysis (see [1], [2], [3], [4], [6], [8], and a series of papers about gradients by T. Branson). In fact, many geometric first-order differential operators are gradients: the Dirac operator, the twistor operator, the exterior derivative, the interior derivative, the conformal Killing operator, the Rarita-Schwinger operator, and so on. Since gradients are defined through fiber-wise orthogonal projections, their principal symbols are related to representation theory for the special orthogonal group $SO(n)$ or the spin group $Spin(n)$ [1], [4], [8]. It is natural to change the group of symmetry, $SO(n)$, to other Lie groups. For the case of the unitary group $U(n)$, the author has discussed Kählerian gradients on Kähler manifolds and given some applications to Kähler geometry [7]. In this paper, we discuss the case of the symplectic group $Sp(n)$, that is, hyper-Kählerian gradients on hyper-Kähler manifolds.

Our main purpose is to construct Bochner-Weitzenböck formulas for hyper-Kählerian gradients. From the definition of them, we can show that

$$\sum_\nu D_\nu^* D_\nu = \nabla^{1,0*}\nabla^{1,0}, \quad \sum_\nu \mathfrak{J}D_\nu^* D_\nu \mathfrak{J}^{-1} = \nabla^{0,1*}\nabla^{0,1}. \tag{1.1}$$

Then our task is to find out independent vectors $(\{a_\nu\}_\nu, \{b_\nu\}_\nu, \{c_\nu\}_\nu, \{d_\nu\}_\nu)$ as many as possible such that

$$\sum_\nu a_\nu D_\nu^* D_\nu + b_\nu \mathfrak{J}D_\nu^* D_\nu \mathfrak{J}^{-1} + c_\nu \mathfrak{J}D_\nu^*\mathfrak{J}^{-1}D_\nu + d_\nu D_\nu^*\mathfrak{J}D_\nu\mathfrak{J}^{-1} = \text{curvature action}. \tag{1.2}$$

We call these identities (1.1) and (1.2) *Bochner-Weitzenböck formulas* or *Bochner identities*. These formulas are useful to make vanishing theorems and eigenvalue estimates [2], [7], [10], [11]. The equation (1.2) means that the principal symbol of the operator on the left vanishes. In other words, there are certain relations among principal symbols of hyper-Kählerian gradients such as the Clifford relation. Our plan to construct Bochner-Weitzenböck formulas is as follows:

1. We relate the principal symbols to the (universal) enveloping algebra $U(\mathfrak{sp}(n,\mathbb{C}))$ of $Sp(n)$.
2. We find out certain relations on $U(\mathfrak{sp}(n,\mathbb{C}))$ through study of Casimir elements.
3. The relations give desired algebraic relations for the principal symbols. Furthermore, they also give some identities for Casimir elements.

As a result, we know that there is a correspondence between identities for Casimir elements of $Sp(n)$ and Bochner-Weitzenböck formulas for hyper-Kählerian gradients.

This paper is organized as follows: In Section 2, we give a review to representation theory for symplectic groups. In Section 3, we introduce Clifford homomorphisms on $Sp(n)$-modules as a generalization of Clifford multiplication on spinor spaces. We relate them to the enveloping algebra of $Sp(n)$. Through study of Casimir elements, we find out the universal Bochner-Weitzenböck formulas, which induce not only identities for Casimir elements but also relations for Clifford homomorphisms. In Section 4, we define hyper-Hermitian gradients on hyper-Hermitian manifolds and prove their conformal covariance. In the last section, by using the universal Bochner-Weitzenböck formulas, we give Bochner-Weitzenböck formulas for hyper-Kählerian gradients on hyper-Kähler manifolds.

2. Representation Theory for Symplectic Groups

Let W be a real $4n$-dimensional vector space with an inner product g. We call (I, J, K) a *hyper-Hermitian structure* on (W, g) if I, J and K are complex structures on W such that $IJ = -JI = K$ and $g(Ix, Iy) = g(Jx, Jy) = g(x, y)$ for any x, y in W. In other words, W is a quaternionic vector space with a compatible inner product g.

We consider the complexification of W, $W \otimes \mathbb{C}$, and extend I, J, K and g complex linearly on $W \otimes \mathbb{C}$. The complex vector space $W \otimes \mathbb{C}$ splits into the direct sum of $W^{1,0}$ and $W^{0,1}$ with respect to I. These vector spaces have Hermitian inner products given by $(u, v) := g(u, \bar{v})$, where $\bar{v}$ is complex conjugate of v. Another complex structure J gives a complex-linear map $J : W^{1,0} \to W^{0,1}$ because of $JI = -IJ$. Then $W^{1,0}$ has an anti-linear map $\mathfrak{J} : W^{1,0} \ni u \mapsto J(\bar{u}) \in W^{1,0}$ such that $\mathfrak{J}^2 = -1$ and $(\mathfrak{J}u, \mathfrak{J}v) = (v, u)$. This means that $W^{1,0}$ has a quaternionic structure $\mathfrak{J}$ compatible with Hermitian inner product. The space $W^{0,1}$ also has such a structure. Moreover $W^{1,0}$ and $W^{0,1}$ have complex symplectic structures. In fact,

$$\Omega^{2,0}(u,v) := (g(Ju,v) + \sqrt{-1}g(Ku,v))/2,\ \Omega^{0,2}(u,v) := (g(Ju,v) - \sqrt{-1}g(Ku,v))/2$$

are complex symplectic forms on $W^{1,0}$ and $W^{0,1}$, respectively.

On a hyper-Hermitian vector space (W, g, I, J, K), we choose an orthonormal basis $\{e_k, Ie_k, Je_k, Ke_k\}_{k=1}^n$ and set

$$\epsilon_k := \frac{1}{\sqrt{2}}(e_k - \sqrt{-1}Ie_k), \quad \epsilon_{-k} := \frac{1}{\sqrt{2}}(Je_k - \sqrt{-1}IJe_k),$$
$$\bar{\epsilon}_k := \frac{1}{\sqrt{2}}(e_k + \sqrt{-1}Ie_k), \quad \bar{\epsilon}_{-k} := \frac{1}{\sqrt{2}}(Je_k + \sqrt{-1}IJe_k),$$

for $k = 1, \dots, n$. These vectors constitute unitary bases $\{\epsilon_\alpha\}_{\alpha=\pm1}^{\pm n}$ of $W^{1,0}$ and $\{\bar{\epsilon}_\alpha\}_{\alpha=\pm1}^{\pm n}$ of $W^{0,1}$ such that

$$\mathfrak{J}(\epsilon_\alpha) = \mathrm{sign}(\alpha)\epsilon_{-\alpha}, \quad \mathfrak{J}(\bar{\epsilon}_\alpha) = \mathrm{sign}(\alpha)\bar{\epsilon}_{-\alpha},$$
$$\Omega^{2,0}(\epsilon_\alpha, \epsilon_\beta) = \mathrm{sign}(\alpha)\delta_{-\alpha\beta}, \quad \Omega^{0,2}(\bar{\epsilon}_\alpha, \bar{\epsilon}_\beta) = \mathrm{sign}(\alpha)\delta_{-\alpha\beta},$$

where $\mathrm{sign}(\alpha)$ is the sign of α,

$$\mathrm{sign}(\alpha) = \begin{cases} 1 & \text{for } \alpha > 0, \\ -1 & \text{for } \alpha < 0. \end{cases}$$

The symplectic group $Sp(n)$ is a group of automorphisms on W preserving hyper-Hermitian structure,

$$Sp(n) := \{A \in SO(W) \mid AI = IA, AJ = JA, KA = AK\}.$$

This group is also realized on $W^{1,0}$ as

$$\begin{aligned} Sp(n) &= \{A \in U(W^{1,0}) \mid \Omega^{2,0}(Au, Av) = \Omega^{2,0}(u,v) \text{ for } u \text{ and } v \text{ in } W^{1,0}\} \\ &= \{A \in U(W^{1,0}) \mid \mathfrak{J}A\mathfrak{J}^{-1} = A\}. \end{aligned}$$

The complexification $Sp(n,\mathbb{C})$ of $Sp(n)$ is called the complex symplectic group, which preserves $\Omega^{2,0}$ on $W^{1,0}$,

$$Sp(n,\mathbb{C}) = \{A \in GL_{\mathbb{C}}(W^{1,0}) \mid \Omega^{2,0}(Au, Av) = \Omega^{2,0}(u,v) \text{ for } u \text{ and } v \text{ in } W^{1,0}\}.$$

There are matrix representations of the Lie algebra $\mathfrak{sp}(n)$ and $\mathfrak{sp}(n,\mathbb{C})$ with our unitary basis $\{\epsilon_\alpha\}_\alpha$,

$$\mathfrak{sp}(n) = \{\begin{pmatrix} A & B \\ -B^* & \bar{A} \end{pmatrix} \mid A + A^* = 0, {}^tB = B\},$$
$$\mathfrak{sp}(n,\mathbb{C}) = \{\begin{pmatrix} A & B \\ C & -{}^tA \end{pmatrix} \mid {}^tC = C, {}^tB = B\}.$$

Then we define canonical elements of $\mathfrak{sp}(n,\mathbb{C})$ by

$$x_{\alpha\beta} := \epsilon_\alpha \otimes \epsilon_\beta^* - \mathrm{sign}(\alpha\beta)\epsilon_{-\beta} \otimes \epsilon_{-\alpha}^* \quad \text{in } W^{1,0} \otimes (W^{1,0})^* = \mathrm{End}(W^{1,0}) \tag{2.1}$$

for $\alpha, \beta = \pm1, \dots, \pm n$. Here, the basis of $\mathfrak{sp}(n,\mathbb{C})$ is $\{x_{\alpha\beta} | \alpha + \beta \geq 0\}$.

Lemma 2.1. 1. *The Lie algebra $\mathfrak{sp}(n,\mathbb{C})$ is spanned by $\{x_{\alpha\beta}\}_{\alpha,\beta=\pm1}^{\pm n}$ with relations*

$$x_{\alpha\beta} = -\mathrm{sign}(\alpha\beta)x_{-\beta-\alpha},$$
$$[x_{\alpha\beta}, x_{\mu\nu}] = \delta_{\beta\mu}x_{\alpha\nu} - \delta_{\alpha\nu}x_{\mu\beta} + \mathrm{sign}(\alpha\beta)(\delta_{-\beta\nu}x_{\mu-\alpha} - \delta_{-\alpha\mu}x_{-\beta\nu}).$$

2. *The natural representation of $\mathfrak{sp}(n,\mathbb{C})$ on $W^{1,0}$ is given by*

$$x_{\alpha\beta}\epsilon_\nu = \delta_{\beta\nu}\epsilon_\alpha + \mathrm{sign}(\nu\beta)\delta_{-\alpha\nu}\epsilon_{-\beta}.$$

Proof. By direct calculations. □

We shall discuss representations of $Sp(n)$ or $Sp(n,\mathbb{C})$. Let V be a finite dimensional irreducible unitary $Sp(n)$-module. We choose $\mathfrak{h} := \mathrm{span}_{\mathbb{R}}\{\sqrt{-1}x_{kk} | 1 \leq k \leq n\}$ as a maximal abelian subalgebra of $\mathfrak{sp}(n)$. Then we decompose V into simultaneous eigenspaces called weight spaces with respect to $\mathfrak{h}$ and have the highest weight $\rho = (\rho^1, \dots, \rho^n)$ in weights for V. Here, ρ^k is the eigenvalue of x_{kk}. This highest weight ρ satisfies *the dominant integral condition*,

$$\rho = (\rho^1, \dots, \rho^n) \in \mathbb{Z}^n, \quad \rho^1 \geq \rho^2 \geq \dots \geq \rho^n \geq 0.$$

Conversely, for a dominant integral weight ρ, we have a unique irreducible unitary $Sp(n)$-module with highest weight ρ up to equivalence. Thus, all the irreducible unitary $Sp(n)$-modules are parametrized by dominant integral weights. So, we denote by (π_ρ, V_ρ) an irreducible unitary representation of $Sp(n)$ with highest weight ρ. For example, the natural representation of $Sp(n)$ on $W^{1,0}$ has highest weight $\mu_1 := (1, 0, \dots, 0)$.

A significant feature common to irreducible $Sp(n)$-modules is that there is a real or quaternionic structure compatible with the action of $Sp(n)$. A *real* (resp. *quaternionic*) *structure* on complex vector space V is said to be an anti-linear automorphism $\mathfrak{J} : V \to V$ with $\mathfrak{J}^2 = 1$ (resp. $\mathfrak{J}^2 = -1$). We shall explain this feature refering to [5]. Let (π_ρ, V_ρ) be as above. Considering the contragradient representation π_ρ^* on dual space V_ρ^*, we know that the weights are $\{-\lambda \mid \lambda \text{ is weight of } (\pi_\rho, V_\rho)\}$ and the lowest one is $-\rho$. But, we can move this lowest weight $-\rho$ to ρ by the Weyl group of $Sp(n)$. Thus, the highest weight of (π_ρ^*, V_ρ^*) is also ρ and there is a $Sp(n)$-isomorphism $V_\rho \simeq V_\rho^*$. Now, we consider a vector space of the $Sp(n)$-invariant bilinear forms on V_ρ, that is, $\mathrm{Hom}_{Sp(n)}(V_\rho, V_\rho^*)$. It follows from Schur's lemma that $\mathrm{Hom}_{Sp(n)}(V_\rho, V_\rho^*)$ is a one-dimensional complex vector space. Then we take a non-zero $Sp(n)$-invariant bilinear form Ω on V_ρ, which is non-degenerate because of irreducibility of V_ρ. We define a symmetric form Ω_+ and a symplectic form Ω_- by $\Omega_\pm(\phi, \psi) := \Omega(\phi, \psi) \pm \Omega(\psi, \phi)$. As $\dim \mathrm{Hom}_{Sp(n)}(V_\rho, V_\rho^*)$ is one, we have either $\Omega_+ = 0$ or $\Omega_- = 0$. Thus, there is a unique non-degenerate symmetric or symplectic form on V_ρ up to a scalar multiple. On the other hands, there is a $Sp(n)$-isomorphism $\bar{V}_\rho \simeq V_\rho^*$ because of unitarity. Here, $(\bar{\pi}_\rho, \bar{V}_\rho)$ is the complex conjugate representation of (π_ρ, V_ρ). As a result, we have an isomorphism $V_\rho \simeq V_\rho^* \simeq \bar{V}_\rho$ as $Sp(n)$-modules. Normalizing the symmetric or symplectic form Ω, we have a unique $Sp(n)$-invariant real or quaternionic structure $\mathfrak{J} : V_\rho \to V_\rho$ such that $(\mathfrak{J}\phi, \mathfrak{J}\psi) = (\psi, \phi)$ for any ϕ and ψ in V_ρ.

Proposition 2.2. *On an irreducible unitary $Sp(n)$-module V_ρ, there is a unique $Sp(n)$-invariant real or quaternionic structure $\mathfrak{J}$ such that $(\mathfrak{J}\phi, \mathfrak{J}\psi) = (\psi, \phi)$ for any ϕ and ψ in V_ρ. Furthermore, if the highest weight ρ satisfies that $\sum_{k=1}^n \rho^k$ is even (resp. odd), then $\mathfrak{J}$ is real (resp. quaternionic).*

Proof. It remains to show that $\sum \rho^k$ is even (resp. odd) iff $\mathfrak{J}$ is real (resp. quaternionic). We consider the natural $Sp(n)$-module $W^{1,0}$ with $Sp(n)$-invariant quaternionic structure $\mathfrak{J}$. The tensor product module $\otimes^{2l} W^{1,0}$ has a real structure $\otimes^{2l}\mathfrak{J}$, and $\otimes^{2l+1}W^{1,0}$ has a quaternionic structure $\otimes^{2l+1}\mathfrak{J}$. Each irreducible $Sp(n)$-module V_ρ can be realized as a component of $\otimes^{2l}W^{1,0}$ or $\otimes^{2l+1}W^{1,0}$. In particular, V_ρ is in $\otimes^{2l}W^{1,0}$ if and only if $\sum \rho^k$ is even. Thus, we have proved the proposition. □

3. Casimir Elements and Universal Bochner-Weitzenböck Formulas

We introduce the enveloping algebra and Casimir elements of the Lie algebra $\mathfrak{sp}(n,\mathbb{C})$. Let $\{X_i\}_{i=1}^s$ be a basis of $\mathfrak{sp}(n,\mathbb{C})$. The enveloping algebra $U(\mathfrak{sp}(n,\mathbb{C}))$ is an associative algebra over $\mathbb{C}$ such that basis is $\{(X_1)^{k_1}\cdots(X_s)^{k_s}|k_i\geq 0\}$ and relations are $X_iX_j - X_jX_i = [X_i, X_j]$ for $i,j=1,\ldots,s$. We denote by $\mathfrak{Z}$ the center of $U(\mathfrak{sp}(n,\mathbb{C}))$, which is characterized as the invariant sub-algebra of $U(\mathfrak{sp}(n,\mathbb{C}))$ under the adjoint action of $\mathfrak{sp}(n,\mathbb{C})$. We call elements in $\mathfrak{Z}$ *Casimir elements.*

We shall construct generators of $\mathfrak{Z}$. Let $\{x_{\alpha\beta} \mid \alpha,\beta = \pm 1,\ldots,\pm n\}$ be the canonical elements of $\mathfrak{sp}(n,\mathbb{C})$ defined in (2.1). For each non-negative integer q, we define an element $x^q_{\alpha\beta}$ in $U(\mathfrak{sp}(n,\mathbb{C}))$ by

$$x^q_{\alpha\beta} := \begin{cases} \sum_{\alpha_1,\alpha_2,\ldots,\alpha_{q-1}=\pm1,\ldots,\pm n} x_{\alpha\alpha_1}x_{\alpha_1\alpha_2}\cdots x_{\alpha_{q-1}\beta} & q\geq 1,\\ \delta_{\alpha\beta} & q=0.\end{cases}$$

It is easy to show that these elements satisfy

$$[x_{\mu\nu}, x^q_{\alpha\beta}] = \delta_{\alpha\nu}x^q_{\mu\beta} - \delta_{\beta\mu}x^q_{\alpha\nu} + \mathrm{sign}(\mu\nu)(\delta_{-\nu\beta}x^q_{\alpha-\mu} - \delta_{-\mu\alpha}x^q_{-\nu\beta}).$$

It follows that $c_q := \sum_\alpha x^q_{\alpha\alpha}$ is an invariant element under the adjoint action of $\mathfrak{sp}(n,\mathbb{C})$, namely, a Casimir element. The following fact for these Casimir elements has been known [13], [14], [15].

Proposition 3.1. *The Casimir elements $\{c_2, c_4,\ldots,c_{2n}\}$ generate the center $\mathfrak{Z}$ algebraically. In particular, the Casimir elements with odd degree are polynomials in $\{c_2,c_4,\ldots,c_{2n}\}$.*

A nice method of calculating eigenvalue of c_q on V_ρ has been shown in [14]. The idea is to use projection operators from $V_\rho\otimes W^{1,0}$ onto irreducible components. The projection operators give homomorphisms among $Sp(n)$-modules, which are just principal symbols of hyper-Kählerian gradients. We call these homomorphisms *Clifford homomorphisms.* Let us clarify the relationship between Clifford homomorphisms and Casimir elements, and calculate the eigenvalues of c_q.

Let V_ρ be an irreducible unitary $Sp(n)$-module with real or quaternionic structure $\mathfrak{J}$ such that $(\mathfrak{J}\phi,\mathfrak{J}\psi) = (\psi,\phi)$. We consider the natural representation $(\pi_{\mu_1}, W^{1,0})$ and the tensor product representation $(\pi_\rho\otimes\pi_{\mu_1}, V_\rho\otimes W^{1,0})$. It follows from a decomposition formula (see [14] or Chapter 7 in [5]) that highest weights of irreducible components in $V_\rho\otimes W^{1,0}$ are

$$\{\rho+\mu_\nu|\ \rho+\mu_\nu \text{ is dominant integral, } \nu=\pm1,\ldots,\pm n\},$$

where we set

$$\mu_k := (\underbrace{0,\dots,0}_{k-1},1,\underbrace{0,\dots,0}_{n-k}) \in \mathbb{Z}^n$$

and $\mu_{-k} := -\mu_k$ for $k = 1,\dots,n$. Define $V_{\rho+\mu_\nu} := \{0\}$ for $\rho+\mu_\nu$ without dominant integral condition, and we can describe the irreducible decomposition as follows:

$$V_\rho \otimes W^{1,0} = \bigoplus_{\nu=\pm 1,\dots,\pm n} V_{\rho+\mu_\nu} = \bigoplus_{k=1,\dots,n} V_{\rho+\mu_k} \oplus V_{\rho-\mu_k}.$$

We adopt a Hermitian inner product and a real or quaternionic structure on each component $V_{\rho+\mu_\nu}$ induced by the ones on $V_\rho \otimes W^{1,0}$. Note that each component has a real (resp. quaternionic) structure if V_ρ has a quaternionic (resp. real) one.

Definition 3.2. We define a linear mapping $p_\nu(u) : V_\rho \to V_{\rho+\mu_\nu}$ for u in $W^{1,0}$ by

$$W^{1,0} \times V_\rho \ni (u,\phi) \mapsto p_\nu(u)\phi := \Pi_\nu(\phi \otimes u) \in V_{\rho+\mu_\nu},$$

where Π_ν is the orthogonal projection from $V_\rho \otimes W^{1,0}$ onto $V_{\rho+\mu_\nu}$. We denote by $p_\nu(u)^*$ the adjoint of $p_\nu(u)$ with respect to inner products on V_ρ and $V_{\rho+\mu_\nu}$. By using real or quaternionic structures, we have other linear mappings

$$\mathfrak{J}p_\nu(u)\mathfrak{J}^{-1} : V_\rho \to V_{\rho+\mu_\nu}, \quad \mathfrak{J}p_\nu(u)^*\mathfrak{J}^{-1} : V_{\rho+\mu_\nu} \to V_\rho.$$

We call these linear mappings $p_\nu(u)$, $p_\nu(u)^*$, $\mathfrak{J}p_\nu(u)\mathfrak{J}^{-1}$, and $\mathfrak{J}p_\nu(u)^*\mathfrak{J}^{-1}$ *Clifford homomorphisms.*

The following lemma is useful.

Lemma 3.3. *The Clifford homomorphisms satisfy*

$$\begin{aligned} \mathfrak{J}p_\nu(u)\mathfrak{J}^{-1} &= p_\nu(\mathfrak{J}u), \quad \mathfrak{J}p_\nu(u)^*\mathfrak{J}^{-1} = p_\nu(\mathfrak{J}u)^*, \\ (\mathfrak{J}p_\nu(u)\mathfrak{J}^{-1})^* &= \mathfrak{J}p_\nu(u)^*\mathfrak{J}^{-1}, \quad \mathfrak{J}\mathfrak{J}p_\nu(u)\mathfrak{J}^{-1}\mathfrak{J}^{-1} = -p_\nu(u). \end{aligned} \tag{3.1}$$

Now, we can relate the Clifford homomorphisms to the enveloping algebra and Casimir elements of $\mathfrak{sp}(n,\mathbb{C})$.

Proposition 3.4. *We assign a constant w_ν to each component $V_{\rho+\mu_\nu}$,*

$$\begin{cases} w_{+k} := -(\rho^k - k + 1) & \text{for } V_{\rho+\mu_k}, \\ w_{-k} := \rho^k - k + 2n + 1 & \text{for } V_{\rho+\mu_{-k}} = V_{\rho-\mu_k}. \end{cases} \tag{3.2}$$

We call this constant w_ν the conformal weight associated to ρ and $\rho+\mu_\nu$. Then, we have the following:

1.

$$\sum_{\alpha=\pm 1,\dots,\pm n} p_\nu(\epsilon_\alpha)^* p_\nu(\epsilon_\alpha) = \frac{\dim V_{\rho+\mu_\nu}}{\dim V_\rho}. \tag{3.3}$$

Here, $\{\epsilon_\alpha\}_\alpha$ is a unitary basis of $W^{1,0}$.

2. *For* $q = 0, 1, \ldots,$

$$\sum_{\nu=\pm1,\ldots,\pm n} w_\nu^q\, p_\nu(\epsilon_\alpha)^* p_\nu(\epsilon_\beta) = \mathrm{sign}(\alpha\beta)\pi_\rho(x^q_{-\alpha-\beta}), \tag{3.4}$$

$$\sum_{\nu=\pm1,\ldots,\pm n} w_\nu^q\, \mathfrak{J} p_\nu(\epsilon_\alpha)^* p_\nu(\epsilon_\beta)\mathfrak{J}^{-1} = \pi_\rho(x^q_{\alpha\beta}), \tag{3.5}$$

$$\sum_{\nu=\pm1,\ldots,\pm n} w_\nu^q\, \mathfrak{J} p_\nu(\epsilon_\alpha)^* \mathfrak{J}^{-1} p_\nu(\epsilon_\beta) = -\mathrm{sign}(\beta)\pi_\rho(x^q_{\alpha-\beta}), \tag{3.6}$$

$$\sum_{\nu=\pm1,\ldots,\pm n} w_\nu^q\, p_\nu(\epsilon_\alpha)^* \mathfrak{J} p_\nu(\epsilon_\beta)\mathfrak{J}^{-1} = -\mathrm{sign}(\alpha)\pi_\rho(x^q_{-\alpha\beta}). \tag{3.7}$$

3. *The eigenvalue of the Casimir element* c_q *on* V_ρ *is*

$$\pi_\rho(c_q) = \pi_\rho(\sum_\alpha x^q_{\alpha\alpha}) = \frac{1}{\dim V_\rho} \sum_{\nu=\pm1,\ldots,\pm n} w_\nu^q \dim V_{\rho+\mu_\nu}. \tag{3.8}$$

Proof. We can prove this proposition with a method given in [7], [8]. So, we give an outline of the proof. First, since Clifford homomorphisms are defined by the orthogonal projection $\Pi_\nu : V_\rho \otimes W^{1,0} \to V_{\rho+\mu_\nu}$, we have

$$\dim V_{\rho+\mu_\nu} = \mathrm{trace}(\Pi_\nu) = \dim V_\rho \times \sum_\alpha p_\nu(\epsilon_\alpha)^* p_\nu(\epsilon_\alpha).$$

Then we have proved (3.3). Next, since the inner product on $V_{\rho+\mu_\nu}$ is given along the tensor inner product on $V_\rho \otimes W^{1,0}$, we show that

$$\sum_\nu p_\nu(\epsilon_\alpha)^* p_\nu(\epsilon_\beta) = (\epsilon_\alpha, \epsilon_\beta) = \delta_{\alpha\beta}. \tag{3.9}$$

Let $c_2 := \sum x_{\alpha\beta}x_{\beta\alpha}$ be the second Casimir element. This Casimir element acts on V_ρ as $\pi_\rho(c_2) = 2(\|\delta+\rho\|^2 - \|\delta\|^2)\mathrm{id}$, where $\|\rho\|^2$ is $\sum_{k=1}^n (\rho^k)^2$ and $\delta = (n, n-1, \ldots, 1)$ is half sum of the positive roots of $\mathfrak{sp}(n, \mathbb{C})$. We define an operator C on $V_\rho \otimes W^{1,0}$ by

$$C := \{(\pi_\rho \otimes \pi_{\mu_1})(c_2) - \mathrm{id} \otimes \pi_{\mu_1}(c_2) - \pi_\rho(c_2) \otimes \mathrm{id}\}/4 = 1/2 \sum \pi_\rho(x_{\alpha\beta}) \otimes \pi_{\mu_1}(x_{\beta\alpha}).$$

This operator acts on irreducible component $V_{\rho+\mu_\nu} \subset V_\rho \otimes W^{1,0}$ as constant multiple $-w_\nu \mathrm{id}$, where

$$-w_\nu = (\|\delta+\rho+\mu_\nu\|^2 - \|\delta+\rho\|^2 - 2n - 1)/2.$$

Then we have

$$\begin{aligned}\sum_\nu -w_\nu p_\nu(\epsilon_\alpha)\phi &= C(\phi \otimes \epsilon_\alpha) = 1/2 \sum \pi_\rho(x_{\beta\nu})\phi \otimes \pi_{\mu_1}(x_{\nu\beta})\epsilon_\alpha \\ &= -\mathrm{sign}(\alpha\beta) \sum_\beta \pi_\rho(x_{-\beta-\alpha})\phi \otimes \epsilon_\beta \\ &= \sum_\nu (-\sum_\beta \mathrm{sign}(\alpha\beta) p_\nu(\epsilon_\beta)\pi_\rho(x_{-\beta-\alpha})\phi),\end{aligned}$$

for $\phi \otimes \epsilon_\alpha$ in $V_\rho \otimes W^{1,0}$. Thus we get

$$w_\nu p_\nu(\epsilon_\alpha) = \sum_\beta \operatorname{sign}(\alpha\beta) p_\nu(\epsilon_\beta)\pi_\rho(x_{-\beta-\alpha}). \tag{3.10}$$

We use this relation q-times and (3.9). Then we have (3.4). For other cases (3.5)-(3.7), we use $\mathfrak{J}(\epsilon_\alpha) = \operatorname{sign}(\alpha)\epsilon_{-\alpha}$ and (3.1). Taking the trace in (3.4), we have the eigenvalue of c_q on V_ρ as (3.8). □

We consider the equation (3.4) for $q = 0, 1, \dots, 2n-1$. Since the conformal weights $\{w_\nu\}_{\nu=\pm1}^{\pm n}$ are different from each other, the Clifford homomorphism $p_\nu(\epsilon_\alpha)^* p_\nu(\epsilon_\beta)$ is realized as a linear combination of $\{\pi_\rho(x^p_{-\alpha-\beta})\}_{p=0}^{2n-1}$. Therefore, if there exist relations for $\{x^q_{\alpha\beta}\}_{q\geq0}$, then we have relations among Clifford homomorphisms. In fact, Proposition 3.1 suggests the existence of such relations. In the rest of this section, we search explicit relations among $\{x^q_{\alpha\beta}\}_{q\geq0}$ and $\{c_q\}_{q\geq0}$. The point is that we try to exchange indices α and β of $x^q_{\alpha\beta}$. To make calculations easy, we define translated elements $\{\hat{x}^q_{\alpha\beta}\}_{q\geq0}$ in $U(\mathfrak{sp}(n,\mathbb{C}))$ by

$$\hat{x}_{\alpha\beta} := x_{\alpha\beta} - (n+1/2)\delta_{\alpha\beta},$$

and

$$\hat{x}^q_{\alpha\beta} := \begin{cases} \sum_{\alpha_1,\alpha_2,\dots,\alpha_{q-1}=\pm1,\dots,\pm n} \hat{x}_{\alpha\alpha_1}\hat{x}_{\alpha_1\alpha_2}\dots\hat{x}_{\alpha_{q-1}\beta} & q \geq 1, \\ \delta_{\alpha\beta} & q = 0, \end{cases}$$

where we remark that

$$\hat{x}^q_{\alpha\beta} = \sum_{p=0}^{q} \binom{q}{p} (-n-1/2)^{q-p} x^p_{\alpha\beta}.$$

The translated elements satisfy

$$\hat{x}_{\alpha\beta} = -\operatorname{sign}(\alpha\beta)\hat{x}_{-\beta-\alpha} - (2n+1)\delta_{\alpha\beta}, \tag{3.11}$$

$$[\hat{x}_{\mu\nu}, \hat{x}^q_{\alpha\beta}] = \delta_{\alpha\nu}\hat{x}^q_{\mu\beta} - \delta_{\beta\mu}\hat{x}^q_{\alpha\nu} + \operatorname{sign}(\mu\nu)(\delta_{-\nu\beta}\hat{x}^q_{\alpha-\mu} - \delta_{-\mu\alpha}\hat{x}^q_{-\nu\beta}), \tag{3.12}$$

$$\sum_\nu \hat{x}^p_{\alpha\nu}\hat{x}^q_{\nu\beta} = \hat{x}^{p+q}_{\alpha\beta}. \tag{3.13}$$

We state the main theorem.

Theorem 3.5 (universal Bochner-Weitzenböck formulas). *The translated element $\hat{x}^q_{\alpha\beta}$ is represented as a linear combination of $\{\hat{x}^p_{-\beta-\alpha}\}_{p=0}^q$ whose coefficients are Casimir elements,*

$$\hat{x}^q_{\alpha\beta} = \operatorname{sign}(\alpha\beta)\{(-1)^q\hat{x}^q_{-\beta-\alpha} - \frac{1-(-1)^q}{2}\hat{x}^{q-1}_{-\beta-\alpha} - \sum_{p=0}^{q-1}(-1)^p\hat{c}_{q-1-p}\hat{x}^p_{-\beta-\alpha}\}. \tag{3.14}$$

Taking the trace in the above equation, we have identities among Casimir elements $\{\hat{c}_p := \sum_\alpha \hat{x}^p_{\alpha\alpha}\}_{p\geq 0}$,

$$2\hat{c}_{2q+1} = -\hat{c}_{2q} - \sum_{p=0}^{2q}(-1)^p \hat{c}_{2q-p}\hat{c}_p, \tag{3.15}$$

for $q = 0, 1, \ldots$.

Proof. It follows from (3.12) and (3.13) that

$$\hat{x}^{q+1}_{\alpha\beta} - \sum_\nu \hat{x}^q_{\nu\beta}\hat{x}_{\alpha\nu} = \sum_\nu [\hat{x}_{\alpha\nu}, \hat{x}_{\nu\beta}] = (2n+1)\hat{x}^q_{\alpha\beta} - \delta_{\alpha\beta}\hat{c}_q - \operatorname{sign}(\alpha\beta)\hat{x}^q_{-\beta-\alpha}.$$

Substituting (3.11) into this equation, we have

$$\hat{x}^{q+1}_{\alpha\beta} = -\delta_{\alpha\beta}\hat{c}_q - \operatorname{sign}(\alpha\beta)\hat{x}^q_{-\beta-\alpha} - \sum_\nu \operatorname{sign}(\nu\alpha)\hat{x}^q_{\nu\beta}\hat{x}_{-\nu-\alpha}. \tag{3.16}$$

This equation means that $\hat{x}^q_{\alpha\beta}$ is a linear combination of $\{\hat{x}^p_{-\beta-\alpha}\}_{p=0}^q$ whose coefficients are Casimir elements given inductively. Then we set

$$\hat{x}^q_{\alpha\beta} := -\operatorname{sign}(\alpha\beta)\sum_{p=0}^q a_{q,p}(\hat{c})\hat{x}^p_{-\beta-\alpha}$$

for $q \geq 0$, where $\{a_{q,p}(\hat{c})\}_{q\geq p\geq 0}$ are polynomials in Casimir elements $\{\hat{c}_0, \hat{c}_1, \ldots\}$. Substituting this into (3.16), we obtain a recursion formula of $\{a_{q,p}(\hat{c})\}_{q\geq p\geq 0}$,

$$a_{q+1,p}(\hat{c}) = \begin{cases} -a_{q,q}(\hat{c}) & \text{for } p = q+1, \\ -a_{q,q-1}(\hat{c}) + 1 & \text{for } p = q, \\ -a_{q,p-1}(\hat{c}) & \text{for } 1 \leq p \leq q-1, \\ \hat{c}_q & \text{for } p = 0. \end{cases}$$

Here, $a_{0,0}(\hat{c}) = -1$, $a_{1,0}(\hat{c}) = 2n+1 = \hat{c}_0 + 1$, $a_{1,1}(\hat{c}) = 1$. We solve this recursion formula and have (3.14). If we take the trace of (3.14) for the case that q is even, then we have trivial identities. When q is odd, we have (3.15). □

Since this theorem induces Bochner-Weitzenböck formulas for hyper-Kählerian gradients in Section 5, we call the equations (3.14) *the universal Bochner-Weitzenböck formulas.*

We shall apply the universal Bochner-Weitzenböck formulas to constructing relations among Clifford homomorphisms. The next lemma follows from Proposition 3.4.

Lemma 3.6. *We define the translated conformal weight* $\hat{w}_\nu$ *by* $\hat{w}_\nu := w_\nu - (n+1/2)$. *Then we have*

$$\begin{aligned}
&\sum_{\nu=\pm1,\dots,\pm n} \hat{w}_\nu^q\, p_\nu(\epsilon_\alpha)^* p_\nu(\epsilon_\beta) = \operatorname{sign}(\alpha\beta)\pi_\rho(\hat{x}^q_{-\alpha-\beta}),\\
&\sum_{\nu=\pm1,\dots,\pm n} \hat{w}_\nu^q\, \mathfrak{J} p_\nu(\epsilon_\alpha)^* p_\nu(\epsilon_\beta)\mathfrak{J}^{-1} = \pi_\rho(\hat{x}^q_{\alpha\beta}),\\
&\sum_{\nu=\pm1,\dots,\pm n} \hat{w}_\nu^q\, \mathfrak{J} p_\nu(\epsilon_\alpha)^* \mathfrak{J}^{-1} p_\nu(\epsilon_\beta) = -\operatorname{sign}(\beta)\pi_\rho(\hat{x}^q_{\alpha-\beta}),\\
&\sum_{\nu=\pm1,\dots,\pm n} \hat{w}_\nu^q\, p_\nu(\epsilon_\alpha)^* \mathfrak{J} p_\nu(\epsilon_\beta)\mathfrak{J}^{-1} = -\operatorname{sign}(\alpha)\pi_\rho(\hat{x}^q_{-\alpha\beta}),
\end{aligned} \tag{3.17}$$

and

$$\pi_\rho(\hat{c}_q) = \pi_\rho(\sum_\alpha \hat{x}^q_{\alpha\alpha}) = \frac{1}{\dim V_\rho} \sum_{\nu=\pm1,\dots,\pm n} \hat{w}_\nu^q \dim V_{\rho+\mu_\nu}.$$

From the universal Bochner-Weitzenböck formulas (3.14) and the equations (3.17), we have identities for Clifford homomorphisms,

$$\begin{aligned}
\sum_\nu \Big\{ &(1-(-1)^q)\hat{w}_\nu^q + \frac{1-(-1)^q}{2}\hat{w}_\nu^{q-1}\\
&+ \sum_{p=0}^{q-1} (-1)^p \pi_\rho(\hat{c}_{q-1-p})\hat{w}_\nu^p \Big\} (p_\nu(u)^* p_\nu(u) + \mathfrak{J} p_\nu(u)^* p_\nu(u)\mathfrak{J}^{-1}) = 0
\end{aligned} \tag{3.18}$$

for $q = 1, 2, \dots$ and any u in $W^{1,0}$. We shall detect independent identities in (3.18). For $s \gg 0$, we consider (3.18) for $q = 1, \dots, 2s$. We represent the coefficients of $\{p_\nu(u)^* p_\nu(u) + \mathfrak{J} p_\nu(u)^* p_\nu(u)\mathfrak{J}^{-1}\}_\nu$ by a $2s \times 2n$ matrix $C(s) \cdot W(s)$, where $C(s)$ is a $2s \times 2s$ matrix,

$$C(s) := \begin{pmatrix}
\pi_\rho(\hat{c}_0)+1 & 2 & 0 & 0 & \dots & \dots & 0\\
\pi_\rho(\hat{c}_1) & -\pi_\rho(\hat{c}_0) & 0 & 0 & \dots & \dots & 0\\
\pi_\rho(\hat{c}_2) & -\pi_\rho(\hat{c}_1) & \pi_\rho(\hat{c}_0)+1 & 2 & 0 & \dots & 0\\
\pi_\rho(\hat{c}_3) & -\pi_\rho(\hat{c}_2) & \pi_\rho(\hat{c}_1) & -\pi_\rho(\hat{c}_0) & 0 & \dots & 0\\
\dots & \dots & \dots & \dots & \dots & \dots & \dots\\
\dots & \dots & \dots & \dots & \dots & \dots & 2\\
\dots & \dots & \dots & \dots & \dots & \dots & -\pi_\rho(\hat{c}_0)
\end{pmatrix},$$

and $W(s)$ is a $2s \times 2n$ matrix,

$$W(s) := \begin{pmatrix}
1 & 1 & \dots & 1\\
\hat{w}_{-n} & \hat{w}_{-n+1} & \dots & \hat{w}_n\\
(\hat{w}_{-n})^2 & (\hat{w}_{-n+1})^2 & \dots & (\hat{w}_n)^2\\
\dots & \dots & \dots & \dots\\
(\hat{w}_{-n})^{2s-1} & (\hat{w}_{-n+1})^{2s-1} & \dots & (\hat{w}_n)^{2s-1}
\end{pmatrix}.$$

Because $\pi_\rho(\hat{c}_0) = 2n$ and $\pi_\rho(\hat{c}_1) = -n(2n+1)$, we should consider only the even cases in (3.18), that is,

$$\sum_\nu \{ \sum_{p=0}^{2q-1} (-1)^p \pi_\rho(\hat{c}_{2q-1-p}) \hat{w}_\nu^p \} (p_\nu(u)^* p_\nu(u) + \mathfrak{J} p_\nu(u)^* p_\nu(u) \mathfrak{J}^{-1}) = 0 \tag{3.19}$$

for $q = 1, \ldots, s$. There is a possibility that these identities are still linear dependent. To simplify a discussion, we assume that the number of irreducible components in $V_\rho \otimes W^{1,0}$ is $2n$. In (3.19), the coefficients of $\{p_\nu(u)^* p_\nu(u) + \mathfrak{J} p_\nu(u)^* p_\nu(u) \mathfrak{J}^{-1}\}_\nu$ are represented by a $s \times 2n$ matrix $D(s) \cdot W(s)$, where $D(s)$ is a $s \times 2s$ matrix,

$$D(s) := \begin{pmatrix} \pi_\rho(\hat{c}_1) & -\pi_\rho(\hat{c}_0) & 0 & 0 & \ldots & \ldots & 0 & 0 \\ \pi_\rho(\hat{c}_3) & -\pi_\rho(\hat{c}_2) & \pi_\rho(\hat{c}_1) & -\pi_\rho(\hat{c}_0) & 0 & \ldots & 0 & 0 \\ \ldots & \ldots & \ldots & \ldots & \ldots & \ldots & \ldots & \ldots \\ \ldots & \ldots & \ldots & \ldots & \ldots & \ldots & \pi_\rho(\hat{c}_1) & -\pi_\rho(\hat{c}_0) \end{pmatrix}.$$

Since the rank of $W(s)$ is $2n$ and the rank of $D(s)$ is s, we have

$$2n - s \leq \operatorname{rank}(D(s) \cdot W(s)) \leq \min\{s, 2n\}.$$

In particular, the rank of $D(n) \cdot W(n)$ is just n. Thus there are at least n independent identities in (3.19). In general, if the number of irreducible components in $V_\rho \otimes W^{1,0}$ is N,

$$N = \#\{\rho + \mu_\nu | \ \rho \text{ is dominant integral}, \nu = \pm 1, \ldots, \pm n\} \leq 2n,$$

then we show that there are at least $[N/2]$ independent identities in (3.19).

On the other hand, from (3.14) and (3.17), we also have

$$\sum_\nu \left\{ (1 + (-1)^q) \hat{w}_\nu^q - \frac{1 - (-1)^q}{2} \hat{w}_\nu^{q-1} \right.$$
$$\left. - \sum_{p=0}^{q-1} (-1)^p \pi_\rho(\hat{c}_{q-1-p}) \hat{w}_\nu^p \right\} (p_\nu(u)^* p_\nu(u) - \mathfrak{J} p_\nu(u)^* p_\nu(u) \mathfrak{J}^{-1}) = 0,$$

for $q = 0, 1, \ldots$. In a similar discussion, we know that there are at least $[\frac{N+1}{2}]$ independent identities in the above. We can also have some independent identities for $\{\mathfrak{J} p_\nu(u)^* \mathfrak{J}^{-1} p_\nu(u)\}_\nu$ and $\{p_\nu(u)^* \mathfrak{J} p_\nu(u) \mathfrak{J}^{-1}\}_\nu$.

Proposition 3.7. *Let V_ρ be an irreducible unitary $Sp(n)$-module such that the number of irreducible components in $V_\rho \otimes W^{1,0}$ is N. We consider the Clifford homomorphisms $\{p_\nu, p_\nu^*, \mathfrak{J} p_\nu \mathfrak{J}^{-1}, \mathfrak{J} p_\nu^* \mathfrak{J}^{-1}\}_{\nu=\pm 1}^{\pm n}$, where we set $p_\nu = 0$ unless $\rho + \mu_\nu$ is dominant integral. Then we have the following independent identities among the Clifford homomorphisms.*

1.

$$\sum_\nu p_\nu(u)^* p_\nu(u) = (u, u).$$

2. *For* $q = 1, \ldots, [N/2]$,

$$\sum_{\nu}\{\sum_{p=0}^{2q-1}(-1)^p \pi_\rho(\hat{c}_{2q-1-p})\hat{w}_\nu^p\}(p_\nu(u)^* p_\nu(u) + \mathfrak{J} p_\nu(u)^* p_\nu(u)\mathfrak{J}^{-1}) = 0.$$

3. *For* $q = 0, 1, \ldots, [\frac{N+1}{2}] - 1$,

$$\sum_{\nu}\{\hat{w}_\nu^{2q} - \sum_{p=0}^{2q-1}(-1)^p \pi_\rho(\hat{c}_{2q-1-p})\hat{w}_\nu^p\}(p_\nu(u)^* p_\nu(u) - \mathfrak{J} p_\nu(u)^* p_\nu(u)\mathfrak{J}^{-1}) = 0,$$

$$\sum_{\nu}\{\hat{w}_\nu^{2q} - \sum_{p=0}^{2q-1}(-1)^p \pi_\rho(\hat{c}_{2q-1-p})\hat{w}_\nu^p\}\mathfrak{J} p_\nu(u)^* \mathfrak{J}^{-1} p_\nu(u) = 0,$$

$$\sum_{\nu}\{\hat{w}_\nu^{2q} - \sum_{p=0}^{2q-1}(-1)^p \pi_\rho(\hat{c}_{2q-1-p})\hat{w}_\nu^p\} p_\nu(u)^* \mathfrak{J} p_\nu(u)\mathfrak{J}^{-1} = 0.$$

4. Hyper-Hermitian gradients and their conformal covariance

A real $4n$-dimensional Riemannian manifold (M, g) is said to be an *almost hyper-Hermitian manifold* if M has three almost complex structures I, J and K such that I and J are orthogonal with respect to the metric g, and $IJ = -JI = K$. If I, J and K are parallel with respect to the Levi-Civita connection, we call (M, g, I, J, K) a *hyper-Kähler manifold*, [9].

We consider an almost hyper-Hermitian manifold (M, g, I, J, K). Some results about hyper-Hermitian structure on vector spaces in Section 2 can be extended globally on M. We decompose the tangent bundle $T(M) \otimes \mathbb{C}$ and the cotangent bundle $T^*(M) \otimes \mathbb{C}$ with respect to I,

$$T(M) \otimes \mathbb{C} = T^{1,0}(M) \oplus T^{0,1}(M), \quad T^*(M) \otimes \mathbb{C} = \Lambda^{1,0}(M) \oplus \Lambda^{0,1}(M).$$

These vector bundles have Hermitian metrics $(u, v) := g(u, \bar{v})$. Another almost complex structure J induces bundle isomorphisms $J : T^{1,0}(M) \to T^{0,1}(M)$ and $J : T^{0,1}(M) \to T^{1,0}(M)$. Then we have a quaternionic structure $\mathfrak{J}$ on each bundle defined by $\mathfrak{J}(u) := J(\bar{u})$. These structure produce bundle isomorphisms,

$$\Lambda^{1,0}(M) \overset{(\cdot,\cdot)}{\simeq} T^{0,1}(M) \overset{J}{\simeq} T^{1,0}(M) \overset{(\cdot,\cdot)}{\simeq} \Lambda^{0,1}(M). \tag{4.1}$$

The structure group of the frame bundle on M reduces to the symplectic group $Sp(n)$. We denote the principal $Sp(n)$-bundle on M by $\mathbf{Sp}(M)$. We consider the Levi-Civita connection ∇^{so} on $T(M)$ and define a connection ∇^{sp} by

$$\nabla^{sp}_X Y := \{\nabla^{so}_X Y - I(\nabla^{so}_X(IY)) - J(\nabla^{so}_X(JY)) - K(\nabla^{so}_X(KY))\}/4.$$

Since $\nabla^{sp} g = 0$ and $\nabla^{sp} I = \nabla^{sp} J = \nabla^{sp} K = 0$, the connection ∇^{sp} is a connection on $\mathbf{Sp}(M)$. Notice that the torsion tensor of ∇^{sp} is zero iff M is a hyper-Kähler manifold.

For an irreducible unitary $Sp(n)$-module V_ρ, we have a Hermitian vector bundle $\mathbf{S}_\rho := \mathbf{Sp}(M) \times_\rho V_\rho$ associated to $\mathbf{Sp}(M)$. We have known that there is a unique $Sp(n)$-invariant real or quaternionic structure $\mathfrak{J}$ on V_ρ. Because of the $Sp(n)$-invariance, we have a real or quaternionic structure $\mathfrak{J}$ on $\mathbf{S}_\rho$ such that $(\mathfrak{J}_x\phi, \mathfrak{J}_x\psi) = (\psi, \phi)$ for ϕ and ψ in $(\mathbf{S}_\rho)_x$. We consider a covariant derivative ∇ on $\mathbf{S}_\rho$ induced from the canonical connection ∇^{sp}. This covariant derivative ∇ splits into the sum of $\nabla^{1,0}$ and $\nabla^{0,1}$,

$$\nabla^{1,0} : \Gamma(\mathbf{S}_\rho) \xrightarrow{\nabla^{1,0}} \Gamma(\mathbf{S}_\rho \otimes \Lambda^{1,0}(M)) = \Gamma(\mathbf{S}_\rho \otimes T^{1,0}(M)),$$
$$\nabla^{0,1} : \Gamma(\mathbf{S}_\rho) \xrightarrow{\nabla^{0,1}} \Gamma(\mathbf{S}_\rho \otimes \Lambda^{0,1}(M)) = \Gamma(\mathbf{S}_\rho \otimes T^{1,0}(M)),$$

where we use isomorphisms (4.1). Because $T^{1,0}(M)$ and $\mathbf{S}_\rho$ are associated vector bundles, we have the irreducible decomposition

$$\mathbf{S}_\rho \otimes T^{1,0}(M) = \bigoplus_{\nu=\pm1,\dots,\pm n} \mathbf{S}_{\rho+\mu_\nu} = \bigoplus_{k=1,\dots,n} \mathbf{S}_{\rho+\mu_k} \oplus \mathbf{S}_{\rho-\mu_k}.$$

We divide $\nabla^{1,0}$ along this decomposition and have first-order differential operators

$$D_\nu : \Gamma(\mathbf{S}_\rho) \xrightarrow{\nabla^{1,0}} \Gamma(\mathbf{S}_\rho \otimes \Lambda^{1,0}(M)) = \Gamma(\mathbf{S}_\rho \otimes T^{1,0}(M)) \xrightarrow{\Pi_\nu} \Gamma(\mathbf{S}_{\rho+\mu_\nu}).$$

for $\nu = \pm1, \dots, \pm n$. We shall express the operator D_ν by using local frame of M. Let $\{\epsilon_\alpha\}_\alpha$ and $\{\bar{\epsilon}_\alpha\}_\alpha$ be local unitary frames of $T^{1,0}(M)$ and $T^{0,1}(M)$ given as (2.1). Then,

$$\begin{aligned} D_\nu(\phi) &= \Pi_\nu(\nabla^{1,0}\phi) = \Pi_\nu\Big(\sum_\alpha \nabla_{\epsilon_\alpha}\phi \otimes \epsilon_\alpha^*\Big) \\ &= \Pi_\nu\Big(\sum_\alpha \nabla_{\epsilon_\alpha}\phi \otimes -\mathrm{sign}(\alpha)\epsilon_{-\alpha}\Big) = -\sum_\alpha \mathrm{sign}(\alpha) p_\nu(\epsilon_{-\alpha})\nabla_{\epsilon_\alpha}\phi. \end{aligned}$$

We twist D_ν by $\mathfrak{J}$ and have another first order differential operator $\mathfrak{J}D_\nu\mathfrak{J}^{-1}$ whose local expression is

$$\mathfrak{J}D_\nu\mathfrak{J}^{-1} = \sum_\alpha p_\nu(\epsilon_\alpha)\nabla_{\bar{\epsilon}_\alpha}.$$

The twisted operator is just the operator $\Pi_\nu \circ \nabla^{0,1}$ as a component of $\nabla^{0,1}$,

$$\mathfrak{J}D_\nu\mathfrak{J}^{-1} : \Gamma(\mathbf{S}_\rho) \xrightarrow{\nabla^{0,1}} \Gamma(\mathbf{S}_\rho \otimes \Lambda^{0,1}(M)) = \Gamma(\mathbf{S}_\rho \otimes T^{1,0}(M)) \xrightarrow{\Pi_\nu} \Gamma(\mathbf{S}_{\rho+\mu_\nu}).$$

We call these first-order differential operators $\{D_\nu, \mathfrak{J}D_\nu\mathfrak{J}^{-1}\}_\nu$ *hyper-Hermitian gradients.* When M is a hyper-Kähler manifold, we call them *hyper-Kählerian gradients.*

In the rest of this section, we see conformal covariance of hyper-Hermitian gradients. Let D_ν be a hyper-Hermitian gradient with respect to the canonical connection ∇^{sp} on an almost hyper-Hermitian manifold (M, g, I, J, K). When we deform the metric g conformally to $g' = e^{2\sigma}g$ for σ in $C^\infty(M)$, the Riemannian manifold (M, g', I, J, K) is also an almost hyper-Hermitian manifold. Then we have another hyper-Hermitian gradient $D'_\nu : \Gamma(\mathbf{S}'_\rho) \to \Gamma(\mathbf{S}'_{\rho+\mu_\nu})$. The conformal deformation gives a principal bundle isomorphism

$$\mathbf{Sp}(M) \ni p = \{\epsilon_\alpha\}_\alpha \mapsto p' = \{\epsilon'_\alpha\}_\alpha = \{e^{-\sigma}\epsilon_\alpha\}_\alpha \in \mathbf{Sp}(M)'.$$

Here, we remark that $(\epsilon'_\alpha)^* = e^\sigma \epsilon^*_\alpha$. This principal bundle isomorphism induces a bundle isomorphism,

$$\psi_\rho : \mathbf{S}_\rho = \mathbf{Sp}(M) \times_\rho V_\rho \ni [p, \phi] \mapsto [p', \phi] \in \mathbf{Sp}(M)' \times_\rho V_\rho = \mathbf{S}'_\rho.$$

Since, precisely speaking, the Clifford homomorphism is defined through not the projection $\Pi_\nu : \mathbf{S}_\rho \otimes T^{1,0}(M) \to \mathbf{S}_{\rho+\mu_\nu}$ but $\Pi_\nu : \mathbf{S}_\rho \otimes \Lambda^{1,0}(M) \to \mathbf{S}_{\rho+\mu_\nu}$, we have

$$p_\nu(\epsilon_\alpha) = e^{-\sigma}\psi_{\rho+\mu_\nu} p_\nu(\epsilon_\alpha)\psi_\rho^{-1} : \mathbf{S}'_\rho \to \mathbf{S}'_{\rho+\mu_\nu}.$$

The following proposition answers the reason why we call w_ν the conformal weight.

Proposition 4.1. *Let D_ν be a hyper-Hermitian gradient with respect to the canonical connection ∇^{sp} on an almost hyper-Hermitian manifold (M, g, I, J, K). Deform the Riemannian metric g conformally to $g' = e^{2\sigma}g$, and consider a hyper-Hermitian gradient D'_ν on (M, g', I, J, K). Then we have*

$$D'_\nu = e^{(-w_\nu/2-1)\sigma}\psi_{\rho+\mu_\nu} \circ D_\nu \circ (e^{(-w_\nu/2)\sigma}\psi_\rho)^{-1}.$$

Here, w_ν is the conformal weight associated to ρ and $\rho + \mu_\nu$.

Proof. The proof is similar to the one for the conformal covariance of the Dirac operator [12]. First we can show that

$$[D_\nu, f] = -p_\nu(\sum \text{sign}(\alpha)(\epsilon_\alpha f)\epsilon_{-\alpha}) \tag{4.2}$$

for f in $C^\infty(M)$. Next, we consider a local connection 1-form ω^β_α on $T^{1,0}(M)$ given by $\nabla^{sp}_V \epsilon_\alpha = \sum \omega^\beta_\alpha(V)\epsilon_\beta$ for each vector field V. It follows from the definition of ∇^{sp} that

$$\begin{aligned}\omega^\beta_\alpha(V) = g(\nabla^{sp}_V \epsilon_\alpha, \bar\epsilon_\beta) &= 1/2\{g(\nabla^{so}_V \epsilon_\alpha, \bar\epsilon_\beta) - \text{sign}(\alpha\beta)g(\nabla^{so}_V \epsilon_{-\beta}, \bar\epsilon_{-\alpha})\}\\ &= -\text{sign}(\alpha\beta)\omega^{-\alpha}_{-\beta}(V).\end{aligned}$$

Thus, ω^β_α is $\mathfrak{sp}(n, \mathbb{C})$-valued. Then the covariant derivative ∇ on $\mathbf{S}_\rho$ is

$$\nabla_V = 1/2\sum \omega^\beta_\alpha(V)\pi_\rho(\epsilon_\beta \otimes \epsilon^*_\alpha - \text{sign}(\alpha\beta)\epsilon_{-\alpha} \otimes \epsilon^*_{-\beta}) = 1/2\sum \omega^\beta_\alpha(V)\pi_\rho(x_{\beta\alpha}).$$

On the other hand, the deformed Levi-Civita connection ∇'^{so} is

$$\nabla'^{so}_V W = \nabla^{so}_V W + (V\sigma)W + (W\sigma)V - g(V, W)\text{grad}(\sigma)$$

for vector fields V and W. As a result, we have

$$\nabla'_V = \psi_\rho \circ \{\nabla_V + 1/2\sum_{\alpha\beta}((\epsilon_\alpha\sigma)g(V, \bar\epsilon_\beta)\pi_\rho(x_{\beta\alpha}) - (\bar\epsilon_\beta\sigma)g(V, \epsilon_\alpha)\pi_\rho(x_{\beta\alpha}))\} \circ \psi_\rho^{-1}.$$

Then,

$$\begin{aligned} D'_\nu &= -\sum \operatorname{sign}(\alpha) p_\nu(\epsilon'_{-\alpha}) \nabla'_{\epsilon'_\alpha} = -\sum \operatorname{sign}(\alpha) p_\nu(\epsilon_{-\alpha}) \nabla'_{\epsilon_\alpha} \\ &= e^{-\sigma} \psi_{\rho+\mu_\nu} \circ \{-\sum_\alpha \operatorname{sign}(\alpha) p_\nu(\epsilon_{-\alpha})(\nabla_{\epsilon_\alpha} + 1/2 \sum_\beta (\epsilon_\beta \sigma) \pi_\rho(x_{\alpha\beta}))\} \circ \psi_\rho^{-1} \\ &= e^{-\sigma} \psi_{\rho+\mu_\nu} \circ \{D_\nu - 1/2 \sum_{\alpha\beta} \operatorname{sign}(\alpha)(\epsilon_\beta \sigma) p_\nu(\epsilon_{-\alpha}) \pi_\rho(x_{\alpha\beta})\} \circ \psi_\rho^{-1} \\ &= e^{-\sigma} \psi_{\rho+\mu_\nu} \circ \{D_\nu - w_\nu/2\, p_\nu(\sum_\beta \operatorname{sign}(\beta)(\epsilon_\beta \sigma) \epsilon_{-\beta})\} \circ \psi_\rho^{-1} \quad \text{(by (3.10))} \\ &= e^{(-w_\mu/2-1)\sigma} \psi_{\rho+\mu_\nu} \circ D_\nu \circ (e^{(-w_\nu/2)\sigma} \psi_\rho)^{-1} \quad \text{(by (4.2))}. \end{aligned}$$

Thus we have proved the proposition. □

5. Bochner-Weitzenböck formulas for hyper-Kählerian gradients

In this section, we assume that (M, g, I, J, K) is a hyper-Kähler manifold, namely, $\nabla^{sp} = \nabla^{so}$. On a hyper-Kähler manifold, the formal adjoint operators of D_ν and $\mathfrak{J} D_\nu \mathfrak{J}^{-1}$ have local expressions as follows:

$$D_\nu^* = \sum_\alpha \operatorname{sign}(\alpha) p_\nu(\epsilon_{-\alpha})^* \nabla_{\bar\epsilon_\alpha} : \Gamma(\mathbf{S}_{\rho+\mu_\nu}) \to \Gamma(\mathbf{S}_\rho),$$

$$(\mathfrak{J} D_\nu \mathfrak{J}^{-1})^* = \mathfrak{J} D_\nu^* \mathfrak{J}^{-1} = -\sum_\alpha p_\nu(\epsilon_\alpha)^* \nabla_{\epsilon_\alpha} : \Gamma(\mathbf{S}_{\rho+\mu_\nu}) \to \Gamma(\mathbf{S}_\rho).$$

Thus, on a hyper-Kähler manifold, the principal symbols of hyper-Kählerian gradients and their adjoints are realized with Clifford homomorphisms. Then we can use results in Section 3.

Proposition 5.1. *We define a second-order derivative $\nabla^2_{V,W}$ by $\nabla_V \nabla_W - \nabla_{\nabla_V W}$ for vector fields V and W. Then hyper-Kählerian gradients satisfy that*

$$\begin{aligned} \sum_\nu \hat{w}_\nu^q D_\nu^* D_\nu &= -\sum_{\alpha,\beta} \pi_\rho(\hat{x}^q_{\alpha\beta}) \nabla^2_{\bar\epsilon_\alpha, \epsilon_\beta}, \\ \sum_\nu \hat{w}_\nu^q \mathfrak{J} D_\nu^* D_\nu \mathfrak{J}^{-1} &= -\sum_{\alpha,\beta} \operatorname{sign}(\alpha\beta) \pi_\rho(\hat{x}^q_{-\alpha-\beta}) \nabla^2_{\epsilon_\alpha, \bar\epsilon_\beta}, \\ \sum_\nu \hat{w}_\nu^q \mathfrak{J} D_\nu^* \mathfrak{J}^{-1} D_\nu &= -\sum_{\alpha,\beta} \operatorname{sign}(\alpha) \pi_\rho(\hat{x}^q_{-\alpha\beta}) \nabla^2_{\epsilon_\alpha, \epsilon_\beta}, \\ \sum_\nu \hat{w}_\nu^q D_\nu^* \mathfrak{J} D_\nu \mathfrak{J}^{-1} &= -\sum_{\alpha,\beta} \operatorname{sign}(\beta) \pi_\rho(\hat{x}^q_{\alpha-\beta}) \nabla^2_{\bar\epsilon_\alpha, \bar\epsilon_\beta}. \end{aligned} \tag{5.1}$$

Here, $\{\epsilon_\alpha\}_\alpha$ is local unitary frame and

$$\hat{x}_{\alpha\beta} := \epsilon_\alpha \otimes \epsilon_\beta^* - \operatorname{sign}(\alpha\beta) \epsilon_{-\beta} \otimes \epsilon_{-\alpha}^* - (n + 1/2) \delta_{\alpha\beta}$$

is a local section of the enveloping algebra bundle $\mathbf{Sp}(M) \times_{\mathrm{Ad}} U(\mathfrak{sp}(n, \mathbb{C}))$.

To obtain Bochner-Weitzenböck formulas for hyper-Kählerian gradients, we need curvature endomorphisms on $\mathbf{S}_\rho$. First, we see the curvature R_T on $T(M)$ defined by $R_T(V,W) := \nabla^2_{V,W} - \nabla^2_{W,V}$. This curvature satisfies

$$R_T(IV,IW)=R_T(V,W),\ \ R_T(JV,JW)=R_T(V,W),\ \ R_T(KV,KW)=R_T(V,W),$$
$$R_T I = I R_T, \quad R_T J = J R_T, \quad R_T K = K R_T.$$

These equations mean that we regard R_T as an endomorphism on $\mathfrak{sp}(n)$. In particular, we have

$$g(R_T(V,W)\epsilon_\alpha, \bar{\epsilon}_\beta) = -\mathrm{sign}(\alpha\beta) g(R_T(V,W)\epsilon_{-\beta}, \bar{\epsilon}_{-\alpha}),$$
$$g(R_T(V,W)\epsilon_\alpha, \epsilon_\beta) = g(R_T(V,W)\bar{\epsilon}_\alpha, \bar{\epsilon}_\beta) = 0,$$
$$R_T(\epsilon_\alpha, \bar{\epsilon}_\beta) = -\mathrm{sign}(\alpha\beta) R_T(\epsilon_{-\beta}, \bar{\epsilon}_{-\alpha}), \quad R_T(\epsilon_\alpha, \epsilon_\beta) = R_T(\bar{\epsilon}_\alpha, \bar{\epsilon}_\beta) = 0$$

for local unitary frame $\{\epsilon_\alpha\}_\alpha$. Notice that the Ricci curvature $\sum_\alpha R_T(\epsilon_\alpha, \bar{\epsilon}_\alpha)$ is zero, and hence, R_T depends only on the conformal Weyl tensor.

We consider the curvature R_ρ on $\mathbf{S}_\rho$,

$$R_\rho(V,W) := \nabla^2_{V,W} - \nabla^2_{W,V} \quad \text{for vector fields } V \text{ and } W.$$

Since the covariant derivative ∇ on $\mathbf{S}_\rho$ is induced from the Levi-Civita connection, the curvature R_ρ is expressed as follows:

$$\begin{aligned} R_\rho(V,W) &= 1/2 \sum_{\alpha,\beta} g(R_T(V,W)\epsilon_\alpha, \bar{\epsilon}_\beta) \pi_\rho(\epsilon_\beta \otimes \epsilon^*_\alpha - \mathrm{sign}(\alpha\beta)\epsilon_{-\alpha} \otimes \epsilon^*_{-\beta}) \\ &= 1/2 \sum_{\alpha,\beta} g(R_T(V,W)\epsilon_\alpha, \bar{\epsilon}_\beta) \pi_\rho(x_{\beta\alpha}). \end{aligned}$$

Since R_ρ satisfies

$$R_\rho(\epsilon_\alpha, \bar{\epsilon}_\beta) = -\mathrm{sign}(\alpha\beta) R_\rho(\epsilon_{-\beta}, \bar{\epsilon}_{-\alpha}),$$

we define a *curvature endomorphism* $\hat{R}^q_\rho$ on $\mathbf{S}_\rho$ for $q = 0, 1, \dots$ by

$$\hat{R}^q_\rho := \sum \pi_\rho(\hat{x}^q_{\beta\alpha}) R_\rho(\epsilon_\alpha, \bar{\epsilon}_\beta).$$

Proposition 5.2. *The curvature endomorphisms $\{\hat{R}^q_\rho\}_q$ are self-adjoint endomorphisms of $\mathbf{S}_\rho$, namely, $((\hat{R}^q_\rho)_x\phi, \psi) = (\phi, (\hat{R}^q_\rho)_x\psi)$ at each x in M. Moreover $\{\hat{R}^q_\rho\}_q$ satisfy that*

$$(1 + (-1)^q)\hat{R}^q_\rho - \frac{1-(-1)^q}{2}\hat{R}^{q-1}_\rho - \sum_{p=0}^{q-1} (-1)^p \pi_\rho(\hat{c}_{q-1-p}) \hat{R}^p_\rho = 0.$$

Proof. We can show the first claim by using the Bianchi identity for R_T. The second claim follows from the universal Bochner-Weitzenböck formulas (3.14). □

For example, we have $\hat{R}^0_\rho = 0$ and $\hat{R}^2_\rho = -n\hat{R}^1_\rho$. Thus, it is enough if we think only $\{\hat{R}^{2q}_\rho\}_q$.

Now, we are in a position to give Bochner-Weitzenböck formulas.

Theorem 5.3. *Let $\{D_\nu, \mathfrak{J}D_\nu\mathfrak{J}^{-1}, D_\nu^*, \mathfrak{J}D_\nu^*\mathfrak{J}^{-1}\}_\nu$ be hyper-Kählerian gradients on a hyper-Kähler manifold. We assume that the number of irreducible components in $\mathbf{S}_\rho \otimes T^{1,0}(M)$ is N. In other words, there are N non-trivial hyper-Kählerian gradients in $\{D_\nu \mid \nu = \pm 1, \dots, \pm n\}$. Then we have the following independent identities for the hyper-Kählerian gradients.*

1.

$$\sum_\nu D_\nu^* D_\nu = \nabla^{1,0*}\nabla^{1,0} = -\sum_\alpha \nabla^2_{\bar{\epsilon}_\alpha, \epsilon_\alpha} = 1/2\nabla^*\nabla, \tag{5.2}$$

$$\sum_\nu \mathfrak{J}D_\nu^* D_\nu \mathfrak{J}^{-1} = \nabla^{0,1*}\nabla^{0,1} = -\sum_\alpha \nabla^2_{\epsilon_\alpha, \bar{\epsilon}_\alpha} = 1/2\nabla^*\nabla, \tag{5.3}$$

where $\nabla^{1,0*}\nabla^{1,0}$ and $\nabla^{0,1*}\nabla^{0,1}$ are the $(1,0)$- and the $(0,1)$- connection Laplacian, and $\nabla^*\nabla$ is the usual connection Laplacian on $\Gamma(\mathbf{S}_\rho)$.

2. *For $q = 1, \dots, [N/2]$,*

$$\sum_\nu \{\sum_{p=0}^{2q-1} (-1)^p \pi_\rho(\hat{c}_{2q-1-p})\hat{w}_\nu^p\}(D_\nu^* D_\nu + \mathfrak{J}D_\nu^* D_\nu \mathfrak{J}^{-1}) = 2\hat{R}_\rho^{2q}. \tag{5.4}$$

3. *For $q = 0, \dots, [\frac{N+1}{2}] - 1$,*

$$\sum_\nu \{2\hat{w}_\nu^{2q} - \sum_{p=0}^{2q-1} (-1)^p \pi_\rho(\hat{c}_{2q-1-p})\hat{w}_\nu^p\}(D_\nu^* D_\nu - \mathfrak{J}D_\nu^* D_\nu \mathfrak{J}^{-1}) = 0, \tag{5.5}$$

$$\sum_\nu \{2\hat{w}_\nu^{2q} - \sum_{p=0}^{2q-1} (-1)^p \pi_\rho(\hat{c}_{2q-1-p})\hat{w}_\nu^p\}\mathfrak{J}D_\nu^* \mathfrak{J}^{-1} D_\nu = 0, \tag{5.6}$$

$$\sum_\nu \{2\hat{w}_\nu^{2q} - \sum_{p=0}^{2q-1} (-1)^p \pi_\rho(\hat{c}_{2q-1-p})\hat{w}_\nu^p\}D_\nu^* \mathfrak{J} D_\nu \mathfrak{J}^{-1} = 0. \tag{5.7}$$

Proof. First, we can prove (5.2) and (5.3) by (5.1). Here, we notice that

$$-\sum_\alpha \nabla^2_{\epsilon_\alpha, \bar{\epsilon}_\alpha} - \sum_\alpha \nabla^2_{\bar{\epsilon}_\alpha, \epsilon_\alpha} = \nabla^*\nabla, \quad \sum_\alpha \nabla^2_{\epsilon_\alpha, \bar{\epsilon}_\alpha} - \sum_\alpha \nabla^2_{\bar{\epsilon}_\alpha, \epsilon_\alpha} = \hat{R}_\rho^0 = 0.$$

Next, by using (3.14) and (5.1), we have

$$\begin{aligned}
&\sum_\nu \hat{w}_\nu^q D_\nu^* D_\nu \\
&= -\sum_{\alpha,\beta} \pi_\rho(\hat{x}^q_{\alpha\beta})(\nabla^2_{\epsilon_\beta, \bar{\epsilon}_\alpha} + R_\rho(\bar{\epsilon}_\alpha, \epsilon_\beta)) = \hat{R}_\rho^q - \sum_{\alpha,\beta} \pi_\rho(\hat{x}^q_{\alpha\beta})\nabla^2_{\epsilon_\beta, \bar{\epsilon}_\alpha} \\
&= \hat{R}_\rho^q + \sum_\nu \{(-1)^q \hat{w}_\nu^q - \frac{1-(-1)^q}{2}\hat{w}_\nu^{q-1} - \sum_{p=0}^{q-1}(-1)^p \pi_\rho(\hat{c}_{q-1-p})\hat{w}_\nu^p\}\mathfrak{J}D_\nu^* D_\nu \mathfrak{J}^{-1}.
\end{aligned}$$

Put this equation between $\mathfrak{J}$ and $\mathfrak{J}^{-1}$, and we have

$$\sum \hat{w}_\nu^q \mathfrak{J} D_\nu^* D_\nu \mathfrak{J}^{-1}$$

$$=\hat{R}_\rho^q + \sum\{(-1)^q \hat{w}_\nu^q - \frac{1-(-1)^q}{2}\hat{w}_\nu^{q-1} - \sum_{p=0}^{q-1}(-1)^p \pi_\rho(\hat{c}_{q-1-p})\hat{w}_\nu^p\} D_\nu^* D_\nu.$$

The above two equations induce (5.4) and (5.5). We can prove (5.6) and (5.7) similarly. The independence of (5.2)-(5.7) is proved by the same discussion as the one in Proposition 3.7. □

References

[1] T. Branson, *Stein-Weiss operators and ellipticity.* J. Funct. Anal. **151**, (1997), 334–383.

[2] T. Branson and O. Hijazi, *Improved forms of some vanishing theorems in Riemannian spin geometry.* Internat. J. Math. **11** (2000), 291–304.

[3] J. Bureš, *The higher spin Dirac operators.* in 'Differential geometry and applications', Masaryk Univ., Brno, (1999), 319–334.

[4] D. Calderbank, P. Gauduchon, M. Herzlich, *Refined Kato inequalities and conformal weights in Riemannian geometry.* J. Funct. Anal. **173** (2000), 214–255.

[5] R. Goodman and N. Wallach, *Representations and invariants of the classical groups.* Encyclopedia of Math. and its Appl. **68**, Cambridge University Press, Cambridge, 1998.

[6] Y. Homma, *Spherical harmonic polynomials for higher bundles.* in 'Int. Conf. on Clifford Analysis, Its Appl. and Related Topics. Beijing', Adv. Appl. Clifford Algebras **11** (S2) (2001), 117–126.

[7] Y. Homma, *Bochner identities for Kählerian gradients.* preprint.

[8] Y. Homma, *Casimir elements and Bochner identities on Riemannian manifolds.* in 'CLIFFORD ALGEBRAS: Applications to Mathematics, Physics, and Engineering' Prog. Math. Phys. **34**, Birkhäuser, (2004), 185–199.

[9] D. Joyce, *Compact Manifolds with Special Holonomy.* Oxford Math. Monographs, Oxford University Press, Oxford, 2000.

[10] W. Kramer, U. Semmelmann and G. Weingart, *The first eigenvalue of the Dirac operator on quaternionic Kähler manifolds.* Comm. Math. Phys. **199** (1998), 327–349.

[11] W. Kramer, U. Semmelmann and G. Weingart, *Eigenvalue estimates for the Dirac operator on quaternionic Kähler manifolds.* Math. Z. **230** (1999), 727–751.

[12] H. B. Lawson and M. L. Michelsohn, *Spin geometry.* Princeton Mathematical Series, **38**. Princeton University Press, Princeton, NJ, 1989.

[13] C. O. Nwachuku and M. A. Rashid, *Eigenvalues of the Casimir operators of the orthogonal and symplectic groups.* J. Math. Phys. **17** (1976), 1611–1616.

[14] S. Okubo, *Casimir invariants and vector operators in simple and classical Lie algebras.* J. Math. Phys. **18**, (1977), 2382–2394.

[15] D. P. Želobenko, *Compact Lie Groups and Their Representations.* Trans. Math. Monographs **40**, American Mathematical Society, Providence, 1973.

Acknowledgement

The author is partially supported by the Grant-in-Aid for Scientific Research (No. 13740120) from the Ministry of Education, Culture, Sports, Science and Technology.

Yasushi Homma
Department of Mathematical Sciences
Waseda University
3-4-1 Ohkubo
Shinjuku-ku
Tokyo 169-8555
Japan
e-mail: homma@gm.math.waseda.ac.jp

Trends in Mathematics: *Advances in Analysis and Geometry*, 209–225

Cohomology Groups of Harmonic Spinors on Conformally Flat Manifolds

Tosiaki Kori

Abstract. We shall investigate various properties of the sheaf of harmonic spinors $\mathcal{N}$ on C^2 and, more generally, on conformally flat spin 4-manifolds. We prove the Runge approximation theorem on C^2, and the vanishing of cohomologies; $H^1(\mathrm{C}^2,\mathcal{N}) = 0$ and $H^1(S^4,\mathcal{N}) = 0$. We shall introduce a concept of divisors of meromorphic spinors on a compact conformally flat spin 4-manifold , and give an analogy of Riemann-Roch theorem.

Mathematics Subject Classification (2000). Primary 58G30; Secondary 53C21.

Keywords. Harmonic spinors, conformally flat manifold, Runge theorem, Riemann-Roch theorem.

1. Introduction

This is a continuation of our previous research on harmonic spinors on conformally flat spin 4-manifolds, [8], [9]. Let (M,S) be a 4-dimensional Riemannian manifold with a compatible spin structure, $S = S^+ \oplus S^-$, $S^\pm$ being the two spin bundles associated to two half-spin representations of $Spin(4)$. Let $D : S^+ \longrightarrow S^-$ be the (half) Dirac operator. A spinor $\varphi \in C^\infty(S^+)$ is called a *harmonic spinor* if $D\varphi = 0$. Let $\mathcal{N}$ be the sheaf of harmonic spinors. Since D is an elliptic operator, we have the exact sequence

$$0 \longrightarrow \mathcal{N} \longrightarrow \mathcal{S}^+ \xrightarrow{D} \mathcal{S}^- \longrightarrow 0, \tag{1.1}$$

where $\mathcal{S}^\pm$ is the sheaf of even (resp. odd) spinors. Therefore

$$\begin{aligned} H^p(M,\mathcal{N}) &= 0 \quad \text{for } p \geq 2, \\ H^1(M,\mathcal{N}) &= coker\{D; C^\infty(M,\mathcal{S}^+) \longrightarrow C^\infty(M,\mathcal{S}^-)\}, \\ H^0(M,\mathcal{N}) &= ker\{D; C^\infty(M,\mathcal{S}^+) \longrightarrow C^\infty(M,\mathcal{S}^-)\}. \end{aligned} \tag{1.2}$$

Research supported by Promotion for Sciences of the ministry of education in Japan (no. 13640224) and Waseda University Grant for special research project (no. 2000A-129).

$H^1(M,\mathcal{N})$ and $H^0(M,\mathcal{N})$ are endowed with the structure of Frechet spaces. When M is compact these are finite dimensional, and we have

$$\dim H^0(M,\mathcal{N}) - \dim H^1(M,\mathcal{N}) = \operatorname{index}(D) = -\frac{1}{24}p_1(M), \tag{1.3}$$

from the Atiyah-Singer index theorem, where $p_1(M)$ is the first Pontryagin number of the 4-manifold M.

We are interested in the vanishing of $H^1(M,\mathcal{N})$. We shall prove it if M is an open subset of C^2 and for $M = S^4$. We conjecture the vanishing of $H^1(M,\mathcal{N})$ for any non-compact conformally flat spin 4-manifold.

Now we shall explain the contents of each sections After the introduction of ingredients on spinor analysis that are studied in [8], we shall prove in Section 3 the Runge approximation theorem, that is, any harmonic spinor on a compact subset K of a domain $G \subset \mathrm{C}^2$ such that $G \setminus K$ has no relatively compact connected component can be approximated uniformly on K by harmonic spinors on G. Several Runge type theorems for Clifford algebra valued functions on a domain in R^n, as well as for Clifford modules, were proved earlier in [2, 3, 11]. We think our argument, which is close to that by Hörmander [7], is worth to be presented. In 3.2 we prove that the Dirac equation $D\varphi = \psi$ has a solution on any open subset G of C^2. Hence we have

$$H^1(G,\mathcal{N}) = 0. \tag{1.4}$$

We can verify the covariance of our theory under conformal transformations on R^4, thus we can extend it to a conformally flat 4-manifold. In particular the above stated properties are independent of the complex structure on R^4. In Section 4 we shall deal with the cohomology of $\mathcal{N}$ on a conformally flat 4-manifold M. We shall see that the cohomology group $H^1(M,\mathcal{N})$ is calculated by a Leray covering. Thus we see that the well known argument to have the classical result $H^1(P^1(\mathrm{C}),\mathcal{O}) = 0$ is valid to prove

$$H^1(S^4,\mathcal{N}) = 0. \tag{1.5}$$

We shall prove that the restriction map

$$H^1(G',\mathcal{N}) \longrightarrow H^1(G,\mathcal{N})$$

has a finite dimensional image for a relatively compact open subset G of an open subset $G' \subset M$. This implies the existence of a non-trivial meromorphic spinor on a relatively compact open subset of M.

Our results should extend to even dimensional conformally flat manifolds. In fact we have already the Runge approximation theorem on a domain in R^n as was proved earlier in [3], so if we adopted it we could in principle obtain the results in Section 4 also on a domain in R^{2n}, but it would be very complicated to write it down because of the 2^n components of spinors.

In Section 5 we shall study the cohomology groups of meromorphic spinors, that is, harmonic spinors with singularity, on compact conformally flat 4-manifolds. We develop a divisor theory for meromorphic spinors. But, because we have no product operation on spinors, this is a linear analogy of classical divisors of meromorphic functions. We shall show a Riemann-Roch type theorem for the cohomological dimensions of meromorphic spinors with prescribed divisor. More precisely, let E be a divisor on a compact conformally flat manifold M and let $\mathcal{L}_E$ be the sheaf of meromorphic spinors having the poles at E of order less than $\deg E$. Then we have

$$\dim H^0(M, \mathcal{L}_E) - \dim H^1(M, \mathcal{L}_E) = \deg E, \tag{1.6}$$

$$H^p(M, \mathcal{L}_E) = 0 \text{ for } p \geq 2. \tag{1.7}$$

2. Preliminaries on the Dirac operator and the Cauchy kernel

Here we shall summarize some ingredients of spinor analysis that are now well known, [1, 2, 10].

2.1. Dirac operator and harmonic spinors

Let $\Delta = \Delta^+ \oplus \Delta^-$ be an irreducible complex representation of the Clifford algebra $Clif(\mathrm{R}^4)$; $Clif(\mathrm{R}^4) \otimes \mathrm{C} = End(\Delta)$. Δ decomposes to irreducible representations $\Delta^\pm$ of $Spin(4)$, each of which has $\dim \Delta^\pm = 2$. Let $S = \mathrm{R}^4 \times \Delta$ be the spinor bundle on R^4. The corresponding bundle S^+ (resp. S^-) is called even (resp. odd) spinor bundle.

We shall choose complex coordinates and look at $\mathrm{R}^4 \simeq \mathrm{C}^2$. Our theory does not depend on the complex structure but on the conformal structure on the manifold. The complex coordinates description is adopted for convenience of notation, though this notation allows us to see things in perspective, for example, our formulas for eigenspinors of the Dirac operator fit for the $SU(2)$ representation theory.

Given a smooth boundary ∂G of a region G we shall denote by $\gamma_{\partial G}$ the Clifford multiplication of the outer unit normal on ∂G. We shall abbreviate it as γ if it is obvious which boundary we are considering. $\gamma_{\partial G}$ changes the chirality:

$$\gamma_{\partial G} : S^+ \oplus S^- \longrightarrow S^- \oplus S^+; \qquad \gamma_{\partial G}^2 = 1. \tag{2.1}$$

Let γ_0 denote the Clifford multiplication of the radial vector $\frac{\partial}{\partial n}$, the unit normal to the unit sphere. The chiral decomposition of γ_0 becomes

$$\gamma_0 = \begin{pmatrix} 0 & \gamma_- \\ \gamma_+ & 0 \end{pmatrix} : \begin{matrix} S^+ \\ \oplus \\ S^- \end{matrix} \longrightarrow \begin{matrix} S^+ \\ \oplus \\ S^- \end{matrix} . \tag{2.2}$$

The Dirac operator is defined by

$$\mathcal{D} = c \circ d, \tag{2.3}$$

where $d : S \longrightarrow S \otimes T^*\mathrm{C}^2 \simeq S \otimes T\mathrm{C}^2$ is the exterior differentiation and $c : S \otimes T\mathrm{C}^2 \longrightarrow S$ is the bundle homomorphism coming from the Clifford multiplication. $\mathcal{D}$ is an elliptic operator.

By means of the decomposition $S = S^+ \oplus S^-$, the Dirac operator is decomposed into chiral components:

$$\mathcal{D} = \begin{pmatrix} 0 & D^\dagger \\ D & 0 \end{pmatrix} : C^\infty\left(\mathrm{C}^2, S^+ \oplus S^-\right) \longrightarrow C^\infty\left(\mathrm{C}^2, S^+ \oplus S^-\right). \tag{2.4}$$

An even (resp. odd) spinor φ is called *harmonic spinor* if $D\varphi = 0$ (resp. $D^\dagger\varphi = 0$). We denote by $\mathcal{N}(U)$ (resp. $\mathcal{N}^\dagger(U)$) the set of even (resp. odd) harmonic spinors on an open set U.

Remark 2.1. In [9] we called $\varphi \in \mathcal{N}(M)$ a *zero mode* spinor on M. The reason why we preferred it was that, on a noncompact manifold M, the condition of harmonicity, $D^\dagger D\varphi = 0$, is not equivalent to $D\varphi = 0$.

The following fundamental properties of harmonic spinors are well known [1].

Theorem 2.2. *A harmonic spinor on a connected open set vanishes identically if it vanishes on an open subset.*

Theorem 2.3. *If U and V are domains in C^2 such that $\overline{V}$ is compact in U, then the restriction map $r^U_V : \mathcal{N}(U) \longrightarrow \mathcal{N}(V)$ is compact.*

The following Stokes' formula holds for $\phi \in C^\infty(\overline{G}, S^+)$ and $\psi \in C^\infty(\overline{G}, S^-)$:

$$\int_G < D\phi, \psi > dV + \int_G < \phi, D^\dagger\psi > dV = \int_{\partial G} < \gamma\,\phi, \psi > d\sigma. \tag{2.5}$$

We shall denote in the sequel

$$(\varphi_1, \varphi_2) = \int_{\mathrm{C}^2} < \varphi_1, \varphi_2 > dV, \quad \text{for } \varphi_1, \varphi_2 \in C^\infty(\mathrm{C}^2, S). \tag{2.6}$$

2.2. Cauchy integral formula

The Cauchy kernel is the Clifford multiplication of the radial component of the gradient of the Newton kernel, [2, 3, 8]. In our description it is defined as follows. We put, for any pair $\zeta \neq z$,

$$\mathcal{K} = \frac{1}{|\zeta - z|^3}\gamma_0(\zeta - z) : C^\infty(\mathrm{C}^2, S) \longrightarrow C^\infty(\mathrm{C}^2, S). \tag{2.7}$$

$\mathcal{K}$ decomposes after $S^+ \oplus S^-$ as

$$\mathcal{K} = \begin{pmatrix} 0 & K^\dagger \\ K & 0 \end{pmatrix}, \tag{2.8}$$

$$K^\dagger(z,\zeta) = \frac{1}{|\zeta - z|^3}\gamma_-(\zeta - z), \qquad K(z,\zeta) = \frac{1}{|\zeta - z|^3}\gamma_+(\zeta - z). \tag{2.9}$$

Proposition 2.4.

$$D_z K^\dagger(z,\zeta) = 0, \qquad D^\dagger_z K(z,\zeta) = 0, \qquad \textit{for} \quad \zeta \neq z. \tag{2.10}$$

Theorem 2.5 (Cauchy's integral formula). *Let G be a domain in C^2 and let $\varphi \in C^\infty(\overline{G}, S^+)$. Then*

$$\varphi(z) = -\frac{1}{2\pi^2}\int_G K^\dagger(z,\zeta)D\varphi(\zeta)dV(\zeta) + \frac{1}{2\pi^2}\int_{\partial G} K^\dagger(z,\zeta)(\gamma\varphi)(\zeta)d\sigma(\zeta), \quad z \in G, \tag{2.11}$$

where $\gamma = \gamma_{\partial G}|S^+$ and $d\sigma$ is the surface measure on ∂G.

These are proved, for example, in Proposition 2.1 and Theorem 2.2 of [8].

2.3. Local solutions

Theorem 2.6. *Given an odd spinor with compact support $\psi \in C_c^\infty(\mathrm{C}^2, S^-)$, there is a solution $\phi \in C^\infty(\mathrm{C}^2, S^+)$ of the equation*

$$D\phi(z) = \psi(z), \qquad z \in \mathrm{C}^2. \tag{2.12}$$

Proof. It is proved in [8] that

$$\phi(z) = \frac{1}{2\pi^2}\int_{\mathrm{C}^2} K^\dagger(z,\zeta)\psi(\zeta)dV(\zeta) \tag{2.13}$$

solves the equation $D\phi = \psi$. □

2.4. Eigen spinors of the tangential Dirac operator

The Dirac operator D has the polar decomposition

$$D = \gamma_+\left(\frac{\partial}{\partial n} - \partial\!\!\!/\right). \tag{2.14}$$

The eigenvalues of the tangential Dirac operator $\partial\!\!\!/$ on $|z| = 1$ are

$$\frac{n}{2}, \quad -\frac{n+3}{2}; \qquad n = 0, 1, \ldots,$$

and the multiplicity of each eigenvalue is equal to $(n+1)(n+2)$.

In [9] we gave a complete orthonormal system of eigenspinors $\{\phi^{\pm(n,m,l)}\}$ of $\partial\!\!\!/$ in $L^2(\{|z| = 1\}, S^+)$:

$$\begin{aligned} \partial\!\!\!/\phi^{(n,m,l)} &= \frac{n}{2}\phi^{(n,m,l)} \\ \partial\!\!\!/\phi^{-(n,m,l)} &= -\frac{n+3}{2}\phi^{-(n,m,l)}, \end{aligned} \tag{2.15}$$

for $l = 0, 1, \ldots, n+1$, $m = 0, 1, \ldots, n$, $n = 0, 1, \ldots$.

$\{\phi^{\pm(n,m,l)}\}$ are extended to $\mathrm{C}^2 \setminus \{0\}$ by the homogeneity relations

$$\phi^{(n,m,l)}(z) = |z|^n \phi^{(n,m,l)}\left(\frac{z}{|z|}\right), \tag{2.16}$$

$$\phi^{-(n,m,l)}(z) = |z|^{-(n+3)} \phi^{-(n,m,l)}\left(\frac{z}{|z|}\right), \tag{2.17}$$

for $z \neq 0$. Then,

$$\begin{aligned} D\phi^{(n,m,l)}(z) &= 0, \qquad \text{on } \mathrm{C}^2. \\ D\phi^{-(n,m,l)}(z) &= 0, \qquad \text{on } \mathrm{C}^2 \setminus \{0\}. \end{aligned} \tag{2.18}$$

2.5. Effects of conformal transformations

Here we look at the effect of conformal transformations on the system $\{\phi^{\pm(n,m,l)}\}$. Let $f : U \longrightarrow \mathbb{R}^4$ be a conformal transformation. f induces a $Spin(4)$-equivariant map $f_\flat$ of $Spin(4)$-principal bundles and it yields a bundle isometry $f' = \Delta(f_\flat) : S \longrightarrow S'$ of the associated spinor bundles. The Dirac operator is conformally covariant, that is, if D' is the Dirac operator corresponding to the metric g'; $f^*g' = e^{2u}g$, then

$$D'_{f(z)} = F \cdot D_z \cdot F^{-1}, \tag{2.19}$$

where u is a smooth function on U and $F = e^{-\frac{3}{2}u}f'$, [6 , 10].

Now let U be a domain containing the disk $\{|z| \le 1\}$. Then we can verify

$$\partial\!\!\!/'_{f(z)} = \pm F\, \partial\!\!\!/_z\, F^{-1} = \pm f'\, \partial\!\!\!/_z\, (f')^{-1}, \qquad \text{on } |z| = 1. \tag{2.20}$$

Hence on the sphere $f(\{|z| = 1\}) \subset f(U)$ the eigenvalues of $\partial\!\!\!/$ are $-\frac{n+3}{2}, \frac{n}{2}$, $n = 0, \pm 1, \ldots$, if f is orientation preserving, while they change to $\frac{n+3}{2}, -\frac{n}{2}$, $n = 0, \pm 1, \ldots$, if f is orientation reversing. The corresponding eigenspinors become $f'\phi^{\pm(n,m,l)}$, that were extended to $\mathbb{R}^4 \setminus f(0)$ by $F\phi^{\pm(n,m,l)}$. In particular, by a coordinate change $T \in SO(4)$ we have the same eigenvalues of $\partial\!\!\!/$ and the eigenspinors are $\phi^{\pm(n,m,l)} \circ T$, and our theory is independent of the choice of the complex structure $\mathbb{C}^2 \simeq \mathbb{R}^4$. By the transformation $f(z) = c + rz$, $r > 0$, we find that the eigenspinors on $|z - c| = r$ are given by

$$r^{\mp(n+\frac{3}{2})}\phi^{\pm(n,m,l)}(z - c).$$

On the other hand by the inversion $f(z) = -\frac{\bar{z}}{|z|}$, we have

$$F\phi^{\pm(n,m,l)}(z) = \overline{|z|^3\gamma(z) \cdot \phi^{\pm(n,m,l)}(z)}. \tag{2.21}$$

Note that $\overline{|z|^3\gamma(z)\cdot\phi^{(n,m,l)}}$ belongs to the eigenvalue $-\frac{n}{2}$.

Having verified the covariance of our theory on $\mathbb{R}^4$ under conformal transformations we can extend it to a manifold which is locally $\mathbb{R}^4$ and patched together by conformal transformations, that is, to a conformally flat 4-manifold. For example, S^4 is obtained by patching up two copies of $\mathbb{C}^2$ together by the inversion $w = f(z) = \frac{\bar{z}}{|z|^2}$. We shall denote these two local coordinates by $\mathbb{C}^2_z$ and $\mathbb{C}^2_w$. f has the conformal weight $u = -\log|z|^2$; $f^*(dwd\bar{w}) = \frac{1}{|z|^4}(dzd\bar{z})$. Therefore an even spinor ϕ on a subset U of S^4 is a pair of $\phi \in C^\infty(U \cap \mathbb{C}^2_z \times \Delta^+)$ and $\widehat{\phi} \in C^\infty(U \cap \mathbb{C}^2_w \times \Delta^-)$ such that

$$\widehat{\phi}(w) = (f'\phi)(f(z)) = \overline{|z|^3(\gamma_+ \cdot \phi)(z)}, \tag{2.22}$$

for $w = f(z)$.

The Cauchy kernel on $\mathbb{C}^2_w$ has the form

$$\widehat{K}^\dagger(w, \eta) = -\overline{|z|^3\gamma_+(z)K^\dagger(z,\zeta)\gamma_+(\zeta)}, \qquad w = f(z),\, \eta = f(\zeta). \tag{2.23}$$

2.6. Expansion of the Cauchy kernel

We proved in [8] that the Cauchy kernel has the following expansion by the spinors $\phi^{\pm(n,m,l)}(z-c)$.

Theorem 2.7. 1. *For* $|z-c| < |\zeta - c|$,

$$\begin{aligned} & K^{\dagger}(z,\zeta)\cdot\gamma_{+}(\zeta-c) \\ = & \sum_{n}\sum_{l=0}^{n+1}\sum_{m=0}^{n} |\zeta-c|^{-(2n+3)}\overline{\phi^{(n,m,l)}(\zeta-c)}\otimes\phi^{(n,m,l)}(z-c). \end{aligned}$$

The convergence is uniform on any compact subset of $\{|z-c| < |\zeta-c|\}$.

2. *For* $|z-c| > |\zeta-c|$,

$$\begin{aligned} & K^{\dagger}(z,\zeta)\cdot\gamma_{+}(\zeta-c) \\ = & -\sum_{n}\sum_{l=0}^{n+1}\sum_{m=0}^{n} |\zeta-c|^{2n+3}\overline{\phi^{-(n,m,l)}(\zeta-c)}\otimes\phi^{-(n,m,l)}(z-c). \end{aligned}$$

The convergence is uniform on any compact subset of $\{|z-c| > |\zeta-c|\}$.

2.7. Meromorphic spinors

The Cauchy integral formula and the expansion of the Cauchy kernel in 2.2 and 2.6 yield the Laurent expansion of a harmonic spinor [2, 8].

Theorem 2.8. *Let* φ *be a smooth even spinor on the annular region* $0 \le r < |z-c| < R \le \infty$ *such that* $D\varphi = 0$. *Then we have the expansion*

$$\varphi(z) = \sum_{(n,m,l)} C_{(n,m,l)}\,\phi^{(n,m,l)}(z-c) + \sum_{(n,m,l)} C_{-(n,m,l)}\,\phi^{-(n,m,l)}(z-c), \quad (2.24)$$

for $r < |z-c| < R$. *The coefficients are uniquely determined by* (n,m,l) *and* c *and are given by*

$$C_{\pm(n,m,l)} = \frac{\rho^{\mp(2n+3)}}{2\pi^2}\int_{|\zeta-c|=\rho} < \varphi(\zeta)\,,\,\phi^{\pm(n,m,l)}(\zeta-c) > \sigma(d\zeta) \qquad (2.25)$$

for any ρ *such that* $r < \rho < R$.

In the expansion of φ in (2.26) the second part

$$\sum_{(n,m,l)} C_{-(n,m,l)}\,\phi^{-(n,m,l)}(z-c) \qquad (2.26)$$

is called *principal part* of φ at c.

Definition 2.9. Let G be a domain in C^2 and let E be a discrete subset of G. A harmonic spinor φ on $G \setminus E$ is said to be *meromorphic* on G if its principal part has only finitely many terms at every point of E. A point of E is called a pole of φ.

Let φ be a meromorphic spinor with a pole at $p \in G$. Let f be a conformal transformation. $f'\varphi$ is expanded around $f(p)$ by a series of $f'\phi^{\pm(n,m,l)}$, and the coefficients are given by the corresponding coefficients $C_{\pm(n,m,l)}$ of the expansion of φ around p. Thus the order of a pole of a meromorphic spinor is invariant by conformal transformations.

S^4 is obtained by patching up C_z^2 and C_w^2 together by the inversion $w = f(z) = \frac{\bar{z}}{|z|^2}$. On C_w^2 the basis vectors of the Laurent expansion at infinity $\widehat{0}$ are

$$\widehat{\phi}^{\pm(m,l,k)} = \overline{|z|^3(\gamma_+ \cdot \phi^{\pm(m,l,k)})(z)},$$

with $\widehat{\phi}^{-(m,l,k)} \in O(|w|^m)$ and $\widehat{\phi}^{(m,l,k)} \in O(|w|^{-(m+3)})$. So, on a neighborhood U of $\widehat{0}$, $\varphi \in \mathcal{N}(U)$ has the Taylor expansion

$$\varphi(w) = \sum C_{-(m,l,k)} \widehat{\phi}^{-(m,l,k)}(w). \tag{2.27}$$

3. Runge's approximation theorem

3.1. Approximation of harmonic spinors on a compact set

First we note the following;

1. For a S^--valued Radon measure μ on $G \subset \mathrm{C}^2$, put
$$K^\dagger\mu(z) = \int K^\dagger(z,\zeta)\mu(d\zeta). \tag{3.1}$$
For $\mu(d\zeta) = \phi(\zeta)\, dV(\zeta)$ with $\phi \in \Gamma(S^-)$, we shall abbreviate to $K^\dagger\phi(z)$. Then
$$K^\dagger\mu(z) \in C^\infty(\mathrm{C}^2 \setminus G, S^+), \qquad DK^\dagger\mu(z) = 0.$$
2. For a ${}^t\overline{S^+}$-valued Radon measure ν on $G \subset \mathrm{C}^2$, put
$$\nu K^\dagger(\zeta) = \int \nu(dz)K^\dagger(z,\zeta). \tag{3.2}$$
$\nu K^\dagger(\zeta)$ belongs to $C^\infty(\mathrm{C}^2 \setminus G, {}^t\overline{S}^-)$, and $D^{\dagger}\, {}^t\overline{(\nu K^\dagger)}(\zeta) = 0$.
3. $$\int \nu K^\dagger(\zeta)\mu(d\zeta) = \int \nu(dz)K^\dagger\mu(z) \tag{3.3}$$
if $Supp[\mu] \cap Supp[\nu] = \phi$.

Theorem 3.1. *Let G be a domain in C^2 and K be a compact subset of G. Then any harmonic spinor defined in a neighborhood of K is approximated uniformly on K by harmonic spinors on G if and only if the open set $G \setminus K$ has no component which is relatively compact in G.*

Proof. Sufficiency: We shall show that every ${}^t\overline{S^+}$-valued Radon measure ν on K which annihilates the harmonic (even) spinors on G annihilates also the harmonic (even) spinors in a neighborhood of K, (then use Hahn-Banach).

Let ν be a ${}^t\overline{S^+}$ valued Radon measure on K that annihilates the harmonic (even) spinors on G.

(1) Put

$$g(\zeta) = \nu K^\dagger(\zeta) = \int \nu(dz) K^\dagger(z,\zeta), \tag{3.4}$$

and $f(\zeta) = {}^t\overline{g}(\zeta)$.

f is well defined and $D^\dagger f(\zeta) = 0$ for $\zeta \in \mathrm{C}^2 \setminus K$. Since $D_z K^\dagger(z,\zeta) = 0$ for $\zeta \neq z$, $f(\zeta) = 0$ on $\mathrm{C}^2 \setminus G$ from the assumption. Hence, by the unique continuation property of (odd) harmonic spinors, $f(\zeta) = 0$ in every component of $\mathrm{C}^2 \setminus K$ which intersects $\mathrm{C}^2 \setminus G$.

Next, since $\phi^\lambda(z)$, $\lambda = (n,m,l)$, is a harmonic spinor on C^2, we have

$$\int \nu(dz)\phi^\lambda(z) = 0.$$

So Theorem 2.7 implies that

$$g(\zeta)\gamma_+(\zeta) = \int \nu(dz) K^\dagger(z,\zeta)\gamma_+(\zeta) = 0,$$

for $|\zeta| > \sup_{z\in K} |z|$. Hence $f(\zeta) = 0$ in the unbounded component of $\mathrm{C}^2 \setminus K$. Since $G \setminus K$ has no component which is relatively compact in G we conclude that $f = 0$ and $g = 0$ on $\mathrm{C}^2 \setminus K$.

(2) Let φ be a harmonic spinor on a neighborhood ω of K, and choose a smooth function u with compact support such that $u = 1$ on K.

Since $u\varphi$ has compact support, we have, from integral formula (2.11),

$$\begin{aligned}\varphi(z) = u(z)\varphi(z) &= -\frac{1}{2\pi^2} K^\dagger D(u\varphi)(z)\\ &= -\frac{1}{2\pi^2} K^\dagger(\, du \cdot \varphi)(z), \quad z \in \omega.\end{aligned}$$

By (3.3),

$$\begin{aligned}\int \nu(dz)\varphi(z) &= -\int \nu(dz)\frac{1}{2\pi^2} K^\dagger(\, du\cdot\varphi)(z)\\ &= -\frac{1}{2\pi^2}\int \nu K^\dagger(\zeta)(\, du\cdot\varphi)(\zeta) dV(\zeta).\end{aligned}$$

Since $g = \nu K^\dagger = 0$ on $\mathrm{C}^2 \setminus K$ and the support of $(du\cdot\varphi)(\zeta)dV(\zeta)$ is contained in $\mathrm{C}^2 \setminus K$, the last integral vanishes. Thus ν annihilates the harmonic spinors on K.

Necessity: We assume that $G \setminus K$ has a component H such that $\overline{H}$ is compact in G. Then the boundary of H is a subset of K and the maximum principle for subharmonic functions yields

$$\sup_{\overline{H}} |\phi| \leq \sup_K |\phi|, \qquad \text{for every harmonic spinor } \phi \text{ on } G, \tag{3.5}$$

where $|\phi| = (<\phi,\phi>)^{\frac{1}{2}}$. Let φ be a harmonic spinor defined in a neighborhood of K. By the assumption we can choose a sequence of harmonic spinors φ_n on G so that $\varphi_n \longrightarrow \varphi$ uniformly on K. (3.5) applied to $\varphi_n - \varphi_m$ implies that φ_n converges uniformly on $\overline{H}$ to a limit Φ. Then $\Phi = \varphi$ on the boundary of H, and

Φ is a harmonic spinor in H and is continuous in $\overline{H}$. In particular, we can choose $\varphi(z) = K^\dagger(z,\zeta)$ for a $\zeta \in H$. Then $\Phi(z) = K^\dagger(z,\zeta)$ on the boundary of H. Hence $K^\dagger(\ \ ,\zeta)$ which is a harmonic spinor in $H \setminus \zeta$ is extended to a harmonic spinor in H. This is a contradiction. □

3.2. Global solutions of $D\phi = \psi$

Theorem 3.2. *Let G be an open set in C^2. Given a $\psi \in C^\infty(G, S^-)$, there is a solution $\phi \in C^\infty(G, S^+)$ of the equation*

$$D\phi(z) = \psi(z), \qquad z \in G. \tag{3.6}$$

Proof. Choose an increasing sequence of compact sets $K_j \subset G$ such that every compact subset of G is contained in some K_j. We may suppose that no component of $G \setminus K_j$ is relatively compact in G. If not take the union of K_j and all components of $G \setminus K_j$ that are relatively compact in G. Let h_j be a smooth function with compact support in G such that $h_j = 1$ in a neighborhood of K_j. Put $f_1 = h_1$, $f_j = h_j - h_{j-1}$. Then f_j has compact support and $f_j = 0$ in a neighborhood of K_{j-1} and $\sum f_j = 1$. From the local existence theorem there exists a $\phi_j \in C^\infty(\mathrm{C}^2, S^+)$ such that

$$D\phi_j = f_j \psi.$$

This means in particular that ϕ_j is a harmonic spinor in a neighborhood of K_{j-1}. By Theorem 3.1 we can find a harmonic spinor φ_j on G so that $\sup_{K_{j-1}} |\varphi_j - \phi_j| < 2^{-j}$. Then the sum

$$\phi = \sum (\varphi_j - \phi_j)$$

is uniformly convergent on every compact subset of G. For each k the sum from $k+1$ to ∞ converges uniformly on K_k to a harmonic spinor in the interior of K_k. Hence $\phi \in C^\infty(G, S^+)$ and we have

$$D\phi = \sum D\phi_j = \sum f_j \psi = \psi.$$

□

4. Cohomology Groups of Harmonic Spinors

4.1. Cohomology on conformally flat manifolds

Let (M, g) be a riemannian 4-manifold with conformally flat metric. We suppose that M has a spin structure and we fix it.

There exists a system of coordinate neighborhoods $(U_\lambda, \chi_\lambda)$ in M such that each $G_\lambda = \chi_\lambda(U_\lambda)$ is a domain in C^2 with local coordinate $\{z_1^\lambda, z_2^\lambda\}$ and such that the transition function $f_{\mu\lambda} = \chi_\mu \chi_\lambda^{-1}$ is a conformal transformation:

$$f_{\mu\lambda}^* \left(dz_1^\mu d\overline{z}_1^\mu + dz_2^\mu d\overline{z}_2^\mu\right) = e^{2u_{\mu\lambda}} \left(dz_1^\lambda d\overline{z}_1^\lambda + dz_2^\lambda d\overline{z}_2^\lambda\right), \tag{4.1}$$

where $u_{\mu\lambda}$ is a smooth function on $G_\mu \cap G_\lambda$. We have also

$$(\chi_\lambda^{-1})^* g = e^{2u_\lambda}(dz_1^\lambda d\overline{z}_1^\lambda + dz_2^\lambda d\overline{z}_2^\lambda), \tag{4.2}$$

with u_λ a smooth function on G_λ, and $u_{\mu\lambda} = u_\lambda - u_\mu$ on $G_\lambda \cap G_\mu$.

The Dirac operator is conformally covariant. Let D be the Dirac operator on (M,g) and let D_λ be the Dirac operator on $G_\lambda \subset \mathrm{C}^2$ that we have discussed hitherto, then

$$D \cdot F_\lambda = F_\lambda \cdot D_\lambda, \tag{4.3}$$

where $F_\lambda = e^{-\frac{3}{2}u_\lambda}(\chi_\lambda)'$, see 2.5.

On each open set U contained in a coordinate neighborhood the Dirac operator

$$D : C^\infty(U, S^+) \longrightarrow C^\infty(U, S^-) \tag{4.4}$$

is surjective from Theorem 3.2. Thus every open covering $\mathcal{U}$ by open subsets contained in coordinate neighborhoods is a Leray covering and we have

$$H^k(M, \mathcal{N}) = H^k(\mathcal{U}, \mathcal{N}) \tag{4.5}$$

for every $k \geq 0$.

The structure of a Fréchet space on $C^\infty(G, S^\pm)$ is defined by the uniform convergence of all the derivatives on compact subsets that are contained in coordinate neighborhoods of G. On the vector subspace of harmonic spinors $\mathcal{N}(G)$ the induced topology coincides with the topology of uniform convergence on compact subsets. We endow $H^1(M,\mathcal{N}) = H^1(\mathcal{U},\mathcal{N})$ with the structure of a Fréchet space in an obvious way.

From (1.2) we know that $H^k(M,\mathcal{N}) = 0$ for $k \geq 2$.

Theorem 4.1.

$$H^1(G, \mathcal{N}) = 0, \tag{4.6}$$

for any open subset G in C^2.

The assertion folows from Theorem 3.2.

Theorem 4.2.

$$H^1(S^4, \mathcal{N}) = 0. \tag{4.7}$$

Proof. Let $U_0 = S^4 \setminus \infty = \mathrm{C}_z^2$ and $U_1 = S^4 \setminus 0$. U_1 is conformally equivalent to C_w^2 by the Kelvin inversion $w = \frac{\bar{z}}{|z|^2}$. It follows from Theorem 4.1 that $\{U_0,\, U_1\}$ is a Leray covering of S^4. Let $f_{01} \in \mathcal{N}(U_0 \cap U_1)$. The Laurent expansion of f_{01} on $U_0 \cap U_1 = \mathrm{C}^2 \setminus 0$ becomes

$$f_{01}(z) = \sum_{m,l,k} C_{(m,l,k)}\, \phi^{(m,l,k)}(z) + \sum_{m,l,k} C_{-(m,l,k)}\, \phi^{-(m,l,k)}(z), \qquad z \neq 0.$$

Put

$$f_0(z) = \sum_{m,l,k} C_{(m,l,k)}\, \phi^{(m,l,k)}(z), \quad \text{and} \quad f_1(z) = \sum_{m,l,k} C_{-(m,l,k)}\, \phi^{-(m,l,k)}(z).$$

Then $f_0 \in \mathcal{N}(U_0)$, while f_1 viewed on the coordinate neighborhood C_w^2, $w = \frac{\bar{z}}{|z|^2}$, is harmonic, see the discussion at the end of 2.7. Hence $f_1 \in \mathcal{N}(U_1)$, and $f_{01} = f_0 - f_1$. Therefore $H^1(S^4, \mathcal{N}) = 0$. □

We would like to pose the following conjecture:

$$H^1(M,\mathcal{N}) = 0,$$

for every non-compact conformally flat spin manifold M.

Theorem 4.3. *Let G' be an open subset of M and G be a relatively compact open subset of G'. Then the restriction map*

$$r^* : H^1(G',\mathcal{N}) \longrightarrow H^1(G,\mathcal{N})$$

has a finite dimensional image.

Proof. This follows from Theorem 2.3 and the Schwartz lemma. The latter says that the image of a homomorphism of Fréchet spaces which is the sum of an epimorphism and a compact morphism has a finite codimension. Let $\mathcal{U}$ be an open covering of G'. Let V be the image of the map

$$r \oplus \delta : Z^1(\mathcal{U},\mathcal{N}) \oplus C^0(\mathcal{U} \cap G,\mathcal{N}) \longrightarrow Z^1(\mathcal{U} \cap G,\mathcal{N}),$$

where δ is the coboundary map and r is the restriction map. Then by the Schwartz lemma,

$$\frac{V}{\mathrm{Im}(r \oplus \delta - r \oplus 0)} = \frac{V}{\delta C^0(\mathcal{U} \cap G,\mathcal{N})} = \mathrm{Im}\, r^*$$

is finite dimensional. □

In particular if M is compact the theorem yields the finiteness of $\dim H^1(M,\mathcal{N})$. But this is obvious from the finiteness of $\dim \mathrm{coker} D$.

4.2. Existence of a non-trivial meromorphic spinor

Let M be a conformally flat 4-manifold and $(U_\lambda, \chi_\lambda)$ be a system of coordinate neighborhoods such that $G_\lambda = \chi_\lambda(U_\lambda)$ is a domain in C^2. A smooth spinor φ on $M \setminus E$, E being a discrete subset, is called a *meromorphic spinor* on M with poles at E if, for each λ, $\varphi_\lambda = (\chi_\lambda)'\varphi$ is a meromorphic spinor on $G_\lambda \subset \mathrm{C}^2$ with poles at $G_\lambda \cap \chi_\lambda(E)$. This is equivalent to saying that a family of meromorphic spinors φ_λ on $G_\lambda \subset \mathrm{C}^2$ such that $\varphi_\mu = (\chi_{\mu\lambda})'\varphi_\lambda$ defines a meromorphic spinor on M.

Theorem 4.4. *Let G be a relatively compact open subset of M. For every point $p \in G$ there is a meromorphic spinor on G which has a pole at p and is smooth on $G \setminus p$.*

Proof. Let (U_1, χ_1) be a coordinate neighborhood of p such that $\chi_1(p) = 0$ and $U_2 = M \setminus \{p\}$. Let

$$d = \dim\, \mathrm{Im}(H^1(M,\mathcal{N}) \longrightarrow H^1(G,\mathcal{N})).$$

We take a set Λ constituting $d+1$ indices from the set of indices (m,l,k); $m \geq 1$, $0 \leq l \leq m$, $0 \leq k \leq m+1$, and we consider spinors $\{\phi^{-(n,m,l)}; (n,m,l) \in \Lambda\}$.

These are meromorphic spinors on U_1 and define a cocycle on $U_1 \cap U_2 = U_1 \setminus \{p\}$. Restricted on G, the cocycles are linearly dependent, hence there is a linear combination

$$\sum_{(n,m,l)\in\Lambda} A_{(n,m,l)}\phi^{-(n,m,l)}$$

that belongs to the coboundary, that is, there exist $\xi_i \in \mathcal{N}(U_i)$; $i = 1.2$, such that

$$\sum_{(n,m,l)\in\Lambda} A_{(n,m,l)}\phi^{-(n,m,l)} = \xi_2 - \xi_1.$$

Therefore a spinor φ which coincides on $U_1 \cap G$ with $\sum_{(n,m,l)\in\Lambda} A_{(n,m,l)}\phi^{-(n,m,l)} + \xi_1$, and which is equal to ξ_2 on $U_2 \cap G$, gives a meromorphic spinor on G with the only pole at p. □

Let $\mathcal{U} = \cup_{i\in I} U_i$ be an open covering of M. A family $(\varphi_i)_{i\in I}$ of meromorphic spinors φ_i on U_i is called a *Mittag-Leffler distribution* on M if the differences $\varphi_{ij} = \varphi_i - \varphi_j$ are harmonic spinors on $U_i \cap U_j$.

The family of differences φ_{ij} defines a cocycle and this cocycle is a coboundary precisely when there exists a meromorphic spinor φ on M such that for each $i \in I$ the difference $\varphi - \varphi_i$ is harmonic on U_i.

From Theorems 4.1 and 4.2 we have the following Mittag-Leffler type theorems.

Theorem 4.5. *Let G be an open subset of $\mathbb{C}^2$. Let $(\varphi_i)_{i\in I}$ be a Mittag-Leffler distribution on G. Then there exists a meromorphic spinor φ on G such that for each $i \in I$ the difference $\varphi - \varphi_i$ is harmonic on U_i.*

Theorem 4.6. *Let $(\varphi_i)_{i\in I}$ be a Mittag-Leffler distribution on S^4. Then there exists a meromorphic spinor φ on S^4 such that for each $i \in I$ the difference $\varphi - \varphi_i$ is harmonic on U_i.*

4.3. Serre duality

For an open subset U, $\mathcal{D}'(U, S^\pm)$ denotes the set of $S^\pm$-valued distributions on U, and $\mathcal{D}'(S^\pm)$ denotes the sheaf of $S^\pm$-valued distributions. For $T \in \mathcal{D}'(U, S^\pm)$ and $\varphi \in C_c^\infty(U, S^\mp)$, respectively, we have, by the definition,

$$DT[\varphi] = -T[D^\dagger\varphi], \quad D^\dagger T[\varphi] = -T[D\varphi],$$

respectively. $\mathcal{D}'(M, S^\pm)$ is a Fréchet space which is dual to $C_c^\infty(M, S^\pm)$, and $D^\dagger$ is the transposition of D.

We have the following exact sequence:

$$0 \longrightarrow \mathcal{N}^\dagger \longrightarrow \mathcal{D}'(S^-) \xrightarrow{D^\dagger} \mathcal{D}'(S^+) \longrightarrow 0. \tag{4.8}$$

Here we used Weyl's lemma to have the kernel $\mathcal{N}^\dagger$.

Let $\mathcal{E}'(M, S^\pm) = \Gamma_c(M, \mathcal{D}'(S^\pm))$ be the space of $S^\pm$-valued distributions with compact supports and let $H_c^k(M, \mathcal{N}^\dagger)$, $k = 0, 1$, be the cohomology groups with compact supports. $\mathcal{E}'(M, S^\pm)$ is the dual of the Fréchet space $C^\infty(M, S^\pm)$. From the short exact sequences (1.1) and (4.10) we have the following exact sequences of cohomology groups:

$$0 \longrightarrow H^0(M,\mathcal{N}) \longrightarrow C^\infty(M,\mathcal{S}^+) \xrightarrow{D} C^\infty(M,\mathcal{S}^-) \longrightarrow H^1(M,\mathcal{N}),$$

$$0 \longleftarrow H^1_c(M,\mathcal{N}^\dagger) \longleftarrow \mathcal{E}'(M,S^+) \xleftarrow{D^\dagger} \mathcal{E}'(M,S^-) \longleftarrow H^0_c(M,\mathcal{N}^\dagger) = 0.$$

$H^0_c(M,\mathcal{N}^\dagger) = 0$ from Theorem 2.2. Suppose that $H^1(M,\mathcal{N}) = 0$. Then D, being surjective, becomes a homomorphism of Fréchet spaces by the Banach theorem.

Theorem 4.7. *Suppose $H^1(M,\mathcal{N}) = 0$. Then the dual of $H^0(M,\mathcal{N})$ is isomorphic to $H^1_c(M,\mathcal{N}^\dagger)$.*

5. Divisors of Meromorphic Spinors

5.1. Divisors and meromorphic spinors

We shall consider the cohomology groups of meromorphic spinors on a compact conformally flat spin manifold M, and prove an analogy of the Riemann-Roch theorem. First we note that there is a meromorphic spinor with a pole at a prescribed point on M, this was proved in Theorem 4.4.

A divisor on a space X is a mapping $E : X \longrightarrow \mathbf{Z}$ such that for any compact subset K there are finitely many points $c \in K$ with $E(c) \neq 0$. With respect to addition the set of all divisors forms an abelian group $Div(X)$. There is a partial ordering on $Div(X)$; for $E, E' \in Div(X)$, set $E \leq E'$ if $E(c) \leq E'(c)$ for every $c \in X$.

Since M is compact, for any $E \in Div(M)$, there are only finitely many $x \in M$ such that $E(x) \neq 0$. Then we define the degree

$$\deg : Div(M) \longrightarrow \mathrm{Z},$$

by $\deg E = \sum_{x \in M} E(x)$.

We shall now define an ordered set of indices to enumerate the basis of spinors of the Laurent expansion: $\phi^{\pm(n,m,l)}$. We introduce a triplet $\lambda = (n,m,l)$, $0 \leq m \leq n$, $0 \leq l \leq n+1$. The lexicographic order for two triplets λ is defined by $\lambda \geq \lambda'$ if either (i)$n > n'$, or (ii)$n = n'$, $m > m'$, or (iii) $n = n'$, $m = m'$ and $l \geq l'$. We introduce also the notation $-\lambda = -(n,m,l)$, and define $-\lambda \geq -\lambda'$ if $\lambda \leq \lambda'$. The smallest positive is $o_+ = (0,0,0)$ and the largest negative is $o_- = -(0,0,0)$.

We denote by $\mathcal{Z}$ the set of all triplets λ and $\mathcal{Z}_{\geq 0_+}$ (resp. $\mathcal{Z}_{\leq 0_-}$) the set of all $\lambda \geq o_+$ (resp. $-\lambda \leq o_-$). We denote $|\pm\lambda| = \pm n$ for $\pm\lambda = \pm(n,m,l)$, and put $\mathcal{Z}_{\pm n} = \{\pm\lambda = \pm(n,m,l);\, 0 \leq m \leq n, 0 \leq l \leq n+1\}$. Note that, by convention, $+0 \neq -0$.

For a meromorphic spinor φ on an open set $G \subset M$ and $c \in G$, the Laurent expansion at c becomes

$$\varphi(z) = \sum_{\lambda \in \mathcal{Z}_{\geq 0}} C_\lambda \phi^\lambda(z-c) + \sum_{-\lambda \in \mathcal{Z}_{\leq 0}} C_{-\lambda} \phi^{-\lambda}(z-c), \qquad r < |z-c| < R. \quad (5.1)$$

We define

$$\mathrm{ord}_c(\varphi) := \begin{cases} 0, & \text{if } C_{-\lambda} = 0 \text{ for all } -\lambda \leq 0_- \text{ and } C_\lambda \neq 0 \text{ for some } \lambda \in \mathcal{Z}_0 \\ k, & \text{if } C_{-\lambda} = 0 \text{ for all } -\lambda \leq 0_- \text{ , } C_\lambda = 0 \text{ for all } |\lambda| \leq k-1, \\ & \quad \text{and } C_\lambda \neq 0 \text{ for some } \lambda \in \mathcal{Z}_k \\ -(k+1), & \text{if } C_{-\lambda} = 0 \text{ for } |-\lambda| \leq -(k+1), \\ & \quad \text{and } C_{-\lambda} \neq 0 \text{ for some } -\lambda \in \mathcal{Z}_{-k} \\ \infty, & \text{if } \varphi \equiv 0 \text{ in a neighborhod of } c \end{cases}$$

The divisor of a meromorphic spinor φ not identically 0 is defined by

$$(\varphi) = \sum_{c \in G} ord_c(\varphi) \cdot c. \quad (5.2)$$

For example, from (2.22) we know that each $\phi^{\pm(n,m,l)}$ gives a meromorphic spinor on S^4. Let $\widehat{0}$ denote the point at infinity: $\mathrm{C}_z^2 \cup \widehat{0} = S^4$. We have

$$\begin{aligned} (\phi^{(n,m,l)}) &= n \cdot 0 - (n+1) \cdot \widehat{0} \\ (\phi^{-(n,m,l)}) &= -(n+1) \cdot 0 + n \cdot \widehat{0}. \end{aligned}$$

Let E be a Divisor. For an open set $U \subset M$ we define

$$\mathcal{L}_E(U) = \{\varphi \in \mathcal{M}(U);\, ord_x(\varphi) \geq -E(x) \quad \text{for } \forall x \in U.\}$$

Example. We have for $E = np$, $p \in M$,

$$\begin{aligned} \mathcal{L}_E(U) &= \{\varphi = \sum_{|-\lambda| \geq -(n-1)} C_{-\lambda} \phi^{-\lambda}(z) + \sum_{\lambda \geq 0_+} C_\lambda \phi^\lambda(z)\} \\ \mathcal{L}_{-E}(U) &= \{\varphi = \sum_{|\lambda| \geq n} C_\lambda \phi^\lambda(z)\} \end{aligned}$$

in a local coordinate around p.

Let $p \in M$. The skyscraper sheaf C_p is defined by

$$\mathrm{C}_p(U) = \begin{cases} \mathrm{C}, & \text{if } p \in U, \\ 0, & \text{if } p \notin U, \end{cases}$$

Let $E \in Div(M)$ and $p \in M$. We look at p also as a divisor, $p \in Div(M)$. Then $E' = E + p \in Div(M)$. We define a sheaf homomorphism

$$\rho : \mathcal{L}_{E'} \longrightarrow \mathrm{C}_p$$

as follows. For an open set U, if $p \neq U$ then ρ_U is the zero homomorphism. If $p \in U$ and $\varphi \in \mathcal{L}_{E'}(U)$, then φ admits a Laurent expansion around p,

$$\varphi(z) = \sum_{|-\lambda| \geq -k} C_{-\lambda} \phi^{-\lambda} + \sum_{\lambda \geq 0_+} C_\lambda \phi^\lambda,$$

where $E(p) = k$, hence $E'(p) = k+1$. Set $\rho_U(\varphi) = C_{-\lambda_0}$ for the smallest $-\lambda_0$ such that $C_{-\lambda_0} \neq 0$ and $|-\lambda_0| = -k$. Then ρ is a sheaf homomorphism and we have the short exact sequence

$$0 \longrightarrow \mathcal{L}_E \longrightarrow \mathcal{L}_{E+p} \longrightarrow \mathrm{C}_p \longrightarrow 0. \tag{5.3}$$

Therefore we have the following exact sequence of cohomologies:

$$\begin{aligned} 0 \longrightarrow H^0(M, \mathcal{L}_E) &\longrightarrow H^0(M, \mathcal{L}_{E+p}) \longrightarrow \mathrm{C} \\ &\longrightarrow H^1(M, \mathcal{L}_E) \longrightarrow H^1(M, \mathcal{L}_{E+p}) \longrightarrow 0. \end{aligned} \tag{5.4}$$

Lemma 5.1. *Suppose M is compact and let $E \leq E'$ be divisors. Then we have an epimorphism*

$$H^1(M, \mathcal{L}_E) \longrightarrow H^1(M, \mathcal{L}_{E'}) \longrightarrow 0.$$

Theorem 5.2. *Suppose E is a divisor on a compact conformally flat four dimensional spin manifold M. Then*

$$\dim H^0(M, \mathcal{L}_E) - \dim H^1(M, \mathcal{L}_E) = \deg E, \tag{5.5}$$

$$H^p(M, \mathcal{L}_E) = 0 \text{ for } p \geq 2. \tag{5.6}$$

Proof. For a divisor $E = 0$, the assertion follows from the Atiyah-Singer index theorem (1.3). In fact, for a conformally flat manifold M, the Weyl tensor vanishes from the definition, and the first Pontryagin number $p_1(M)$ is zero.

Let $E \in Div(M)$ and $E' = E + p$. Let $V = Im(H^0(M, \mathcal{L}_{E'}) \longrightarrow \mathrm{C})$ and $W = \mathrm{C}/V$. Then $\dim V + \dim W = 1 = \deg E' - \deg E$. From the exact sequence (5.4) we have,

$$\begin{aligned} \dim H^0(M, \mathcal{L}_{E'}) &= \dim H^0(M, \mathcal{L}_E) + \dim V, \\ \dim H^1(M, \mathcal{L}_{E'}) &= \dim H^1(M, \mathcal{L}_E) - \dim W. \end{aligned}$$

Hence

$$\begin{aligned} &\dim H^0(M, \mathcal{L}_{E'}) - \dim H^1(M, \mathcal{L}_{E'}) - \deg E' \\ &\quad = \dim H^0(M, \mathcal{L}_E) - \dim H^1(M, \mathcal{L}_E) - \deg E. \end{aligned}$$

Thus (5.5) holds for E (resp. E') if it holds for E' (resp. E). In particular (5.5) is true for every divisor $E' \geq 0$. Any divisor may be written as

$$E = p_1 + \ldots + p_m - p_{m+1} - \ldots - p_n.$$

Hence the first assertion is proved by induction. As for the second part, we know $H^p(M, \mathcal{N}) = 0$ for $p \geq 2$, that is, $H^p(M, \mathcal{L}_0) = 0$ for $p \geq 2$. Then, by the same argument using the exact sequence (5.4) as in the first part, we can prove

$$H^p(M, \mathcal{L}_E) = 0 \text{ for } p \geq 2,$$

for every divisor E. □

References

[1] B. Booss-Bavnbeck and K.P. Wojciechowski, *Elliptic boundary value problems for Dirac operators.* Birkhäuser Verlag, Basel-Boston, 1993.

[2] F. Brackx, R. Delanghe and F. Sommen, *Clifford analysis.* Pitman advanced publishing program, (1982).

[3] R. Delanghe and F. Brackx, *Runge's theorem in hypercomplex function theory,* J. Approx. Theory **29** (1980), 200–211.

[4] O. Forster, *Lectures on Riemann surfaces,* Springer-Verlag, 1981.

[5] R. Gunning and H. Rossi, *Analytic functions of several complex variables,* Prentice-Hall, Inc., 1965.

[6] N. Hitchin, *Harmonic Spinors,* Adv. in Math. **14** (1974).

[7] L. Hörmander, *An Introduction to Complex Analysis in Several Variables,* D. Van Norstrand, 1967.

[8] T. Kori, *Spinor analysis on* $\mathbf{C}^2$ *and on conformally flat 4-manifolds,* Japanese J. of Math. **28-1** (2002), 1–30.

[9] T. Kori, *Index of the Dirac operator on* S^4 *and the infinite dimensional Grassmannian on* S^3, Japanese J. of Math. **22-1** (1996), 1–36.

[10] H.B. Lawson and M.L. Michelsohn, *Spin Geometry,* Princeton Univ. Press, 1989.

[11] J. Ryan, *Runge approximation theorems in complex Clifford analysis together with some of their applications,* J. of Functional Anal. **70** No.2 (1987), 221–253.

Tosiaki Kori
Department of Mathematics
School of Science and Engineering
Waseda University
3-4-1 Okubo, Shinjuku-ku
Tokyo
Japan
e-mail: `kori@waseda.jp`

Trends in Mathematics: *Advances in Analysis and Geometry*, 227–255

Spin Geometry, Clifford Analysis, and Joint Seminormality

Mircea Martin

Abstract. The first part of this article studies the integral and maximal operators associated with fundamental solutions of Dirac operators on Clifford bundles. The main goal is to obtain explicit estimates for integral transforms of this kind in terms of the corresponding maximal functions. As direct consequences of such estimates one derives several quantitative Hartogs–Rosenthal type theorems concerning monogenic approximation on compact sets.

The second part illustrates a Clifford analysis approach to the theory of seminormal systems of Hilbert space operators. The four existing concepts of joint seminormality are reevaluated by assuming that the remainders in some Bochner–Kodaira identities are semidefinite, and a new concept is introduced based on a Bochner–Weitzenböck identity. A rather general singular integral model of jointly seminormal pairs of systems of self-adjoint operators that involves Riesz transforms is presented, and a Putnam type commutator inequality for that model is proved.

Mathematics Subject Classification (2000). Primary: 42B20, 47B20; Secondary: 44A35, 47A13, 47A30.

Keywords. Clifford algebras, Dirac operators, hyponormal operators, Putnam's inequality, Riesz transforms.

1. Introduction

The primary tools used in developing Clifford analysis and spin geometry are Clifford algebras and Dirac operators. Though the same objects have played key parts in some other important areas of mathematical research such as harmonic analysis, index theory, K-theory, and noncommutative geometry, we should point out that Clifford analysis and spin geometry address a great deal of specific problems, interesting in their own right, and employ several powerful techniques devised or discovered over the years as means of solving those problems.

For an excellent historical perspective on the development of Clifford analysis, we refer to the monograph of Brackx, Delanghe, and Sommen [BDS], and all the references and comments therein. More additional material is presented in the books by Delanghe, Sommen, and Souček [DSS], Gürlebeck and Sprößig [GS1,2], and Mitrea [Mi]. Other recent contributions can be found in the volumes edited by Ryan [Ry1], and by Ryan and Sprößig [RyS].

Relevant ideas, results, and methods in spin geometry are discussed in the monographs of Berline, Getzler, and Vergne [BGV], Cnops [Cn], Friedrich [F], Gilbert and Murray [GM], Gilkey, Leahy, and Park [GLP], Lawson and Michelson [LM], and Roe [Ro].

The goal of this article is to show that Clifford analysis and spin geometry provide an appropriate framework for defining and studying seminormal systems of Hilbert space operators. Significant applications of Clifford analysis techniques to the spectral theory of commuting tuples of operators are due to McIntosh [Mc1,2], McIntosh and Pryde [McP], and McIntosh, Pryde, and Ricker [McPR]. Our approach elaborates on ways of implementing analogues of spin geometry techniques in multidimensional operator theory. Specifically, we associate to each n-tuple of bounded linear operators on a Hilbert space a Clifford bundle equipped with a Dirac type operator. The corresponding Dirac and Bochner Laplacians are involved in two Bochner–Kodaira identities and a Bochner–Weitzenböck identity. Motivated by Bochner's method in spin geometry, it is natural to assume that the remainders in such identities are semidefinite operators. It turns out that this assumption yields the existing classes of seminormal tuples, as well as a new one.

For the convenience of the reader, we recall that the theory of seminormal, i.e., hyponormal or cohyponormal operators originated in connection with pairs (X, Y) of self-adjoint operators satisfying the commutator condition $\sqrt{-1}[X, Y] \geq 0$, which in terms of the operator $T = X + \sqrt{-1}Y$ and its adjoint $T^* = X - \sqrt{-1}Y$ reduces to $[T^*, T] \geq 0$. The early contributions to the field are presented in the monograph by Putnam [Pm1]. Subsequent monographs by Clancey [Cl], Xia [X2], Martin and Putinar [MnPr], and Conway [Co] give accounts of the development of the one-variable theory at later stages. Steps towards a multivariable theory are indicated in Athavale [At], Cho, Curto, Huruya, and Zelazko [CCHZ], Curto [Cu], Curto and Jian [CJ], Curto, Muhly, and Xia [CMX], Douglas, Paulsen, and Yan [DPY], Martin [Mn1], [Mn4], Martin and Salinas [MnSa], McCullough and Paulsen [McCP], and Xia [X3], to mention just a few of the contributors in this area.

One of the basic results in the single variable theory of seminormal operators is a theorem proved by Putnam [Pm2] that relates the norm of the self-commutator of a seminormal operator to the area of its spectrum. For subnormal operators, Axler and Shapiro [AS] derived Putnam's result from an inequality in rational approximation due to Alexander [Al1,2]. As highlights of our article, we generalize both Putnam's theorem and Alexander's inequality to higher dimensions, and the generalizations rely essentially on Clifford analysis.

The article has two sections. Section 2 introduces Dirac operators on Euclidean spaces or Clifford bundles, Bochner–Weitzenböck and Bochner–Kodaira identities, fundamental solutions, and the integral and maximal operators associated with fundamental solutions.

Theorems A and C proved in this section point out in specific forms that the integral operators are dominated by the maximal operators. The two other results, Theorems B and D, are quantitative Hartogs–Rosenthal type theorems concerning monogenic approximation on compact sets.

In Section 3, we develop a Clifford analysis and spin geometry approach to joint seminormality. We present a multidimensional generalization of the Hilbert transform model of seminormal operators, that is referred to as a Riesz transform model. The main result in this section, Theorem E, is a Putnam type inequality for Riesz transform models derived from Theorem C.

The article is in part expository and mainly self-contained. Additional references and comments are included within each section.

2. Integral and Maximal Operators in Spin Geometry

We start with a brief review centered on the geometric construction and basic properties of Dirac operators on Clifford bundles. Thorough accounts of the general theory can be found, for instance, in [BGV], [GM], [LM], or [Ro].

Next we introduce the integral and maximal operators associated with the fundamental solution of a Dirac operator. Our goal is to show that the integral operators are pointwise dominated and, therefore, controlled by the corresponding maximal operators. The inequality that quantifies the link involves a universal constant that can be computed and, presumably, has a geometric meaning.

We conclude this section with consequences of that inequality concerning monogenic approximation on compact sets.

2.1. Clifford Algebras and Dirac Operators on Euclidean Spaces

The *real Clifford algebra* $\mathfrak{A}_m = \mathfrak{A}(\mathbb{R}^m)$ associated with the Euclidean space $\mathbb{R}^m$, $m \geq 1$, can be described as the universal unital real C^*-algebra with generators e_k, $1 \leq k \leq m$, subject to the relations

$$e_k e_l + e_l e_k = -2\delta_{kl} e_0, \qquad 1 \leq k, l \leq m, \tag{2.1}$$

$$e_k^* = -e_k, \qquad 1 \leq k \leq m, \tag{2.2}$$

where e_0 is the identity of $\mathfrak{A}_m$ and δ_{kl} equals 1 or 0 as $k = l$ or $k \neq l$. The involution defined by (2.2) is also called *Clifford conjugation.* The generators can be identified with the standard orthonormal basis for $\mathbb{R}^m$ and thus one gets an embedding of $\mathbb{R}^m$ into $\mathfrak{A}_m$.

We next let $\mathfrak{S}$ denote a unitary $\mathfrak{A}_m$-module, that is, a real or complex Hilbert space upon which the algebra $\mathfrak{A}_m$ acts such that every generator e_k of $\mathfrak{A}_m$, $1 \leq k \leq m$, determines a skew-adjoint unitary operator, and e_0 corresponds to the identity operator. The space $\mathcal{E}(\mathbb{R}^m, \mathfrak{S})$ of smooth $\mathfrak{S}$-valued functions on $\mathbb{R}^m$ is an

$\mathfrak{A}_m$-module under pointwise Clifford algebra action. Therefore, it makes sense to define a first-order linear differential operator $D_{\text{euc}} : \mathcal{E}(\mathbb{R}^m, \mathfrak{S}) \to \mathcal{E}(\mathbb{R}^m, \mathfrak{S})$ by setting

$$D_{\text{euc}} = e_1\partial_1 + \cdots + e_m\partial_m, \tag{2.3}$$

where $\partial_k = \partial/\partial x_k$, $1 \leq k \leq m$. The so defined operator D_{euc} is called the *Euclidean Dirac operator* on $\mathcal{E}(\mathbb{R}^m, \mathfrak{S})$.

We will refer to functions u in $\mathcal{E}(\mathbb{R}^m, \mathfrak{S})$ that satisfy the equation $D_{\text{euc}}u = 0$ as *monogenic*, or *Clifford analytic*, functions. By squaring D_{euc} and using (2.1) we get the operator $\Delta_{\text{euc}} = D_{\text{euc}}^2$, which turns out to be the standard Laplace operator on $\mathbb{R}^m$ extended to vector-valued functions. Consequently, each monogenic functions is harmonic.

The most important example of a monogenic $\mathfrak{A}_m$-valued function is provided by the *Cauchy kernel* Φ on $\mathbb{R}^m$ defined as

$$\Phi(x) = \frac{1}{\omega_m} \cdot \frac{\bar{x}}{|x|^m}, \qquad x \in \mathbb{R}^m_0 = \mathbb{R}^m \setminus \{0\}, \tag{2.4}$$

where ω_m is the area of the unit sphere $\mathbb{S}^{m-1}$ in $\mathbb{R}^m$, and $\bar{x}$ and $|x|$ stand for the Clifford conjugate and the Euclidean norm of a vector x in $\mathbb{R}^m$, respectively. Actually, Φ is the fundamental solution of the Dirac operator D_{euc}, that is,

$$D_{\text{euc}}(\Phi * u) = u, \tag{2.5}$$

for any compactly supported function u in $\mathcal{E}(\mathbb{R}^m, \mathfrak{S})$, where $\mathfrak{S}$ is an $\mathfrak{A}_m$-module and $\Phi * u$ denotes the convolution product given by

$$\Phi * u(x) = \int_{\mathbb{R}^m} \Phi(x-y)u(y)dy, \qquad x \in \mathbb{R}^m. \tag{2.6}$$

In some parts of our article we are going to use the *complex Clifford algebras* $\mathfrak{A}^c_m = \mathfrak{A}(\mathbb{C}^m)$ defined as complexifications of their real counterparts, that is, $\mathfrak{A}^c_m = \mathfrak{A}_m \otimes \mathbb{C}$, $m \geq 1$. As a matter of fact, we will be interested only in the case when m is even. Some specific constructions pertaining to this case are summarized below.

If $m = 2n$, with $n \geq 1$, a more convenient set of generators, ε_i, $1 \leq i \leq n$, for $\mathfrak{A}^c_{2n}$ as a complex C^*-algebra can be introduced by setting

$$\varepsilon_i = \frac{1}{2}\left(e_i - \sqrt{-1}e_{n+i}\right), \qquad 1 \leq i \leq n. \tag{2.7}$$

We easily get

$$e_i = \varepsilon_i - \varepsilon_i^*, \quad e_{n+i} = \sqrt{-1}(\varepsilon_i + \varepsilon_i^*), \qquad 1 \leq i \leq n, \tag{2.8}$$

and the following two important properties,

$$\varepsilon_i\varepsilon_j + \varepsilon_j\varepsilon_i = 0 \quad \varepsilon_i\varepsilon_j^* + \varepsilon_j^*\varepsilon_i = \delta_{ij}e_0, \qquad 1 \leq i,j \leq n, \tag{2.9}$$

which are equivalent to relations (2.1).

Moreover, algebra $\mathfrak{A}^c_{2n}$ has a unique irreducible representation on the so called *space of complex spinors* $\mathfrak{S}_n$. As a complex Hilbert space,

$$\mathfrak{S}_n = \Lambda^*(\mathbb{C}^n) = \bigoplus_{p=0}^{n} \Lambda^p(\mathbb{C}^n), \tag{2.10}$$

where $\Lambda^*(\mathbb{C}^n)$ is the complex exterior algebra over $\mathbb{C}^n$. To each vector $\sigma \in \mathbb{C}^n$ one associates the operator $\varepsilon(\sigma)$ of left exterior multiplication by σ on $\mathfrak{S}_n$. If $\{\sigma_1, \ldots, \sigma_n\}$ is the standard orthonormal basis for $\mathbb{C}^n$, it can be checked easily that the operators $\varepsilon_i = \varepsilon(\sigma_i)$ and their adjoints $\varepsilon_i^* = \varepsilon(\sigma_i)^*$, $1 \leq i \leq n$, satisfy conditions (2.9). Thus one gets a realization of $\mathfrak{A}^c_{2n}$ within the C^*-algebra $\mathcal{L}(\mathfrak{S}_n)$ of all linear operators on $\mathfrak{S}_n$. Actually, $\mathfrak{A}^c_{2n} = \mathcal{L}(\mathfrak{S}_n)$.

Next, we identify $\mathbb{R}^{2n}$ with $\mathbb{C}^n$ and assuming that $\mathfrak{S}$ is an $\mathfrak{A}^c_{2n}$-module , consider the Dirac operator D_{euc} defined by (2.3) as acting on $\mathcal{E}(\mathbb{C}^n, \mathfrak{S})$. Using (2.8) we have

$$D_{\text{euc}} = \mathcal{D}_{\text{euc}} + \mathcal{D}^*_{\text{euc}} \tag{2.11}$$

with

$$\mathcal{D}_{\text{euc}} = \sum_{i=1}^{n} \varepsilon_i \left(\partial_i + \sqrt{-1}\partial_{n+i}\right), \quad \mathcal{D}^*_{\text{euc}} = \sum_{i=1}^{n} \varepsilon_i^* \left(-\partial_i + \sqrt{-1}\partial_{n+i}\right). \tag{2.12}$$

Since $\mathcal{D}^2_{\text{euc}} = 0$ and $\mathcal{D}^{*2}_{\text{euc}} = 0$, for the Laplace operator $\Delta_{\text{euc}} = D^2_{\text{euc}}$ we obtain the identity

$$\Delta_{\text{euc}} = \mathcal{D}_{\text{euc}}\mathcal{D}^*_{\text{euc}} + \mathcal{D}^*_{\text{euc}}\mathcal{D}_{\text{euc}}. \tag{2.13}$$

In the special case when $\mathfrak{S} = \mathfrak{S}_n$, the space $\mathcal{E}(\mathbb{C}^n, \mathfrak{S}_n)$ can be thought of as the space of all smooth complex differential forms of type $(0, p)$, $0 \leq p \leq n$, over $\mathbb{C}^n$, and (2.11) becomes $D_{\text{euc}} = \sqrt{2}(\bar{\partial} + \bar{\partial}^*)$.

2.2. Dirac Operators on Clifford Bundles

The previous constructions can be generalized to Riemannian manifolds. If M is such a manifold, every cotangent space $T^*_x M$, $x \in M$, has an inner product induced by the Riemannian metric, so it makes sense to take the Clifford algebras $\mathfrak{A}(T^*_x M)$ for each $x \in M$, and form an algebra bundle over M. The resulting bundle will be denoted by $\mathfrak{A}(M)$ and referred to as the *Clifford algebra bundle* of M. The cotangent bundle T^*M is naturally identified with a subbundle of $\mathfrak{A}(M)$.

Suppose next that E is a smooth Riemannian vector bundle over M equipped with a smooth action $\Gamma : \mathfrak{A}(M) \otimes E \to E$ of the Clifford algebra bundle $\mathfrak{A}(M)$ that makes each fiber E_x a unitary $\mathfrak{A}(T^*_x M)$-module, for any $x \in M$.

We denote the space of all smooth sections, or compactly supported smooth sections, from M to E by $\mathcal{E}(M, E)$ and $\mathcal{D}(M, E)$, respectively.

Assume now that E has a linear connection $\nabla : \mathcal{E}(M, E) \to \mathcal{E}(M, T^*M \otimes E)$, and, in addition, suppose that ∇ preserves the metric on E and is compatible with the Levi-Civita connection on M and the Clifford action Γ. Such bundles will be called *Clifford bundles* over M, and each Clifford bundle E has a canonically associated first-order linear differential operator $D : \mathcal{E}(M,E) \to \mathcal{E}(M,E)$ called the *Dirac operator* on E and defined as

$$\mathcal{E}(M,E) \xrightarrow{\nabla} \mathcal{E}(M,T^*M \otimes E) \hookrightarrow \mathcal{E}(M,\mathfrak{A}(M) \otimes E) \xrightarrow{\Gamma} \mathcal{E}(M,E). \tag{2.14}$$

The Dirac operator D is elliptic and formally self-adjoint. Its square, $\Delta = D^2$, is referred to as the *Dirac Laplacian* on E.

There is another naturally defined Laplacian Δ^c on E, called the *Bochner*, or *connection Laplacian*, and given by $\Delta^c = \nabla^*\nabla$, where $\nabla^* : \mathcal{E}(M,T^*M \otimes E) \to \mathcal{E}(M,E)$ stands for the formal adjoint of ∇. The two Laplacians are related by an equation of the form

$$\Delta = \Delta^c + \mathcal{R}, \tag{2.15}$$

where the remainder $\mathcal{R}$ is an operator of order zero that can be explicitly expressed in terms of the curvature operator of E. A complete proof and more details may be found in [LM, Section II.8], where (2.15) is called the *general Bochner identity*, in [BGV, Section 3.5], where an extension of (2.15) is referred to as the *Lichnerowicz formula*, or in [GM, Chapter 5], where (2.15) is called the *Bochner-Weitzenböck identity*.

We next turn our attention to complex manifolds. Specifically, we assume that M is a Kähler manifold and let E be a holomorphic bundle of left modules over $\mathfrak{A}^c(M) = \mathfrak{A}(M) \otimes_{\mathbb{R}} \mathbb{C}$ furnished with a Hermitian metric. In addition, we assume that the Hermitian metric and the associated canonical connection make E a Clifford bundle over M.

The complexified cotangent bundle $T^*M \otimes \mathbb{C}$ splits into two subbundles, $T^*M \otimes \mathbb{C} = T^{0,1}M \oplus T^{1,0}M$. Consequently, the canonical connection ∇ on E splits into two pieces, namely,

$$\nabla = \nabla^{0,1} + \nabla^{1,0}. \tag{2.16}$$

In its turn, the Dirac operator D on E can be expressed as

$$D = \mathcal{D} + \mathcal{D}^*, \tag{2.17}$$

where $\mathcal{D}$ is a first-order differential operator on E and $\mathcal{D}^*$ is its formal adjoint. Moreover, $\mathcal{D}^2 = 0$ and $(\mathcal{D}^*)^2 = 0$.

As a result of the preceding splittings, we get

$$\mathcal{D}\mathcal{D}^* + \mathcal{D}^*\mathcal{D} = 2(\nabla^{0,1})^*\nabla^{0,1} + \mathcal{R}^{0,1}, \tag{2.18}$$

and

$$\mathcal{D}\mathcal{D}^* + \mathcal{D}^*\mathcal{D} = 2(\nabla^{1,0})^*\nabla^{1,0} + \mathcal{R}^{1,0}. \tag{2.19}$$

We will refer to (2.18) and (2.19) as the *general Bochner-Kodaira identities*. Actually, (2.15) is just the average of (2.18) and (2.19). Specifically,

$$\begin{aligned} \Delta &= \mathcal{D}\mathcal{D}^* + \mathcal{D}^*\mathcal{D}, \\ \Delta^c &= (\nabla^{0,1})^*\nabla^{0,1} + (\nabla^{1,0})^*\nabla^{1,0}, \\ 2\mathcal{R} &= \mathcal{R}^{0,1} + \mathcal{R}^{1,0}. \end{aligned} \tag{2.20}$$

The two remainders $\mathcal{R}^{0,1}$ and $\mathcal{R}^{1,0}$ are operators of order zero on E, and likewise $\mathcal{R}$ they can be computed using the Clifford action and the curvature operator on E. We refer the reader to [LM, Sections IV.8 and IV.11] for complete details.

As a concluding remark we recall that the special features of the remainders $\mathcal{R}$ in (2.15), and $\mathcal{R}^{0,1}$ and $\mathcal{R}^{1,0}$ in (2.18) and (2.19) provide the basis of a powerful technique discovered by Bochner (see [Bo], [BoY], [G]) that yields various vanishing theorems under appropriate positivity assumptions on the curvature operator of the vector bundle E. The last sections in [LM, Chapters II and IV] and [BGV, Chapter 3] offer a great deal of relevant examples.

2.3. Fundamental Solutions and Related Operators

Suppose E is a Clifford bundle over a Riemannian manifold M of dimension $m \geq 2$, and let $D : \mathcal{E}(M, E) \to \mathcal{E}(M, E)$ be the associated Dirac operator. On M we take the distance function and the volume measure derived from the Riemannian metric.

We also introduce the vector bundle $\mathcal{L}(E) = \mathcal{L}(E, E)$ over $M \times M$ whose fiber at a point $(x, y) \in M \times M$ equals the space $\mathcal{L}(E_y, E_x)$ consisting of linear maps from E_y into E_x. The norm in E_x and the operator norm in $\mathcal{L}(E_y, E_x)$ are denoted by $\|\cdot\|_x$ and $\|\cdot\|_{x,y}$.

A notable property of the Dirac operator D is that both D and its transpose, D', have the *unique continuation property.* This can be proved, for instance, by applying a result of Aronszajn [Ar] to the Dirac Laplacian $\Delta = D^2$.

For more details on the unique continuation property we refer to the monograph by Tarkhanov [T], where this property is called *the uniqueness condition for the Cauchy problem in the small*, in short, $(U)_S$. As pointed out there, property $(U)_S$ implies the existence of a right fundamental solution for D, i.e., a smooth section $\Phi : M \times M \setminus \Delta(M) \to \mathcal{L}(E)$ where $\Delta(M) = \{(x, x) : x \in M\}$, such that

(i) $\Phi(x, y) : E_y \to E_x$ is a linear map for every $(x, y) \in M \times M \setminus \Delta(M)$;
(ii) the singular integral operator $\mathfrak{J} : \mathcal{D}(M, E) \to \mathcal{E}(M, E)$ given by

$$\mathfrak{J}u(x) = \lim_{\rho \downarrow 0} \int_{M \setminus B(x,\rho)} \Phi(x, y)u(y)dy, \qquad x \in M, \tag{2.21}$$

where $B(x, \rho)$ stands for the closed ball in M centered at $x \in M$ with radius $\rho > 0$ is well defined;
(iii) $\mathfrak{J}$ has the property

$$D\mathfrak{J}u = u, \qquad u \in \mathcal{D}(M, E). \tag{2.22}$$

In other words, $\mathfrak{J}$ is a right inverse to D.

Next, for every $x \in M$ and $t \in (0, \infty)$ we introduce the set $\Omega[x, t]$ defined as

$$\Omega[x, t] = \{y \in M : y \neq x, \ \|\Phi(x, y)\|_{x,y} \geq t\} \cup \{x\}, \tag{2.23}$$

and let $\mu_x : (0, \infty) \to (0, \infty)$ be the function given by

$$\mu_x(t) = \operatorname{vol} \Omega[x, t], \qquad t \in (0, \infty). \tag{2.24}$$

We will assume that μ_x has finite values on $(0,\infty)$ and the quantity $\lambda = \lambda(\Phi)$ defined by

$$\lambda = \sup_{x \in M} \sup_{t \in (0,\infty)} t^{m/(m-1)} \mu_x(t), \tag{2.25}$$

where $m = \dim M$, is finite.

Actually, based on some norm estimates for the fundamental solution Φ proved in Calderbank [Ca], the last assumption is always met if M is a compact manifold, because $\|\Phi(x,y)\|_{x,y}[\mathrm{dist}(x,y)]^{m-1}$, $(x,y) \in M \times M \setminus \Delta(M)$, is uniformly bounded. The property is still true if M is noncompact with an Euclidean metric on the complement of a compact subset, and the restriction of the Clifford bundle E to that complement is isomorphic to a product bundle with a constant metric in each fiber. For instance, if $M = \mathbb{R}^m$, $m \geq 2$, and $E = M \times \mathfrak{S}$, with $\mathfrak{S}$ an $\mathfrak{A}_m$-module, then, as pointed out in Section 2.1, for the Euclidean Dirac operator we have

$$\Phi(x,y) = \frac{1}{\omega_m} \cdot \frac{\bar{x} - \bar{y}}{|x-y|^m}, \qquad x, y \in \mathbb{R}^m, \; x \neq y, \tag{2.26}$$

whence,

$$\|\Phi(x,y)\|_{x,y} = \frac{1}{\omega_m} \cdot \frac{1}{|x-y|^{m-1}}, \qquad x, y \in \mathbb{R}^m, \; x \neq y. \tag{2.27}$$

Consequently, the corresponding sets $\Omega[x,t]$ defined by (2.23) are Euclidean balls centered at $x \in \mathbb{R}^m$ of radius $\rho = (t\omega_m)^{-1/(m-1)}$. Therefore, the functions μ_x given by (2.24) are of the form

$$\mu_x(t) = \lambda t^{-m/(m-1)}, \qquad x \in \mathbb{R}^m, \; t \in (0,\infty), \tag{2.28}$$

where

$$\lambda = \frac{1}{(\omega_m)^{m/(m-1)}} \cdot \mathrm{vol}(\mathbb{B}^m), \tag{2.29}$$

with $\mathbb{B}^m$ the unit ball in $\mathbb{R}^m$.

Let us now return to the general case of a Dirac operator D on a Clifford bundle E with fundamental solution Φ, and introduce the *maximal operator* $\mathfrak{M}$ on $\mathcal{D}(M,E)$ that associates to each compactly supported section $u \in \mathcal{D}(M,E)$ the scalar-valued function $\mathfrak{M}u$ defined as

$$\mathfrak{M}u(x) = \sup_{t \in (0,\infty)} \frac{1}{\mathrm{vol}\,\Omega[x,t]} \int_{\Omega[x,t]} \|u(y)\|_y dy, \; x \in M. \tag{2.30}$$

If D is the Euclidean Dirac operator, then (2.30) is just the definition of the classical *Hardy-Littlewood maximal operators* (see [S1,2]).

We are now in a position to formulate the main result of this section that relates the integral operator (2.21) and the maximal operator (2.30). Before doing that we need to introduce the uniform norm $\|\cdot\|_\infty$ and the L^p-norms $\|\cdot\|_p$, $1 \leq p < \infty$, on $\mathcal{D}(M,E)$ given by

$$\|u\|_\infty = \sup_{x \in M} \|u(x)\|_x, \qquad u \in \mathcal{D}(M,E), \tag{2.31}$$

and

$$\|u\|_p = \left[\int_M \|u(x)\|_x^p dx\right]^{1/p}, \qquad u \in \mathcal{D}(M,E). \tag{2.32}$$

The Banach spaces that result by completing $\mathcal{D}(M,E)$ in the norms (2.31) and (2.32) are denoted by $L^\infty(M,E)$ and $L^p(M,E)$.

Theorem A. *If $D : \mathcal{E}(M,E) \to \mathcal{E}(M,E)$ is the Dirac operator on a Clifford bundle E with fundamental solution Φ, then there exists a constant $A = A(\Phi)$ such that*

$$\|\mathfrak{I}u(x)\|_x \leq A[\mathfrak{M}u(x)]^{(m-1)/m} \cdot \|u\|_1^{1/m} \tag{2.33}$$

for all $u \in \mathcal{D}(M,E)$ and $x \in M$.

Moreover, a possible value of A in (2.33) is given by

$$A = m\lambda^{(m-1)/m}, \tag{2.34}$$

with λ as in (2.25).

In particular, operator $\mathfrak{I}$ can be extended to $L^1(M,E) \cap L^\infty(M,E)$ and

$$\|\mathfrak{I}u\|_\infty \leq m\lambda^{(m-1)/m}\|u\|_\infty^{(m-1)/m}\|u\|_1^{1/m}, \tag{2.35}$$

for each $u \in L^1(M,E) \cap L^\infty(M,E)$.

2.4. Proof of Theorem A

We continue with a proof of Theorem A. Suppose that $u \in \mathcal{D}(M,E)$, $u \neq 0$, and $x \in M$ are fixed, and let $\mu_{u,x} : (0,\infty) \to [0,\infty)$ be the decreasing function defined by

$$\mu_{u,x}(t) = \int_{\Omega[x,t]} \|u(y)\|_y dy. \tag{2.36}$$

Using (2.36) and (2.24) we next rewrite the definition (2.30) of the maximal operator as

$$\mathfrak{M}u(x) = \sup_{t \in (0,\infty)} \frac{\mu_{u,x}(t)}{\mu_x(t)}. \tag{2.37}$$

In addition, from (2.25) we get

$$\mu_x(t) \leq \lambda t^{-m/(m-1)}, \qquad t \in (0,\infty). \tag{2.38}$$

Suppose that $\tau \in (0,\infty)$ is given and define two functions $k_-, k_+ : M \to [0,\infty)$ by setting

$$k_-(y) = \begin{cases} \|\Phi(x,y)\|_{x,y} - \tau, & \text{if } y \in \Omega[x,\tau] \\ 0, & \text{if } y \in M \setminus \Omega[x,\tau] \end{cases} \tag{2.39}$$

$$k_+(y) = \begin{cases} \tau, & \text{if } y \in \Omega[x,\tau] \\ \|\Phi(x,y)\|_{x,y}, & \text{if } y \in M \setminus \Omega[x,\tau]. \end{cases} \tag{2.40}$$

Since obviously $\|\Phi(x,y)\|_{x,y} = k_-(y) + k_+(y)$ and

$$\|\mathfrak{I}u(x)\|_x \leq \int_M \|\Phi(x,y)\|_{x,y}\|u(y)\|_y dy, \tag{2.41}$$

we get

$$\|\mathfrak{J}u(x)\|_x \le \int_{\Omega[x,\tau]} k_-(y)\|u(y)\|_y dy + \int_M k_+(y)\|u(y)\|_y dy. \tag{2.42}$$

We claim that

$$\int_{\Omega[x,\tau]} k_-(y)\|u(y)\|_y dy \le (m-1)\lambda \cdot \tau^{-1/(m-1)}\mathfrak{M}u(x). \tag{2.43}$$

To prove (2.43) we use (2.39) and express the integral over $\Omega[x,\tau]$ as a Stieltjes integral associated with the decreasing function $\mu_{u,x}$ defined in (2.36), that is,

$$\int_{\Omega[x,\tau]} k_-(y)\|u(y)\|_y dy = -\int_\tau^\infty (t-\tau)d\mu_{u,x}(t). \tag{2.44}$$

From (2.36) and (2.37) we get

$$\mu_{u,x}(t) \le \mu_x(t)\mathfrak{M}u(x) \le \lambda t^{-m/(m-1)}\mathfrak{M}u(x), \tag{2.45}$$

whence

$$\lim_{t\to\infty} t\mu_{u,x}(t) = 0.$$

An integration by parts in (2.44) now yields

$$\int_{\Omega[x,\tau]} k_-(y)\|u(y)\|_y dy = \int_\tau^\infty \mu_{u,x}(t)dt,$$

which in conjunction with (2.45) gives

$$\int_{\Omega[x,\tau]} k_-(y)\|u(y)\|_y dy \le \lambda\mathfrak{M}u(x)\int_\tau^\infty t^{-m/(m-1)}dt,$$

an inequality equivalent to (2.43).

On the other hand, from (2.23) and (2.40) we clearly have $k_+(y) \le \tau$ for all $y \in M$. Therefore, the second integral in (2.42) has the estimate

$$\int_M k_+(y)\|u(y)\|_y dy \le \tau\|u\|_1. \tag{2.46}$$

By (2.42), (2.43), and (2.46) we get

$$\|\mathfrak{J}u(x)\|_x \le (m-1)\lambda\tau^{-1/(m-1)}\mathfrak{M}u(x) + \tau\|u\|_1. \tag{2.47}$$

Finally, to minimize the right-hand side of (2.47) we choose

$$\tau = \left[\lambda \cdot \frac{\mathfrak{M}u(x)}{\|u\|_1}\right]^{(m-1)/m}.$$

A direct calculation shows that for this value of τ inequality (2.47) takes the form (2.33) with a constant A as in (2.34).

The last statement and inequality (2.35) in Theorem A follow from (2.33) by observing that

$$\mathfrak{M}u(x) \le \|u\|_\infty, \qquad \text{for any } x \in M.$$

The proof of Theorem A is complete. □

As a matter of fact, the previous proof shows that Theorem A is a result about integral and maximal operators associated to arbitrary *measurable* kernels $\Phi : M \times M \setminus \Delta(M) \to \mathcal{L}(E)$ with the property that (2.25) yields a finite quantity $\lambda = \lambda(\Phi)$. This observation makes it possible to apply Theorem A to kernels that are not necessarily fundamental solutions. A situation of this kind will be presented in Section 3. We would like to mention that Theorem A generalizes some estimates proved in Martin [Mn2,3] for the Euclidean Cauchy kernel, and an inequality in one-variable complex analysis due to Ahlfors and Beurling [AB]. At the same time, when applied to other kinds of kernels, Theorem A generalizes an inequality proved by Hedberg [He] for Riesz potentials, an inequality for convolution operators with homogeneous kernels due to Martin and Szeptycki [MnSz], as well as an estimate for solid Bochner–Martinelli–Koppelman transforms in several variable complex analysis established in Martin [Mn5]. For some other related results we refer to Lieb [L].

Inequality (2.33) in Theorem A is not the only pointwise estimate that relates the integral and maximal operators associated with the fundamental solution of the Dirac operator. A slight addition to the proof leads to a generalization of Theorem A. To make the point, suppose that $1 \leq p < m$. We are going to show that under the assumptions in Theorem A there exists a constant $A_p = A_p(\Phi)$ such that

$$\|\mathfrak{J}u(x)\|_x \leq A_p[\mathfrak{M}u(x)]^{(m-p)/m}\|u\|_p^{p/m}, \qquad u \in \mathcal{D}(M,E),\ x \in M. \tag{2.48}$$

A possible value of A_p is given by

$$A_p = \frac{m}{p}\left[\frac{p(m-1)}{m-p}\cdot\lambda\right]^{(m-1)/m}. \tag{2.49}$$

The case $p = 1$ reduces to Theorem A. If $1 < p < m$, we proceed as follows. Let $q = p/(p-1)$ be the conjugate exponent to p and then, by applying Hölder's inequality to the second integral in (2.42), one gets

$$\int_M k_+(y)\|u(y)\|_y dy \leq \|k_+\|_q\|u\|_p. \tag{2.50}$$

Next, using (2.40) we have

$$\|k_+\|_q^q = \int_{\Omega[x,\tau]} \tau^q dy + \int_{M\setminus\Omega[x,\tau]} \|\Phi(x,y)\|_{x,y}^q dy.$$

The first integral in the equation above equals $\tau^q\mu_x(\tau)$, with μ_x given by (2.24). The second integral can be expressed as a Stieltjes integral associated with the decreasing function μ_x, namely,

$$\int_{M\setminus\Omega[x,\tau]} \|\Phi(x,y)\|_{x,y}^q dy = -\int_0^\tau t^q d\mu_x(t).$$

Combining the last two equations and using an integration by parts we conclude that

$$\|k_+\|_q^q \le q \int_0^\tau t^{q-1}\mu_x(t)dt.$$

Finally, by (2.38) we have, after some simple calculations,

$$\|k_+\|_q \le \left(\frac{mp}{m+p-1}\cdot\lambda\right)^{(p-1)/p} \tau^{(m+p-1)/mp}. \tag{2.51}$$

From (2.50) and (2.51) we obtain the following substitute for (2.46),

$$\int_M k_+(y)\|u(y)\|_y dy \le \left(\frac{mp}{m+p-1}\cdot\lambda\right)^{(p-1)/p} \tau^{(m+p-1)/mp}\|u\|_p. \tag{2.52}$$

Therefore, we can replace (2.47) with

$$\|\mathfrak{J}u(x)\|_x \le (m-1)\lambda\tau^{-1/(m-1)}\mathfrak{M}u(x) + \left(\frac{mp}{m+p-1}\cdot\lambda\right)^{(p-1)/p} \tau^{(m+p-1)/mp}\|u\|_p. \tag{2.53}$$

To minimize the right-hand side of (2.53) we choose

$$\tau = \left(\frac{m+p-1}{mp}\cdot\lambda\right)^{(m-1)/m}\left[\frac{\mathfrak{M}u(x)}{\|u\|_p}\right]^{p(m-1)/m}.$$

It remains to observe that for this value of τ inequality (2.53) reduces to (2.48) with a constant A_p as in (2.49).

2.5. Monogenic Approximation on Compact Sets

The second main result in this section is a quantitative Hartogs-Rosenthal theorem for Dirac operators on Clifford bundles that is derived from Theorem A. We use the same notations as in Section 2.3.

In addition, let $\Omega \subseteq M$ be a compact set and denote by $\mathcal{C}(\Omega, E)$ the Banach space of all continuous sections from Ω to the restriction of the bundle E to Ω, with the uniform norm given by

$$\|u\|_{\Omega,\infty} = \sup_{x\in\Omega}\|u(x)\|_x, \qquad u \in \mathcal{C}(\Omega, E).$$

We also introduce the norm $\|\cdot\|_{\Omega,p}$, $1 \le p < \infty$, defined as

$$\|u\|_{\Omega,p} = \left[\int_\Omega \|u(x)\|_x^p dx\right]^{1/p}, \qquad u \in \mathcal{C}(\Omega, E).$$

We associate to the Dirac operator D on E and each compact set $\Omega \subseteq M$ the solution subspace $\mathcal{S}_D(\Omega, E)$ of $\mathcal{C}(\Omega, E)$ defined as the uniform closure in $\mathcal{C}(\Omega, E)$ of the set consisting of restrictions to Ω of sections u from $\mathcal{E}(M, E)$ such that $Du = 0$ on certain open neighborhoods of Ω in M that may vary from a section to another.

Suppose that Φ is a right fundamental solution of D. Operator $\mathfrak{J}$ defined by (2.21) makes sense for more general sections u, not necessarily smooth. In particular, we can define an operator $\mathfrak{I}_\Omega : \mathcal{E}(M, E) \to \mathcal{C}(\Omega, E)$ by setting $\mathfrak{I}_\Omega(\cdot) = \mathfrak{I}(\chi_\Omega \cdot)|_\Omega$, with χ_Ω the characteristic function of Ω, that is,

$$\mathfrak{I}_\Omega u(x) = \int_\Omega \Phi(x, y)u(y)dy, \qquad u \in \mathcal{E}(M, E),\ x \in \Omega,$$

where the integral has the same meaning as in (2.21).

The next result is a preliminary step in estimating the distance to the solution subspace $\mathcal{S}_D(\Omega, E)$.

Lemma 1. *If $u \in \mathcal{E}(M, E)$, then*

$$\operatorname{dist}_{\mathcal{C}(\Omega,E)}[u, \mathcal{S}_D(\Omega, E)] \leq \|\mathfrak{I}_\Omega(Du)\|_{\Omega,\infty}. \tag{2.54}$$

Proof. Let χ_n, $n \geq 1$, be a sequence of smooth compactly supported functions on M taking values in $[0, 1]$, with each χ_n equal to 1 on an open relatively compact neighborhood U_n of Ω in M, such that $\operatorname{supp} \chi_{n+1} \subseteq \operatorname{supp} \chi_n$ and

$$\bigcap_{n\geq 1} \operatorname{supp} \chi_n = \Omega. \tag{2.55}$$

Suppose $u \in \mathcal{E}(M, E)$ is given and let $u_n \in \mathcal{E}(M, E)$ be defined as

$$u_n = u - \mathfrak{I}(\chi_n Du), \qquad n \geq 1.$$

By (2.22) we get

$$Du_n = Du - \chi_n Du = (1 - \chi_n)Du,$$

whence $Du_n|_{U_n} = 0$, that is, $u_n|_\Omega \in \mathcal{S}_D(\Omega, E)$. Consequently,

$$\operatorname{dist}_{\mathcal{C}(\Omega,E)}[u, \mathcal{S}_D(\Omega, E)] \leq \|u - u_n\|_{\Omega,\infty} = \|\mathfrak{I}(\chi_n Du)\|_{\Omega,\infty}.$$

Inequality (2.54) follows from (2.55) and the last estimate by taking the limit. The proof is complete. □

Combining Lemma 1 and Theorem A we get the following estimate of the distance in $\mathcal{C}(\Omega, E)$ from a smooth global section u to the solution subspace $\mathcal{S}_D(\Omega, E)$.

Theorem B. *Suppose $D : \mathcal{E}(M, E) \to \mathcal{E}(M, E)$ is the Dirac operator on a Clifford bundle E. If $\Omega \subseteq M$ is a compact set and $u \in \mathcal{E}(M, E)$, then*

$$\operatorname{dist}_{\mathcal{C}(\Omega,E)}[u, \mathcal{S}_D(\Omega, E)] \leq A\|Du\|_{\Omega,\infty}^{(m-1)/m} \|Du\|_{\Omega,1}^{1/m}, \tag{2.56}$$

where A is as in (2.34) in Theorem A.

In particular,

$$\operatorname{dist}_{\mathcal{C}(\Omega,E)}[u, \mathcal{S}_D(\Omega, E)] \leq m[\lambda(\Phi)]^{(m-1)/m}(\operatorname{vol}\Omega)^{1/m}\|Du\|_{\Omega,\infty}. \tag{2.57}$$

Since the space consisting of restrictions to Ω of sections from $\mathcal{E}(M, E)$ is dense in $\mathcal{C}(\Omega, E)$, from Theorem B we get the next Hartog-Rosenthal type theorem.

Corollary. *If $\Omega \subseteq M$ is a compact set of volume zero, then $\mathcal{S}_D(\Omega, E) = \mathcal{C}(\Omega, E)$.*

A different proof of this corollary for the Euclidean Dirac operator can be found in Brackx, Delanghe, and Sommen [BDS].

As a concluding remark, we would like to point out that Theorem B generalizes Alexander's inequality [Al1,2] and an inequality due to Putinar [P], both in rational approximation, to the setting of monogenic approximation in arbitrary dimension.

2.6. Additional Results

The tools we developed so far make it possible to find norm estimates for the extension of the integral operator $\mathfrak{I}_\Omega$ to the Lebesgue spaces $L^p(\Omega, E)$ defined as completions of $\mathcal{C}(\Omega, E)$ in the norms $\|\cdot\|_{\Omega,p}$, $1 < p < \infty$.

We start by observing that the proof of Theorem A in Section 2.4 actually gives a stronger uniform estimate than (2.35), namely,

$$\int_M \|\Phi(x,y)\|_{x,y}\|u(y)\|_y dy \leq m[\lambda(\Phi)]^{(m-1)/m}\|u\|_\infty^{(m-1)/m}\|u\|_1^{1/m}. \tag{2.58}$$

If we choose $u \in L^\infty(M,E) \cap L^1(M,E)$, with $\|u(y)\|_y = 1$ for each $y \in \Omega$, and $\|u(y)\|_y = 0$, for each $y \in M \setminus \Omega$, then from (2.58) we get

$$\int_\Omega \|\Phi(x,y)\|_{x,y} dy \leq m[\lambda(\Phi)]^{(m-1)/m}(\operatorname{vol}\Omega)^{1/m}. \tag{2.59}$$

Suppose next that the adjoint kernel $\Phi^* : M \times M \setminus \Delta(M) \to \mathcal{L}(E)$ is such that the corresponding quantity $\lambda(\Phi^*)$ defined as in (2.25) is bounded. Since $\|\Phi^*(x,y)\|_{y,x} = \|\Phi(x,y)\|_{x,y}$ if $x \neq y$, we get the next companion to (2.59),

$$\int_\Omega \|\Phi(x,y)\|_{x,y} dx \leq m[\lambda(\Phi^*)]^{(m-1)/m}(\operatorname{vol}\Omega)^{1/m}. \tag{2.60}$$

For the fundamental solution of the Euclidean Dirac operator defined by (2.26) we clearly have

$$\lambda(\Phi^*) = \lambda(\Phi).$$

Though this might be the case in general, nevertheless we can replace $\lambda(\Phi)$ and $\lambda(\Phi^*)$ in (2.59) and (2.60) with

$$\lambda(\Phi, \Phi^*) = \max[\lambda(\Phi), \lambda(\Phi^*)]. \tag{2.61}$$

We are now in a position to refine the last statement in Theorem A.

Theorem C. *Suppose that $\Omega \subseteq M$ is a compact set and $u \in L^p(\Omega, E)$, $1 < p < \infty$. Then*

$$\|\mathfrak{I}_\Omega u\|_{\Omega,p} \leq m[\lambda(\Phi,\Phi^*)]^{(m-1)/m}(\operatorname{vol}\Omega)^{1/m}\|u\|_{\Omega,p}. \tag{2.62}$$

Proof. Let $q = p/(p-1)$ be the conjugate exponent to p. Using the canonical pairing $(\cdot,\cdot)$ that relates $L^p(\Omega, E)$ and $L^q(\Omega, E)$ we get

$$(\mathfrak{I}_\Omega u, v) = \int_\Omega \int_\Omega \langle \Phi(x,y)u(y), v(x)\rangle_x dy dx,$$

for any $v \in L^q(\Omega, E)$, whence, by applying Cauchy's inequality,

$$|(\mathfrak{I}_\Omega u, v)| \leq \int_\Omega \int_\Omega \|\Phi(x,y)\|_{x,y} \|u(y)\|_y \|v(x)\|_x dy dx. \tag{2.63}$$

Assume that $v \in L^q(\Omega, E)$ is fixed and let f and g be the scalar-valued functions on $\Omega \times \Omega \setminus \Delta(\Omega)$ given by

$$f(x,y) = \|\Phi(x,y)\|_{x,y}^{1/p} \|u(y)\|_y,$$
$$g(x,y) = \|\Phi(x,y)\|_{x,y}^{1/q} \|v(x)\|_x.$$

Inequality (2.63) amounts to

$$|(\mathfrak{I}_\Omega u, v)| \leq \int_\Omega \int_\Omega f(x,y) g(x,y) dy dx.$$

By Hölder's inequality we get

$$|(\mathfrak{I}_\Omega u, v)| \leq \|f\|_p \|g\|_q, \tag{2.64}$$

where

$$\|f\|_p = \left[\int_\Omega \int_\Omega f^p(x,y) dy dx\right]^{1/p}, \quad \|g\|_q = \left[\int_\Omega \int_\Omega g^q(x,y) dy dx\right]^{1/q}.$$

Moreover, from (2.60) we have

$$\begin{aligned} \|f\|_p^p &= \int_\Omega \|u(y)\|_y^p \left[\int_\Omega \|\Phi(x,y)\|_{x,y} dx\right] dy, \\ &\leq m[\lambda(\Phi, \Phi^*)]^{(m-1)/m} (\operatorname{vol} \Omega)^{1/m} \|u\|_{\Omega,p}^p, \end{aligned}$$

and similarly, from (2.59) we obtain

$$\begin{aligned} \|g\|_q^q &= \int_\Omega \|v(x)\|_x^q \left[\int_\Omega \|\Phi(x,y)\|_{x,y} dy\right] dx, \\ &\leq m[\lambda(\Phi, \Phi^*)]^{(m-1)/m} (\operatorname{vol} \Omega)^{1/m} \|v\|_{\Omega,q}^q. \end{aligned}$$

The last two estimates in conjunction with (2.64) lead to

$$|(\mathfrak{I}_\Omega u, v)| \leq m[\lambda(\Phi, \Phi^*)]^{(m-1)/m} (\operatorname{vol} \Omega)^{1/m} \|u\|_{\Omega,p} \|v\|_{\Omega,q},$$

an inequality that completes the proof of Theorem C. □

The comments made in connection with Theorem A remain true for Theorem C, namely, the estimate (2.62) takes place for arbitrary measurable kernels with the property that quantity (2.61) is finite.

Theorem C can be used to estimate the distance in $L^p(\Omega, E)$, $1 < p < \infty$, from a smooth section $u \in \mathcal{E}(M, E)$ to the subspace $S_D^p(\Omega, E)$ obtained as the closure in $L^p(\Omega, E)$ of the solution subspace $\mathcal{S}_D(\Omega, E)$. The next result complements inequality (2.57) in Theorem B and follows from Theorem C in the same way Theorem B was derived from Theorem A.

Theorem D. *Suppose $D : \mathcal{E}(M,E) \to \mathcal{E}(M,E)$ is the Dirac operator on a Clifford bundle E. If $\Omega \subseteq M$ is a compact set and $u \in \mathcal{E}(M,E)$, then*

$$\operatorname{dist}_{L^p(\Omega,E)}[u, S_D^p(\Omega,E)] \le m[\lambda(\Phi,\Phi^*)]^{(m-1)/m}(\operatorname{vol}\Omega)^{1/m}\|Du\|_{\Omega,p}, \tag{2.65}$$

where $1 < p < \infty$.

3. Clifford Analysis Techniques in Multidimensional Operator Theory

This section outlines a Clifford analysis approach to multidimensional operator theory, with a particular emphasis on the study of seminormal systems of Hilbert space operators. The concept of joint seminormality is introduced in a geometric setting in connection with Bochner–Weizenböck and Bochner–Kodaira type identities, and Bochner's method.

We will next present a Riesz transform model of pairs of jointly seminormal self-adjoint systems of operators, that in dimension one reduces to the Hilbert transform model of a single seminormal operator. The main result in this section is a Putnam type inequality for the Riesz transform model derived from Theorem C. That result and, more importantly, the way we prove it, illustrate an interesting connection between spin geometry and multidimensional operator theory.

3.1. Bochner–Weizenböck and Bochner–Kodaira Identities in Multidimensional Operator Theory

To begin with, we assume that $\mathcal{H}$ is a complex infinite dimensional Hilbert space and denote by $\mathcal{L}(\mathcal{H})$ the C^*-algebra of all bounded linear operators on $\mathcal{H}$. If $T = (T_1, T_2, \ldots, T_n)$, $n \ge 1$, is an n-tuple of operators in $\mathcal{L}(\mathcal{H})$, we set $T^* = (T_1^*, T_2^*, \ldots, T_n^*)$, where T_i^* stands for the adjoint of T_i, $1 \le i \le n$, and define the real and imaginary parts of T as

$$X = \operatorname{Re}(T) = (X_1, X_2, \ldots, X_n), \quad X_i = \operatorname{Re}(T_i) = \frac{T_i + T_i^*}{2}, \qquad 1 \le i \le n,$$

and

$$Y = \operatorname{Im}(T) = (Y_1, Y_2, \ldots, Y_n), \quad Y_i = \operatorname{Im}(T_i) = \frac{T_i - T_i^*}{2\sqrt{-1}}, \qquad 1 \le i \le n.$$

Next, we take the $\mathfrak{A}_{2n}^c$-module of *$\mathcal{H}$-valued spinors* $\mathfrak{S} = \mathfrak{S}_n \otimes \mathcal{H}$, where $\mathfrak{S}_n$ is the space of complex spinors introduced in Section 2.1, and form the product bundle $E = \mathbb{C}^n \times \mathfrak{S}$. The corresponding vector bundle $\mathcal{L}(E)$ is isomorphic to the product bundle $\mathbb{C}^n \times \mathcal{L}(\mathfrak{S})$, and the spaces of smooth sections $\mathcal{E}(\mathbb{C}^n, E)$ and $\mathcal{D}(\mathbb{C}^n, E)$ can be identified with the spaces of smooth functions $\mathcal{E}(\mathbb{C}^n, \mathfrak{S})$ and $\mathcal{D}(\mathbb{C}^n, \mathfrak{S})$, respectively.

To the n-tuple T we associate a Dirac operator on the Clifford bundle E defined as

$$D = D_{\text{euc}} + D(X,Y), \tag{3.1}$$

where D_{euc} is the standard Euclidean Dirac operator on E and, using notations already introduced in Section 2.1,

$$D(X,Y) = \sqrt{-1}\sum_{i=1}^{n}(e_i \otimes X_i + e_{n+i} \otimes Y_i) \in \mathfrak{A}_{2n}^{c} \otimes \mathcal{L}(\mathcal{H}). \tag{3.2}$$

Based on a result due to Gilkey [Gi], there exist a unique linear connection ∇ and a unique endomorphism $\mathcal{R}$ on E such that

$$\Delta = \Delta^c + \mathcal{R}, \tag{3.3}$$

where $\Delta = D^2$ and $\Delta^c = \nabla^*\nabla$ are the Dirac and the Bochner Laplacians.

We will refer to (3.3) as the *Bochner–Weitzenböck identity associated to T*. Both the linear connection ∇ and the remainder $\mathcal{R}$ in (3.3) depend on T.

Let us assume for a while that T is a commuting tuple, that is, its entries have the property

(c) $\quad [T_i, T_j] = 0, \qquad 1 \leq i, j \leq n,$

where $[T_i, T_j] = T_iT_j - T_jT_i$ is the commutator of T_i and T_j, $1 \leq i, j \leq n$. A direct calculation shows that in terms of the real and imaginary parts of T condition (c) amounts to

(c′) $\quad [X_i, X_j] = [Y_i, Y_j], \qquad 1 \leq i, j \leq n,$

and

(c″) $\quad [X_i, Y_j] = [X_j, Y_i], \qquad 1 \leq i, j \leq n.$

With this additional assumption, we are now in a position to set up the Bochner–Kodaira identities (2.18) and (2.19) introduced in Section 2.2. We start by splitting the Dirac operator (3.1) as in (2.17). That is done according to (2.11) for D_{euc} and writing operator (3.2) as

$$D(X,Y) = \mathcal{D}(T) + \mathcal{D}^*(T), \tag{3.4}$$

where

$$\mathcal{D}(T) = \sqrt{-1}\sum_{i=1}^{n} \varepsilon_i \otimes T_i, \quad \mathcal{D}^*(T) = -\sqrt{-1}\sum_{i=1}^{n} \varepsilon_i^* \otimes T_i^*, \tag{3.5}$$

with ε_i, $1 \leq i \leq n$, defined by (2.7).

Condition (c) above guarantees—in fact it is equivalent to—the next important property,

$$\mathcal{D}(T)^2 = 0, \quad \mathcal{D}^*(T)^2 = 0. \tag{3.6}$$

The linear connection ∇ used in (3.3) naturally decomposes as in (2.16) and, consequently, we get the two identities (2.18) and (2.19), as well as the three equations in (2.20). Actually, after some cancellations, the resulting relations only involve operators in $\mathfrak{A}_{2n}^{c} \otimes \mathcal{L}(\mathcal{H})$, and we end up with the following equations,

$\mathbf{BK}_{\mathrm{L}}(T)$: $\quad \Delta(T) = e_0 \otimes \Delta_{\mathrm{L}}^{c}(T) + \mathcal{R}_{\mathrm{L}}(T),$

$\mathbf{BK}_{\mathrm{R}}(T)$: $\quad \Delta(T) = e_0 \otimes \Delta_{\mathrm{R}}^{c}(T) + \mathcal{R}_{\mathrm{R}}(T),$

called the *left and right Bochner-Kodaira identities associated to* T, where

$$\Delta(T) = \mathcal{D}(T)\mathcal{D}^*(T) + \mathcal{D}^*(T)\mathcal{D}(T), \tag{3.7}$$

$$\Delta^c_{\mathrm{L}}(T) = \sum_{i=1}^{n} T_i^* T_i, \qquad \Delta^c_{\mathrm{R}}(T) = \sum_{i=1}^{n} T_i T_i^*, \tag{3.8}$$

and the two remainders are given by

$$\mathcal{R}_{\mathrm{L}}(T) = \sum_{i,j=1}^{n} \varepsilon_i \varepsilon_j^* \otimes [T_i, T_j^*], \qquad \mathcal{R}_{\mathrm{R}}(T) = \sum_{i,j=1}^{n} \varepsilon_i^* \varepsilon_j \otimes [T_i^*, T_j]. \tag{3.9}$$

We proceed with two observations. First, we notice that the two just mentioned Bochner–Kodaira identities are true for not necessarily commuting tuples. Second, we point out that the left identity $\mathrm{BK}_{\mathrm{L}}(T)$ for T coincides, up to a unitary equivalence, with the right identity $\mathrm{BK}_{\mathrm{R}}(T^*)$ for the adjoint tuple T^*. To be specific, we let θ denote the unitary element in $\mathfrak{A}^c_{2n}$ given by

$$\theta = \prod_{i=1}^{n} (\varepsilon_i + \varepsilon_i^*) = (-\sqrt{-1})^n \prod_{i=1}^{n} e_{n+i},$$

and observe that

$$\theta^* \varepsilon_i \theta = (-1)^{n-1} \varepsilon_i^*, \qquad 1 \le i \le n. \tag{3.10}$$

It remains to check, by direct calculations and a repeated use of (3.10), the following equations in $\mathfrak{A}^c_{2n} \otimes \mathcal{L}(\mathcal{H})$,

(i) $\qquad (\theta \otimes I)^* \mathcal{D}(T)(\theta \otimes I) = (-1)^{n-1} \mathcal{D}^*(T^*),$

$\qquad\;\; (\theta \otimes I)^* \mathcal{D}^*(T)(\theta \otimes I) = (-1)^{n-1} \mathcal{D}(T^*),$

whence

$$(\theta \otimes I)^* \Delta(T)(\theta \otimes I) = \Delta(T^*),$$

as well as

(ii) $\qquad (\theta \otimes I)^* e_0 \otimes \Delta^c_{\mathrm{L}}(T)(\theta \otimes I) = e_0 \otimes \Delta^c_{\mathrm{R}}(T^*),$

(iii) $\qquad (\theta \otimes I)^* \mathcal{R}_{\mathrm{L}}(T)(\theta \otimes I) = \mathcal{R}_{\mathrm{R}}(T^*).$

The last observations also indicate that the right identity $\mathrm{BK}_{\mathrm{R}}(T)$ is equivalent to $\mathrm{BK}_{\mathrm{L}}(T^*)$. Therefore, instead of dealing with $\mathrm{BK}_{\mathrm{L}}(T)$ and $\mathrm{BK}_{\mathrm{R}}(T)$, we can choose just one equation as a basic form of the Bochner–Kodaira identity. In the case of a single operator T, that is, when $n = 1$, the standard definition of the self-commutator $\mathcal{C}(T)$ of T is $\mathcal{C}(T) = [T^*, T]$, and for that reason it seems that the natural choice should be the right identity, whose remainder displays the same kind of commutators. Consequently, we rewrite $\mathrm{BK}_{\mathrm{R}}(T)$ simply as

BK(T): $\qquad \Delta(T) = e_0 \otimes \Delta^c(T) + \mathcal{R}(T),$

where $\Delta^c(T) = \Delta^c_{\mathrm{R}}(T)$ and $\mathcal{R}(T) = \mathcal{R}_{\mathrm{R}}(T)$, and replace $\mathrm{BK}_{\mathrm{L}}(T)$ with

BK(T^*): $\qquad \Delta(T^*) = e_0 \otimes \Delta^c(T^*) + \mathcal{R}(T^*).$

The splitting (3.4) is primarily motivated by the use of complex variables. If one decides on using real variables, as in Clifford analysis, then one needs to set up identities in terms of the self-adjoint tuples X and Y. To this end, we return to equation (3.2) and decompose $D(X,Y)$ by setting

$$D(X,Y) = D'(X) + D''(Y), \tag{3.11}$$

where

$$D'(X) = \sqrt{-1}\sum_{i=1}^{n} e_i \otimes X_i, \quad D''(Y) = \sqrt{-1}\sum_{i=1}^{n} e_{n+i} \otimes Y_i. \tag{3.12}$$

Assuming that X and Y satisfy the commuting conditions (c′) and (c″), or the equivalent condition (3.6), we get

$$\Delta(T) = D(X,Y)^2 = D'(X)^2 + D'(X)D''(Y) + D''(Y)D'(X) + D''(Y)^2,$$

and

$$\Delta(T^*) = D(X,-Y)^2 = D'(X)^2 - D'(X)D''(Y) - D''(Y)D'(X) + D''(Y)^2.$$

We next introduce the following combinations of the operators involved in equations BK(T) and BK(T^*) above,

$$\Delta(X,Y) = \frac{1}{2}[\Delta(T) - \Delta(T^*)], \tag{3.13}$$

$$\Delta^c(X,Y) = \frac{1}{2}[\Delta^c(T) - \Delta^c(T^*)], \tag{3.14}$$

and

$$\mathcal{R}(X,Y) = \frac{1}{2}[\mathcal{R}(T) - \mathcal{R}(T^*)]. \tag{3.15}$$

As a result, from BK(T) and BK(T^*) we derive the equation

$$\mathbf{BW}(X,Y): \qquad \Delta(X,Y) = e_0 \otimes \Delta^c(X,Y) + \mathcal{R}(X,Y),$$

that will be referred to as the *Bochner–Weitzenböck identity for the pair of self-adjoint tuples X and Y.*

We have the following explicit equation for the operator (3.13),

$$\Delta(X,Y) = D'(X)D''(Y) + D''(Y)D'(X).$$

In addition, for the operators (3.14) and (3.15) we get

$$\Delta^c(X,Y) = -\sqrt{-1}\sum_{i=1}^{n}[X_i, Y_i],$$

and

$$\mathcal{R}(X,Y) = \sqrt{-1}\sum_{i,j=1}^{n} \varepsilon_i^*\varepsilon_j \otimes ([X_i, Y_j] + [X_j, Y_i]). \tag{3.16}$$

The last equation can be simplified by using condition (c″). We would also like to notice that identity BW(X,Y) only requires condition (c″).

We now introduce two definitions of seminormal systems of operators that catch the geometric significance of the concept of joint seminormality. The terminology proposed below is the same as in Martin [Mn1, Mn4] and Martin and Salinas [MnSa].

Definition 1. An n-tuple T of operators on $\mathcal{H}$ is called *seminormal* provided the remainder $\mathcal{R}(T)$ in $\mathrm{BK}(T)$, or the remainder $\mathcal{R}(T^*)$ in $\mathrm{BK}(T^*)$, is semidefinite as an operator in $\mathfrak{A}_{2n}^c \otimes \mathcal{L}(\mathcal{H})$.

If $\mathcal{R}(T) \geq 0$, or $\mathcal{R}(T) \leq 0$, then T is said to be *right hyponormal*, or *right cohyponormal*, respectively.

If $\mathcal{R}(T^*) \geq 0$, or $\mathcal{R}(T^*) \leq 0$, then T is called *left cohyponormal*, or *left hyponormal*, respectively.

In the case when $n = 1$, there is no difference between left or right seminormality and Definition 1 records the standard definitions of seminormal, hyponormal, and cohyponormal operators. It can be proved that the concepts of joint hyponormality and t-hyponormality introduced by Athavale [At] and Xia [X3] correspond in our terminology to left hyponormality and right hyponormality (see, for instance, [Mn1]).

Definition 2. A pair (X, Y) of self-adjoint n-tuples of operators on $\mathcal{H}$ satisfying the commuting condition (c$''$) is called *jointly seminormal* provided the remainder $\mathcal{R}(X, Y)$ in $\mathrm{BW}(X, Y)$ is semidefinite as an operator in $\mathfrak{A}_{2n}^c \otimes \mathcal{L}(\mathcal{H})$.

If $\mathcal{R}(X, Y) \geq 0$, or $\mathcal{R}(X, Y) \leq 0$, then the pair (X, Y) is said to be *jointly hyponormal*, or *jointly cohyponormal*, respectively.

From Definition 1 we easily get that the adjoint T^* of a right hyponormal, or cohyponormal, tuple T is left cohyponormal, or hyponormal. For more details on seminormal tuples of operators, including some spectral properties of *commuting* right hyponormal or left cohyponormal tuples derived from the Bochner–Kodaira identities $\mathrm{BK}(T)$ or $\mathrm{BK}(T^*)$ we refer to [Mn1], [Mn4], and [MnSa], and the references therein.

We also note that the self-adjoint pair (X, Y) corresponding to a commuting tuple T that is simultaneously right and left hyponormal, or cohyponormal, turns out to be jointly hyponormal, or cohyponormal, as it follows from Definitions 1 and 2 and equation (3.15) for $\mathcal{R}(X, Y)$. At the same time, it is quite apparent that a pair (X, Y) is jointly cohyponormal if and only if the pair $(X, -Y)$ is jointly hyponormal.

We end this section with a test for *remainder-like* positive semidefinite operators in $\mathfrak{A}_{2n}^c \otimes \mathcal{L}(\mathcal{H}) = \mathcal{L}(\mathfrak{S}_n \otimes \mathcal{H})$. To state it, we assume that $\mathcal{R} : \mathfrak{S}_n \otimes \mathcal{H} \to \mathfrak{S}_n \otimes \mathcal{H}$ is an operator of the form

$$\mathcal{R} = \sum_{i,j=1}^{n} \varepsilon_i^* \varepsilon_j \otimes R_{ij}, \qquad R_{ij} \in \mathcal{L}(\mathcal{H}),\ 1 \leq i, j \leq n,$$

and associate to it the $n \times n$ operator matrix $\mathcal{C} = (R_{ij})_{i,j=1}^n$, regarded as an operator on $\mathbb{C}^n \otimes \mathcal{H}$. Actually, if we use the orthogonal decomposition (2.10) of

$\mathfrak{S}_n$ and identify $\Lambda^{n-1}(\mathbb{C}^n)$ with $\mathbb{C}^n$, then matrix $\mathcal{C}$ represents the restriction of operator $\mathcal{R}$ to its invariant subspace $\Lambda^{n-1}(\mathbb{C}^n)\otimes\mathcal{H}$. As the next result points out, that restriction completely characterizes operator $\mathcal{R}$.

Lemma 2. *With the notation introduced above, the following two statements are equivalent:*

(i) $\mathcal{R}\geq 0$ *as an operator on* $\mathfrak{S}_n\otimes\mathcal{H}$*;*
(ii) $\mathcal{C}\geq 0$ *as an operator on* $\mathbb{C}^n\otimes\mathcal{H}$*.*

For a complete proof, we refer to Proposition 2.6 and Lemma 2.7 in Martin [Mn4].

In particular, from Lemma 2 and equation (3.16), we conclude that a self-adjoint pair (X,Y) satisfying condition (c$''$) is jointly hyponormal if and only if the $n\times n$ operator matrix

$$\mathcal{C}(X,Y)=\sqrt{-1}([X_i,Y_j]+[X_j,Y_i])_{i,j=1}^n \tag{3.17}$$

is positive semidefinite. We will refer to matrix (3.17) as the *commutator matrix* of the pair (X,Y).

3.2. Riesz Transform Models of Jointly Hyponormal Self-Adjoint Pairs

In this section, we introduce a singular integral model of jointly hyponormal pairs of self-adjoint tuples of operators that generalizes to higher dimension the Hilbert transform model discovered by Xia [X1] and Pincus [Pi] for operators with one-dimensional self-commutators, and afterwards set up in full generality by Kato [Ka] and Muhly [My].

We first recall that the Riesz transforms of a complex-valued function φ in the Schwartz space $\mathcal{S}(\mathbb{R}^n,\mathbb{C})$ are defined by the principal value integrals

$$R_i\varphi(x)=\frac{\Gamma((n+1)/2)}{\pi^{(n+1)/2}}\text{p.v.}\int_{\mathbb{R}^n}\frac{x_i-y_i}{|x-y|^{n+1}}\varphi(y)dy,\qquad 1\leq i\leq n,\ x\in\mathbb{R}^n. \tag{3.18}$$

We let k_i denote the kernels

$$k_i(x)=\frac{\Gamma((n+1)/2)}{\pi^{(n+1)/2}}\cdot\frac{x_i}{|x|^{n+1}},\qquad 1\leq i\leq n,\ x\in\mathbb{R}^n_0=\mathbb{R}^n\setminus\{0\}, \tag{3.19}$$

and write (3.18) as $R_i\varphi(x)=k_i*\varphi(x)$, $1\leq i\leq n$.

In addition, we introduce the operators M_i on $\mathcal{S}(\mathbb{R}^n,\mathbb{C})$ defined as

$$M_i\varphi(x)=x_i\varphi(x),\qquad 1\leq i\leq n,\ x=(x_1,x_2,\dots,x_n)\in\mathbb{R}^n. \tag{3.20}$$

A simple calculation shows that the commutators $[M_i,R_j]$, $1\leq i,j\leq n$, are singular integral convolution operators, namely,

$$[M_i,R_j]\varphi(x)=k_{ij}*\varphi(x),\qquad \varphi\in\mathcal{S}(\mathbb{R}^n,\mathbb{C}),\ x\in\mathbb{R}^n, \tag{3.21}$$

where the kernels k_{ij} are given by

$$k_{ij}(x)=\frac{\Gamma((n+1)/2)}{\pi^{(n+1)/2}}\cdot\frac{x_ix_j}{|x|^{n+1}},\qquad x\in\mathbb{R}^n_0. \tag{3.22}$$

In particular, we have $[M_i,R_j]=[M_j,R_i]$, $1\leq i,j\leq n$.

We next form the commutator $n \times n$ matrix $\mathcal{C} = ([M_i, R_j])_{i,j=1}^n$, and consider the corresponding operator on the Schwartz space $\mathcal{S}(\mathbb{R}^n, \mathbb{C}^n)$ of $\mathbb{C}^n$-valued functions on $\mathbb{R}^n$. On $\mathcal{S}(\mathbb{R}^n, \mathbb{C})$ we take the inner product inherited from the standard Lebesgue space $L^2(\mathbb{R}^n, \mathbb{C})$ of square integrable complex-valued functions on $\mathbb{R}^n$, and then convert $\mathbb{C}^n \otimes \mathcal{S}(\mathbb{R}^n, \mathbb{C})$ into an inner product space in the usual way. We are now in a position to formulate the following result.

Lemma 3. *The operator $\mathcal{C} = ([M_i, R_j])_{i,j=1}^n$ is positive semidefinite, that is,*

$$\langle \mathcal{C}u, u\rangle \geq 0$$

for any $u \in \mathcal{S}(\mathbb{R}^n, \mathbb{C}^n) \equiv \mathbb{C}^n \otimes \mathcal{S}(\mathbb{R}^n, \mathbb{C})$.

A complete proof of Lemma 3 based on a Fourier transform argument can be found in Martin [Mn4]. To make a point, we just want to notice that by taking the Fourier transforms of the kernels (3.19) we get

$$(R_i\varphi)\hat{}\,(\xi) = -\sqrt{-1}\xi_i|\xi|^{-1}\hat{\varphi}(\xi), \qquad \varphi \in \mathcal{S}(\mathbb{R}^n, \mathbb{C}),\ \xi \in \mathbb{R}^n_0,$$

for each $1 \leq i \leq n$, a set of equations that immediately leads to several basic properties of the Riesz transforms. For instance, one gets that each R_i, $1 \leq i \leq n$, extends to a skew-adjoint operator on $L^2(\mathbb{R}^n, \mathbb{C})$, such that $\sum_{i=1}^n R_i^* R_i$ equals the identity on $L^2(\mathbb{R}^n, \mathbb{C})$, and $\sum_{i=1}^n \|R_i\varphi\|^2 = \|\varphi\|^2$ for any $\varphi \in L^2(\mathbb{R}^n, \mathbb{C})$. More properties of Riesz transforms are discussed in Stein's monographs [S1,2].

The Fourier transforms of the kernels k_{ij} defined by (3.22) are equal to

$$\hat{k}_{ij}(\xi) = \delta_{ij}|\xi|^{-1} - \xi_i\xi_j|\xi|^{-3}, \qquad 1 \leq i, j \leq n,\ \xi \in \mathbb{R}^n_0.$$

Lemma 3 proves essential in developing the Riesz transform model of a jointly seminormal pair (X, Y).

The two self-adjoint n-tuples involved in that model consist of operators on a direct integral Hilbert space. To introduce it, suppose $\mathfrak{H}$ is an infinite dimensional separable complex Hilbert space and

$$\{0\} = \mathfrak{H}_0 \subset \mathfrak{H}_1 \subset \mathfrak{H}_2 \subset \mathfrak{H}_3 \subset \cdots \subset \mathfrak{H}$$

is a chain of subspaces such that

(i) $\dim \mathfrak{H}_k = k$, $k \geq 0$,
(ii) $\bigcup_{k \geq 0} \mathfrak{H}_k$ spans $\mathfrak{H}$.

For every $x \in \mathbb{R}^n$, let $\mathfrak{H}(x)$ be one of the subspaces $\mathfrak{H}_k$, $k \geq 0$, such that

(iii) $\Omega_k = \{x \in \mathbb{R}^n : \mathfrak{H}(x) = \mathfrak{H}_k\}$, $k \geq 0$, are Borel sets, and
(iv) $\Omega = \bigcup_{k \geq 1} \Omega_k$ is compact.

Next we form the direct integral Hilbert space

$$\mathcal{H} = \int_{\mathbb{R}^n}^{\oplus} \mathfrak{H}(x)dx = \int_{\Omega}^{\oplus} \mathfrak{H}(x)dx$$

consisting of classes of weakly measurable functions $\varphi \in L^2(\mathbb{R}^n, \mathfrak{H})$ such that

$$\varphi(x) \in \mathfrak{H}(x) \qquad \text{a.e.}$$

Let $P : \mathbb{R}^n \to \mathcal{L}(\mathfrak{H})$ be the projection-valued map that assigns to every $x \in \mathbb{R}^n$ the orthogonal projection onto $\mathfrak{H}(x)$.

Finally, we take $n+1$ operator-valued functions $a_1, a_2, \ldots, a_n, b \in L^\infty(\mathbb{R}^n, \mathcal{L}(\mathfrak{H}))$ such that

$$\begin{aligned} a_i(x) &= P(x)a_i(x)P(x) = a_i(x)^* \quad \text{a.e.}, \qquad 1 \le i \le n, \\ b(x) &= P(x)b(x)P(x) \quad \text{a.e.} \end{aligned}$$

We will refer to the function b as the *primary parameter*, and to the functions $a_1, a_2, \ldots, a_n$ as the *secondary parameters*. Based on the previous equations, we define $n+1$ operators $A_1, A_2, \ldots, A_n, B \in \mathcal{L}(\mathcal{H})$ as

$$\begin{aligned} A_i\varphi(x) &= a_i(x)\varphi(x), \qquad \varphi \in \mathcal{H},\ 1 \le i \le n,\ x \in \Omega, \\ B\varphi(x) &= b(x)\varphi(x), \qquad \varphi \in \mathcal{H},\ x \in \Omega. \end{aligned}$$

We now introduce the self-adjoint n-tuples $X = (X_1, X_2, \ldots, X_n)$ and $Y = (Y_1, Y_2, \ldots, Y_n)$ of operators on $\mathcal{H}$ by setting $X_i = M_i \mid \mathcal{H}$, $1 \le i \le n$, with M_i given by (3.20), and $Y_i = A_i - \sqrt{-1}B^*PR_iPB$, $1 \le i \le n$, where R_i are the Riesz operators defined by (3.18). The operators M_i and R_i are regarded as acting on $\mathfrak{H}$-valued functions.

We claim that the so-defined pair (X, Y) is jointly hyponormal, that is, the corresponding $n \times n$ commutator matrix $\mathcal{C}(X, Y)$ given by (3.17) is positive semi-definite. To prove our claim we first observe that

$$[X_i, Y_j] = [X_j, Y_i] = -\sqrt{-1}B^*P[M_i, R_j]PB, \qquad 1 \le i, j \le n. \tag{3.23}$$

Substituting these equations into (3.17) we get

$$\mathcal{C}(X, Y) = 2\left(B^*P[M_i, R_j]PB\right)_{i,j=1}^n. \tag{3.24}$$

It remains to observe that whenever $u \in \mathbb{C}^n \otimes \mathcal{H} \subset \mathbb{C}^n \otimes L^2(\mathbb{R}^n, \mathfrak{H})$ we get $\langle \mathcal{C}(X, Y)u, u\rangle = 2\langle \mathcal{C}_\Omega(I \otimes B)u, (I \otimes B)u\rangle$, where $\mathcal{C}_\Omega$ is the compression of the operator $\mathcal{C}$ in Lemma 3 to $\mathbb{C}^n \otimes L^2(\Omega, \mathfrak{H})$. Since u can be approximated by restrictions to Ω of functions from $\mathbb{C}^n \otimes \mathcal{S}(\mathbb{R}^n, \mathfrak{H})$, Lemma 3 shows that $\mathcal{C}_\Omega \ge 0$, hence $\mathcal{C}(X, Y) \ge 0$.

The reader should be aware that in the case when $n = 1$ the previous model coincides with the singular integral model of pure hyponormal operators. If $n = 1$, there is just one Riesz operator, the Hilbert transform. For details and other properties of the Hilbert transform model, we refer to [Cl], [Ka], [My], [Ps], [PX], [PXX], and [X1,2].

3.3. Putnam-Type Inequalities for Riesz Transform Models

In this section, we prove a norm estimate of the remainder in the Bochner–Weitzenböck identity associated to the Riesz transform model. This estimate is a multidimensional form of Putnam's theorem. To illustrate the point, we first recall that Putnam's theorem states that if T is a seminormal operator, then

$$\|[T^*, T]\| \le \frac{1}{\pi}\text{area}(\operatorname{spec} T), \tag{3.25}$$

where $\operatorname{spec} T$ denotes the spectrum of T.

We will refer to (3.25) as the *one-dimensional form of Putnam's inequality* for a seminormal operator T. Set $X = \operatorname{Re}(T)$ and $Y = \operatorname{Im}(T)$. A well known result due also to Putnam and proved, for instance, in [Pm1], states that the projections of $\operatorname{spec} T$ onto the real and imaginary axes are equal to $\operatorname{spec} X$ and $\operatorname{spec} Y$, respectively. Therefore, (3.25) leads to the *Cartesian form of Putnam's inequality,*

$$\|\mathcal{C}(X,Y)\| \leq \frac{2}{\pi}\|Y\| \cdot \operatorname{length}(\operatorname{spec} X), \tag{3.26}$$

where we used the projection property recalled above in conjunction with $\operatorname{spec} Y \subseteq [-\|Y\|, \|Y\|]$.

The original proof of (3.25) in Putnam [Pm2] uses the Cartesian form (3.26) as a starting point and Putnam's cut-down technique, so (3.25) and (3.26) can be derived one from the other.

There is also a Cartesian form of Putnam's inequality for the Hilbert transform model. Using the notation in 3.2, that amounts to

$$\|\mathcal{C}(X,Y)\| \leq \frac{2}{\pi} \operatorname*{ess\,sup}_{x\in\Omega} \|b(x)\|^2 \cdot \operatorname{length}(\Omega), \tag{3.27}$$

where b is the primary parameter of the model. The proof of (3.27) is straightforward, and (3.27) turns out to be a refined version of (3.26), because $\Omega = \operatorname{spec} X$ and $\operatorname{ess\,sup}_{x\in\Omega} \|b(x)\|^2 = \|B\|^2 \leq \|Y\|$. For more details we refer to [Cl, Chapter 2]. However, it is important to notice that every pure seminormal operator is unitarily equivalent to a Hilbert transform model, therefore (3.27) implies (3.26) and also (3.25) for arbitrary seminormal operators.

The main result in this section generalizes (3.27) for the Riesz transform model introduced in 3.2. The notation used subsequently is the same as in 3.2.

Theorem E. *Let X and Y be the self-adjoint tuples of the Riesz transform model. Then the remainder $\mathcal{R}(X,Y)$ in identity $BW(X,Y)$ satisfies the inequality*

$$\|\mathcal{R}(X,Y)\| \leq A_n\|B\|^2(\operatorname{vol}\Omega)^{1/n}, \tag{3.28}$$

with

$$A_n = 2n \cdot \frac{\Gamma((n+1)/2)}{\pi^{(n+1)/2}}[\operatorname{vol}(\mathbb{B}^n)]^{(n-1)/n} \tag{3.29}$$

and

$$\|B\| = \operatorname*{ess\,sup}_{x\in\Omega} \|b(x)\|, \tag{3.30}$$

where b is the primary parameter.

The comments made right before Lemma 2 in Section 3.1 show that for the commutator matrix $\mathcal{C}(X,Y)$ given by (3.24) we clearly have

$$\|\mathcal{C}(X,Y)\| \leq \|\mathcal{R}(X,Y)\|. \tag{3.31}$$

In effect, by (3.28) and (3.31) we get a norm estimate for $\mathcal{C}(X,Y)$ similar to (3.27), that in fact reduces to (3.27) when $n = 1$.

Proof of Theorem E. We begin the proof by taking the $\mathfrak{A}^c_{2n}$-module $\mathfrak{S} = \mathfrak{S}_n \otimes \mathcal{H}$ and the Clifford bundle $E = \mathbb{R}^n \times \mathfrak{S}$. On E we consider the kernel Φ that assigns to each $(x, y) \in \mathbb{R}^n \times \mathbb{R}^n$, $x \neq y$, the operator $\Phi(x, y) : \mathfrak{S} \to \mathfrak{S}$ given by

$$\Phi(x,y) = \sum_{i,j=1}^{n} \varepsilon_i^* \varepsilon_j \otimes \frac{(x_i - y_i)(x_j - y_j)}{|x-y|^{n+1}}. \tag{3.32}$$

Operator (3.32) is in fact an element of the Clifford algebra $\mathfrak{A}^c_{2n}$. Specifically, by setting

$$\varepsilon(x-y) = \frac{1}{|x-y|^{(n+1)/2}} \sum_{i=1}^{n} (x_i - y_i)\varepsilon_i \in \mathfrak{A}^c_{2n} \tag{3.33}$$

we have

$$\Phi(x,y) = \varepsilon(x-y)^* \varepsilon(x-y). \tag{3.34}$$

Using (3.33) and (3.34) one can compute the operator norm of $\Phi(x, y)$. A direct calculation that is left to our reader leads to

$$\|\Phi(x,y)\| = \|\varepsilon(x-y)\|^2 = \frac{1}{|x-y|^{n-1}}, \qquad (x,y) \in \mathbb{R}^n \times \mathbb{R}^n, \ x \neq y. \tag{3.35}$$

The Hilbert space $L^2(\Omega, E)$, where $\Omega \subset \mathbb{R}^n$ is the compact set involved in the Riesz transform model, coincides with $\mathfrak{S}_n \otimes L^2(\Omega, \mathfrak{H})$, and the integral operator $\mathfrak{I}_\Omega$ on $L^2(\Omega, E)$ associated with Φ becomes under this identification an integral operator on $\mathfrak{S}_n \otimes L^2(\Omega, \mathfrak{H})$.

Eventually, we are going to apply Theorem C to operator $\mathfrak{I}_\Omega$ in the case when $m = n$, $M = \mathbb{R}^n$, and $p = 2$. Using (3.35) we obviously get that the sets $\Omega[x, t]$ defined as in (2.23) are Euclidean balls of radius $\rho = t^{-1/(n-1)}$ and, consequently, the constants $\lambda(\Phi)$ and $\lambda(\Phi^*)$ needed in Theorem C are both given by

$$\lambda(\Phi) = \lambda(\Phi^*) = \mathrm{vol}(\mathbb{B}^n). \tag{3.36}$$

With all the prerequisites at hand, we finally observe that by using (3.16) and (3.23) in conjunction with (3.21), (3.22), and (3.32) above, we can express the remainder $\mathcal{R}(X, Y) \in \mathfrak{A}^c_{2n} \otimes \mathcal{L}(\mathcal{H})$ associated with the Riesz transform model as

$$\mathcal{R}(X,Y) = \frac{2\Gamma((n+1)/2)}{\pi^{(n+1)/2}} \cdot (e_0 \otimes B)^* (e_0 \otimes P) \mathfrak{I}_\Omega (e_0 \otimes P)(e_0 \otimes B), \tag{3.37}$$

where, as in Section 3.2, P is the projection operator from $L^2(\Omega, \mathfrak{H})$ onto the direct integral Hilbert space $\mathcal{H}$, and B is the operator on $\mathcal{H}$ corresponding to the primary parameter.

To conclude the proof, suppose next that $u \in \mathfrak{S}_n \otimes \mathcal{H}$ is given and notice that by (3.37) and because $\mathcal{R}(X, Y)$ is positive semidefinite we have

$$\begin{aligned} \langle \mathcal{R}(X,Y)u, u\rangle &= \frac{2\Gamma((n+1)/2)}{\pi^{(n+1)/2}} \langle \mathfrak{I}_\Omega (e_0 \otimes B)u, (e_0 \otimes B)u\rangle \\ &\le \frac{2\Gamma((n+1)/2)}{\pi^{(n+1)/2}} \|\mathfrak{I}_\Omega (e_0 \otimes B)u\| \|(e_0 \otimes B)u\|. \end{aligned}$$

By (2.62) in Theorem C and based on (3.36) we obtain

$$\|\mathfrak{I}_\Omega(e_0 \otimes B)u\| \le n[\mathrm{vol}(\mathbb{B}^n)]^{(n-1)/n}(\mathrm{vol}\,\Omega)^{1/n}\|(e_0 \otimes B)u\|.$$

On the other hand,

$$\|(e_0 \otimes B)u\| \le \|B\|\|u\|.$$

Combining the last three estimates we get

$$\langle \mathcal{R}(X,Y)u,u\rangle \le A_n\|B\|^2(\mathrm{vol}\,\Omega)^{1/n}\|u\|^2, \tag{3.38}$$

with A_n and $\|B\|$ as in (3.29) and (3.30).

Since $u \in \mathfrak{S}_n \otimes \mathcal{H}$ is an arbitrary element and $\mathcal{R}(X,Y) \ge 0$, inequality (3.28) is equivalent to (3.38). The proof of Theorem E is complete. □

Acknowledgments

This article is based on my plenary talk at the *International Conference on Clifford Analysis and Its Applications* held in Macao, China, as a satellite conference to ICM-2002. I wish to thank Alan McIntosh, Tao Qian, and all their colleagues for organizing such an excellent conference. Special thanks are also due to Swanhild Bernstein, David Calderbank, Sirkka-Liisa Eriksson-Bique, Klaus Gürlebeck, Thomas Hempfling, Sören Kraußhar, Alan McIntosh, Tao Qian, John Ryan, Frank Sommen, and Wolfgang Sprößig for stimulating and enjoyable conversations.

References

[AB] Ahlfors, L. and Beurling, A., Conformal invariants and function theoretic null sets, *Acta Math* **83** (1950), 101–129.

[Al1] Alexander, H., Projections of polynomial hulls, *J. Funct. Anal.* **13** (1973), 13–19.

[Al2] Alexander, H., On the area of the spectrum of an element of a uniform algebra. In *Complex Approximation Proceedings, Quebec, July 3–8, 1978*, Birkhäuser, 1980, 3–12.

[Ar] Aronszajn, N., A unique continuation theorem for solutions of elliptic partial differential equations or inequalities of second order, *J. Math. Pures Appl.* **36** (1957), 235–249.

[At] Athavale, A., On joint hyponormal operators, *Proc. Amer. Math. Soc.* **103** (1988), 417–423.

[AS] Axler, S. and Shapiro, J. H., Putnam's theorem, Alexander's spectral area estimate, and VMO, *Math. Ann.* **271** (1985), 161–183.

[BGV] Berline, H., Getzler, E., and Vergne, M., *Heat Kernels and Dirac Operators*, Grundlehren der Mathematischen Wissenschaften, **298**, Springer Verlag, New York, 1992.

[Bo] Bochner, S., Curvature and Betti numbers. I; II, *Ann. of Math.* **49** (1948), 379–390; **50** (1949), 79–93.

[BoY] Bochner, S. and Yano, K. *Curvature and Betti Numbers*, Ann. of Math. Studies **32**, Princeton University Press, Princeton, 1953.

[BDS] Brackx, F., Delanghe, R., and Sommen, F., *Clifford Analysis*, Pitman Research Notes in Mathematics Series **76**, 1982.

[Ca] Calderbank, D. M. J., Dirac operators and Clifford analysis on manifolds with boundary, Odense Universitet, Preprint 53, 1997.

[Cl] Clancey, K. F., *Seminormal Operators*, Lecture Notes in Math. **742**, Springer Verlag, 1979.

[CCHZ] Cho, M., Curto, R. E., Huruya, T., and Zelazko, W., Cartesian form of Putnam's inequality for doubly commuting n-tuples, *Indiana Univ. Math. J.* **49** (2000), 1437–1448.

[Cn] Cnops, J., *An Introduction to Dirac Operators on Manifolds*, Progress in Mathematical Physiscs, Birkhäuser, Basel, 2002.

[Co] Conway, J. B., *The Theory of Subnormal Operators*, Mathematical Surveys and Monographs, **36**, American Mathematical Society, Providence, RI, 1991.

[Cu] Curto, R. E. Joint hyponormality: a bridge between hyponormality and subnormality, in *Operator Theory, Operator Algebras and Applications*, W. B. Arveson and R. G. Douglas (eds.); Part 2, *Proc. Sympos. Pure Math.* **51** (1990), 69–91.

[CJ] Curto, R. E. and Jian, R. A matricial identity involving the self-commutator of a commuting n-tuple, *Proc. Amer. Math. Soc.* **121** (1994), 461–464.

[CMX] Curto, R. E., Muhly, P. A., and Xia, J., Hyponormal pairs of commuting operators, in *Operator Theory: Advances and Applications* **35**, Birkhäuser Verlag, 1988, 1–22.

[DSS] Delanghe, R., Sommen, F., and Souček, V., *Clifford Algebra and Spinor-Valued Functions*, Kluwer Academic Publishers, 1992.

[DPY] Douglas, R. G., Paulsen, V., and Yan, K., Operator theory and algebraic geometry, *Bull. Amer. Math. Soc.* **20** (1988), 67–71.

[F] Friedrich, T., *Dirac Operators in Riemannian Geometry*, Graduate Studies in Mathematics **25**, Amer. Math. Soc., Providence, RI, 2000.

[GM] Gilbert, J. E. and Murray, M. A. M., *Clifford Algebras and Dirac Operators in Harmonic Analysis*, Cambridge Studies in Advanced Mathematics, **26**, Cambridge University Press, 1991.

[Gi] Gilkey, P. B., The spectral geometry of a Riemann manifold, *J. Diff. Geom.* **10** (1975), 601–608.

[GLP] Gilkey, P. B., Leahy, J. V., and Park, JH., *Spectral Geometry, Riemannian Submersions, and the Gromov-Lawson Conjecture*, CRC Press, Boca Raton, FL, 1999.

[G] Goldberg, R., *Curvature and Homology*, Academic Press, New York, London, 1962.

[GS1] Gürlebeck, K. and Sprössig, W., *Quaternionic Analysis and Elliptic Boundary Value Problems*, Birkhäuser, Basel, 1990.

[GS2] Gürlebeck, K. and Sprössig, W., *Quaternionic and Clifford Calculus for Physicists and Engineers*, John Wiley & Sons, New York, 1997.

[He] Hedberg, L., On certain convolution inequalities, *Proc. Amer. Math. Soc.* **36** (1972), 505–510.

[Ka] Kato, T., Smooth operators and commutators, *Studia Math.* **31** (1968), 531–546.

[LM] Lawson, H. B. and Michelson, M.-L., *Spin Geometry*, Princeton Mathematical Series, **38**, Princeton University Press, Princeton, 1989.

[L] Lieb, E. H., Sharp constants in the Hardy-Littlewood-Sobolev and related inequalities, *Annals of Math.* **118** (1983) 349–379.

[Mn1] Martin, M., Joint seminormality and Dirac operators, *Integral Equations Operator Theory* **30** (1998), 101–121.

[Mn2] Martin, M., Higher-dimensional Ahlfors-Beurling inequalities in Clifford analysis, *Proc. Amer. Math. Soc.* **126** (1998), 2863–2871.

[Mn3] Martin, M., Convolution and maximal operator inequalities in Clifford analysis, in *Clifford Algebras and Their Applications in Mathematical Physics, Vol. 2: Clifford Analysis*, Progress in Physics **9**, Birkhäuser Verlag, Basel, 2000, 95–113.

[Mn4] Martin, M., Self-commutator inequalities in higher dimension, *Proc. Amer. Math. Soc.* **130** (2002), 2971–2983.

[Mn5] Martin, M., Uniform approximation by closed forms in several complex variables, preprint, 2002.

[MnPr] Martin M. and Putinar, M., *Lectures on Hyponormal Operators*, Operator Theory: Advances and Applications **39**, Birkhäuser Verlag, Basel, 1989.

[MnSa] Martin, M. and Salinas, N., Weitzenböck type formulas and joint seminormality, *Contemporary Math.*, Amer. Math. Soc. **212** (1998), 157–167.

[MnSz] Martin, M. and Szeptycki, P., Sharp inequalities for convolution operators with homogeneous kernels and applications, *Indiana Univ. Math. J.* **46** (1997), 975–988.

[McCP] McCullough, S. and Paulsen, V., A note on joint hyponormality, *Proc. Amer. Math. Soc.* **107** (1989), 187–195.

[Mc1] McIntosh, A., Operators which have a H_∞ functional calculus, in *Miniconference on Operator Theory and Partial Differential Equations, 1986*, Proceedings of the Centre for Mathematical Analysis, ANU, Canberra **14** (1986), 210–231.

[Mc2] McIntosh, A., Clifford algebras, Fourier theory, singular integrals, and harmonic functions on Lipschitz domains, in *Clifford Algebras in Analysis and Related Topics*, J. Ryan (Ed.), CRC Press, Boca Raton, FL, 1995.

[McP] McIntosh, A. and Pryde, A., A functional calculus for several commuting operators, *Indiana Univ. Math. J.* **36** (1987), 421–439.

[McPR] McIntosh, A., Pryde, A., and Ricker, W., Comparison of joint spectra for certain classes of commuting operators, *Studia Math.* **88** (1988), 23–36.

[Mi] Mitrea, M., *Singular Integrals, Hardy Spaces, and Clifford Wavelets*, Lecture Notes in Mathematics, **1575**, Springer-Verlag, Heidelberg, 1994.

[My] Muhly, P. S., A note on commutators and singular integrals, *Proc. Amer. Math. Soc.* **54** (1976), 117–121.

[Ps] Pincus, J. D., Commutators and systems of singular integral equations, *Acta Math.* **121** (1968), 219–249.

[PX] Pincus, J. D. and Xia, D., The analytic model of a hyponormal operator with rank one self-commutator, *Integral Equations Operator Theory* **4** (1981), 134–150.

[PXX] Pincus, J. D., Xia, D., and Xia, J., The analytic model of a hyponormal operator with rank one self-commutator, *Integral Equations Operator Theory* **7** (1984), 516–535.

[P] Putinar, M., Extreme hyponormal operators, *Operator Theory: Advances and Applications* **28** (1988), 249–265.

[Pm1] Putnam, C. R., *Commutation Properties of Hilbert Space Operators and Related Topics*, Springer-Verlag, Berlin, Heidelberg, New York, 1967.

[Pm2] Putnam, C. R., An inequality for the area of hyponormal spectra, *Math. Z.* **116** (1970), 323–330.

[Ro] Roe, J., *Elliptic Operators, Topology, and Asymptotic Methods*, 2nd ed., Pitman Research Notes, **395**, Longman, 1998.

[Ry1] Ryan, J. (Ed.), *Clifford Algebras in Analysis and Related Topics*, CRC Press, Boca Raton, FL, 1995.

[Ry2] Ryan, J., Dirac operators, conformal transformations and aspects of classical harmonic analysis, *Journal of Lie Theory* **8** (1998), 67–82.

[RyS] Ryan, J. and Sprößig, W. (Eds.), *Clifford Algebras and Their Applications in Mathematical Physics, Volume 2: Clifford Analysis*, Progress in Physics **19**, Birkhäuser, Basel, 2000.

[S1] Stein, E. M., *Singular Integrals and Differentiability Properties of Functions*, Princeton Univ. Press, Princeton, NJ,, 1970.

[S2] Stein, E. M., *Harmonic Analysis: Real-Variable Methods, Orthogonality, and Oscillatory Integrals*, Princeton Univ. Press, Princeton, NJ, 1993.

[T] Tarkhanov, N. N., *The Cauchy Problem for Solutions of Elliptic Equations*, Akademie Verlag, Berlin, 1995.

[X1] D. Xia, On non-normal operators. I, *Chinese J. Math.* **3**(1963), 232–246; II, *Acta Math. Sinica* **21** (1987), 103–108.

[X2] D. Xia, *Spectral Theory of Hyponormal Operators*, Birkhäuser Verlag, Basel-Boston-Stuttgart, 1983.

[X3] D. Xia, *On some classes of hyponormal tuples of commuting operators*, in Operator Theory: Advances and Applications **48**, Birkhäuser Verlag, 1990, 423–448.

Mircea Martin
Department of Mathematics
Baker University
Baldwin City, 66006 Kansas
USA
e-mail: `mircea.martin@bakeru.edu`

Trends in Mathematics: *Advances in Analysis and Geometry*, 257–286

A Mean Value Laplacian for Strongly Kähler–Finsler Manifolds

Zhong Chunping and Zhong Tongde

Abstract. It is well known that the Laplace operator plays an important role in the theory of harmonic integrals and the Bochner technique both in Riemannian and Kähler manifolds. In recent years, under the initiation of S.S. Chern, the global differential geometry of real and complex Finsler manifolds has gained a great development ([1], [2], [3], [4]). A lot of results about the Laplacian and its applications have been obtained in a real Finsler manifold ([5], [6]). But up to now there are no results for the Laplacian and its applications in a complex Finsler manifold. The key point in the theory of the Bochner technique and harmonic integrals is to define a suitable Laplace operator. In the case of Finsler manifolds the difficulty is that the Finsler metric depends on the fibre coordinates. Using the idea that the Laplacian on Euclidean space or on a Riemannian manifold measures the average value of a function around a point, P. Centore ([7]) generalizes the Laplacian on a Riemannian manifold to a real Finsler manifold. Considering a complex manifold as a real manifold, there is a one-to-one correspondence between the real coordinates and the complex coordinates ([8]). In this paper, we use the mean value idea to define the Laplacian on a strongly Kähler–Finsler manifold, first for functions and then for forms, and we derive some remarkable properties for the Laplacian for functions and extend the Laplacian to arbitrary forms. Indeed our Laplacian on strongly Kähler–Finsler manifolds generalizes the Kählerian Laplacian. And it is worth to remark that using the osculating Kähler metric – which we obtain in the following to define the pointwise and global inner product when we define the Hodge–Laplace operator of (p, q)-forms – is more natural than using the fundamental tensor of the Finsler metric and can avoid many complicated calculations.

Mathematics Subject Classification (2000). 32C10; 53B40; 53C55.

Keywords. Strongly Kähler-Finsler manifold, mean value Laplacian, Hodge–Laplace operator.

This project is supported by the Natural Science Foundation of China (No.10271097).

1. Background and fundamental results

Consider a real $2n$-dimensional manifold V_{2n} of class C^r covered by a system of coordinate neighborhoods.

If we denote the coordinates of a point p in two different coordinate neighborhoods by (x^i) and $(\tilde{x}^i)$, respectively, then there exists relations of the form

$$\tilde{x}^i = \tilde{x}^i(x^k), \tag{1.1}$$

where the function $\tilde{x}^i(x^k)$ are of class C^r with non-vanishing Jacobian, and the Latin indices run over the range $1, 2, \cdots, n, \bar{1}, \bar{2}, \cdots, \bar{n}$.

Now,if we put

$$z^\alpha = x^\alpha + ix^{\bar{\alpha}}, \quad \bar{z}^\alpha = x^\alpha - ix^{\bar{\alpha}}, \tag{1.2}$$

where the Greek indices run over the range $1, 2, \cdots, n$, then we have a one-to-one correspondence

$$(z^\alpha, \bar{z}^\alpha) \rightleftharpoons (x^i),$$

and $(z^\alpha, \bar{z}^\alpha)$ may be considered to be coordinates of a point in our $2n$-dimensional manifold V_{2n}, and equation (1.1) may be always written as relations

$$\tilde{z}^\alpha = \tilde{z}^\alpha(z, \bar{z}), \quad \bar{\tilde{z}}^\alpha = \bar{\tilde{z}}^\alpha(z, \bar{z}) \tag{1.3}$$

formally.

Now if we can cover the whole manifold V_{2n} entirely by a system of complex coordinates $(z^\alpha, \bar{z}^\alpha)$ in such a way that, in the intersection of two complex coordinates neighborhoods, $(z^\alpha, \bar{z}^\alpha)$ and $(\tilde{z}^\alpha, \bar{\tilde{z}}^\alpha)$, $\tilde{z}^\alpha$ are complex analytic functions with non-vanishing Jacobian of the complex coordinates z^α, then we say that the manifold V_{2n} admits a complex analytic structure, and we call the manifold a complex analytic (or simply analytic) manifold of real dimension $2n$ and of complex dimension n.

In this case, equation (1.3) takes the form

$$\tilde{z}^\alpha = \phi^\alpha(z), \quad \bar{\tilde{z}}^\alpha = \bar{\phi}^\alpha(\bar{z}), \tag{1.4}$$

where $\bar{\phi}^\alpha$ denotes the complex conjugate of the function $\phi^\alpha(z)$. Also, if we put

$$\bar{z}^\alpha = z^{\bar{\alpha}}$$

and assume that the barred Greek indices take the values $\bar{1}, \bar{2}, \cdots, \bar{n}$, then for $(z^\alpha, \bar{z}^\alpha)$ we can write z^i, $i = 1, 2, \cdots, n, \bar{1}, \bar{2}, \cdots, \bar{n}$, and write the transformation (1.4) as

$$\tilde{z}^i = f^i(z^k). \tag{1.5}$$

The Jacobian of (1.4) is then, as is easily seen,

$$\det\left(\frac{\partial \tilde{z}^i}{\partial z^j}\right) > 0,$$

and thus the manifold is always orientable.

In a complex analytic manifold M of complex dimension n vectors, tensors, affine connections, etc., are defined with respect to the coordinate transformations (1.5) having the special form (1.4), in just the same way as in real case ([8]).

For example, for a contravariant vector ξ^i the law of transformation of its components,

$$\tilde{\xi}^i = \frac{\partial \tilde{z}^i}{\partial z^r}\xi^r,$$

separates to

$$\tilde{\xi}^\alpha = \frac{\partial \tilde{z}^\alpha}{\partial z^\lambda}\xi^\lambda \text{ and } \tilde{\xi}^{\bar{\alpha}} = \frac{\partial \bar{\tilde{z}}^\alpha}{\partial \bar{z}^\lambda}\xi^{\bar{\lambda}}. \tag{1.6}$$

by virtue of the special form (1.4) of the transformation (1.5). Thus, the $2n$ components of ξ separate completely to the blocks ξ^α and $\xi^{\bar{\alpha}}$.

Equation (1.6) shows that if ξ^i are components of a contravariant vector, then $(\xi^\alpha, 0)$ and $(0, \xi^{\bar{\alpha}})$ are also components of contravariant vectors. Moreover, taking the complex conjugate of equation (1.6), we get

$$\overline{\tilde{\xi}^{\bar{\alpha}}} = \frac{\partial \tilde{z}^\alpha}{\partial z^\lambda}\overline{\xi^{\bar{\lambda}}} \text{ and } \overline{\tilde{\xi}^{\alpha}} = \frac{\partial \bar{\tilde{z}}^\alpha}{\partial \bar{z}^\lambda}\overline{\xi^{\lambda}},$$

which shows that, if ξ^i are components of a contravariant vector, then $(\overline{\xi^{\bar{\alpha}}}, \overline{\xi^\alpha})$ are also components of a contravariant vector.

For tensors,the same arguments apply.

The volume form of $\mathbb{C}^n = \mathbb{R}^{2n}$ is given by $dV = dx^1 \wedge dy^1 \wedge \cdots \wedge dx^n \wedge dy^n$, where, as usual, $z^\alpha = x^\alpha + iy^\alpha, \alpha = 1, 2, \cdots, n$. dV can also be expressed as

$$dV = \frac{(-1)^{n(n-1)/2}}{(2i)^n} d\bar{z}^1 \wedge \cdots \wedge d\bar{z}^n \wedge dz^1 \wedge \cdots \wedge dz^n,$$

or

$$dV = C_n d\bar{z} \wedge dz, \tag{1.7}$$

where $C_n = \frac{(-1)^{n(n-1)/2}}{(2i)^n}, dz = dz^1 \wedge \cdots \wedge dz^n$.

Let M be a compact complex manifold of dimension n with local coordinates z^α around a point z. If $v \in T_z^{1,0}M$, then $v = v^\alpha \frac{\partial}{\partial z^\alpha}$, where the v^α are coordinates for the tangent bundle canonically induced from the z^α on the base manifold. F is a complex Finsler metric, i.e., a continuous function $F : T^{1,0}M \longrightarrow \mathbb{R}^+$ satisfying ([4])

(i) $G = F^2$ is smooth on $\tilde{M}, \tilde{M} = T^{1,0}M/o(M)$;
(ii) $F(v) > 0$ for all $v \in \tilde{M}$;
(iii) $F(\lambda v) = |\lambda| F(v)$ for all $v \in T^{1,0}M$ and $\lambda \in \mathbb{C}$.

Condition (iii) is essential in the following, because it guarantees that the unit ball

$$I = \{v \in T_z^{1,0}M \mid F(v) \leq 1\}$$

is centered at 0, i.e.,

(iii)′ F is symmetric:$F(z; -v) = F(z; v)$.

In addition to the unit ball, one can consider a ball, in the tangent space $T_z^{1,0}M$ of any radius $\varepsilon > 0$. The indicatrix I_ε of radius ε will be defined to be this (solid) ball:

$$I_\varepsilon = \{v \in T_z^{1,0}M \mid F(v) \leq \varepsilon\}.$$

Apart from balls I_ε in the tangent space, we will see that we can talk about balls $\sum_\varepsilon$ on the manifold, and we must be careful to distinguish the two.

Now we denote

$$G_{\alpha\bar\beta}(z;v)=\frac{\partial^2 G}{\partial v^\alpha \partial \bar v^\beta},\quad G^{\bar\delta\beta}G_{r\bar\delta}=\delta_r^\beta,\quad \mathcal{G}=\det{}^2(G_{\alpha\bar\beta}). \tag{1.8}$$

By the property (iii) of F, we have

$$G(z;v)=G_{\alpha\bar\beta}(z;v)v^\alpha\bar v^\beta. \tag{1.9}$$

In the following, we assume that M is a strongly Kähler–Finsler manifold ([4]), that is,

$$\Gamma^\alpha_{\mu;\nu}=\Gamma^\alpha_{\nu;\mu}, \tag{1.10}$$

where

$$\Gamma^\alpha_{\beta;\mu}=G^{\bar\tau\alpha}(G_{\beta\bar\tau;\mu}-G_{\beta\bar\tau\gamma}\Gamma^\gamma_\mu),\quad \Gamma^\alpha_\mu=G^{\bar\tau\alpha}G_{\bar\tau;\mu}.$$

Let $f:M\longrightarrow\mathbb{R}$ be a smooth function with derivatives $f_\alpha=\frac{\partial f}{\partial z^\alpha}$, $f_{\alpha\bar\beta}=\frac{\partial^2 f}{\partial z^\alpha\partial\bar z^\beta}$, etc. We are looking for a Laplacian $\triangle f:M\longrightarrow\mathbb{R}$.

Now let us remember the mean value property of a harmonic function f in the complex plane $\mathbb{C}^1$. It is well known that the Laplace operator $\triangle$ in $\mathbb{C}^1$ is defined by

$$\triangle=\frac{\partial^2}{\partial x^2}+\frac{\partial^2}{\partial y^2}=4\frac{\partial^2}{\partial z\partial\bar z},$$

where $z=x+iy$. A smooth function f in a domain $D\subset\mathbb{C}^1$ is said to be harmonic, if it satisfies

$$\triangle f=0,$$

and then f has the important mean value property, that is,

$$f(a)=\frac{1}{2\pi}\int_0^{2\pi}f(a+re^{i\theta})d\theta,$$

holds for $\{z:|z-a|\le r\}\subset D$.

In general, in the Euclidean space $\mathbb{R}^n$ we say that a function f is harmonic, or has zero Laplacian, if

$$f(x)=\text{average value of } f \text{ on a solid sphere (of any radius) around } x=\frac{\int_{\sum_\varepsilon}fdV}{\int_{\sum_\varepsilon}dV},$$

where

$$\textstyle\sum_\varepsilon:=\{y\in\mathbb{R}^n\mid d(x,y)\le\varepsilon\},\quad dV:=dx^1\wedge\cdots\wedge dx^n.$$

Thus

$$\frac{\int_{\sum_\varepsilon}fdV}{\int_{\sum_\varepsilon}dV}-f(x)=0$$

for harmonic functions in Euclidean space. To make sense of the expression

$$\frac{\int_{\sum_\varepsilon}fdV}{\int_{\sum_\varepsilon}dV}$$

for a general function, we must "infinitesimalize," i.e., consider the above quantity as $\varepsilon \to 0$. Furthermore, we know from the more usual expression $\triangle f = \sum_{i=1}^{n} f_{ii}$ that we are only interested in the 2nd-order term of this expression, so we should divide by ε^2 as $\varepsilon \to 0$. In addition, we need a normalizing factor of $2(n+2)$, and we obtain

$$\triangle f(x) = 2(n+2) \lim_{\varepsilon \to 0} \frac{1}{\varepsilon^2} \left(\frac{\int_{\sum_\varepsilon} f dV}{\int_{\sum_\varepsilon} dV} - f(x) \right).$$

This expression can be easily extended to a Riemannian manifold. Since for a Riemannian manifold with positive definite metric

$$ds^2 = g_{ij} dx^i dx^j, i, j = 1, 2, \cdots, n,$$

one can choose the volume form to be

$$dV = \sqrt{g} dx, g = \det(g_{ij}),$$

we can define

$$\triangle f(x) = 2(n+2) \lim_{\varepsilon \to 0} \frac{1}{\varepsilon^2} \left(\frac{\int_{\sum_\varepsilon} f \sqrt{g} dV}{\int_{\sum_\varepsilon} \sqrt{g} dV} - f(x) \right).$$

The main purpose of this paper is to generalize this formula to strongly Kähler–Finsler manifolds. Let M be a strongly Kähler–Finsler manifold. Then we have a distance function between pairs of points, and this function satisfies certain properties. For $z \in M$, let $U \subset M$ be a coordinate chart with $z \in U$. Then define a function $\rho_z : U \longrightarrow \mathbb{R}$, where $\rho_z(w)$ is the Finsler distance from w to z, ρ_z satisfying

(i) $\rho_z(z) = 0$;
(ii) $\rho_z > 0$ on $U \backslash \{z\}$;
(iii) ρ_z is continuous on U;
(iv) ρ_z is differentiable on $U \backslash \{z\}$.

Define $\sum_\varepsilon = \{v \in M \mid \rho_z(v) \leq \varepsilon\}$, the ball of radius ε around z. These balls are defined intrinsically on the complex Finsler manifold and will be essential for our Laplacian.

Now we can use F to define a nowhere-zero volume form. Note that as a metric space, (M, F) is equipped with a natural Hausdorff measure. Choose the volume form which generates the $2n$-dimensional measure, i.e., take

$$dV = \left(\frac{\kappa_{2n}}{\int_I d\bar{v} \wedge dv} \right) d\bar{z} \wedge dz, \tag{1.11}$$

where

$$\kappa_{2n} = \text{volume of the unit sphere in } \mathbb{R}^{2n}.$$

$$I = \{v \in T_p^{1,0} M \mid F(v) \leq 1\} = \text{the solid unit ball in the tangent space.}$$

$$\begin{aligned} C_n d\bar{v} \wedge dv &= C_n d\bar{v}^1 \wedge \cdots \wedge d\bar{v}^n \wedge dv^1 \wedge \cdots \wedge dv^n, \quad C_n = (-1)^{n(n-1)/2}/(2i)^n \\ &= \text{Euclidean volume form in tangent space with coordinates } v^\alpha. \end{aligned}$$

This generalizes the Riemannian volume form and is written in coordinates as

$$dV(p) := \kappa_{2n}\Big(\int_{\{v\in T_p^{1,0}M \mid F(v)\le 1\}} d\bar v^1\wedge\cdots\wedge d\bar v^n\wedge dv^1\wedge\cdots dv^n\Big)^{-1} \times d\bar z^1\wedge\cdots\wedge d\bar z^n\wedge dz^1\wedge\cdots\wedge dz^n.$$

We have

$$dV(\bar v_1,\cdots,\bar v_n,v_1,\cdots,v_n) = \kappa_{2n}\det(\bar v_\alpha^\beta)\det(v_\alpha^\beta)\Big(\int_{\{v\in T_p^{1,0}M \mid F(v)\le 1\}} d\bar v^1\wedge\cdots\wedge d\bar v^n\wedge dv^1\wedge\cdots dv^n\Big)^{-1},$$

where

$$v_\alpha = v_\alpha^\beta\frac{\partial}{\partial z^\beta},\quad \bar v_\alpha = \bar v_\alpha^\beta\frac{\partial}{\partial \bar z^\beta}.$$

Proposition 1.1. *dV is a well-defined (n,n)-form.*

Proof. dV is an (n,n)-form if, when we change coordinates to $\tilde z^\alpha = \tilde z^\alpha(z)$, generating a new expression

$$d\tilde V = \kappa_{2n}\Big(\int_{\{\tilde v\in T_p^{1,0}M \mid F(\tilde v)\le 1\}} d\bar{\tilde v}^1\wedge\cdots\wedge d\bar{\tilde v}^n\wedge d\tilde v^1\wedge\cdots d\tilde v^n\Big)^{-1} \times d\bar{\tilde z}^1\wedge\cdots\wedge d\bar{\tilde z}^n\wedge d\tilde z^1\wedge\cdots\wedge d\tilde z^n,$$

and a new coordinate expression $\tilde v_\alpha$ for v_α, then

$$dV(\bar v_1,\cdots,\bar v_n,v_1,\cdots,v_n) = d\tilde V(\bar{\tilde v}_1,\cdots,\bar{\tilde v}_n,\tilde v_1,\cdots,\tilde v_n).$$

i) Write

$$\tilde v_\alpha^\gamma\frac{\partial}{\partial \tilde z^\gamma} = v_\alpha^\gamma\frac{\partial}{\partial z^\gamma}.$$

Then

$$\tilde v_\alpha^\gamma\frac{\partial z^\beta}{\partial \tilde z^\gamma}\frac{\partial}{\partial z^\beta} = v_\alpha^\beta\frac{\partial}{\partial z^\beta},$$

so

$$\tilde v_\alpha^\gamma\frac{\partial z^\beta}{\partial \tilde z^\gamma} = v_\alpha^\beta,$$

$$\det(\tilde v_\alpha^\gamma)\det\Big(\frac{\partial z^\beta}{\partial \tilde z^\gamma}\Big) = \det(v_\alpha^\beta). \tag{1.12}$$

ii)Calculate

$$
\begin{aligned}
&dV(\bar{v}_1,\cdots,\bar{v}_n,v_1,\cdots,v_n)\\
&=\kappa_{2n}\det(\bar{v}^\beta_\alpha)\det(v^\beta_\alpha)\Big(\int_{\{v\in T^{1,0}_pM \,|\, F(v)\le 1\}} d\bar{v}^1\wedge\cdots\wedge d\bar{v}^n\wedge dv^1\wedge\cdots dv^n\Big)^{-1}\\
&=\kappa_{2n}\det(\bar{\tilde{v}}^\gamma_\alpha)\det\Big(\frac{\partial\bar{z}^\beta}{\partial\bar{\tilde{z}}^\gamma}\Big)\det(\tilde{v}^\gamma_\alpha)\det\Big(\frac{\partial z^\beta}{\partial\tilde{z}^\gamma}\Big)\\
&\quad\times\Big(\int_{\{\tilde{v}\in T^{1,0}_pM \,|\, F(\tilde{v})\le 1\}}\det\Big(\frac{\partial\bar{z}^\sigma}{\partial\bar{\tilde{z}}^\tau}\Big)\det\Big(\frac{\partial z^\beta}{\partial\tilde{z}^\tau}\Big)d\bar{\tilde{v}}^1\wedge\cdots\wedge d\bar{\tilde{v}}^n\wedge d\tilde{v}^1\wedge\cdots d\tilde{v}^n\Big)^{-1}\\
&\quad\text{(using (1.12))}\\
&=\kappa_{2n}\det(\bar{\tilde{v}}^\gamma_\alpha)\det(\tilde{v}^\gamma_\alpha)\Big(\int_{\{\tilde{v}\in T^{1,0}_pM \,|\, F(\tilde{v})\le 1\}} d\bar{\tilde{v}}^1\wedge\cdots\wedge d\bar{\tilde{v}}^n\wedge d\tilde{v}^1\wedge\cdots d\tilde{v}^n\Big)^{-1}\\
&=d\tilde{V}(\bar{\tilde{v}}_1,\cdots,\bar{\tilde{v}}_n,\tilde{v}_1,\cdots,\tilde{v}_n).
\end{aligned}
$$

□

Now we denote the volume form by

$$dV = C_n\omega(z)d\bar{z}\wedge dz,\ \ \omega(z)=\kappa_{2n}\Big(C_n\int_I d\bar{v}\wedge dv\Big)^{-1}. \tag{1.13}$$

With this volume form, we can define $\triangle f$ on a strongly Kähler-Finsler manifold:

$$\triangle f(z) = 2(2n+2)\lim_{\varepsilon\to 0}\frac{1}{\varepsilon^2}\left(\frac{\int_{\sum_\varepsilon} f\omega(z)d\bar{z}\wedge dz}{\int_{\sum_\varepsilon}\omega(z)d\bar{z}\wedge dz} - f(z)\right). \tag{1.14}$$

Before working out the coordinate expression for $\triangle f(z)$ we give some estimations.

2. Some estimates

A complex Finsler manifold has an intrinsically defined exponential map, which sends one-dimensional subspaces of a tangent space isometrically onto geodesics. We shall express the exponential map *exp* in local coordinates, and compute some expressions to be used in later calculations. A geodesic in a complex Finsler manifold is a parametrized path $z(t) = (z^1(t), z^2(t),\cdots,z^n(t))$, which satisfies the differential equation

$$\ddot{z}^\alpha = -\Gamma^\alpha_{\beta;\mu}(z,\dot{z})\dot{z}^\beta\dot{z}^\mu,$$

where a dot above the z indicates differentiation with respect to t. This equation allows us to write *exp* in local coordinates around a point $p\in M$ as

$$exp : T^{1,0}_pM \longrightarrow M,$$

$$z^\alpha(exp(v)) = v^\alpha - \frac{1}{2}\Gamma^\alpha_{\beta;\mu}(v)v^\beta v^\mu - \frac{1}{6}\Gamma^\alpha_{\beta;\mu\gamma}v^\beta v^\mu v^\gamma + \frac{1}{3}\Gamma^\alpha_{\beta;\mu}\Gamma^\mu_{\delta;\lambda}v^\beta v^\delta v^\lambda + o(|v|^4), \tag{2.1}$$

where we are writing $v\in T^{1,0}_pM$ as $v = v^\alpha\frac{\partial}{\partial z^\alpha}$, and $\Gamma^\alpha_{\beta;\mu\gamma} = \frac{\partial}{\partial z^\gamma}\Gamma^\alpha_{\beta;\mu}$.

In a Riemannian space, *exp* is C^∞ everywhere on T_pM; in a complex Finsler space, *exp* is C^∞ everywhere except at the origin $(p,0)$, at which it is only C^1([4]).

This deficiency, however, will not cause us any trouble. We will use the exponential map to move integrals from the manifold to the tangent space, where the structure of F is exhibited more clearly. In particular, exp^{-1} takes the sphere ε on M to the indicatrix of "radius" ε on $T_p^{1,0}M$, and on $T_p^{1,0}M$ indicatrices of different radius have a natural scaling property.

Indeed, $T_p^{1,0}M$ can be thought of as a manifold on its own, and exp as a map between the manifolds $T_p^{1,0}M$ and M. Since exp is invertible, we can pull back forms on M to forms on $T_p^{1,0}M$, and we will calculate pullbacks (in local coordinates) for some special cases. To begin with, let v^α be the coordinate functions on $T_p^{1,0}M$ canonically induced by the coordinate functions z^α on M. Then forms on the manifold will be written as wedge products of 1-forms dv^α. Take the exterior derivative of both sides of (2.1). (This is allowed because exp is C^1 at p, although higher derivatives do not generally exist.)

$$\begin{aligned}
dz^\alpha=&\frac{\partial}{\partial v^\tau}\Big[v^\alpha-\frac{1}{2}\Gamma^\alpha_{\beta;\mu}(v)v^\beta v^\mu-\frac{1}{6}\Gamma^\alpha_{\beta;\mu\gamma}v^\beta v^\mu v^\gamma+\frac{1}{3}\Gamma^\alpha_{\beta;\mu}\Gamma^\mu_{\delta;\lambda}v^\beta v^\delta v^\lambda+o(|v|^4)\Big]dv^\tau\\
=&dv^\alpha-\frac{1}{2}\Gamma^\alpha_{\beta;\mu}\delta^\beta_\tau v^\mu dv^\tau-\frac{1}{2}\Gamma^\alpha_{\beta;\mu}v^\beta\delta^\mu_\tau dv^\tau-\frac{1}{2}\frac{\partial\Gamma^\alpha_{\beta;\mu}}{\partial v^\tau}v^\beta v^\mu dv^\tau-\frac{1}{3}\Gamma^\alpha_{\tau;\mu\gamma}v^\mu v^\gamma dv^\tau\\
&-\frac{1}{6}\Gamma^\alpha_{\beta;\mu\tau}v^\beta v^\mu dv^\tau+\frac{1}{3}\Gamma^\alpha_{\tau;\mu}\Gamma^\mu_{\delta;\lambda}v^\delta v^\lambda dv^\tau+\frac{2}{3}\Gamma^\alpha_{\beta;\mu}\Gamma^\mu_{\tau;\lambda}v^\beta v^\lambda dv^\tau+o(|v|^3)\\
=&dv^\alpha-\Gamma^\alpha_{\tau;\mu}v^\mu dv^\tau+C^\alpha_{\tau\beta\mu}v^\beta v^\mu dv^\tau+o(|v|^3),
\end{aligned}$$

where we write

$$C^\alpha_{\tau\beta\mu}=-\frac{1}{2}\frac{\partial\Gamma^\alpha_{\beta;\mu}}{\partial v^\tau}-\frac{1}{3}\Gamma^\alpha_{\tau;\mu\beta}-\frac{1}{6}\Gamma^\alpha_{\beta;\mu\tau}+\frac{1}{3}\Gamma^\alpha_{\tau;\ell}\Gamma^\ell_{\beta;\mu}+\frac{2}{3}\Gamma^\alpha_{\beta;\ell}\Gamma^\ell_{\tau;\mu}. \tag{2.2}$$

To work with (n,n) forms, we also need to calculate $dz=dz^1\wedge\cdots\wedge dz^n$ in terms of $dv=dv^1\wedge\cdots\wedge dv^n$.

$$\begin{aligned}
dz=&(dv^1-\Gamma^1_{\tau;\mu}v^\mu dv^\tau+C^1_{\tau\beta\mu}v^\beta v^\mu dv^\tau)\wedge(dv^2-\Gamma^2_{\tau;\mu}v^\mu dv^\tau+C^2_{\tau\beta\mu}v^\beta v^\mu dv^\tau)\\
&\wedge\cdots\wedge(dv^n-\Gamma^n_{\tau;\mu}v^\mu dv^\tau+C^n_{\tau\beta\mu}v^\beta v^\mu dv^\tau)\\
=&\Big[1-\Gamma^\beta_{\beta;\mu}v^\mu+\Big(\Gamma^\rho_{\rho;\mu}\Gamma^\sigma_{\sigma;\beta}-\Gamma^\rho_{\sigma;\mu}\Gamma^\sigma_{\rho;\beta}+C^\theta_{\theta\beta\mu}\Big)v^\beta v^\mu\Big]dv\\
=&\Big[1-\Gamma^\beta_{\beta;\mu}v^\mu+\Big(\Gamma^\rho_{\rho;\mu}\Gamma^\sigma_{\sigma;\beta}-\frac{1}{3}\Gamma^\rho_{\sigma;\mu}\Gamma^\sigma_{\rho;\beta}-\frac{1}{3}\Gamma^\alpha_{\alpha;\mu\beta}-\frac{1}{6}\Gamma^\alpha_{\beta;\mu\alpha}\\
&+\frac{1}{3}\Gamma^\alpha_{\alpha;\gamma}\Gamma^\gamma_{\beta;\mu}-\frac{1}{2}\frac{\partial\Gamma^\alpha_{\beta;\mu}}{\partial v^\alpha}\Big)v^\beta v^\mu\Big]dv,
\end{aligned}$$

where we are neglecting terms of higher order than 2. Because exp is only C^1, the above expression is only continuous and not necessarily differentiable at p.

The exponential map allows us to change coordinates so that the spheres $\sum_\varepsilon$ on M become indicatrices of radius ε on $T_p^{1,0}M$. Thus any integral $\int_{\sum_\varepsilon}\omega$ (of any (n,n) form ω) becomes an integral $\int_{I_\varepsilon}exp^*\omega$ on the tangent space, and, if $exp^*\omega(v)$ (where $exp^*\omega=exp^*\omega(v)d\bar{v}\wedge dv$) is homogenous we can use the scaling property of the I_ε to write $\int_{I_\varepsilon}exp^*\omega$ as $\varepsilon^k\int_I exp^*\omega$ for some appropriate exponent k. In particular, when we come to integrate $\int_{\sum_\varepsilon} f\omega$ for a function f, we will replace f by its power series around p to get

$$\begin{aligned}\int_{\sum_\varepsilon} f\omega =& f(p)\int_{\sum_\varepsilon}\omega + f_\alpha(p)\int_{\sum_\varepsilon} z^\alpha\omega + f_{\bar\alpha}(p)\int_{\sum_\varepsilon}\bar z^\alpha\omega\\ &+\frac{1}{2}f_{\alpha\bar\beta}(p)\int_{\sum_\varepsilon} z^\alpha\bar z^\beta\omega + \int_{\sum_\varepsilon} o(|z|^3)\omega,\end{aligned}\tag{2.3}$$

where$|z| = \sqrt{\sum |z^\alpha|^2}$. After moving all these new integrals to the tangent space, we will see that we do in fact have homogenous functions, so we can write them all as integrals over I, which is the same set as I_1. We now determine the exponent k given above for integrals of the form $\int_{\sum_\varepsilon} z^{|\alpha|}dV$, where α is a multi-exponent for the z^α's.

Case A. $\int_{\sum_\varepsilon} dV$. This amounts to finding the volume of the sphere of radius ε in the complex Finsler manifold. We develop a power series in terms of ε for this volume.

$$\mathrm{Vol}\Big(\sum_\varepsilon\Big) = \int_{\sum_\varepsilon} dV = C_n\int_{\sum_\varepsilon}\omega(z)d\bar z\wedge dz,$$

where $\omega(z) = \kappa_{2n}\Big(C_n\int_I d\bar v\wedge dv\Big)^{-1}$ are the coefficients of the volume form as given above. Replace $\omega(z)$ by its Taylor series around p:

$$\begin{aligned}
=&C_n\int_{\sum_\varepsilon}\Big[\omega(p)+\omega_\alpha(p)z^\alpha+\omega_{\bar\alpha}(p)\bar z^\alpha+\frac{1}{2}\omega_{\alpha\bar\beta}(p)z^\alpha\bar z^\beta + o(|z|^3)\Big]d\bar z\wedge dz\\
=&C_n\int_{I_\varepsilon}\Big[\omega(p)+\omega_\alpha(p)\Big(v^\alpha-\frac{1}{2}\Gamma^\alpha_{\beta;\mu}v^\beta v^\mu\Big)+\omega_{\bar\alpha}(p)\Big(v^{\bar\alpha}-\frac{1}{2}\Gamma^{\bar\alpha}_{\bar\beta;\bar\mu}v^{\bar\beta}v^{\bar\mu}\Big)\\
&+\frac{1}{2}\omega_{\alpha\bar\beta}(p)v^\alpha\bar v^\beta\Big]\Big(1-\Gamma^\beta_{\beta;\mu}+C_{\beta;\mu}v^\beta v^\mu\Big)\Big(1-\Gamma^{\bar\beta}_{\bar\beta;\bar\mu}+\overline{C_{\beta;\mu}}\bar v^\beta\bar v^\mu\Big)d\bar v\wedge dv\\
=&C_n\int_{I_\varepsilon}\Big[\omega(p)+\Big(\omega_\mu(p)-\omega(p)\Gamma^\beta_{\beta;\mu}\Big)v^\mu+\Big(\omega_{\bar\mu}(p)-\omega(p)\Gamma^{\bar\beta}_{\bar\beta;\bar\mu}\Big)v^{\bar\mu}\\
&\times\Big(-\frac{1}{2}\Gamma^\alpha_{\beta;\mu}\omega_\alpha(p)-\omega_\beta(p)\Gamma^\lambda_{\lambda;\mu}+\omega(p)C_{\beta\mu}\Big)v^\beta v^\mu\\
&\times\Big(-\frac{1}{2}\Gamma^{\bar\alpha}_{\bar\beta;\bar\mu}\omega_{\bar\alpha}(p)-\omega_{\bar\beta}(p)\Gamma^{\bar\lambda}_{\bar\lambda;\bar\mu}+\omega(p)\overline{C_{\beta\mu}}\Big)v^{\bar\beta}v^{\bar\mu}+\frac{1}{2}\omega_{\alpha\bar\beta}(p)v^\alpha\bar v^\beta\Big]d\bar v\wedge dv.
\end{aligned}$$

Here we are ignoring terms of order higher than 2 and abbreviating

$$C_{\beta;\mu}:=\Gamma^\rho_{\rho;\mu}\Gamma^\sigma_{\sigma;\beta}-\frac{1}{3}\Gamma^\rho_{\sigma;\mu}\Gamma^\sigma_{\rho;\beta}-\frac{1}{3}\Gamma^\alpha_{\alpha;\mu\beta}-\frac{1}{6}\Gamma^\alpha_{\beta;\mu\alpha}+\frac{1}{3}\Gamma^\alpha_{\alpha;\gamma}\Gamma^\gamma_{\beta;\mu}-\frac{1}{2}\frac{\partial\Gamma^\alpha_{\beta;\mu}}{\partial v^\alpha}.$$

Notice that I_ε is symmetric on $T^{1,0}_pM$. This is an advantage of working in the tangent space because symmetry ensures that $C_n\int_{I_\varepsilon} v^\mu d\bar v\wedge dv = 0$, or, more generally, $C_n\int_{I_\varepsilon} h(v)d\bar v\wedge dv = 0$ if h is any odd function. In the line above we see the integral $C_n\int_{I_\varepsilon}\Gamma^\beta_{\beta;\mu}v^\mu d\bar v\wedge dv$. Consider equation (1.10). Since $F^2(v)$ is an even function, and since $G_{\alpha\bar\beta} = \frac{\partial^2 G(z;v)}{\partial v^\alpha\partial\bar v^\beta}$ is obtained by taking two derivatives of an even function, $G_{\alpha\bar\beta}(v)$ must be also even. Then $G^{\bar\beta\alpha}(v)$ is even, and derivatives of the form $\frac{\partial G_{\beta\bar\tau}}{\partial z^\mu}(v)$, with respect to coordinates z^μ in the manifold, are also even (as functions of the tangent space coordinates v). Furthermore $G_{\beta\bar\tau\gamma}$ is odd, $G_{\bar\tau;\mu}$ is odd, so $G_{\beta\bar\tau\gamma}\Gamma^\gamma_\mu$ is even. We see from equation (1.10) that $\Gamma^\alpha_{\beta;\mu}$ must then be even. The function v^μ (on the tangent space) is odd, so the product $\Gamma^\beta_{\beta;\mu}(v)v^\mu$ of an even function and an odd function must be odd. Thus $C_n\int_{I_\varepsilon}\Gamma^\beta_{\beta;\mu}v^\mu d\bar v\wedge dv=0$, therefore

$$\begin{aligned}&\mathrm{Vol}(\textstyle\sum_\varepsilon)\\&=C_n\int_{I_\varepsilon}\omega(p)d\bar v\wedge dv+C_n\int_{I_\varepsilon}\Big[\Big(-\frac{1}{2}\Gamma^\alpha_{\beta;\mu}\omega_\alpha(p)-\omega_\beta(p)\Gamma^\lambda_{\lambda;\mu}+\omega(p)C_{\beta\mu}\Big)v^\beta v^\mu\\&\quad+\Big(-\frac{1}{2}\Gamma^{\bar\alpha}_{\bar\beta\bar\mu}\omega_{\bar\alpha}(p)-\omega_{\bar\beta}(p)\Gamma^{\bar\lambda}_{\bar\lambda;\bar\mu}+\omega(p)\overline{C_{\beta\mu}}\Big)v^{\bar\beta}v^{\bar\mu}+\frac{1}{2}\omega_{\alpha\bar\beta}(p)v^\alpha v^{\bar\beta}\Big]d\bar v\wedge dv.\end{aligned}\tag{2.4}$$

Note that the first term in (2.4) is the integral of a constant. Because of the scaling property $I_{r\varepsilon}=rI_\varepsilon$, we can change our region of integration to I_1, via

$$C_n\int_{I_\varepsilon}\omega(p)d\bar v\wedge dv=C_n\varepsilon^{2n}\int_{I_1}\omega(p)d\bar v\wedge dv=C_n\varepsilon^{2n}\int_I\omega(p)d\bar v\wedge dv,$$

where n is the dimension of M, and, of course, $I=I_1$. Similarly, the second term of (2.4) is the integral of a function homogeneous of degree 2, so we can again change our region of integration to I, but this time with a factor of ε^{2n+4}:

$$\begin{aligned}&=C_n\varepsilon^{2n}\int_I\omega(p)d\bar v\wedge dv+C_n\varepsilon^{2n+4}\int_I\Big[\Big(-\frac{1}{2}\Gamma^\alpha_{\beta;\mu}\omega_\alpha(p)-\omega_\beta(p)\Gamma^\lambda_{\lambda;\mu}+\omega(p)C_{\beta\mu}\Big)v^\beta v^\mu\\&\quad+\Big(-\frac{1}{2}\Gamma^{\bar\alpha}_{\bar\beta;\bar\mu}\omega_{\bar\alpha}(p)-\omega_{\bar\beta}(p)\Gamma^{\bar\lambda}_{\bar\lambda;\bar\mu}+\omega(p)\overline{C_{\beta\mu}}\Big)v^{\bar\beta}v^{\bar\mu}+\frac{1}{2}\omega_{\alpha\bar\beta}(p)v^\alpha v^{\bar\beta}\Big]d\bar v\wedge dv.\end{aligned}\tag{2.5}$$

Recall that $\omega(p)$ is given by $\omega(p)=\frac{\kappa_{2n}}{C_n\int_I d\bar v\wedge dv}$, where neither the numerator nor the denominator depends on v. Thus we can take outside the integral:

$$=C_n\varepsilon^{2n}\frac{\kappa_{2n}}{C_n\int_I d\bar v\wedge dv}\int_I d\bar v\wedge dv+0\cdot\varepsilon^{2n+2}+C_{2n+4}\cdot\varepsilon^{2n+4}+o(\varepsilon^{2n+6}),$$

where C_{2n+4} will be written explicitly later. In summary,

$$\mathrm{Vol}\Big(\sum_\varepsilon\Big)=\kappa_{2n}\varepsilon^{2n}+C_{2n+4}\cdot\varepsilon^{2n+4}+o(\varepsilon^{2n+6}).\tag{2.6}$$

Case B. $\int_{\sum_\varepsilon}z^\alpha dV$. The function z^α is "odd" around the origin, which is the center of $\sum_\varepsilon$, so we would expect the integral to be near zero. In fact, this integral has order ε^{2n+2} instead of ε^{2n+1} which seems natural. Again, we work this out by transferring $\int_{\Sigma_\varepsilon} z^\alpha dV$ to $T_p^{1,0}M$ and using the symmetry of the indicatrix there:

$$\begin{aligned}
&\int_{\Sigma_\varepsilon} z^\alpha dV = C_n \int_{\Sigma_\varepsilon} z^\alpha \omega(z) d\bar{z} \wedge dz \\
=&C_n \int_{\Sigma_\varepsilon} \Big[z^\alpha \omega(p) + \omega_\beta(p) z^\alpha z^\beta + \omega_{\bar{\beta}}(p) z^\alpha \bar{z}^\beta + \frac{1}{2}\omega_{\beta\bar{\mu}}(p) z^\alpha z^\beta \bar{z}^\mu + o(|z|^4)\Big] d\bar{z}\wedge dz \\
=&C_n \int_{\Sigma_\varepsilon} \Big[\Big(v^\alpha - \frac{1}{2}\Gamma^\alpha_{\beta;\mu}(v) v^\beta v^\mu \Big)\omega(p) \\
&+ \omega_\beta(p)\Big(v^\alpha - \frac{1}{2}\Gamma^\alpha_{\gamma;\mu}(v)v^\gamma v^\mu\Big)\Big(v^\beta - \frac{1}{2}\Gamma^\beta_{\rho;\theta}(v)v^\beta v^\theta\Big) \\
&+ \omega_{\bar{\beta}}(p)\Big(v^\alpha - \frac{1}{2}\Gamma^\alpha_{\gamma;\mu}(v)v^\gamma v^\mu\Big)\Big(\overline{v^\beta} - \frac{1}{2}\Gamma^{\bar{\beta}}_{\bar{\rho};\bar{\theta}}(v)\bar{v}^\beta \bar{v}^\theta\Big) \\
&+ \frac{1}{2}\omega_{\beta\bar{\mu}}(p)\Big(v^\alpha - \frac{1}{2}\Gamma^\alpha_{\gamma;\mu}(v)v^\gamma v^\mu\Big)\Big(v^\beta - \frac{1}{2}\Gamma^\beta_{\lambda;\delta}(v)v^\lambda v^\delta\Big)\Big(\bar{v}^\mu - \frac{1}{2}\Gamma^{\bar{\mu}}_{\bar{\tau};\bar{\nu}}(v)\bar{v}^\tau \bar{v}^\nu\Big)\Big] \\
&\times \Big(1 - \Gamma^\beta_{\beta;\mu}(v)v^\mu + C_{\beta\mu}v^\beta v^\mu\Big)\Big(1 - \Gamma^{\bar{\beta}}_{\bar{\beta};\bar{\mu}}(v)\bar{v}^\mu + \overline{C_{\beta\mu}}\bar{v}^\beta\bar{v}^\mu\Big) d\bar{v}\wedge dv \\
=&C_n \int_{I_\varepsilon} \Big[\omega(p)v^\alpha + d_{\alpha\beta}v^\alpha v^\beta + d_{\alpha\bar{\beta}}v^\alpha \bar{v}^\beta + o(|z|^3)\Big] d\bar{v}\wedge dv,
\end{aligned}$$

for some coefficient $d_{\alpha\beta}, d_{\alpha\bar{\beta}}$. Again, oddness of z^α ensures that

$$\int_{I_\varepsilon} \omega(p) v^\alpha d\bar{v} \wedge dv$$

vanishes, and homogeneity allows us to evaluate I_ε solely on I by adding an appropriate factor:

$$\begin{aligned}
=&C_n d_{\alpha\beta}\left(\int_I v^\alpha v^\beta d\bar{v}\wedge dv\right)\varepsilon^{2n+2} + C_n d_{\alpha\bar{\beta}}\left(\int_I v^\alpha \bar{v}^\beta d\bar{v}\wedge dv\right)\varepsilon^{2n+2} + d_{zn+3}\varepsilon^{2n+3} \\
&+ o(\varepsilon^{2n+4}),
\end{aligned}$$

for some constant d_{2n+3}. The important result here is that we don't need any terms involving ε^{2n+1}, so

$$\left|\int_{\Sigma_\varepsilon} z^\alpha dV\right| \le d_{2n+2}\varepsilon^{2n+2} + o(\varepsilon^{2n+4}).$$

By estimations of Case A and Case B, there are two conclusions, the first conclusion assures that integrals of the form $\int_{\Sigma_\varepsilon} z^\alpha dV$ are actually of the same order of magnitude as the integrals $\int_{\Sigma_\varepsilon} z^\alpha z^\beta dV$, and the second conclusion allows us to ignore higher-order terms when they appear in integrals.

Remark 2.1. Expression (2.6) of Case A is interesting because of its parallels with the Riemannian case. The first term in both the complex Finsler and Riemannian case is $\kappa_{2n}\varepsilon^{2n}$, which is simply the Euclidean volume of a sphere of radius ε in $\mathbb{R}^{2n}$, so even in a complex Finsler space, volume measurement doesn't differ dramatically from the Euclidean case. Also, in both cases the coefficient of ε^{2n+2} vanishes. The most interesting coefficient is C_{2n+4}, which measures how our space differs from Euclidean space. In the Riemannian case, this coefficient (up to a constant) is actually the scalar curvature([7]).

Since expression (2.6) is invariant by its very construction, this clearly suggests thinking of the second coefficient as a scalar curvature for strongly Kähler–Finsler manifolds ([8]).

Definition 2.2. Define the *Puiseux curvature* P at z of the strongly Kähler–Finsler manifold (M, F) to be

$$P = \alpha_{2n} C_{2n+4},$$

where α_{2n} is a constant.

By (2.5) and replacing $\omega(p)$ with $\omega(p) = \frac{\kappa_{2n}}{C_n \int_I d\bar{v}\wedge dv}$ we can get an expression for the Puiseux curvature:

$$\begin{aligned}
P &= \alpha_{2n} C_{2n+4} \\
&= \alpha_{2n}\kappa_{2n} \int_I \Bigg\{ \Big[-\frac{1}{2}\Gamma^{\alpha}_{\beta;\mu}\frac{\partial}{\partial z^{\alpha}}\Big(\int_I d\bar{v}\wedge dv\Big)^{-1} + \frac{1}{2}\frac{\partial}{\partial z^{\alpha}}\frac{\partial}{\partial z^{\beta}}\Big(\int_I d\bar{v}\wedge dv\Big)^{-1} \\
&\quad - \frac{\partial}{\partial z^{\beta}}\Big(\int_I d\bar{v}\wedge dv\Big)^{-1}\Gamma^{\lambda}_{\lambda;\mu} + \Big(\int_I d\bar{v}\wedge dv\Big)^{-1}\Big(\Gamma^{\rho}_{\rho;\mu}\Gamma^{\sigma}_{\sigma;\beta} - \frac{1}{3}\Gamma^{\rho}_{\sigma;\mu}\Gamma^{\sigma}_{\rho;\beta} - \frac{1}{3}\Gamma^{\alpha}_{\alpha;\mu\beta} \\
&\quad - \frac{1}{6}\Gamma^{\alpha}_{\beta;\mu\alpha} + \frac{1}{3}\Gamma^{\alpha}_{\alpha;\gamma}\Gamma^{\gamma}_{\beta;\mu} - \frac{1}{2}\frac{\partial\Gamma^{\alpha}_{\beta;\mu}}{\partial v^{\alpha}}\Big)\Big] v^{\beta}v^{\mu} + \Big[-\frac{1}{2}\Gamma^{\bar{\alpha}}_{\bar{\beta};\bar{\mu}}\frac{\partial}{\partial \bar{z}^{\alpha}}\Big(C_n\int_I d\bar{v}\wedge dv\Big)^{-1} \\
&\quad + \frac{1}{2}\frac{\partial}{\partial \bar{z}^{\alpha}}\frac{\partial}{\partial \bar{z}^{\beta}}\Big(\int_I d\bar{v}\wedge dv\Big)^{-1} - \frac{\partial}{\partial \bar{z}^{\beta}}\Big(\int_I d\bar{v}\wedge dv\Big)^{-1}\Gamma^{\bar{\alpha}}_{\bar{\lambda};\bar{\mu}} + \Big(\int_I d\bar{v}\wedge dv\Big)^{-1} \\
&\quad \times \Big(\Gamma^{\bar{\rho}}_{\bar{\rho};\bar{\mu}}\Gamma^{\bar{\sigma}}_{\bar{\sigma};\bar{\beta}} - \frac{1}{3}\Gamma^{\bar{\rho}}_{\bar{\sigma};\bar{\mu}}\Gamma^{\bar{\sigma}}_{\bar{\rho};\bar{\beta}} - \frac{1}{3}\Gamma^{\bar{\alpha}}_{\bar{\alpha};\bar{\mu}\bar{\beta}} - \frac{1}{6}\Gamma^{\bar{\alpha}}_{\bar{\beta};\bar{\mu}\bar{\alpha}} + \frac{1}{3}\Gamma^{\bar{\alpha}}_{\bar{\alpha};\bar{\gamma}}\Gamma^{\bar{\gamma}}_{\bar{\beta};\bar{\mu}} - \frac{1}{2}\frac{\partial\Gamma^{\bar{\alpha}}_{\bar{\beta};\bar{\mu}}}{\partial \bar{v}^{\alpha}}\Big)\Big] v^{\bar{\beta}}v^{\bar{\mu}} \\
&\quad + \frac{1}{2}\frac{\partial}{\partial z^{\alpha}}\frac{\partial}{\partial \bar{z}^{\beta}}\Big(\int_I d\bar{v}\wedge dv\Big)^{-1} v^{\alpha}v^{\bar{\beta}} \Bigg\} d\bar{v}\wedge dv.
\end{aligned}$$

Lemma 2.3. *$P : M \longrightarrow \mathbb{R}$ given by $z \longmapsto$ Puiseux curvature at z is C^{∞}.*

Proof. The expression for C_{2n+4} involves integration over I, but an integral over I is equal to an integral over $I\backslash\{0\}$. If C_{2n+4} is seen as an integration over $I\backslash\{0\}$, then all limits of integration and integrands are smooth functions of z, so P is smooth. □

3. Mean value Laplacian and its coordinate expression

In this section, we use the constructions of Section 2 to show that the Laplacian $\triangle$ for a function exists, and to derive a coordinate expression for $\triangle$, which will make clear that $\triangle$ is a differential operator.

Firstly, we will state formally the definition of the Laplacian.

Definition 3.1. Let M be a complex manifold with a strongly Kähler–Finsler metric F, and let $f : M \longrightarrow \mathbb{R}$ be a smooth function. Let $dV = C_n\omega(z)d\bar{z} \wedge dz$, where $\omega(z) = \frac{\kappa_{2n}}{C_n \int_I d\bar{v}\wedge dv}$ is the coefficient of the canonical volume form.

Define the *Laplacian* $\triangle$ at a point p by

$$\triangle f(z) = 2(2n+2)\lim_{\varepsilon\to 0}\frac{1}{\varepsilon^2}\left(\frac{\int_{\Sigma_\varepsilon} f dV}{\int_{\Sigma_\varepsilon} dV} - f(z)\right), \tag{3.1}$$

or if we write $dV = C_n\omega(z)d\bar{z} \wedge dz$,

$$\triangle f(z) = 2(2n+2)\lim_{\varepsilon\to 0}\frac{1}{\varepsilon^2}\left(\frac{\int_{\Sigma_\varepsilon} f\omega(z)d\bar{z}\wedge dz}{\int_{\Sigma_\varepsilon} \omega(z)d\bar{z}\wedge dz} - f(z)\right). \tag{3.2}$$

Theorem 3.2. *Let M be a complex manifold with a strongly Kähler–Finsler metric F, and let $f : M \longrightarrow \mathbb{R}$ be a smooth function. Let p be a point in M. Then $\triangle f(p)$ exists and is a linear, 2nd-order differential operator with the coordinate expression:*

$$\begin{aligned}\triangle f =& (2n+2)\left[\frac{\int_I v^\alpha\bar{v}^\beta d\bar{v}\wedge dv}{\int_I d\bar{v}\wedge dv} f_{\alpha\bar{\beta}}\right.\\ &+ f_\alpha\left(\frac{2\omega_{\bar{\beta}}}{\omega}\frac{\int_I v^\alpha\bar{v}^\beta d\bar{v}\wedge dv}{\int_I d\bar{v}\wedge dv} - \frac{2\int_I \Gamma^{\bar{\lambda}}_{\bar{\lambda};\bar{\beta}} v^\alpha\bar{v}^\beta d\bar{v}\wedge dv}{\int_I d\bar{v}\wedge dv}\right)\\ &\left.+ f_{\bar{\alpha}}\left(\frac{2\omega_{\beta}}{\omega}\frac{\int_I \bar{v}^\alpha v^\beta d\bar{v}\wedge dv}{\int_I d\bar{v}\wedge dv} - \frac{2\int_I \Gamma^{\lambda}_{\lambda;\beta} \bar{v}^\alpha v^\beta d\bar{v}\wedge dv}{\int_I d\bar{v}\wedge dv}\right)\right].\end{aligned}$$

Proof. Write f as a Taylor series around p,

$$f(z) = f(p) + f_\alpha(p)z^\alpha + f_{\bar{\alpha}}(p)\bar{z}^\alpha + \frac{1}{2}f_{\alpha\bar{\beta}}(p)z^\alpha\bar{z}^\beta + o(|z|^3).$$

Substitute this expression into the definition:

$$\begin{aligned}&2(2n+2)\lim_{\varepsilon\to 0}\frac{1}{\varepsilon^2}\left[\frac{\int_{\Sigma_\varepsilon} f dV}{\int_{\Sigma_\varepsilon} dV} - f(z)\right]\\ =& 2(2n+2)\lim_{\varepsilon\to 0}\frac{1}{\varepsilon^2}\left[\frac{\int_{\Sigma_\varepsilon}\left(f(p) + f_\alpha(p)z^\alpha + f_{\bar{\alpha}}(p)\bar{z}^\alpha + \frac{1}{2}f_{\alpha\bar{\beta}}(p)z^\alpha\bar{z}^\beta + o(|z|^3)\right)dV}{\int_{\Sigma_\varepsilon} dV}\right.\\ &\left.- \frac{\int_{\Sigma_\varepsilon} f(p)dV}{\int_{\Sigma_\varepsilon} dV}\right]\end{aligned}$$

$$=2(2n+2)\lim_{\varepsilon\to 0}\frac{1}{\varepsilon^2}\frac{\int_{\sum_\varepsilon}\Big(f_\alpha(p)z^\alpha+f_{\bar\alpha}(p)\bar z^\alpha+\frac{1}{2}f_{\alpha\bar\beta}(p)z^\alpha\bar z^\beta+o(|z|^3)\Big)dV}{\int_{\sum_\varepsilon}dV}$$

$$=2(2n+2)\lim_{\varepsilon\to 0}\frac{1}{\varepsilon^2}\frac{\int_{\sum_\varepsilon}\Big(f_\alpha(p)z^\alpha+f_{\bar\alpha}(p)\bar z^\alpha+\frac{1}{2}f_{\alpha\bar\beta}(p)z^\alpha\bar z^\beta+o(|z|^3)\Big)\omega(z)d\bar z\wedge dz}{\int_{\sum_\varepsilon}\omega(z)d\bar z\wedge dz}$$

$$=2(2n+2)\lim_{\varepsilon\to 0}\frac{1}{\varepsilon^2}\int_{I_\varepsilon}\Big[f_\alpha(p)(v^\alpha-\frac{1}{2}\Gamma^\alpha_{\beta;\mu}v^\beta v^\mu)+f_{\bar\alpha}(p)(\bar v^\alpha-\frac{1}{2}\Gamma^{\bar\alpha}_{\bar\beta;\bar\mu}\bar v^\beta\bar v^\mu)$$
$$+\frac{1}{2}f_{\alpha\bar\beta}(p)(v^\alpha-\frac{1}{2}\Gamma^\alpha_{\lambda;\delta}v^\lambda v^\delta)(\bar v^\beta-\frac{1}{2}\Gamma^{\bar\beta}_{\bar\lambda;\bar\delta}v^{\bar\lambda}v^{\bar\delta})\Big]$$
$$\times\frac{\Big(\omega(p)+\omega_\alpha(p)(v^\alpha-\frac{1}{2}\Gamma^\alpha_{\lambda;\delta}v^\lambda v^\delta)+\omega_{\bar\alpha}(p)(\bar v^\alpha-\frac{1}{2}\Gamma^{\bar\alpha}_{\bar\lambda;\bar\delta}\bar v^\lambda\bar v^\delta)\Big)(1-\Gamma^\rho_{\rho;\theta}v^\theta)(1-\Gamma^{\bar\rho}_{\bar\rho;\bar\theta}\bar v^\theta)d\bar v\wedge dv}{\int_{I_\varepsilon}\Big(\omega(p)+\omega_\alpha(p)(v^\alpha-\frac{1}{2}\Gamma^\alpha_{\lambda;\delta}v^\lambda v^\delta)+\omega_{\bar\alpha}(p)(\bar v^\alpha-\frac{1}{2}\Gamma^{\bar\alpha}_{\bar\lambda;\bar\delta}\bar v^\lambda\bar v^\delta)\Big)(1-\Gamma^\rho_{\rho;\theta}v^\theta)(1-\Gamma^{\bar\rho}_{\bar\rho;\bar\theta}\bar v^\theta)d\bar v\wedge dv}$$

$$=2(2n+2)\lim_{\varepsilon\to 0}\left[\varepsilon^2\int_{I_\varepsilon}\Big(\omega(p)+\omega_\alpha v^\alpha+\omega_{\bar\alpha}\bar v^\alpha+o(|v|^2)\Big)d\bar v\wedge dv\right]^{-1}$$
$$\times\int_{I_\varepsilon}\Big[f_\alpha(p)\omega(p)v^\alpha+\frac{1}{2}f_{\alpha\bar\beta}(p)\omega(p)v^\alpha\bar v^\beta+f_\alpha(p)\omega_{\bar\beta}(p)v^\alpha\bar v^\beta-f_\alpha(p)\omega(p)\Gamma^{\bar\lambda}_{\bar\lambda;\bar\beta}v^\alpha\bar v^\beta$$
$$+f_{\bar\alpha}(p)\omega(p)\bar v^\alpha+f_{\bar\alpha}(p)\omega_\beta(p)\bar v^\alpha v^\beta-f_{\bar\alpha}(p)\omega(p)\Gamma^\rho_{\rho;\beta}\bar v^\alpha v^\beta\Big]d\bar v\wedge dv.$$

Again, eliminating integrands homogeneous of degree 1, and ignoring terms of higher order than $2n+2$, and moving to I, we have

$$=2(2n+2)\lim_{\varepsilon\to 0}\varepsilon^{2n+2}\left[\frac{\int_I\Big(\frac{1}{2}f_{\alpha\bar\beta}(p)\omega(p)+f_\alpha(p)\omega_{\bar\beta}(p)-f_\alpha(p)\omega(p)\Gamma^{\bar\lambda}_{\bar\lambda;\bar\beta}\Big)v^\alpha\bar v^\beta d\bar v\wedge dv}{\varepsilon^{2n+2}\omega(p)\int_I d\bar v\wedge dv}\right.$$
$$\left.+\frac{\Big(f_{\bar\alpha}(p)\omega_\beta(p)-f_{\bar\alpha}(p)\omega(p)\Gamma^\lambda_{\lambda;\beta}\Big)\bar v^\alpha v^\beta d\bar v\wedge dv}{\varepsilon^{2n+2}\omega(p)\int_I d\bar v\wedge dv}\right].$$

Separating into coefficient of $f_{\alpha\bar\beta}$ and f_α, $f_{\bar\alpha}$,

$$\triangle f=(2n+2)\Big[\frac{\int_I v^\alpha\bar v^\beta d\bar v\wedge dv}{\int_I d\bar v\wedge dv}f_{\alpha\bar\beta}+f_\alpha\Big(\frac{2\omega_{\bar\beta}}{\omega}\frac{\int_I v^\alpha\bar v^\beta d\bar v\wedge dv}{\int_I d\bar v\wedge dv}-\frac{2\int_I\Gamma^{\bar\lambda}_{\bar\lambda;\bar\beta}v^\alpha\bar v^\beta d\bar v\wedge dv}{\int_I d\bar v\wedge dv}\Big)$$
$$+f_{\bar\alpha}\Big(\frac{2\omega_\beta}{\omega}\frac{\int_I\bar v^\alpha v^\beta d\bar v\wedge dv}{\int_I d\bar v\wedge dv}-\frac{2\int_I\Gamma^\lambda_{\lambda;\beta}\bar v^\alpha v^\beta d\bar v\wedge dv}{\int_I d\bar v\wedge dv}\Big)\Big].$$

This expression clearly shows that $\triangle$ is a differential operator. □

4. Mean value Laplacian generalizes the Kählerian Laplacian

Firstly,we remark that for the leading coefficient in the coordinate expression of $\triangle$, we have

Lemma 4.1. *Let M be a strongly Kähler–Finsler manifold of dimension n with Finsler metric F. Define*

$$K^{\bar{\beta}\alpha} = (2n+2)\frac{\int_I v^\alpha \bar{v}^\beta d\bar{v} \wedge dv}{\int_I d\bar{v} \wedge dv}. \tag{4.1}$$

Then $K^{\bar{\beta}\alpha}$ are the components of a Hermitian symmetric, twice-contravariant, positive-definite tensor.

Proof. Clearly, $\overline{K^{\bar{\beta}\alpha}} = K^{\bar{\alpha}\beta}$, so these components are Hermitian symmetric. To see that

$$K^{\bar{\beta}\alpha} = (2n+2)\frac{\int_I v^\alpha \bar{v}^\beta d\bar{v} \wedge dv}{\int_I d\bar{v} \wedge dv}$$

are the components of a Hermitian tensor, make a coordinate change

$$\tilde{z}^\alpha = \tilde{z}^\alpha(z)$$

on the manifold. Then we have the following relations on the tangent space:

$$\tilde{v}^\alpha = \frac{\partial z^\alpha}{\partial \tilde{z}^\beta} v^\beta,$$
$$d\tilde{v} = \det\left(\frac{\partial z^\alpha}{\partial \tilde{z}^\beta}\right) dv.$$

In the $\tilde{z}^\alpha$ coordinate system,

$$\begin{aligned}\tilde{K}^{\bar{\beta}\alpha} &= (2n+2)\frac{\int_I \tilde{v}^\alpha \tilde{\bar{v}}^\beta d\tilde{\bar{v}} \wedge d\tilde{v}}{\int_I d\tilde{\bar{v}} \wedge d\tilde{v}} \\ &= (2n+2)\frac{\int_I \frac{\partial z^\alpha}{\partial \tilde{z}^\gamma}\frac{\partial \bar{z}^\beta}{\partial \bar{\tilde{z}}^\delta} v^\gamma \bar{v}^\delta \left|\det\left(\frac{\partial z^\alpha}{\partial \tilde{z}^\beta}\right)\right|^2 d\bar{v} \wedge dv}{\int_I \left|\det\left(\frac{\partial z^\alpha}{\partial \tilde{z}^\beta}\right)\right|^2 d\bar{v} \wedge dv}.\end{aligned}$$

Since none of the terms $\frac{\partial z^\alpha}{\partial \tilde{z}^\gamma}\frac{\partial \bar{z}^\beta}{\partial \bar{\tilde{z}}^\delta}$ and $\left|\det\left(\frac{\partial z^\alpha}{\partial \tilde{z}^\beta}\right)\right|^2$ depends on v and $\bar{v}$, we can take them all outside the integral signs, and cancel the determinant, to get

$$\tilde{K}^{\bar{\beta}\alpha} = (2n+2)\frac{\partial z^\alpha}{\partial \tilde{z}^\gamma}\frac{\partial \bar{z}^\beta}{\partial \bar{\tilde{z}}^\delta}\frac{\int_I v^\gamma \bar{v}^\delta d\bar{v} \wedge dv}{\int_I d\bar{v} \wedge dv} = \frac{\partial z^\alpha}{\partial \tilde{z}^\gamma}\frac{\partial \bar{z}^\beta}{\partial \bar{\tilde{z}}^\delta} K^{\bar{\delta}\gamma},$$

i.e., $\tilde{K}^{\bar{\beta}\alpha}$ are the components of a twice-contravariant Hermitian tensor.

To see that this tensor is positive-definite, simply use the Hermitian symmetry to choose a coordinate system $\tilde{z}$ in which $\tilde{K}^{\bar{\beta}\alpha}$ is diagonal, that is, one for which $\tilde{K}^{\bar{\beta}\alpha} = 0$ if $\alpha \neq \beta$. Then

$$\tilde{K}^{\bar{\alpha}\alpha} = (2n+2)\frac{\int_I \tilde{v}^\alpha \bar{\tilde{v}}^\alpha d\bar{\tilde{v}} \wedge d\tilde{v}}{\int_I d\bar{\tilde{v}} \wedge d\tilde{v}} > 0,$$

so $K^{\bar{\alpha}\alpha}$ is positive-definite. □

Since a symmetric,positive-definite,twice-covariant Hermitian tensor is a metric tensor, we can see the $K^{\bar{\beta}\alpha}$ as the (inverse) components of a Kähler metric. Thus we have canonically associated to our strongly Kähler-Finsler manifold a Kähler metric.

Definition 4.2. The metric

$$K^{\bar{\beta}\alpha} = (2n+2)\frac{\int_I v^\alpha \bar{v}^\beta d\bar{v} \wedge dv}{\int_I d\bar{v} \wedge dv}$$

on a strongly Kähler–Finsler manifold (M, F) is called the (inverse of the) *osculating* Kähler metric.

Lemma 4.3. *Let M be a Kähler manifold of complex dimension n with metric $G_{\alpha\bar{\beta}}$. Then*

$$G^{\bar{\beta}\alpha} = \frac{1}{2}(2n+2)\frac{\int_I v^\alpha \bar{v}^\beta d\bar{v} \wedge dv}{\int_I d\bar{v} \wedge dv}. \tag{4.2}$$

Proof. To simplify the calculations, choose coordinates such that

$$G^{\bar{\beta}\alpha}(p) = G_{\alpha\bar{\beta}}(p) = \delta_{\alpha\beta}(p), \tag{4.3}$$

i.e., such that I at p is the unit sphere D in $\mathbb{R}^{2n}$. Then, by symmetry of D,

$$K^{\bar{\beta}\alpha} = 0, \quad \text{if } \alpha \neq \beta, \tag{4.4}$$

and

$$K^{\bar{1}1} = K^{\bar{2}2} = \cdots = K^{\bar{n}n}. \tag{4.5}$$

Let $v^\alpha = \xi_\alpha + i\xi_{n+\alpha}$, $\alpha = 1, \ldots, n$. Then we calculate

$$K^{\bar{n}n} = (2n+2)\frac{\int_D (\xi_n^2 + \xi_{2n}^2) d\xi_1 d\xi_2 \cdots d\xi_{2n}}{\int_D d\xi_1 d\xi_2 \cdots d\xi_{2n}} = 2(2n+2)\frac{\int_D \xi_{2n}^2 d\xi_1 d\xi_2 \cdots d\xi_{2n}}{\int_D d\xi_1 d\xi_2 \cdots d\xi_{2n}},$$

where

$$\begin{aligned}
&2(2n+2)\frac{\int_D \xi_{2n}^2 d\xi_1 d\xi_2 \cdots d\xi_{2n}}{\int_D d\xi_1 d\xi_2 \cdots d\xi_{2n}} \\
=&2(2n+2)\frac{\int_{-1}^1 \xi_{2n}^2 \Big(\int_{-\sqrt{1-\xi_{2n}^2}}^{\sqrt{1-\xi_{2n}^2}} \int \cdots \int d\xi_1 \cdots d\xi_{2n-2} d\xi_{2n-1}\Big) d\xi_{2n}}{\int_{-1}^1 \Big(\int_{-\sqrt{1-\xi_{2n}^2}}^{\sqrt{1-\xi_{2n}^2}} \int \cdots \int d\xi_1 \cdots d\xi_{2n-2} d\xi_{2n-1}\Big) d\xi_{2n}} \\
=&2(2n+2)\frac{\int_{-1}^1 \xi_{2n}^2 \Big((2n-1)\text{-sphere of radius } \sqrt{1-\xi_{2n}^2}\Big) d\xi_{2n}}{\int_{-1}^1 \Big((2n-1)\text{-sphere of radius } \sqrt{1-\xi_{2n}^2}\Big) d\xi_{2n}} \\
=&2(2n+2)\frac{\int_{-1}^1 \xi_{2n}^2 \kappa_{2n-1} (\sqrt{1-\xi_{2n}^2})^{2n-1} d\xi_{2n}}{\int_{-1}^1 \kappa_{2n-1} (\sqrt{1-\xi_{2n}^2})^{2n-1} d\xi_{2n}}.
\end{aligned}$$

Let $\xi_{2n} = \sin\theta$ and integrate:

$$\begin{aligned}
&=2(2n+2)\frac{2\int_0^{\frac{\pi}{2}} \sin^2\theta \cos^{2n-1}\theta \cos\theta d\theta}{2\int_0^{\frac{\pi}{2}} \cos^{2n}\theta d\theta} \\
&=2(2n+2)\frac{\frac{-1}{2n+2}\cos^{2n+1}\theta \sin\theta\Big|_0^{\frac{\pi}{2}} + \frac{1}{2n+2}\int_0^{\frac{\pi}{2}} \cos^{2n}\theta d\theta}{\int_0^{\frac{\pi}{2}} \cos^{2n}\theta d\theta} \\
&=2,
\end{aligned}$$

by (4.1) to (4.5),

$$K^{\bar{\beta}\alpha} = 2G^{\bar{\beta}\alpha}.$$

We know that $G^{\bar{\beta}\alpha}$ is a twice-contravariant Hermitian tensor, and since lemma 2 proved that $K^{\bar{\beta}\alpha}$ is a twice-contravariant Hermitian tensor, it follows that $G^{\bar{\beta}\alpha} = \frac{1}{2}K^{\bar{\beta}\alpha}$ in any coordinate system, i.e.,

$$G^{\bar{\beta}\alpha} = \frac{1}{2}(2n+2)\frac{\int_I v^\alpha \bar{v}^\beta d\bar{v}\wedge dv}{\int_I d\bar{v}\wedge dv}. \qquad \square$$

As remarked before, the terms

$$(2n+2)\frac{\int_I v^\alpha \bar{v}^\beta d\bar{v}\wedge dv}{\int_I d\bar{v}\wedge dv}$$

appear as the leading coefficients in $\triangle$. If our space is Kählerian, they will also appear in the first-order coefficients, and this fact suggests us to prove the following theorem, which states that our mean-value Laplacian generalizes the Kählerian Laplacian.

Theorem 4.4. *Let (M,F) be Kähler manifold with its canonical Laplacian* ([8])

$$L(f) = 2G^{\bar{\beta}\alpha}\frac{\partial^2 f}{\partial z^\alpha \partial \bar{z}^\beta}. \tag{4.6}$$

Define

$$\triangle f = (2n+2)\lim_{\varepsilon\to 0}\frac{1}{\varepsilon^2}\left(\frac{\int_{\sum_\varepsilon} f dV}{\int_{\sum_\varepsilon} dV} - f\right).$$

Then

$$\triangle f = L(f). \tag{4.7}$$

Proof. Consider the coordinate expression (3.3) for

$$\begin{aligned}
\triangle f = (2n+2)\Big[&\frac{\int_I v^\alpha \bar{v}^\beta d\bar{v}\wedge dv}{\int_I d\bar{v}\wedge dv} f_{\alpha\bar{\beta}} + f_\alpha\Big(\frac{2\omega_{\bar{\beta}}}{\omega}\frac{\int_I v^\alpha \bar{v}^\beta d\bar{v}\wedge dv}{\int_I d\bar{v}\wedge dv} - \frac{2\int_I \Gamma^{\bar{\lambda}}_{\bar{\lambda};\bar{\beta}} v^\alpha \bar{v}^\beta d\bar{v}\wedge dv}{\int_I d\bar{v}\wedge dv}\Big) \\
&+ f_{\bar{\alpha}}\Big(\frac{2\omega_\beta}{\omega}\frac{\int_I \bar{v}^\alpha v^\beta d\bar{v}\wedge dv}{\int_I d\bar{v}\wedge dv} - \frac{2\int_I \Gamma^\lambda_{\lambda;\beta} \bar{v}^\alpha v^\beta d\bar{v}\wedge dv}{\int_I d\bar{v}\wedge dv}\Big)\Big].
\end{aligned}$$

Since the Christoffel symbols $\Gamma^{\bar{\lambda}}_{\bar{\lambda};\bar{\beta}}$ don't depend on tangent vectors in a Kähler manifold, we can take them outside the integral signs, which leaves a common factor $(2n+2)\int_I v^\alpha \bar{v}^\beta d\bar{v}\wedge dv / \int_I d\bar{v}\wedge dv$:

$$\triangle f = (2n+2)\frac{\int_I v^\alpha \bar{v}^\beta d\bar{v}\wedge dv}{\int_I d\bar{v}\wedge dv}\left[f_{\alpha\bar{\beta}} + f_\alpha\left(\frac{2\omega_{\bar{\beta}}}{\omega} - 2\Gamma^{\bar{\lambda}}_{\bar{\lambda};\bar{\beta}}\right) + f_{\bar{\alpha}}\left(\frac{2\omega_\beta}{\omega} - 2\Gamma^{\lambda}_{\lambda;\beta}\right)\right]. \quad (4.8)$$

In a Kähler manifold, the volume form $dV = \omega(z) d\bar{z}\wedge dz$ is

$$dV = \sqrt{\mathcal{G}}\, d\bar{z}\wedge dz,$$

so

$$\frac{\omega_{\bar{\beta}}}{\omega} = \frac{1}{\sqrt{\mathcal{G}}}\frac{\partial\sqrt{\mathcal{G}}}{\partial \bar{z}^\beta}.$$

There is, however, the Kähler identity

$$\Gamma^{\bar{\lambda}}_{\bar{\lambda};\bar{\beta}} = G^{\bar{\lambda}\varepsilon}\frac{\partial G_{\varepsilon\bar{\lambda}}}{\partial \bar{z}^\beta} = \frac{1}{\sqrt{\mathcal{G}}}\frac{\partial\sqrt{\mathcal{G}}}{\partial \bar{z}^\beta},$$

so the coefficients of f_α and $f_{\bar{\alpha}}$ in expression (4.8) are zero, leaving

$$\triangle f = (2n+2)\frac{\int_I v^\alpha \bar{v}^\beta d\bar{v}\wedge dv}{\int_I d\bar{v}\wedge dv} f_{\alpha\bar{\beta}}. \quad (4.9)$$

Now use Lemma 4.3,which asserts the identity

$$G^{\bar{\beta}\alpha} = \frac{1}{2}(2n+2)\frac{\int_I v^\alpha \bar{v}^\beta d\bar{v}\wedge dv}{\int_I d\bar{v}\wedge dv},$$

to convert (4.9) to

$$\triangle f = 2G^{\bar{\beta}\alpha} f_{\alpha\bar{\beta}} = L(f). \qquad \square$$

Remark 4.5. Although the Laplacian we have defined does not result from any Kählerian metric, it shares some important properties with Kählerian Laplacians. In particular, $\triangle$ is "self-adjoint" in the sense that it satisfies a version of Green's Theorem for complex Finsler manifolds, and as a result of this, it can be written in a divergence form very similar to the divergence form of the Kählerian Laplacian. Furthermore, our definition can be extended to the Hodge–Laplace operator of arbitrary (p,q) forms. It is worth to remark that using the osculating Kähler metric $K^{\bar{\beta}\alpha}$ to define the pointwise and global inner product when we define the Hodge–Laplace operator of (p,q) forms is more natural than using the Finsler metric $G^{\bar{\beta}\alpha}$ and can avoid many complicate calculation.

5. Properties of the Laplacian for functions on strongly Kähler–Finsler manifolds

In this section, we will prove some remarkable properties of the mean value Laplacian $\triangle$ defined in Section 3.

Definition 5.1. A linear operator $\mathcal{D}$ is *self-adjoint* on a complex Finsler manifold (M, F), if

$$\int_M (u\mathcal{D}(v))\omega(w)d\bar{w} \wedge dw = \int_M (v\mathcal{D}(u))\omega(w)d\bar{w} \wedge dw \tag{5.1}$$

for any compactly supported functions u and v.

Lemma 5.2. *Let M be a strongly Kähler–Finsler manifold with Finsler metric F and $\triangle$ be defined as above. Then $\triangle$ is self-adjoint, i.e.,*

$$\int_M (u \triangle v - v \triangle u)\omega(w)d\bar{w} \wedge dw = 0. \tag{5.2}$$

Proof.

$$\begin{aligned}
&\int_M (u \triangle v - v \triangle u)\omega(w)d\bar{w} \wedge dw \\
=&2(2n+2)\int_M \left[u(w) \lim_{\varepsilon\to 0} \frac{\int_{\sum_\varepsilon(w)}(v(z)-v(w))\omega(z)d\bar{z}\wedge dz}{\varepsilon^2 \int_{\sum_\varepsilon(w)} \omega(z)d\bar{z}\wedge dz} \right. \\
&\left. - v(w) \lim_{\varepsilon\to 0} \frac{\int_{\sum_\varepsilon(w)}(u(z)-u(w))\omega(z)d\bar{z}\wedge dz}{\varepsilon^2 \int_{\sum_\varepsilon(w)} \omega(z)d\bar{z}\wedge dz} \right] \omega(w)d\bar{w}\wedge dw
\end{aligned}$$

(where $\sum_\varepsilon(w)$ (resp.$\sum_\varepsilon(z)$) is the ball of radius ε around w(resp.z))

$$\begin{aligned}
=&2(2n+2)\lim_{\varepsilon\to 0}\frac{1}{\varepsilon^2}\int_M \int_{\sum_\varepsilon(w)} \Big(\frac{1}{\int_{\sum_\varepsilon(w)} \omega(z)d\bar{z}\wedge dz}\Big) \\
&\times [u(w)v(z) - v(w)u(z) - u(w)v(w) + u(w)v(w)] \\
&\times \omega(z)\omega(w)d\bar{z}\wedge dz\wedge d\bar{w}\wedge dw \\
=&2(2n+2)\lim_{\varepsilon\to 0}\frac{1}{\varepsilon^2}\left[\int_M \int_{\sum_\varepsilon(w)} \Big(\frac{1}{\int_{\sum_\varepsilon(w)} \omega(z)d\bar{z}\wedge dz}\Big) \right. \\
&\times u(w)v(z)\omega(z)\omega(w)d\bar{z}\wedge dz\wedge d\bar{w}\wedge dw \\
&\left. - \int_M \int_{\sum_\varepsilon(w)} \Big(\frac{1}{\int_{\sum_\varepsilon(w)} \omega(z)d\bar{z}\wedge dz}\Big) v(w)u(z)\omega(z)\omega(w)d\bar{z}\wedge dz\wedge d\bar{w}\wedge dw \right] \\
=&2(2n+2)\lim_{\varepsilon\to 0}\frac{1}{\varepsilon^2}\left[\int_M \int_{\sum_\varepsilon(w)} \Big(\frac{1}{\int_{\sum_\varepsilon(w)} \omega(z)d\bar{z}\wedge dz}\Big) \right. \\
&\times u(w)v(z)\omega(z)\omega(w)d\bar{z}\wedge dz\wedge d\bar{w}\wedge dw \\
&\left. - \int_M \int_{\sum_\varepsilon(z)} \Big(\frac{1}{\int_{\sum_\varepsilon(w)} \omega(z)d\bar{z}\wedge dz}\Big) v(w)u(z)\omega(z)\omega(w)d\bar{w}\wedge dw\wedge d\bar{z}\wedge dz \right]
\end{aligned}$$

(because, along the "diagonal" of $M \times M$,

$$\int_M \int_{\sum_\varepsilon(z)} f(z,w)d\bar{w}\wedge dw\wedge d\bar{z}\wedge dz = \int_M \int_{\sum_\varepsilon(w)} f(z,w)d\bar{z}\wedge dz\wedge d\bar{w}\wedge dw \Big)$$

$$
\begin{aligned}
=&2(2n+2)\lim_{\varepsilon\to 0}\frac{1}{\varepsilon^2}\Bigg[\int_M\int_{\sum_\varepsilon(w)}\Big(\frac{1}{\int_{\sum_\varepsilon(w)}\omega(z)d\bar z\wedge dz}\Big)\\
&\times u(w)v(z)\omega(z)\omega(w)d\bar z\wedge dz\wedge d\bar w\wedge dw\\
&-\int_M\int_{\sum_\varepsilon(w)}\Big(\frac{1}{\int_{\sum_\varepsilon(z)}\omega(w)d\bar w\wedge dw}\Big)u(w)v(z)\omega(z)\omega(w)d\bar z\wedge dz\wedge d\bar w\wedge dw\Bigg],
\end{aligned}
$$

(renaming the parameters in the second term)

$$
\begin{aligned}
=&2(2n+2)\lim_{\varepsilon\to 0}\frac{1}{\varepsilon^2}\int_M\Bigg[\int_{\sum_\varepsilon(w)}u(w)v(z)\omega(z)\omega(w)\Bigg]\\
&\times\Bigg[\frac{1}{\int_{\sum_\varepsilon(w)}\omega(z)d\bar z\wedge dz}-\frac{1}{\int_{\sum_\varepsilon(z)}\omega(w)d\bar w\wedge dw}\Bigg]d\bar z\wedge dz\wedge d\bar w\wedge dw\\
=&2(2n+2)\lim_{\varepsilon\to 0}\frac{1}{\varepsilon^2}\int_M\Bigg[\int_{\sum_\varepsilon(w)}u(w)v(z)\omega(z)\omega(w)\Bigg]\\
&\times\Bigg[\frac{1}{\kappa_{2n}\varepsilon^{2n}+C_{2n+4}(w)\varepsilon^{2n+4}+\cdots}-\frac{1}{\kappa_{2n}\varepsilon^{2n}+C_{2n+4}(z)\varepsilon^{2n+4}+\cdots}\Bigg]\\
&\times d\bar z\wedge dz\wedge d\bar w\wedge dw,
\end{aligned}
$$

(using (2.6))

$$
\begin{aligned}
=&2(2n+2)\lim_{\varepsilon\to 0}\frac{1}{\varepsilon^2}\int_M\Bigg[\frac{1}{\varepsilon^{2n}}\int_{\sum_\varepsilon(w)}u(w)v(z)\omega(z)\omega(w)\Bigg]\\
&\times\Bigg[\frac{\kappa_{2n}(C_{2n+4}(z)-C_{2n+4}(w))\varepsilon^{2n+4}+\cdots}{\kappa_{2n}^2\varepsilon^{2n}}\Bigg]d\bar z\wedge dz\wedge d\bar w\wedge dw\\
=&2(2n+2)\lim_{\varepsilon\to 0}\frac{\varepsilon^{2n+4}}{\varepsilon^{2n+4}}\frac{1}{\kappa_{2n}}\int_M\Bigg[\frac{1}{\varepsilon^{2n-2}}\int_{\sum_\varepsilon(w)}u(w)v(z)\omega(z)\omega(w)\Bigg]\\
&\times[C_{2n+4}(z)-C_{2n+4}(w)]d\bar z\wedge dz\wedge d\bar w\wedge dw\\
=&0.
\end{aligned}
$$

The last line we obtained by letting $\varepsilon\to 0$. When this happens, $z\to w, C_{2n+4}$ is just the Puiseux curvature defined in Section 2 (Definition 2.2.), where we also proved that it is smooth. Since C_{2n+4} is smooth,we have $C_{2n+4}(z)\to C_{2n+4}(w)$. □

Self adjointness induces an equivalent condition on the coefficients of the differential operator $\mathcal{D}$. For a second-order, linear differential operator $\mathcal{D}$ with Hermitian symmetric leading coefficients, i.e.,

$$
\mathcal{D}(f)=A^{\bar\beta\alpha}f_{\alpha\bar\beta}+B^\alpha f_\alpha,\quad \overline{A^{\bar\beta\alpha}}=A^{\bar\alpha\beta}, \tag{5.3}
$$

this is indeed the case. For the Laplace–Beltrami operator of the Kähler manifold $L(f)$, we can write

$$L(f) = \frac{1}{\sqrt{\mathcal{G}}} \frac{\partial}{\partial z^\alpha}\left(f_{\bar\beta} \mathcal{G}^{\bar\beta\alpha} \sqrt{\mathcal{G}}\right) \tag{5.4}$$

and give a name to this convenient form. Notice that in the case of a Kähler manifold if we write $L(f)$ in the form (5.3) we have $B^\alpha = 0$.

Definition 5.3. We will say that an operator $\mathcal{D}$ is *in divergence form* if

$$\mathcal{D}(f) = \frac{1}{\omega} \frac{\partial}{\partial z^\alpha}\left(f_{\bar\beta} h^{\bar\beta\alpha} \omega\right), \tag{5.5}$$

where $\omega(z) d\bar z \wedge dz$ is the canonical Finsler volume form dV of a strongly Kähler–Finsler manifold, and $h^{\bar\beta\alpha}$ is some Hermitian symmetric, contravariant 2-tensor on M.

Lemma 5.4. *If $\mathcal{D}$ is a second-order linear differential operator of the form*

$$\mathcal{D}f = A^{\bar\beta\alpha} f_{\alpha\bar\beta} + B^\alpha f_\alpha$$

with Hermitian symmetric leading coefficients, then

$$\mathcal{D} \textit{ is self-adjoint} \iff \mathcal{D} \textit{ is in divergence form.}$$

Proof. $\Longleftarrow$: Assume that $\mathcal{D}$ is in divergence form, i.e., that

$$\mathcal{D}(f) = \frac{1}{\omega} \frac{\partial}{\partial z^\alpha}\left(f_{\bar\beta} h^{\bar\beta\alpha} \omega\right)$$

for some $h^{\bar\beta\alpha}$. Let u and v be any compactly supported smooth functions on the manifold M. Then

$$\begin{aligned}
\int_M u\mathcal{D}(v)\omega(z) d\bar z \wedge dz &= \int_M u \frac{1}{\omega} \frac{\partial}{\partial z^\alpha}\left(v_{\bar\beta} h^{\bar\beta\alpha}\omega\right)\omega(z) d\bar z \wedge dz \\
&= -\int_M u_\alpha v_{\bar\beta} h^{\bar\beta\alpha} \omega(z) d\bar z \wedge dz \\
&= \int_M v \frac{\partial}{\partial \bar z^\beta}\left(u_\alpha h^{\bar\beta\alpha}\omega\right) d\bar z \wedge dz \\
&= \int_M v \frac{1}{\omega} \frac{\partial}{\partial \bar z^\beta}\left(u_\alpha h^{\bar\beta\alpha}\omega\right)\omega d\bar z \wedge dz \\
&= \int_M v\mathcal{D}(u)\omega(z) d\bar z \wedge dz.
\end{aligned}$$

This is just the definition of self-adjointness.

$\Longrightarrow$: Let $\mathcal{D}$ be written as in expression (5.3). Assume self-adjointness and choose two arbitrary compactly supported functions u and v. Then

$$\begin{aligned}
0 &= \int_M (u\mathcal{D}v - v\mathcal{D}u)\omega(z)d\bar{z}\wedge dz \\
&= \int_M (uA^{\bar{\beta}\alpha}v_{\alpha\bar{\beta}} + uB^{\alpha}v_{\alpha} - vA^{\bar{\beta}\alpha}u_{\alpha\bar{\beta}} - vB^{\alpha}u_{\alpha})\omega(z)d\bar{z}\wedge dz \qquad (*)\\
&= \int_M \Big[- v_{\alpha}\Big(\frac{\partial A^{\bar{\beta}\alpha}}{\partial \bar{z}^{\beta}}u\omega + A^{\bar{\beta}\alpha}u_{\bar{\beta}}\omega + A^{\bar{\beta}\alpha}u\omega_{\bar{\beta}}\Big) \qquad (**)\\
&\quad + u_{\alpha}\Big(\frac{\partial A^{\bar{\beta}\alpha}}{\partial \bar{z}^{\beta}}v\omega + A^{\bar{\beta}\alpha}v_{\bar{\beta}}\omega + A^{\bar{\beta}\alpha}v\omega_{\bar{\beta}}\Big) \qquad (***)\\
&\quad + uB^{\alpha}v_{\alpha}\omega - vB^{\alpha}u_{\alpha}\omega\Big] d\bar{z}\wedge dz.
\end{aligned}$$

To achieve the three lines above,we used integration by parts on the first term $(*)$ (integrating $v_{\alpha\bar{\beta}}d\bar{z}\wedge dz$) and on the third term in $(*)$ (integrating $u_{\alpha\bar{\beta}}d\bar{z}\wedge dz$). We can now use the Hermitian symmetric of the coefficient $A^{\bar{\beta}\alpha}$ to cancel the second in $(**)$ with the second term in $(***)$.

$$= \int_M (u_{\alpha}v - v_{\alpha}u)\Big[\frac{\partial A^{\bar{\beta}\alpha}}{\partial \bar{z}^{\beta}} + A^{\bar{\beta}\alpha}\frac{\partial}{\partial \bar{z}^{\beta}}\Big(\ln\omega\Big) - B^{\alpha}\Big]\omega(z)d\bar{z}\wedge dz$$

$$0 = \int_M v^2\frac{\partial}{\partial z^{\alpha}}\Big(\frac{u}{v}\Big)\Big[\frac{\partial A^{\bar{\beta}\alpha}}{\partial \bar{z}^{\beta}} + A^{\bar{\beta}\alpha}\frac{\partial}{\partial \bar{z}^{\beta}}\Big(\ln\omega\Big) - B^{\alpha}\Big]\omega(z)d\bar{z}\wedge dz.$$

Because any function can be written as $v^2\frac{\partial}{\partial z^{\alpha}}(\frac{u}{v})$,therefore we have

$$B^{\alpha} = \frac{\partial A^{\bar{\beta}\alpha}}{\partial \bar{z}^{\beta}} + A^{\bar{\beta}\alpha}\frac{\partial}{\partial \bar{z}^{\beta}}\Big(\ln\omega\Big). \qquad (5.6)$$

Furthermore,by the following Sub-Lemma (Lemma 5.5) the conclusion follows. □

Lemma 5.5 (Sub-Lemma). *Recall that $\mathcal{D}f = A^{\bar{\beta}\alpha}f_{\alpha\bar{\beta}} + B^{\alpha}f_{\alpha}$. In this notation,*

$$\mathcal{D}f = \frac{1}{\omega}\frac{\partial}{\partial z^{\alpha}}\Big(f_{\bar{\beta}}h^{\bar{\beta}\alpha}\omega\Big) \iff B^{\alpha} = \frac{\partial A^{\bar{\beta}\alpha}}{\partial \bar{z}^{\beta}} + A^{\bar{\beta}\alpha}\frac{\partial}{\partial \bar{z}^{\beta}}\Big(\ln\omega\Big) \text{ and } A^{\bar{\beta}\alpha} = h^{\bar{\beta}\alpha}.$$

Proof. $\Longrightarrow$:

$$\begin{aligned}
\mathcal{D}f = A^{\bar{\beta}\alpha}f_{\alpha\bar{\beta}} + B^{\alpha}f_{\alpha} &= \frac{1}{\omega}\frac{\partial}{\partial z^{\alpha}}\Big(f_{\bar{\beta}}h^{\bar{\beta}\alpha}\omega\Big)\\
&= f_{\alpha\bar{\beta}}h^{\bar{\beta}\alpha} + f_{\alpha}\Big[\frac{\partial h^{\bar{\beta}\alpha}}{\partial \bar{z}^{\beta}} + h^{\bar{\beta}\alpha}\frac{\partial}{\partial \bar{z}^{\beta}}\Big(\ln\omega\Big)\Big],
\end{aligned}$$

by assumption $A^{\bar{\beta}\alpha} = h^{\bar{\beta}\alpha}$, it follows that

$$B^{\alpha} = \frac{\partial h^{\bar{\beta}\alpha}}{\partial \bar{z}^{\beta}} + h^{\bar{\beta}\alpha}\frac{\partial}{\partial \bar{z}^{\beta}}\Big(\ln\omega\Big) = \frac{\partial A^{\bar{\beta}\alpha}}{\partial \bar{z}^{\beta}} + A^{\bar{\beta}\alpha}\frac{\partial}{\partial \bar{z}^{\beta}}\Big(\ln\omega\Big).$$

$\Longleftarrow$: Assume

$$B^{\alpha} = \frac{\partial A^{\bar{\beta}\alpha}}{\partial \bar{z}^{\beta}} + A^{\bar{\beta}\alpha}\frac{\partial}{\partial \bar{z}^{\beta}}\Big(\ln\omega\Big).$$

Let $h^{\bar\beta\alpha} = A^{\bar\beta\alpha}$, then

$$\begin{aligned}\frac{1}{\omega}\frac{\partial}{\partial z^\alpha}\Big(f_{\bar\beta}h^{\bar\beta\alpha}\omega\Big) &= \frac{1}{\omega}\frac{\partial}{\partial z^\alpha}\Big(f_{\bar\beta}A^{\bar\beta\alpha}\omega\Big)\\ &= A^{\bar\beta\alpha}f_{\alpha\bar\beta} + f_\alpha\frac{\partial A^{\bar\beta\alpha}}{\partial\bar z^\beta} + f_\alpha A^{\bar\beta\alpha}\frac{\partial}{\partial\bar z^\beta}\Big(\ln\omega\Big)\\ &= A^{\bar\beta\alpha}f_{\alpha\bar\beta} + B^\alpha f_\alpha\\ &= \mathcal{D}f.\end{aligned}$$

□

(5.6) now implies (by Lemma 5.5) that

$$\mathcal{D}f = \frac{1}{\omega}\frac{\partial}{\partial z^\alpha}\Big(f_{\bar\beta}h^{\bar\beta\alpha}\omega\Big).$$

for some $h^{\bar\beta\alpha}$. In fact, we can see from the calculations that we must have $h^{\bar\beta\alpha} = A^{\bar\beta\alpha}$.

Recall the the coordinate expression for $\triangle f$,

$$\begin{aligned}\triangle f =&(2n+2)\Bigg[\frac{\int_I v^\alpha\bar v^\beta d\bar v\wedge dv}{\int_I d\bar v\wedge dv}f_{\alpha\bar\beta}\\ &+ f_\alpha\left(\frac{2\omega_{\bar\beta}}{\omega}\frac{\int_I v^\alpha\bar v^\beta d\bar v\wedge dv}{\int_I d\bar v\wedge dv} - \frac{2\int_I\Gamma^{\bar\lambda}_{\bar\lambda;\bar\beta}v^\alpha\bar v^\beta d\bar v\wedge dv}{\int_I d\bar v\wedge dv}\right)\\ &+ f_{\bar\alpha}\left(\frac{2\omega_\beta}{\omega}\frac{\int_I \bar v^\alpha v^\beta d\bar v\wedge dv}{\int_I d\bar v\wedge dv} - \frac{2\int_I\Gamma^{\lambda}_{\lambda;\beta}\bar v^\alpha v^\beta d\bar v\wedge dv}{\int_I d\bar v\wedge dv}\right)\Bigg].\end{aligned}$$

Lemma 5.2 shows that $\triangle$ is self-adjoint, Lemma 4.3 gives conditions on the coefficients of $\triangle$. In particular, the $A^{\bar\beta\alpha}$ in Lemma 5.4 must just be

$$A^{\bar\beta\alpha} = K^{\bar\beta\alpha} = (2n+2)\frac{\int_I v^\alpha\bar v^\beta d\bar v\wedge dv}{\int_I d\bar v\wedge dv},$$

i.e., the coefficients of the osculating Hermitian metric. The B^α of course are just

$$B^\alpha = (2n+2)\left[\frac{2\omega_{\bar\beta}}{\omega}\frac{\int_I v^\alpha\bar v^\beta d\bar v\wedge dv}{\int_I d\bar v\wedge dv} - \frac{2\int_I\Gamma^{\bar\lambda}_{\bar\lambda;\bar\beta}v^\alpha\bar v^\beta d\bar v\wedge dv}{\int_I d\bar v\wedge dv}\right]. \tag{5.7}$$

Lemma 5.5 gives the condition

$$\begin{aligned}B^\alpha &= \frac{\partial A^{\bar\beta\alpha}}{\partial\bar z^\beta} + A^{\bar\beta\alpha}\frac{\partial}{\partial\bar z^\beta}\Big(\ln\omega\Big)\\ &= \frac{\partial}{\partial\bar z^\beta}\left[(2n+2)\frac{\int_I v^\alpha\bar v^\beta d\bar v\wedge dv}{\int_I d\bar v\wedge dv}\right] + (2n+2)\frac{\int_I v^\alpha\bar v^\beta d\bar v\wedge dv}{\int_I d\bar v\wedge dv}\frac{\partial}{\partial\bar z^\beta}\Big(\ln\omega\Big).\end{aligned}$$

Using $\omega = \frac{\kappa_{2n}}{\int_I d\bar{v}\wedge dv}$,

$$\begin{aligned}
\frac{1}{(2n+2)}B^\alpha =& \frac{\frac{\partial}{\partial \bar{z}^\beta}\int_I v^\alpha \bar{v}^\beta d\bar{v}\wedge dv}{\int_I d\bar{v}\wedge dv} \\
&- \frac{\int_I v^\alpha \bar{v}^\beta d\bar{v}\wedge dv \cdot \frac{\partial}{\partial \bar{z}^\beta}\int_I d\bar{v}\wedge dv}{\left(\int_I d\bar{v}\wedge dv\right)^2} + \frac{\int_I v^\alpha \bar{v}^\beta d\bar{v}\wedge dv}{\int_I d\bar{v}\wedge dv}\frac{\partial}{\partial \bar{z}^\beta}\left[\ln\left(\frac{\kappa_{2n}}{\int_I d\bar{v}\wedge dv}\right)\right] \\
=& \frac{\frac{\partial}{\partial \bar{z}^\beta}\int_I v^\alpha \bar{v}^\beta d\bar{v}\wedge dv}{\int_I d\bar{v}\wedge dv} - \frac{\int_I v^\alpha \bar{v}^\beta d\bar{v}\wedge dv \cdot \frac{\partial}{\partial \bar{z}^\beta}\int_I d\bar{v}\wedge dv}{\left(\int_I d\bar{v}\wedge dv\right)^2} \\
&- \frac{\int_I v^\alpha \bar{v}^\beta d\bar{v}\wedge dv}{\int_I d\bar{v}\wedge dv}\frac{1}{\int_I d\bar{v}\wedge dv}\frac{\partial}{\partial \bar{z}^\beta}\int_I d\bar{v}\wedge dv \\
=& -2\frac{\partial}{\partial \bar{z}^\beta}\left[\ln\left(\int_I d\bar{v}\wedge dv\right)\right]\frac{\int_I v^\alpha \bar{v}^\beta d\bar{v}\wedge dv}{\int_I d\bar{v}\wedge dv} + \frac{\frac{\partial}{\partial \bar{z}^\beta}\int_I v^\alpha \bar{v}^\beta d\bar{v}\wedge dv}{\int_I d\bar{v}\wedge dv}. \qquad (5.8)
\end{aligned}$$

Of course from (5.7) we already know

$$\frac{1}{(2n+2)}B^\alpha = \frac{2\omega_{\bar{\beta}}}{\omega}\frac{\int_I v^\alpha \bar{v}^\beta d\bar{v}\wedge dv}{\int_I d\bar{v}\wedge dv} - \frac{2\int_I \Gamma^{\bar{\lambda}}_{\bar{\lambda};\beta} v^\alpha \bar{v}^\beta d\bar{v}\wedge dv}{\int_I d\bar{v}\wedge dv},$$

and if we again use $\omega = \frac{\kappa_{2n}}{\int_I d\bar{v}\wedge dv}$, then

$$\frac{1}{(2n+2)}B^\alpha = -2\frac{\partial}{\partial \bar{z}^\beta}\left[\ln\left(\int_I d\bar{v}\wedge dv\right)\right]\frac{\int_I v^\alpha \bar{v}^\beta d\bar{v}\wedge dv}{\int_I d\bar{v}\wedge dv} - \frac{2\int_I \Gamma^{\bar{\lambda}}_{\bar{\lambda};\beta} v^\alpha \bar{v}^\beta d\bar{v}\wedge dv}{\int_I d\bar{v}\wedge dv}. \qquad (5.9)$$

Equating (5.7) and (5.8), cancelling the common term, and multiplying by $\int_I d\bar{v}\wedge dv$ gives

$$\frac{\partial}{\partial \bar{z}^\beta}\int_I v^\alpha \bar{v}^\beta d\bar{v}\wedge dv = -2\int_I \Gamma^{\bar{\lambda}}_{\bar{\lambda};\beta} v^\alpha \bar{v}^\beta d\bar{v}\wedge dv. \qquad (5.10)$$

This, then, must be an identity in strongly Kähler–Finsler manifolds.

If we go back to Lemma 5.4, we can see a way to write $\triangle f$ in divergence form directly. The proof of Lemma 5.4 necessitates $h^{\bar{\beta}\alpha} = A^{\bar{\beta}\alpha} = K^{\bar{\beta}\alpha}$, the coefficient of the osculating Hermitian metric, and this is all needed to prove.

Theorem 5.6. *Let M be a complex manifold with a strongly Kähler–Finsler metric F, and let $f : M \to \mathbb{R}$ be a smooth function. Let $dV = \omega(z)d\bar{z}\wedge dz$, where $\omega(z) = \frac{\kappa_{2n}}{\int_I d\bar{v}\wedge dv}$ is the coefficient of the canonical volume form, and $K^{\bar{\beta}\alpha} = (2n+2)\int_I v^\alpha \bar{v}^\beta d\bar{v}\wedge dv / \int_I d\bar{v}\wedge dv$ is the osculating Hermitian metric. Then the Laplacian*

$$\triangle f(p) = 2(2n+2)\lim_{\varepsilon\to 0}\frac{1}{\varepsilon^2}\left(\frac{\int_{\Sigma_\varepsilon} f dV}{\int_{\Sigma_\varepsilon} dV} - f(p)\right)$$

can also be written as

$$\triangle f(p) = \frac{1}{\omega(p)}\frac{\partial}{\partial z^\alpha}\left[f_{\bar{\beta}}K^{\bar{\beta}\alpha}\omega(p)\right]. \qquad (5.11)$$

Thus $\triangle$ is a self-adjoint operator (that is, $\int_M (u\triangle v - v\triangle u)\omega(w)d\bar{w}\wedge dw = 0$) in divergence form. □

This provides an interesting interpretation of our Laplacian: we can see $\triangle$ as the Laplacian of the osculating Hermitian manifold with respect to a volume form which comes from the strongly Kähler–Finsler metric.

6. The Laplacian for differential forms

So far, we have defined the mean-value Laplacian for functions on a strongly Kähler–Finsler manifold. Now we will extend our definition to arbitrary (p,q)-forms. The key point is defining a pointwise inner product and a global inner product.

Assume M is a compact complex Finsler manifold with a strongly Kähler–Finsler metric F. $\Omega^{p,q}(M)$ (or $\Omega^{p,q}$) denotes the $C^\infty(p,q)$-forms on M. The elements of $\Omega^{p,q}$ in local coordinates are

$$\varphi(z) = \frac{1}{p!q!}\varphi_{\alpha_1\cdots\alpha_p\bar{\beta}_1\cdots\bar{\beta}_q} dz^{\alpha_1}\wedge\cdots\wedge dz^{\alpha_p}\wedge d\bar{z}^{\beta_1}\wedge\cdots\wedge d\bar{z}^{\beta_q},$$

$$\psi(z) = \frac{1}{p!q!}\psi_{\lambda_1\cdots\lambda_p\bar{\mu}_1\cdots\bar{\mu}_q} dz^{\lambda_1}\wedge\cdots\wedge dz^{\lambda_p}\wedge d\bar{z}^{\mu_1}\wedge\cdots\wedge d\bar{z}^{\mu_q}.$$

Define the pointwise inner product of $\varphi(z)$ and $\psi(z)$ by

$$\begin{aligned}\langle\varphi,\psi\rangle(z) = &\int_I\int_I\cdots\int_I\times\int_I\int_I\cdots\int_I \varphi(v)\bar{\psi}(v)\Big(\frac{(2n+2)d\bar{v}_1\wedge dv_1}{\int_I d\bar{v}\wedge dv}\Big)\\ &\times\Big(\frac{(2n+2)d\bar{v}_2\wedge dv_2}{\int_I d\bar{v}\wedge dv}\Big)\cdots\Big(\frac{(2n+2)d\bar{v}_p\wedge dv_p}{\int_I d\bar{v}\wedge dv}\Big)\\ &\times\Big(\frac{(2n+2)d\bar{v}_1\wedge dv_1}{\int_I d\bar{v}\wedge dv}\Big)\Big(\frac{(2n+2)d\bar{v}_2\wedge dv_2}{\int_I d\bar{v}\wedge dv}\Big)\times\cdots\times\Big(\frac{(2n+2)d\bar{v}_q\wedge dv_q}{\int_I d\bar{v}\wedge dv}\Big),\end{aligned}\tag{6.1}$$

where

$$\begin{aligned}&\varphi(v_1,v_2,\cdots,v_p;\bar{v}_1,\cdots,\bar{v}_q)\bar{\psi}(v_1,v_2,\cdots,v_p;\bar{v}_1,\cdots,\bar{v}_q)\\ &=\frac{1}{p!q!}\varphi_{\alpha_1\cdots\alpha_p\bar{\beta}_1\cdots\bar{\beta}_q}\bar{\psi}_{\lambda_1\cdots\lambda_p\bar{\mu}_1\cdots\bar{\mu}_q}\\ &\quad\times v_1^{\alpha_1}v_2^{\alpha_2}\cdots v_p^{\alpha_p}\bar{v}_1^{\beta_1}\bar{v}_2^{\beta_2}\cdots\bar{v}_q^{\beta_q}\bar{v}_1^{\lambda_1}\bar{v}_2^{\lambda_2}\cdots\bar{v}_p^{\lambda_p}v_1^{\mu_1}v_2^{\mu_2}\cdots v_q^{\mu_q}.\end{aligned}$$

If we substitute this into (6.1),then we can perform the integrations over $v_1,v_2,\ldots,v_p$ and $v_1,v_2,\ldots,v_q$ separately, and since the coefficients $\varphi_{\alpha_1\cdots\alpha_p\bar{\beta}_1\cdots\bar{\beta}_q}(z)$ and $\psi_{\lambda_1\cdots\lambda_p\bar{\mu}_1\cdots\bar{\mu}_q}(z)$ do not depend on tangent vectors, we can take them outside the integral to get (by Lemma 4.1)

$$\begin{aligned}\langle\varphi,\psi\rangle(z) =&\frac{1}{p!q!}\varphi_{\alpha_1\cdots\alpha_p\bar{\beta}_1\cdots\bar{\beta}_q}\bar{\psi}_{\lambda_1\cdots\lambda_p\bar{\mu}_1\cdots\bar{\mu}_q}(2n+2)^{2(p+q)}\times\frac{\int_I v_1^{\alpha_1}\bar{v}_1^{\lambda_1}d\bar{v}_1\wedge dv_1}{\int_I d\bar{v}\wedge dv}\cdots\\ &\frac{\int_I v_p^{\alpha_p}\bar{v}_p^{\lambda_p}d\bar{v}_p\wedge dv_p}{\int_I d\bar{v}\wedge dv}\frac{\int_I v_1^{\mu_1}\bar{v}_1^{\beta_1}d\bar{v}_1\wedge dv_1}{\int_I d\bar{v}\wedge dv}\cdots\frac{\int_I v_q^{\mu_q}\bar{v}_q^{\beta_q}d\bar{v}_q\wedge dv_q}{\int_I d\bar{v}\wedge dv}\\ =&\frac{1}{p!q!}\varphi_{\alpha_1\cdots\alpha_p\bar{\beta}_1\cdots\bar{\beta}_q}\bar{\psi}_{\lambda_1\cdots\lambda_p\bar{\mu}_1\cdots\bar{\mu}_q}K^{\bar{\lambda}_1\alpha_1}\cdots K^{\bar{\lambda}_p\alpha_p}K^{\bar{\beta}_1\mu_1}\cdots K^{\bar{\beta}_q\mu_q}\end{aligned}$$

$$= \frac{1}{p!q!}\varphi_{\alpha_1\cdots\alpha_p\bar{\beta}_1\cdots\bar{\beta}_q}\overline{\psi^{\bar{\alpha}_1\cdots\bar{\alpha}_p\beta_1\cdots\beta_q}}, \tag{6.2}$$

where

$$\overline{\psi^{\bar{\alpha}_1\cdots\bar{\alpha}_p\beta_1\cdots\beta_q}} = \bar{\psi}_{\lambda_1\cdots\lambda_p\bar{\mu}_1\cdots\bar{\mu}_q}K^{\bar{\lambda}_1\alpha_1}\cdots K^{\bar{\lambda}_p\alpha_p}K^{\bar{\beta}_1\mu_1}\cdots K^{\bar{\beta}_q\mu_q}.$$

When (M,F) is a Kähler manifold, then $K^{\bar{\lambda}_1\alpha_1} = 2\mathcal{G}^{\bar{\lambda}_1\alpha_1},\cdots,K^{\bar{\beta}_q\mu_q} = 2\mathcal{G}^{\bar{\beta}_q\mu_q}$ (Lemma 4.3), and the pointwise inner product (6.1),(6.2) reduces to the case of a Kähler manifold.

We define this pointwise inner product in a particular coordinate system, so we should check that this definition is independent of coordinates. This follows from writing

$$\widetilde{\varphi}_{k_1k_2\cdots k_p\bar{\ell}_1\bar{\ell}_2\cdots\bar{\ell}_q} = \frac{\partial\tilde{z}^{\alpha_1}}{\partial z^{k_1}}\frac{\partial\tilde{z}^{\alpha_2}}{\partial z^{k_2}}\cdots\frac{\partial\tilde{z}^{\alpha_p}}{\partial z^{k_p}}\frac{\partial\bar{\tilde{z}}^{\bar{\beta}_1}}{\partial\bar{z}^{\ell_1}}\frac{\partial\bar{\tilde{z}}^{\bar{\beta}_2}}{\partial\bar{z}^{\ell_2}}\cdots\frac{\partial\bar{\tilde{z}}^{\bar{\beta}_q}}{\partial\bar{z}^{\ell_q}}\varphi_{\alpha_1\cdots\alpha_p\bar{\beta}_1\cdots\bar{\beta}_q}, \tag{6.3}$$

$$\widetilde{\psi}_{s_1s_2\cdots s_p\bar{t}_1\bar{t}_2\cdots\bar{t}_q} = \frac{\partial\tilde{z}^{\lambda_1}}{\partial z^{s_1}}\frac{\partial\tilde{z}^{\lambda_2}}{\partial z^{s_2}}\cdots\frac{\partial\tilde{z}^{\lambda_p}}{\partial z^{s_p}}\frac{\partial\bar{\tilde{z}}^{\bar{\mu}_1}}{\partial\bar{z}^{t_1}}\frac{\partial\bar{\tilde{z}}^{\bar{\mu}_2}}{\partial\bar{z}^{t_2}}\cdots\frac{\partial\bar{\tilde{z}}^{\bar{\mu}_q}}{\partial\bar{z}^{t_q}}\psi_{\lambda_1\cdots\lambda_p\bar{\mu}_1\cdots\bar{\mu}_q}, \tag{6.4}$$

$$\widetilde{K}^{\bar{s}_1k_1} = \frac{\partial\bar{z}^{s_1}}{\partial\bar{\tilde{z}}^{\lambda_1}}\frac{\partial z^{k_1}}{\partial\tilde{z}^{\alpha_1}}K^{\bar{\lambda}_1\alpha_1},\cdots,\widetilde{K}^{\bar{\ell}_qt_q} = \frac{\partial\bar{z}^{\ell_q}}{\partial\bar{\tilde{z}}^{\beta_q}}\frac{\partial z^{t_q}}{\partial\tilde{z}^{\mu_q}}K^{\bar{\beta}_q\mu_q}, \tag{6.5}$$

in a tilde coordinate system $\tilde{z}^\alpha = \tilde{z}^\alpha(z)$. Substituting (6.3), (6.4) and (6.5) into (6.2) gives

$$\begin{aligned}&\varphi_{\alpha_1\cdots\alpha_p\bar{\beta}_1\cdots\bar{\beta}_q}\bar{\psi}_{\lambda_1\cdots\lambda_p\bar{\mu}_1\cdots\bar{\mu}_q}K^{\bar{\lambda}_1\alpha_1}\cdots K^{\bar{\lambda}_p\alpha_p}K^{\bar{\beta}_1\mu_1}\cdots K^{\bar{\beta}_q\mu_q}\\ =&\widetilde{\varphi}_{k_1k_2\cdots k_p\bar{\ell}_1\bar{\ell}_2\cdots\bar{\ell}_q}\bar{\widetilde{\psi}}_{s_1s_2\cdots s_p\bar{t}_1\bar{t}_2\cdots\bar{t}_q}\widetilde{K}^{\bar{s}_1k_1}\widetilde{K}^{\bar{s}_2k_2}\cdots\widetilde{K}^{\bar{s}_pk_p}\widetilde{K}^{\bar{\ell}_1t_1}\widetilde{K}^{\bar{\ell}_2t_2}\cdots\widetilde{K}^{\bar{\ell}_qt_q},\end{aligned}$$

so that the pointwise inner product (6.1) or (6.2) is invariant on the manifold, this invariance depending solely on the tensor (density) character of

$$K^{\bar{\lambda}_1\alpha_1}\cdots K^{\bar{\lambda}_p\alpha_p}K^{\bar{\beta}_1\mu_1}\cdots K^{\bar{\beta}_q\mu_q}\omega(z).$$

There are many other possible densities, too ([5], [9]).

Associated to the osculating Kähler metric we have the Kähler form

$$\omega = iK_{\alpha\bar{\beta}}dz^\alpha d\bar{z}^\beta$$

and

$$\omega^n = 2^n n! K C_n d\bar{z}dz, K = \det(K_{\alpha\bar{\beta}}).$$

By the pointwise inner products (6.1) or (6.2) we can obviously define the global inner product $(\cdot,\cdot)$ on $\Omega^{p,q}(M)$:

$$\begin{aligned}(\varphi,\psi) =& \int_M \langle\varphi,\psi\rangle(z)\frac{\omega^n}{n!}\\ =&2^nC_n\int_M \varphi_{\alpha_1\cdots\alpha_p\bar{\beta}_1\cdots\bar{\beta}_q}\bar{\psi}_{\lambda_1\cdots\lambda_p\bar{\mu}_1\cdots\bar{\mu}_q}K^{\bar{\lambda}_1\alpha_1}\cdots K^{\bar{\lambda}_p\alpha_p}\\ &\times K^{\bar{\beta}_1\mu_1}\cdots K^{\bar{\beta}_q\mu_q}Kd\bar{z}\wedge dz.\end{aligned} \tag{6.6}$$

Now we define the Laplace operator for $\Omega^{p,q}(M)$. First we define a co-differential $\delta : \Omega^p(M) \to \Omega^{p-1}(M)$, where $\Omega^p(M)$ is the space of $(p,0)$-forms on M by requiring (for any $(p,0)$-form φ and $(p-1,0)$-form ψ)

$$(\varphi, d\psi) = (\delta\varphi, \psi). \tag{6.7}$$

As usual, define $\delta f = 0$ for any function f. Now simply define, for any $(p,0)$-form φ,

$$\Delta\varphi = d\delta\varphi + \delta d\varphi.$$

This is our Laplacian on $(p,0)$-forms.

We should check that this new definition is compatible with the old one for the case of 0-forms, or functions. We have

Theorem 6.1. *Let $f : M \to \mathbb{R}$ be a smooth function. Our two definitions of Laplacians agree, i.e.,*

$$\Delta f = \delta df = \frac{1}{\omega(z)} \frac{\partial}{\partial \bar{z}^\beta}\Big[f_\alpha K^{\bar{\beta}\alpha}\omega(z)\Big].$$

Proof. We will calculate explicitly $\delta : \Omega^1(M) \to \Omega^0(M)$. Let φ be a $(1,0)$-form. Condition (6.7) requires

$$(\varphi, d\psi) = (\delta\varphi, \psi),$$

where ψ is a $(0,0)$-form, i.e., ψ is a function $\psi : M \to \mathbb{R}$. Using (6.6), the definition of the global inner product, we write

$$(\varphi, d\psi) = C_n \int_M \varphi_\alpha (d\psi)_{\bar{\beta}} K^{\bar{\beta}\alpha}\omega(z) d\bar{z} \wedge dz.$$

Since $d\psi$ is just the exterior derivative of ψ, $(d\psi)_{\bar{\beta}} = \psi_{\bar{\beta}}$, it follows

$$\begin{aligned}(\varphi, d\psi) &= C_n \int_M \varphi_\alpha \psi_{\bar{\beta}} K^{\bar{\beta}\alpha}\omega(z) d\bar{z} \wedge dz \\ &= C_n \int_M \psi \frac{\partial}{\partial \bar{z}^\beta}\Big[\varphi_\alpha K^{\bar{\beta}\alpha}\omega(z)\Big] d\bar{z} \wedge dz \\ &= C_n \int_M \frac{1}{\omega(z)} \psi \frac{\partial}{\partial \bar{z}^\beta}\Big[\varphi_\alpha K^{\bar{\beta}\alpha}\omega(z)\Big] d\bar{z} \wedge dz \\ &= \Big(\frac{1}{\omega(z)} \frac{\partial}{\partial \bar{z}^\beta}\Big[\varphi_\alpha K^{\bar{\beta}\alpha}\omega(z)\Big], \psi\Big).\end{aligned}$$

Since (6.7) determines δ uniquely, it must be that

$$\delta\varphi = \frac{1}{\omega(z)} \frac{\partial}{\partial \bar{z}^\beta}\Big[\varphi_\alpha K^{\bar{\beta}\alpha}\omega(z)\Big].$$

Now let $\varphi = df$. Then $\varphi_\alpha = f_\alpha$, and so

$$\delta\varphi = \delta df = \frac{1}{\omega(z)} \frac{\partial}{\partial \bar{z}^\beta}\Big[f_\alpha K^{\bar{\beta}\alpha}\omega(z)\Big],$$

or equivalently,

$$\delta\varphi = \delta df = \frac{1}{\omega(z)} \frac{\partial}{\partial z^\alpha}\Big[f_{\bar{\beta}} K^{\bar{\beta}\alpha}\omega(z)\Big],$$

which was to be shown. □

Theorem 6.2. *The Laplace operator $\triangle$ for $(p,0)$-forms is self-adjoint.*

Proof. Let φ, ψ be two $(p,0)$-forms. Then

$$\begin{aligned}(\varphi, \triangle\psi) &= (\varphi, (d\delta + \delta d)\psi)\\ &= (\varphi, d\delta\psi) + (\varphi, \delta d\psi)\\ &= (\delta\varphi, \delta\psi) + (d\varphi, d\psi)\\ &= (d\delta\varphi, \psi) + (\delta d\varphi, \psi)\\ &= ((d\delta + \delta d)\varphi, \psi)\\ &= (\triangle\varphi, \psi).\end{aligned}$$

□

In terms of integrals,we would write

$$\begin{aligned}&C_n \int_M \varphi_{\alpha_1\alpha_2\cdots\alpha_p}(\triangle\psi)_{\bar\beta_1\bar\beta_2\cdots\bar\beta_p} K^{\bar\beta_1\alpha_1}K^{\bar\beta_2\alpha_2}\cdots K^{\bar\beta_p\alpha_p}\omega(z)d\bar z\wedge dz\\ =&C_n \int_M \psi_{\alpha_1\alpha_2\cdots\alpha_p}(\triangle\varphi)_{\bar\beta_1\bar\beta_2\cdots\bar\beta_p} K^{\bar\beta_1\alpha_1}K^{\bar\beta_2\alpha_2}\cdots K^{\bar\beta_p\alpha_p}\omega(z)d\bar z\wedge dz,\end{aligned}$$

which is a direct generalization of the condition given for functions.

In the following we define the Hodge–Laplace operator for (p,q)-forms on $\Omega^{p,q}(M)$. Fist we define the operator $* : \Omega^{p,q}(M) \to \Omega^{n-q,n-p}(M)$ by the relation

$$\varphi \wedge *\bar\psi = \langle\varphi, \psi\rangle dV(M).$$

Similar to the proof of Theorem 2.1 in [10] we can prove

Theorem 6.3. *There is a linear map $* : \Omega^{p,q}(M) \to \Omega^{n-q,n-p}(M)$ such that*

(1) $\varphi \wedge *\bar\psi = \langle\varphi, \psi\rangle dV(M)$;
(2) $\overline{*\psi} = *\bar\psi$;
(3) $**\varphi = (-1)^{p+q}\varphi$. □

Then the inner product $(\cdot,\cdot)$ of (6.6) can also be expressed as

$$(\varphi, \psi) = \int_M \varphi \wedge *\bar\psi. \tag{6.8}$$

Next we define the adjoint of $\bar\partial, \partial$ and d with respect to $(\cdot,\cdot)$ on a compact strongly Kähler–Finsler manifold:

$$\begin{aligned}&\bar\partial^* : \Omega^{(p,q)}(M) \to \Omega^{(p,q-1)}(M),\\ &\partial^* : \Omega^{(p,q)}(M) \to \Omega^{(p-1,q)}(M),\\ &\delta : \Omega^{p}(M) \to \Omega^{(p-1)}(M),\\ &(\bar\partial\varphi, \psi) = (\varphi, \bar\partial^*\psi),\ (\partial\varphi, \psi) = (\varphi, \partial^*\psi),\ (d\varphi, \psi) = (\varphi, \delta\psi).\end{aligned}$$

Theorem 6.4. *Let (M,F) be a compact strongly Kähler–Finsler manifold. Then*

$$\bar\partial^* = -*\partial*,\quad \partial^* = -*\bar\partial*,\quad \delta = -*d*.$$

Proof. We only prove the first equation, the remaining proofs are similar. Take $\varphi \in \Omega^{(p,q)}(M)$, $\psi \in \Omega^{(p,q+1)}(M)$, then $\chi = \varphi \wedge *\bar{\psi} \in \Omega^{(n,n-1)}(M)$, so that

$$d\chi = (\partial + \bar{\partial})\chi = \bar{\partial}\chi,$$

therefore

$$0 = \int_M \bar{\partial}\chi = \int_M \bar{\partial}(\varphi \wedge *\bar{\psi}) = \int_M \bar{\partial}\varphi \wedge *\bar{\psi} + (-1)^{p+q} \int_M \varphi \wedge \bar{\partial}(*\bar{\psi}),$$

and we have

$$\begin{aligned}(\bar{\partial}\varphi, \psi) &= \int_M \bar{\partial}\varphi \wedge *\bar{\psi} = (-1) \int_M \varphi \wedge (-1)^{p+q}\bar{\partial} * \bar{\psi} = (-1) \int_M \varphi \wedge * * \bar{\partial} * \bar{\psi} \\ &= (\varphi, - * \partial * \psi) = (\varphi, \bar{\partial}^* \psi),\end{aligned}$$

that is, $\bar{\partial}^* = - * \partial *$. □

Definition 6.5. Let

$$\triangle = d\delta + \delta d, \quad \Box = \bar{\partial}\bar{\partial}^* + \bar{\partial}^*\bar{\partial}, \quad \overline{\Box} = \partial\partial^* + \partial^*\partial,$$

where $d = \partial + \bar{\partial}$, $\delta = \partial^* + \bar{\partial}^*$, operators $\triangle, \Box : \Omega^{p,q}(M) \longrightarrow \Omega^{p,q}(M)$ are called *Hodge–Laplace operator.*

Evidently,we have

$$\begin{aligned}\triangle &= d\delta + \delta d = (\partial + \bar{\partial})(\bar{\partial}^* + \partial^*) + (\bar{\partial}^* + \partial^*)(\partial + \bar{\partial}) \\ &= \Box + \overline{\Box} + \partial\bar{\partial}^* + \bar{\partial}^*\partial + \overline{\partial\bar{\partial}^* + \bar{\partial}^*\partial}.\end{aligned}$$

Definition 6.6. Let (M, F) be compact strongly Kähler–Finsler manifolds and $\varphi \in \Omega^{p,q}(M)$. If $\Box\varphi = 0$, then φ is called a *harmonic* (p, q)*-form.*

Theorem 6.7. *Let* (M, F) *be a compact strongly Kähler–Finsler manifold. Then* $\Box\varphi = 0$ *if and only if* $\bar{\partial}\varphi = \bar{\partial}^*\varphi = 0$.

Proof.

$$\begin{aligned}(\Box\varphi, \varphi) &= (\bar{\partial}\bar{\partial}^*\varphi, \varphi) + (\bar{\partial}^*\bar{\partial}\varphi, \varphi) \\ &= (\bar{\partial}^*\varphi, \bar{\partial}^*\varphi) + (\bar{\partial}\varphi, \bar{\partial}\varphi).\end{aligned}$$

□

Theorem 6.8. $\Box$ *is self-adjoint, that is,* $(\Box\varphi, \psi) = (\varphi, \Box\psi)$.

Proof.

$$\begin{aligned}(\Box\varphi, \psi) &= (\bar{\partial}\bar{\partial}^*\varphi, \psi) + (\bar{\partial}^*\bar{\partial}\varphi, \psi) \\ &= (\bar{\partial}^*\varphi, \bar{\partial}^*\psi) + (\bar{\partial}\varphi, \bar{\partial}\psi) \\ &= (\varphi, (\bar{\partial}\bar{\partial}^* + \bar{\partial}^*\bar{\partial})\psi) \\ &= (\varphi, \Box\psi).\end{aligned}$$

□

References

[1] S.S. Chern, *Finsler geometry is just Riemannian geometry without the quadratic restriction.* AMS Notices **43** (1996), 959–963.

[2] D. Bao, S.S. Chern and Z. Shen (eds.), *Finsler geometry* (Proceedings of the Joint Summer Research Conference on Finsler Geometry, July 16–20,1995, Seattle, Washington). Cont. Math. Vol. **196** (1996) Amer. Math. Soc. Providence, RI.

[3] D. Bao, S.S. Chern and Z. Shen, *An introduction to Riemann-Finsler geometry.* New York: Springer-Verlag, New York, 2000.

[4] M. Abate and G. Patrizio, *Finsler metrics - A global approach.* Lecture Notes in Math, Vol **1591**, Springer-Verlag, 1994.

[5] D. Bao and B. Lackey, *A Hodge decomposition theorem for Finsler spaces.* C.R. Acad. Sci. Paris **323** (1996), 51–56.

[6] P.L. Antonelli and B. Lackey (eds.), *The Theory of Finslerian Laplacians and applications.* MAIA **459**, Kluwer Academic Publishers, 1998.

[7] P. Centore, *A mean-value Laplacian for Finsler spaces.* The theory of Finslerian Laplacians and applications, MAIA **459**,Kluwer Academic Publishers, 1998, 151–186.

[8] K. Yano and S. Bochner, *Curvature and Betti Numbers.* Princeton Univ. Press, 1953.

[9] Zhong Chunping and Zhong Tongde, *Hodge–Laplace operator on complex Finsler manifolds.* Proceedings of the ICM 2002 Satellite Conference on Geometric Function Theory in Several Complex Variables, USTC. Hefei, Anhui, P.R. of China, Aug. 30–Sep. 2, 2002. Edited by C.H. Fitzgerald and Sheng Gong. World Scientific. (to appear).

[10] J. Morrow, K. Kodaira,*Complex manifolds*, New York: Holt, Rinehart and Winston, Inc, 1971.

Zhong Chunping
School of Mathematical Sciences
Xiamen University
Xiamen, 361005
P.R. China
e-mail: `chpzhong@sina.com`

Zhong Tongde
Institute of Mathematics
Xiamen University
Xiamen, 361005
P.R. China

Part C
Applications

Trends in Mathematics: *Advances in Analysis and Geometry*, 289–300

Non-commutative Determinants and Quaternionic Monge-Ampère Equations

Semyon Alesker

Abstract. First we give a survey of the notions and the properties of non-commutative determinants of Moore and Dieudonne, especially in the quaternionic case, with a particular emphasis to applications in quaternionic analysis. Then we review the theory of plurisubharmonic functions and Monge-Ampere equations in quaternionic variables, following [4] and [5].

Mathematics Subject Classification (2000). 15A33, 32U05, 35J60.

Keywords. Moore determinant, Monge-Ampére equation, plurisubharmonic function.

0. Introduction

This paper has two goals. First, we give a survey of the notions and properties of non-commutative determinants of Dieudonnè and Moore especially in the quaternionic case and with special emphasis to applications in quaternionic analysis. In fact the Moore determinant is usually defined only for quaternionic hyperhermitian matrices (though the quaternions can be replaced by more general fields with some rather special properties; we do not discuss here such generalizations). Our exposition follows mostly [4]. Some references are given in Section 1 of this paper. For more references see [4]. Another goal is to discuss the theory of plurisubharmonic functions of quaternionic variables and of quaternionic Monge-Ampère equations. The notion of plurisubharmonic function of quaternionic variables was introduced by the author in [4] and independently by G. Henkin [24]. The quaternionic Monge-Ampère equation was introduced and studied by the author in [5] where in particular the existence and uniqueness of the solution of the Dirichlet problem was proved (under appropriate assumptions). This material is discussed in Section 2. Some references to the works in quaternionic analysis are given in Section 2. In Section 3 we state some open questions.

We believe that the theory of non-commutative determinants should be very relevant for Clifford and quaternionic analysis. The quaternionic Monge-Ampère equations and plurisubharmonic functions of quaternionic variables are examples of such applications which seems to be interesting by itself. The standard reference for Clifford analysis is the book by Brackx, Delanghe, and Sommen [11].

1. Linear algebra

In this section we will discuss the determinants of Moore and Dieudonné especially in the context of quaternionic matrices. The most general theory of non-commutative determinants (or quasideterminants) is due to I. Gelfand and V. Retakh. It was introduced in [20], see the recent survey [22] for further developments and applications. Though both the Dieudonné and Moore determinants can be expressed through the Gelfand-Retakh quasideterminants [21], [22] we will discuss a more elementary traditional approach to them.

1.1. The Dieudonné determinant

The Dieudonné determinant of quaternionic matrices takes values in non-negative real numbers and behaves exactly like the absolute value of the usual determinant of real or complex matrices from all points of view (algebraic and analytic). However we would like to present first purely algebraic theory of this determinant over an arbitrary non-commutative field following the original approach of Dieudonné [17] (see also [6]).

Let F be an infinite field, not necessarily commutative. Let $M_n(F)$ denote the ring of $n \times n$-matrices with entries in F. Let $GL_n(F)$ denote the group of invertible $n \times n$-matrices. An *elementary* matrix is a matrix which has units on the diagonal and at most one non-zero element out of the diagonal. Let E_n denote the subgroup of $GL_n(F)$ generated by all elementary matrices. Set also $F^*_{ab} := F^*/[F^*, F^*]$ the abelinization of the multiplicative group of F (here F^* denotes the multiplicative group of F, and $[F^*, F^*]$ denotes its commutator subgroup).

1.1.1. Theorem (Dieudonné). *Let $n \geq 2$. The group E_n is a normal subgroup of $GL_n(F)$. For the quotient-group $GL_n(F)/E_n$ there exists a natural isomorphism $D : GL_n(F)/E_n \longrightarrow F^*_{ab}$.*
This isomorphism D is uniquely defined by the property that for any invertible diagonal matrix $X = \begin{bmatrix} x_1 & & 0 \\ & \ddots & \\ 0 & & x_n \end{bmatrix}$, $D(X) = \prod_i x_i \bmod [F^*, F^*]$.

1.1.2. Definition (Dieudonné determinant). The Dieudonné determinant is a map

$$D : M_n(F) \longrightarrow F^*_{ab} \cup \{0\}$$

defined as follows: if X is an invertible matrix then $D(X)$ is as in Theorem 1.1.1; if X is not invertible then $D(X) := 0$.

Note also that it is convenient to define the Dieudonné determinant of elements of F, i.e. 1×1-matrices, as $D(0) = 0$ and for $x \neq 0$ as $D(x) := x \bmod [F^*, F^*]$. Let us state some basic general properties of the Dieudonné determinant. For the proofs we again refer to [6].

1.1.3. Theorem. (i) $D(Id) = 1$.

(ii) *For* $X, Y \in M_n(F)$,

$$D(XY) = D(X)D(Y).$$

(iii) *For any block-matrix* $A = \begin{bmatrix} X & 0 \\ 0 & Y \end{bmatrix}$, *with* X, Y *being square matrices,*

$$D(A) = D(X)D(Y).$$

(iv) *If one interchanges two rows or two columns of the matrix then the Dieudonné determinant is multiplied by* $-1 \bmod [F^*, F^*]$.

Now let us consider in more details the case of the quaternionic field $F = \mathbb{H}$. The commutator subgroup $[\mathbb{H}^*, \mathbb{H}^*]$ coincides with the subgroup of quaternions of absolute value 1. Thus we can identify $\mathbb{H}^*_{ab}$ with the multiplicative group $\mathbb{R}_{>0}$ by

$$q \bmod [\mathbb{H}^*, \mathbb{H}^*] \leftrightarrow |q| := \sqrt{q\bar{q}}.$$

So in the quaternionic case the Dieudonné determinant maps

$$D : M_n(\mathbb{H}) \longrightarrow \mathbb{R}_{\geq 0}.$$

Let us discuss its properties in greater detail.

1.1.4. Theorem. (i) *For any complex* $(n \times n)$*-matrix* X *considered as quaternionic matrix, the Dieudonné determinant* $D(X)$ *is equal to the absolute value of the usual determinant of* X.

(ii) *For any quaternionic matrix* X,

$$D(X) = D(X^t) = D(X^*),$$

where X^t *and* X^* *denote the transposed and quaternionic conjugate matrices of* X, *respectively.*

(iii) $D(X \cdot Y) = D(X)D(Y)$.

The following result is a weak version of the decomposition of the determinant in rows (columns).

1.1.5. Theorem. *Let* $A = \begin{bmatrix} a_{11} & \dots & a_{1n} \\ \dots & \dots & \dots \\ a_{n1} & \dots & a_{nn} \end{bmatrix}$ *be a quaternionic matrix. Then*

$$D(A) \leq \sum_{i=1}^{n} |a_{1i}| D(M_{1i}).$$

Similar inequalities hold for any other row or column.

(In this theorem $|a|$ denotes the absolute value of a quaternion a, and M_{pq} denotes the minor of the matrix A obtained from it by deleting the p-th row and q-th column).

In a sense, the Dieudonné determinant provides the theory of *absolute value* of the determinant. However it is not always sufficient and we loose many of the algebraic properties of the usual determinant. The notion of the Moore determinant provides such a theory, but only on the class of quaternionic *hyperhermitian* matrices. These notions will be discussed in the next subsection.

1.2. The Moore determinant

Let V be a right vector space over quaternions.

1.2.1. Definition. A *hyperhermitian semilinear form* on V is a map $a : V \times V \longrightarrow \mathbb{H}$ satisfying the following properties:

(a) a is additive with respect to each argument;
(b) $a(x, y \cdot q) = a(x, y) \cdot q$ for any $x, y \in V$ and any $q \in \mathbb{H}$;
(c) $a(x, y) = \overline{a(y, x)}$ for any $x, y \in V$.

1.2.2. Example. Let $V = \mathbb{H}^n$ be the standard coordinate space considered as right vector space over $\mathbb{H}$. Fix a *hyperhermitian* $n \times n$-matrix $(a_{ij})_{i,j=1}^n$, i.e. $a_{ij} = \bar{a}_{ji}$, where $\bar{x}$ denotes the usual quaternionic conjugation of x. For $x = (x_1, \dots, x_n)$, $y = (y_1, \dots, y_n)$ define

$$A(x, y) = \sum_{i,j} \bar{x}_i a_{ij} y_j$$

(note the order of the terms!). Then A defines a hyperhermitian semilinear form on V.

In general one has the following standard claims.

1.2.3. Claim. *Fix a basis in a finite dimensional right quaternionic vector space V. Then there is a natural bijection between hyperhermitian semilinear forms on V and $n \times n$-hyperhermitian matrices.*

This bijection is in fact described in the previous Example 1.2.2.

1.2.4. Claim. *Let A be the matrix of the given hyperhermitian form in the given basis. Let C be transition matrix from this basis to another one. Then the matrix A' of the given form in the new basis is equal to*

$$A' = C^* AC.$$

1.2.5. Remark. Note that for any hyperhermitian matrix A and for any matrix C the matrix C^*AC is also hyperhermitian. In particular the matrix C^*C is always hyperhermitian.

1.2.6. Definition. A hyperhermitian semilinear form a is called *positive definite* if $a(x, x) > 0$ for any non-zero vector x.

Let us fix a positive definite hyperhermitian form $(\cdot,\cdot)$ on our quaternionic right vector space V. The space with such a fixed form will be called *hyperhermitian* space.

For any quaternionic linear operator $\phi : V \longrightarrow V$ in a hyperhermitian space V one can define the adjoint operator $\phi^* : V \longrightarrow V$ in the usual way, i.e. $(\phi x, y) = (x, \phi^* y)$ for any $x, y \in V$. Then if one fixes an orthonormal basis in the space V then the operator ϕ is selfadjoint if and only if its matrix in this basis is hyperhermitian.

1.2.7. Claim. *For any selfadjoint operator in a hyperhermitian space there exists an orthonormal basis such that its matrix in this basis is diagonal and real.*

The proof is standard. Now we are going to define the Moore determinant of hyperhermitian matrices. The definition below is different from the original one due to Moore [29] but equivalent to it.

First note that every hyperhermitian $n \times n$-matrix A defines a hyperhermitian semilinear form on the coordinate space $\mathbb{H}^n$. It also can be considered as a *symmetric* bilinear form on $\mathbb{R}^{4n}$ (which is the realization of $\mathbb{H}^n$). Let us denote its $4n \times 4n$-matrix by ${}^{\mathbf{R}}A$. Let us consider the entries of A as formal variables (each quaternionic entry corresponds to four commuting real variables). Then $\det({}^{\mathbf{R}}A)$ is a homogeneous polynomial of degree $4n$ in $n(2n-1)$ real variables. Let us denote by Id the identity matrix. One has the following result.

1.2.8. Theorem. *There exists a polynomial P defined on the space of all hyperhermitian $n \times n$-matrices such that for any hyperhermitian $n \times n$-matrix A one has $\det({}^{\mathbf{R}}A) = P^4(A)$ and $P(Id) = 1$. P is defined uniquely by these two properties. Furthermore P is homogeneous of degree n and has integer coefficients.*

Thus for any hyperhermitian matrix A the value $P(A)$ is a real number, and it is called the *Moore determinant* of the matrix A. The explicit formula for the Moore determinant was given by Moore [29] (see also [7]). From now on the Moore determinant of a matrix A will be denoted by $\det A$. This notation should not cause any confusion with the usual determinant of real or complex matrices due to part (i) of the next theorem.

1.2.9. Theorem. (i) *The Moore determinant of any complex hermitian matrix considered as quaternionic hyperhermitian matrix is equal to its usual determinant.*

(ii) *For any hyperhermitian matrix A and any matrix C*

$$\det(C^* A C) = \det A \cdot \det(C^* C).$$

1.2.10. Example. (a) Let $A = diag(\lambda_1, \dots, \lambda_n)$ be a diagonal matrix with real λ_i's. Then A is hyperhermitian and has Moore determinant $\det A = \prod_i \lambda_i$.

(b) A general hyperhermitian 2×2 matrix A has the form

$$A = \begin{bmatrix} a & q \\ \bar{q} & b \end{bmatrix},$$

where $a, b \in \mathbb{R}$, $q \in \mathbb{H}$. Then $\det A = ab - q\bar{q} = (ab - \bar{q}q)$.

Let us introduce more notation. Let A be any hyperhermitian $n \times n$-matrix. For any non-empty subset $I \subset \{1, \dots, n\}$ the minor $M_I(A)$ of A which is obtained by deleting the rows and columns with indexes from the set I, is clearly hyperhermitian. For $I = \{1, \dots, n\}$ let $\det M_{\{1,\dots,n\}} = 1$.

1.2.11. Claim. *Let A be a non-negative (resp. positive) definite hyperhermitian matrix. Then* $\det A \geq 0$ (*resp.* $\det A > 0$).

The following theorem is a quaternionic generalization of the standard Sylvester criterion.

1.2.12. Theorem (Sylvester criterion). *A hyperhermitian $n \times n$-matrix A is positive definite if and only if the Moore determinants of all the left upper minors of A are positive.*

Let us define now the mixed discriminant of hyperhermitian matrices in analogy with the case of real symmetric matrices [2].

1.2.13. Definition. Let $A_1, \dots, A_n$ be hyperhermitian $n \times n$-matrices. Consider the homogeneous polynomial in real variables $\lambda_1, \dots, \lambda_n$ of degree n equal to $\det(\lambda_1 A_1 + \dots + \lambda_n A_n)$. The coefficient of the monomial $\lambda_1 \cdot \dots \cdot \lambda_n$ divided by $n!$ is called the *mixed discriminant* of the matrices $A_1, \dots, A_n$, and it is denoted by $det(A_1, \dots, A_n)$.

Note that the mixed discriminant is symmetric with respect to all variables, and linear with respect to each of them, i.e.

$$det(\lambda A_1' + \mu A_1'', A_2, \dots, A_n) = \lambda \cdot det(A_1', A_2, \dots, A_n) + \mu \cdot det(A_1'', A_2, \dots, A_n)$$

for any *real* λ, μ. Note also that $det(A, \dots, A) = \det A$. One has the following generalization of Aleksandrov's inequalities for mixed discriminants [2].

1.2.14. Theorem. (i) *The mixed discriminant of positive (resp. non-negative) definite matrices is positive (resp. non-negative).*

(ii) *Fix positive definite hyperhermitian $n \times n$-matrices $A_1, \dots, A_{n-2}$. On the real linear space of hyperhermitian $n \times n$-matrices consider the bilinear form*

$$B(X, Y) := det(X, Y, A_1, \dots, A_{n-2}).$$

Then B is a non-degenerate quadratic form, and its signature has one plus and the rest are minus signs.

1.2.15. Corollary. *Let $A_1, \dots, A_{n-1}$ be positive definite hyperhermitian $n \times n$-matrices. Then for any hyperhermitian matrix X,*

$$det(A_1, \dots, A_{n-1}, X)^2 \geq det(A_1, \dots, A_{n-1}, A_{n-1}) \cdot det(A_1, \dots, A_{n-2}, X, X),$$

and the equality is satisfied if and only if the matrix X is proportional to A_{n-1}.

Finally let us give an explicit formula for the Moore determinant (which was the original definition by Moore [29]). Let $A = (a_{i,j})_{i,j=1}^n$ be a hyperhermitian $n \times n$-matrix. Let σ be a permutation of $\{1, \dots, n\}$. Write σ as a product of disjoint cycles such that each cycle starts with with the smallest number. Since disjoint cycles commute we may write

$$\sigma = (k_{11} \dots k_{1j_1})(k_{21} \dots k_{2j_2}) \dots (k_{m1} \dots k_{mj_m})$$

where for each i we have $k_{i1} < k_{ij}$ for all $j > 1$, and $k_{11} > k_{21} > \dots > k_{m1}$. This expression is unique. Let $sgn(\sigma)$ be the parity of σ.

1.2.16. Theorem. *The Moore determinant of A is equal to*

$$\det A = \sum_{\sigma} sgn(\sigma) a_{k_{11},k_{12}} \dots a_{k_{1j_1},k_{11}} a_{k_{21},k_{22}} \dots a_{k_{mj_m},k_{m1}}$$

where the sum runs over all permutations σ.

2. Quaternionic Monge-Ampère equations

First we introduce the class of plurisubharmonic functions on $\mathbb{H}^n$ following [4]. Note that this notion was also introduced independently by G. Henkin [24]. Let Ω be a domain in $\mathbb{H}^n$.

2.1.1. Definition. A real valued function $u : \Omega \longrightarrow \mathbb{R}$ is called quaternionic plurisubharmonic if it is upper semi-continuous and its restriction to any right *quaternionic* line is subharmonic.

Recall that upper semi-continuity means that $u(x_0) \geq \limsup_{x \longrightarrow x_0} u(x)$ for any $x_0 \in \Omega$. We will denote by $P(\Omega)$ the class of plurisubharmonic functions in the domain Ω.

Also we will call a C^2-smooth function $u : \Omega \longrightarrow \mathbb{R}$ to be *strictly plurisubharmonic* if its restriction to any right quaternionic line is strictly subharmonic (i.e. the Laplacian is strictly positive).

2.1.2. Definition. An open bounded domain $\Omega \subset \mathbb{H}^n$ with a smooth boundary $\partial\Omega$ is called strictly pseudoconvex if for every point $z_0 \in \partial\Omega$ there exists a neighborhood $\mathcal{O}$ and a smooth strictly psh function h on $\mathcal{O}$ such that $\Omega \cap \mathcal{O} = \{h < 0\}$, $h(z_0) = 0$, and $\nabla h(z_0) \neq 0$.

We will write a quaternion q in the usual form

$$q = t + x \cdot i + y \cdot j + z \cdot k,$$

where t, x, y, z are real numbers, and i, j, k satisfy the usual relations

$$i^2 = j^2 = k^2 = -1,\ ij = -ji = k,\ jk = -kj = i,\ ki = -ik = j.$$

The Dirac-Weyl operator $\frac{\partial}{\partial \bar{q}}$ is defined as follows. For any $\mathbb{H}$-valued function f

$$\frac{\partial}{\partial \bar{q}} f := \frac{\partial f}{\partial t} + i \frac{\partial f}{\partial x} + j \frac{\partial f}{\partial y} + k \frac{\partial f}{\partial z}.$$

Let us also define the operator $\frac{\partial}{\partial q}$:

$$\frac{\partial}{\partial q}f := \overline{\frac{\partial}{\partial \bar{q}}\bar{f}} = \frac{\partial f}{\partial t} - \frac{\partial f}{\partial x}i - \frac{\partial f}{\partial y}j - \frac{\partial f}{\partial z}k.$$

Remark. The operator $\frac{\partial}{\partial \bar{q}}$ is called sometimes the Cauchy-Riemann-Moisil-Fueter operator since it was introduced by Moisil in [28] and used by Fueter [18], [19] to define the notion of quaternionic analyticity. For further results on quaternionic analyticity we refer e.g. to [1], [11], [30], [31], [37], and for applications to mathematical physics to [23]. In Clifford analysis one often calls the above operator the Dirac operator. But in fact it was used earlier by J.C. Maxwell in [27], vol. II, pp. 570–576, where he has applied the quaternions to electromagnetism.

Now we can write the quaternionic Monge-Ampère equation with respect to a C^2-smooth psh function u on Ω:

$$\det(\frac{\partial^2 u}{\partial q_i \partial \bar{q}_j}) = f,$$

where f is a given function. Note that the matrix $(\frac{\partial^2 u}{\partial q_i \partial \bar{q}_j})$ is quaternionic hyperhermitian (since u is real valued), det means the Moore determinant of this matrix. Note also that since the function u is psh, the matrix $(\frac{\partial^2 u}{\partial q_i \partial \bar{q}_j})$ is non-negative definite, and hence its Moore determinant is non-negative by Claim 1.2.11.

Let B denote the Euclidean ball in $\mathbb{H}^n$.

2.1.3. Theorem. *Let $f \in C^\infty(\bar{B})$, $f > 0$. Let $\phi \in C^\infty(\partial B)$. There exists a unique psh function $u \in C^\infty(\bar{B})$ which is a solution of the Dirichlet problem*

$$\det(\frac{\partial^2 u}{\partial q_i \partial \bar{q}_j}) = f,$$

$$u|_{\partial B} = \phi.$$

The real version of this result was proved for arbitrary strictly convex bounded domains in $\mathbb{R}^n$ by Caffarelli, Nirenberg, and Spruck [12]. The complex version of it was proved for arbitrary strictly pseudoconvex bounded domains in $\mathbb{C}^n$ by Caffarelli, Kohn, Nirenberg, and Spruck [13] and Krylov [26]. Our method is a modification of the method of the paper [13]. Note also that in the case $n = 1$ the problem is reduced to the classical Dirichlet problem for the Laplacian in $\mathbb{R}^4$ (which is a linear problem); it was solved in the XIXth century. Note also that *interior* regularity of the solution of the Dirichlet problem for real Monge-Ampère equations was proved earlier by A. Pogorelov, and the proof was briefly described in [32]–[34]. The complete proof was published in [35] and [14], [15]. In [15] Cheng and Yau gave a different proof of interior regularity; they also studied some related geometric problems on flat manifolds.

Now we will state results on existence and uniqueness of generalized solutions of the Dirichlet problem in general strictly pseudoconvex bounded domains in $\mathbb{H}^n$. First we will formulate a result that clarifies in which sense the solutions are generalized.

2.1.4. Theorem ([4]). *For any function* $u \in C(\Omega) \cap P(\Omega)$ *one can define a non-negative measure* $\det(\frac{\partial^2 u}{\partial q_i \partial \bar{q}_j})$ *which is defined uniquely by the following two properties:*

(a) *if* $u \in C^2(\Omega)$, *then it has the obvious meaning;*
(b) *if* $u_N \longrightarrow u$ *uniformly on compact subsets in* Ω, *and* $u_N, u \in C(\Omega) \cap P(\Omega)$, *then*

$$\det(\frac{\partial^2 u_N}{\partial q_i \partial \bar{q}_j}) \xrightarrow{w} \det(\frac{\partial^2 u}{\partial q_i \partial \bar{q}_j}),$$

where the convergence of measures is understood in the weak sense.

Remarks. (a) It is easy to see that if $u_N \longrightarrow u$ uniformly on compact subsets, and $u_N \in C(\Omega) \cap P(\Omega)$, then $u \in C(\Omega) \cap P(\Omega)$.

(b) Note that the real analogue of this result was proved by A.D. Aleksandrov [3], and the complex analogue by Chern, Levine, and Nirenberg [16].

2.1.5. Theorem ([5]). *Let* $\Omega \subset \mathbb{H}^n$ *be a bounded quaternionic strictly pseudoconvex domain. Let* $f \in C(\bar{\Omega})$, $f \geq 0$. *Let* $\phi \in C(\partial\Omega)$. *Then there exists a unique function* $u \in C(\bar{\Omega})$ *which is psh in* Ω *such that*

$$\det(\frac{\partial^2 u}{\partial q_i \partial \bar{q}_j}) = f \text{ in } \Omega,$$

$$u|_{\partial\Omega} \equiv \phi.$$

Remark. The real analogue of this result was proved by A.D. Aleksandrov [3], and the complex one by E. Bedford and B.A. Taylor [10].

Finally we would like to state the minimum principle analogous to the complex case proved by Bedford and Taylor [10]. This minimum principle is heavily used in the proof of the above results.

2.1.6. Theorem. *Let* Ω *be a bounded open set in* $\mathbb{H}^n$. *Let* u, v *be continuous functions on* $\bar{\Omega}$ *which are plurisubharmonic in* Ω. *Assume that*

$$\det(\frac{\partial^2 u}{\partial q_i \partial \bar{q}_j}) \leq \det(\frac{\partial^2 v}{\partial q_i \partial \bar{q}_j}) \text{ in } \Omega.$$

Then

$$\min\{u(z) - v(z) | z \in \bar{\Omega}\} = \min\{u(z) - v(z) | z \in \partial\Omega\}.$$

3. Questions and comments

We would like to state few questions closely related to the material of this paper.

Question 1. Find a geometric (or any other) interpretation of the quaternionic Monge-Ampère equation (or of an appropriate modification of it).

Remind that the (modified) real Monge-Ampère equations appear in construction of convex hypersurfaces in $\mathbb{R}^n$ with the prescribed conditions on curvature. For this material we refer to [9], [35]. One of the main applications of (modified) complex Monge-Ampère equations is the construction of Kähler metrics on complex manifolds. After the proof of the Calabi-Yau theorem [38], [39] and the Aubin-Yau theorem [8], [39] they became one of the key tools in complex differential geometry, see e.g. [9], [25] for further discussion.

Question 2. (due to L. Polterovich.) Find a geometric characterization of quaternionic strictly pseudoconvex domains. (Note that we have not defined the notion of quaternionic pseudoconvex domain in the non-strict sense.)

Question 3. (due to G. Henkin.) This question is closely related to the previous one. Let $\Omega \subset \mathbb{H}^n$ be a domain which admits an exhaustion by level sets of plurisubharmonic function; in other words, there exists a plurisubharmonic function $h : \Omega \longrightarrow \mathbb{R}$ such that for any number c the set $\{h \leq c\}$ is compact. (Note that in the classical complex situation this property is one of the equivalent definitions of the pseudoconvex domain.) It was observed by G. Henkin [24] that if h is a strictly plurisubharmonic Morse function then Ω admits a homotopy retraction onto a compact subset of dimension at most $\frac{3}{4}\dim_{\mathbb{R}}\Omega = 3n$ (indeed the Morse index of every critical point of such a function is bounded from above by $3n$). This implies that the boundary $\partial\Omega$ is connected provided $n > 1$. These properties are analogous to the corresponding properties of pseudoconvex domains in the complex spaces (where the constant $\frac{3}{4}$ is replaced by $\frac{1}{2}$). It would be of interest to understand the relation between the class of domains with this property and the class of strictly pseudoconvex domains in the sense of this paper.

Question 4. Generalize Theorem 2.1.3 on the existence of the *regular* solution (under suitable assumptions on regularity of the initial data) to arbitrary strictly pseudoconvex bounded domains with smooth boundary (and not only for the Euclidean ball).

Question 5. Generalize the theory of plurisubharmonic functions and Monge-Ampère equations to the octonionic case, say to the case of the octonionic plane $\mathbb{O}^2$. The hyperhermitian matrices of size 2 have the form $\begin{bmatrix} a & q \\ \bar{q} & b \end{bmatrix}$ with $a, b \in \mathbb{R}$, $q \in \mathbb{O}$. The octonionic analogue of the Moore determinant is $\det \begin{bmatrix} a & q \\ \bar{q} & b \end{bmatrix} = ab - q\bar{q}$ $(= ab - \bar{q}q)$.

References

[1] W.W. Adams, C.A. Berenstein, P. Loustaunau, I. Sabadini, I, D.C. Struppa, Regular functions of several quaternionic variables and the Cauchy-Fueter complex. *J. Geom. Anal.* **9** (1999), no. 1, 1–15.

[2] A.D. Aleksandrov, Die gemischte Diskriminanten und die gemischte Volumina. *Math. Sbornik* **3** (1938), 227–251.

[3] A.D. Aleksandrov, Dirichlet's problem for the equation $Det\,||z_{ij}|| = \varphi(z_1, \cdots, z_n, z, x_1, \cdots, x_n)$. I. (Russian) *Vestnik Leningrad. Univ. Ser. Mat. Meh. Astr.* **13** (1958), no. 1, 5–24.

[4] S. Alesker, Non-commutative linear algebra and plurisubharmonic functions of quaternionic variables. *Bull. Sci. Math.* **127** (2003), no. 1, 1–35. also: math.CV/0104209

[5] S. Alesker, Quaternionic Monge-Ampère equations. *J. Geom. Anal.* **13** (2003), no. 2, 183–216. also: math.CV/0208005.

[6] E. Artin, *Geometric algebra.* Interscience Publishers, Inc., New York-London, 1957.

[7] H. Aslaksen,Quaternionic determinants. *Math. Intelligencer* **18** (1996), no. 3, 57–65.

[8] T. Aubin, Equations du type de Monge-Ampère sur les variétés kähleriennes compactes. *C.R. Acad. Sci. Paris* **283** (1976), 119–121.

[9] T. Aubin, *Nonlinear analysis on manifolds. Monge-Ampère equations.* Grundlehren der Mathematischen Wissenschaften [Fundamental Principles of Mathematical Sciences], 252. Springer-Verlag, New York, 1982.

[10] E. Bedford, B.A. Taylor, The Dirichlet problem for a complex Monge-Ampère equation. *Invent. Math.* **37** (1976), no. 1, 1–44.

[11] F. Brackx, R. Delanghe, F. Sommen, *Clifford analysis.* Research Notes in Mathematics, 76. Pitman (Advanced Publishing Program), Boston, MA, 1982.

[12] L. Caffarelli, L. Nirenberg, J. Spruck, The Dirichlet problem for nonlinear second-order elliptic equations. I. Monge-Ampère equation. *Comm. Pure Appl. Math.* **37** (1984), no. 3, 369–402.

[13] L. Caffarelli, J.J. Kohn, L. Nirenberg, J. Spruck, The Dirichlet problem for nonlinear second-order elliptic equations. II. Complex Monge-Ampère, and uniformly elliptic, equations. *Comm. Pure Appl. Math.* **38** (1985), no. 2, 209–252.

[14] S.Y. Cheng, S.-T. Yau, On the regularity of the Monge-Ampère equation $\det(\partial^2 u/\partial x_i \partial s x_j) = F(x,u)$. *Comm. Pure Appl. Math.* **30** (1977), no. 1, 41–68.

[15] S.Y. Cheng, S.-T. Yau, The real Monge-Ampére equation and affine flat structures. *Proceedings of the 1980 Beijing Symposium on Differential Geometry and Differential Equations*, Vol. 1, 2, 3 (Beijing, 1980), 339–370, Science Press, Beijing, 1982.

[16] S.S. Chern, H.I. Levine, L. Nirenberg, Intrinsic norms on a complex manifold. *Global Analysis (Papers in Honor of K. Kodaira)*, 119–139 Univ. Tokyo Press, Tokyo, 1969.

[17] J. Dieudonné, Les déterminants sur un corps non commutatif. (French) *Bull. Soc. Math. France* **71** (1943), 27–45.

[18] R. Fueter, Die Funktionentheotie der Differentialgleichungen $\Delta u = 0$ und $\Delta\Delta u = 0$ mit vier reelen Variablen. *Comment. Math. Helv.* **7** (1935), 307–330.

[19] R. Fueter, Über die analytische Darstellung der regulären Functionen einer Quaternionenvariablen. *Comment. Math. Helv.* **8** (1936), 371–378.

[20] I. Gelfand, V. Retakh, Determinants of matrices over noncommutative rings. (Russian) *Funktsional. Anal. i Prilozhen.* **25** (1991), no. 2, 13–25, 96; translation in *Funct. Anal. Appl.* **25** (1991), no. 2, 91–102.

[21] I. Gelfand, V. Retakh, R.L. Wilson, Quaternionic quasideterminants and determinants. math.QA/0206211.

[22] I. Gelfand, S. Gelfand, V. Retakh, R.L. Wilson, Quasideterminants. math.QA/0208146.

[23] F. Gürsey, H.C. Tze, Complex and quaternionic analyticity in chiral and gauge theories. I. *Ann. Physics* **128** (1980), no. 1, 29–130.

[24] G. Henkin, Private communication.

[25] D. Joyce, *Compact manifolds with special holonomy.* Oxford Mathematical Monographs. Oxford University Press, Oxford, 2000.

[26] N.V. Krylov, Smoothness of the value function for a controlled diffusion process in a domain, *Izvestiya Akademii Nauk SSSR, seriya matematicheskaya,* Vol. **53**, No. 1 (1989), 66–96 in Russian; English translation in *Math. USSR Izvestija*, Vol. **34**, No. 1 (1990)

[27] J.C. Maxwell, *The Scientific Letters and Papers of James Clerk Maxwell*, Volume II, 1862–1873, edited by P. M. Harman, Cambridge Univ. Press, 1995.

[28] G.C. Moisil, Sur les quaternions monogènes. *Bull. Sci. Math.* **55** (1931), 168–194.

[29] E.H. Moore, On the determinant of an hermitian matrix of quaternionic elements. *Bull. Amer. Math. Soc.* **28** (1922), 161–162.

[30] V.P. Palamodov, Holomorphic synthesis of monogenic functions of several quaternionic variables. *J. Anal. Math.* **78** (1999), 177–204.

[31] D. Pertici, Quaternion regular functions and domains of regularity. *Boll. Un. Mat. Ital.* B (7) **7** (1993), no. 4, 973–988.

[32] A.V. Pogorelov, The regularity of the generalized solutions of the equation $\det(\partial^2 u/\partial x^i \partial x^j) = \varphi(x^1, x^2, \dots, x^n) > 0$. (Russian) *Dokl. Akad. Nauk SSSR* **200** (1971), 534-537.

[33] A.V. Pogorelov, The Dirichlet problem for the multidimensional analogue of the Monge-Ampère equation. (Russian) *Dokl. Akad. Nauk SSSR* **201** (1971), 790–793.

[34] A.V. Pogorelov, A regular solution of the n-dimensional Minkowski problem. *Dokl. Akad. Nauk SSSR* **199**, 785–788 (Russian); translated as *Soviet Math. Dokl.* **12** (1971), 1192–1196.

[35] A.V. Pogorelov, *Mnogomernoe uravnenie Monzha-Ampera* $\det\|z_{ij}\| = \phi(z_1, \cdots, z_n, z, x_1, \cdots, x_n)$. *(Russian) [The multidimensional Monge-Ampère equation* $\det\|z_{ij}\| = \phi(z_1, \cdots, z_n, z, x_1, \cdots, x_n)$*]* "Nauka", Moscow, 1988.

[36] J. Rauch, B.A. Taylor, The Dirichlet problem for the multidimensional Monge-Ampère equation. *Rocky Mountain J. Math.* **7** (1977), no. 2, 345–364.

[37] A. Sudbery, Quaternionic analysis. *Math. Proc. Cambridge Philos. Soc.* **85** (1979), no. 2, 199–224.

[38] S.T. Yau, On Calabi's conjecture and some new results in algebraic geometry. *Proceedings of the National Academy of Sciences of the U.S.A.* **74** (1977), 1798–1799.

[39] S.T. Yau, On the Ricci curvature of a compact Kähler manifold and the complex Monge-Ampère equation. I. *Comm. Pure Appl. Math.* **31** (1978), no. 3, 339–411.

Semyon Alesker
Department of Mathematics
Tel Aviv University, Ramat Aviv
69978 Tel Aviv
Israel

Trends in Mathematics: *Advances in Analysis and Geometry*, 301–310

Galpern–Sobolev Type Equations with Non-constant Coefficients

K. Gürlebeck and W. Sprößig

Abstract. We study Galpern–Sobolev equations by the help of a quaternionic operator calculus. Some previous work will be extended to the case of variable coefficients. Rothe's time discretization method is used to reduce the problem to a series of stationary problems. Solving the resulting stationary problems by means of quaternionic analysis we obtain integral representation formulas for the solution of the Galpern–Sobolev equation. The truncation error and the stability of the method is studied, too.

Mathematics Subject Classification (2000). Primary 30G35; Secondary 35G15.

Keywords. Galpern–Sobolev equations, representation formulas, operator calculus, finite difference schemes.

1. Introduction

Let $\Omega \subset \mathbb{R}^3$ be a bounded domain with a piecewise smooth boundary Γ. A fluid flow with a low Reynolds number, i.e. the inertia term of forces is smaller than the viscous term, can be modelled by the well-known Navier–Stokes equations:

$$\partial_t u + \frac{1}{\rho}\,\mathrm{grad}\,p - \nu\Delta u = \frac{f}{\rho} \text{ in } \Omega \tag{1.1}$$

$$\mathrm{div}\,u = 0 \text{ in } \Omega \tag{1.2}$$

$$u = g \text{ on } \Gamma \tag{1.3}$$

$$u(0,x) = u_0 \text{ in } \Omega\,. \tag{1.4}$$

The coefficient ν describes the diffusion of momentum and is called *kinematic viscosity*. ρ is the *density* of the fluid. $\eta = \rho\nu$ is the so-called *dynamic viscosity*. Under a given body force f the pressure $p = p(t,x)$ and the velocity $u = u(t,x)$ of the fluid flow is searched. On the boundary the velocity data g are given. Because of Gauß theorem we have

$$\int_\Gamma g(y)\cdot n(y)d\Gamma_y = 0 \tag{1.5}$$

where n is the unit vector of outer normal to Γ at point y. Condition (1.2) expresses the incompressibility.

The case of a steady incompressible fluid, i.e. $\partial_t u = 0$ was treated by hypercomplex methods in [3] and [4]. For $g = 0$ representation formulas for the solution of (1.1)-(1.4) were obtained as

$$u + \frac{1}{\eta}\, T_\Omega \mathbf{Q} p = T_\Omega \mathbf{Q} T_\Omega\, \frac{f}{\eta}\,, \tag{1.6}$$

where T_Ω is the Teodorescu transform and $I - \mathbf{Q}$ a generalized Bergman projection. In the books [3], [4], [10] this alternative strategy for solving linear and nonlinear boundary value problems of stationary equations in mathematical physics was presented. With this special hypercomplex operator calculus the problems of existence, uniqueness, regularity, representation of solutions and the construction of well-adapted numerical methods can be studied in the framework of one and the same theory.

Recently, first attempts were made to include time-dependent problems into this concept. As in [10] time-harmonic cases are considered it was described in [5] how Rothe's semi-discretisation method can be applied also in the Clifford analysis context. This can be done by working in the Pauli algebra as well as the Dirac algebra. Using complex distances G. Kaiser [8] obtained another interesting generalization of the quaternionic operator calculus and applied this to Maxwell's equations. Without discussing details it can be stated that all these approaches to time-dependend problems require to work in complex quaternions or complex Clifford algebras. The Dirac or generalized Cauchy–Riemann operator has to be replaced by a disturbed Dirac operator.

As a model case in [5] a version of the so-called *Galpern–Sobolev equation* (cf. [9], [13]) was studied. It is an equation of the following type:

$$u_t - \eta \Delta u_t - \nu \Delta u = f(x, u, \nabla u) \tag{1.7}$$

where $u = u(x,t), x \in \Omega \subset \mathbb{R}^3, t \geq 0$, and η, ν are non-negative constants. Several physical phenomena lead to boundary value problems or Cauchy problems of this partial differential equations, e.g for $n = 1$; non-steady flow of second-order fluids and the theory of seepage of homogenous fluids through a fissured rock. Here u represents the absolute value of the velocity of the fluid and η characterizes the fissured rock (increasing η corresponds to a decreasing degree of fissuring). Existence- and uniquess theory the reader can find for instance in [9].

Dealing with this problem in [5] it was shown that by discretizing the time we can reduce the original problem approximately to a sequence of steady state problems. These problems could be factorized by applying the hypercomplex operator calculus, working in the complex quaternions. The factorization leads to two first order differential equation with a disturbed Dirac operator. Finally, for these disturbed Dirac equations and the discretized problem integral representations for the solutions could be obtained.

In this paper we generalize the approach to differential equations with variable coefficients. For comparison we study again an equation of Galpern–Sobolev-type as follows:

Let $\Omega \subset \mathbb{R}^3$ be a bounded domain with a sufficiently smooth boundary Γ and $[0,T]$ a finite time-segment. Further, let $\underline{f} : \Omega \times [0,T] \to \mathbb{R}^3$ and $\underline{u} : \Omega \times [0,T] \to \mathbb{R}^3$. We consider the differential equation

$$\partial_t(\mathbf{u} + \eta\beta^{-1}D\,\beta\,D\mathbf{u}) + \nu\alpha^{-1}D\,\alpha\,D\,\mathbf{u} = \mathbf{f}(x,t) \quad \text{in } \Omega \times (0,T] \tag{1.8}$$

with the boundary condition

$$\mathbf{u}(x,t) = \mathbf{g}(x,t) \qquad \text{on } \partial\Omega \times (0,T] \tag{1.9}$$

and the initial condition

$$\mathbf{u}(x,0) = \mathbf{h}(x) \qquad \text{on } \Omega\,. \tag{1.10}$$

The coefficients η and ν are positive constants and the positive functions $\alpha, \beta \in C^\infty(\overline{\Omega})$. The operator D is defined by

$$D\,\mathbf{u} = -\ \operatorname{div}\ \mathbf{u} +\ \operatorname{rot}\ \mathbf{u}.$$

Remark 1.1. Galpern–Sobolev problems can be also considered in domains in $\mathbb{R}^n$. For $n = 1$ we get the well-known Benjamin–Bona–Mahony equation, which (cf. [15]) that is seen as an alternative model of the KdV equations. The case $\alpha, \beta = const$ reproduces equation (1.7).

2. Preliminaries

To study our problem more detailed it is necessary to introduce some basic notation. We use here the notations from [1] because the properties of the used integral operators are proved therein.

Let $\mathbb{H}$ be the set of real quaternions. This means that elements of $\mathbb{H}$ are the form

$$a = \sum_{k=0}^{3} a_k i_k, \quad \text{where} \quad \{a_k | k \in \mathbb{N}_3^0 := \mathbb{N}_3 \cup \{0\}; \mathbb{N}_3 := \{1,2,3\}\} \subset \mathbb{R};$$

$i_0 = 1$ the unit; i_1, i_2, i_3 are called imaginary units, and they define the arithmetic rules in $\mathbb{H}$; by definition $i_k^2 = -i_0$, $k \in \mathbb{N}_3$; $i_1 i_2 = -i_2 i_1 = i_3$; $i_2 i_3 = -i_3 i_2 = i_1$; $i_3 i_1 = -i_1 i_3 = i_2$.

Natural operations of addition and multiplication in $\mathbb{H}$ turn it into a skew-field. The main involution in $\mathbb{H}$ is called quaternionic conjugation. This involution is defined in the following way:

$$\overline{i_0} := i_0; \quad \overline{i_k} := -i_k; \quad \text{for} \quad k \in \mathbb{N}_3,$$

and it extends onto $\mathbb{H}$ by $\mathbb{R}$-linearity, i.e., for $a \in \mathbb{H}$

$$\overline{a} := \sum_{k=0}^{3} \overline{a_k i_k} = \sum_{k=0}^{3} a_k \overline{i_k} = a_0 - \sum_{k=1}^{3} a_k i_k.$$

Note that

$$\overline{a}a = a\overline{a} = \sum_{k=0}^{3} a_k^2 = |a|_{\mathbf{R}^4}^2 =: |a|_{\mathbf{H}}^2$$

Therefore for $a \in \mathbb{H} \setminus \{0\}$ the quaternion

$$a^{-1} := \frac{1}{a\overline{a}}\ \overline{a} = \frac{1}{|a|^2}\overline{a}$$

is an inverse to a. We notice also a very important property of the quaternionic conjugation: for any $a, b \in \mathbb{H}$

$$\overline{ab} = \overline{b} \cdot \overline{a}.$$

For our purposes we need a notion of a complex quaternion. The set of complex quaternions $\mathbb{H}(\mathbb{C})$ consists of the elements

$$a = \sum_{k=0}^{3} a_k i_k$$

where $\{i_k\}$ is as above,

$$\{a_k | k \in \mathbb{N}_3^0\} \subset \mathbb{C},$$

the set of usual complex numbers with the imaginary unit i. By definition

$$i \cdot i_k = i_k \cdot i\ , \quad k \in \mathbb{N}_0^3\ .$$

Arithmetic rules are defined in $\mathbb{H}(\mathbb{C})$ just as in $\mathbb{H}$. It is obvious that $\mathbb{H}$ is a real subalgebra in $\mathbb{H}(\mathbb{C})$. Each element $a \in \mathbb{H}(\mathbb{C})$ can be represented in the form

$$a = a^1 + a^2 \cdot i = a^1 + i \cdot a^2\ ,$$

where $a^1, a^2 \in \mathbb{H}$. Quaternionic conjugation on $\mathbb{H}(\mathbb{C})$ acts on quaternionic units only, not on i.
Let Ω be a domain in $\mathbb{R}^3$, we shall consider $\mathbb{H}(\mathbb{C})$ - valued functions defined in Ω:

$$f\ :\ \Omega \longrightarrow \mathbb{H}(\mathbb{C}).$$

Notation $C^p(\Omega; \mathbb{H}(\mathbb{C})), p \in \mathbb{N} \cup \{0\}$, has the usual component-wise meaning. On $C^1(\Omega; \mathbb{H}(\mathbb{C}))$ the Dirac operator D is defined by:

$$D(f) \quad := \quad \sum_{k=1}^{3} i_k \frac{\partial f}{\partial x_k} =: \sum_{k=1}^{3} i_k \cdot \partial_k f.$$

D is a right-linear operator with respect to scalars from $\mathbb{H}(\mathbb{C})$. Let $\triangle$ be the 3-dimensional Laplace operator $\triangle := \sum_{k=1}^{3} \partial_k^2$. Then on $C^2(\Omega; \mathbb{H}(\mathbb{C}))$ we have analogously to the complex case the factorization:

$$\triangle = -D^2.$$

Let $\lambda \in \mathbb{C} \setminus \{0\}$, and let α be its complex square root: $\alpha \in \mathbb{C}$, $\alpha^2 = \lambda$, $\operatorname{Re} \alpha > 0$. The 3-dimensional Helmholtz operator

$$\triangle_\lambda := \triangle + \lambda$$

acts on $C^2(\Omega; \mathbb{H}(\mathbb{C}))$. For the stated above α the operators

$$D_\alpha \quad := \quad \alpha + D \quad ; \quad \overline{D}_\alpha := \alpha - D$$

are called left mutually conjugate generalized Cauchy-Riemann operators. The following equalities hold:

$$\Delta_\lambda = D_\alpha \cdot \overline{D}_\alpha = \overline{D}_\alpha \cdot D_\alpha.$$

They give factorizations of the Helmholtz operator with a complex (in particular, real) parameter λ. Elements of the set $\mathfrak{M}_\alpha(\Omega; \mathbb{H}(\mathbb{C})) := \ker D_\alpha$ are referred to as left-α-hyperholomorphic functions.

Let Θ_λ be a fundamental solution of the Helmholtz operator in $\mathbb{R}^3$:

$$\Theta_\lambda(x) := -\frac{1}{4\pi} \cdot \frac{1}{|x|} e^{-i\alpha|x|} ,$$

Then the fundamental solution of the operator D_α is given by

$$e_\alpha(x) := \overline{D}_\alpha[\Theta_\lambda](x) = \frac{1}{|x|^2} \left(\alpha|x|^2 + (i\alpha|x| + 1) \cdot \sum_{k=1}^{3} i_k x_k \right) \Theta_\lambda(x) . \qquad (2.1)$$

3. Time discretisation

Let us now start to deduce an implicit discretisation method. We abbreviate the meshwidth $\tau := T/N$ and $\mathbf{u}_k = \mathbf{u}(\cdot, k\tau)$. The time derivative $\partial_t \mathbf{u}$ at $k\tau$ is approximated by the forward difference

$$\frac{\mathbf{u}(x, (k+1)\tau - \mathbf{u}(x, k\tau)}{\tau} \qquad (k = 0, 1, ...N - 1).$$

Then we obtain from equation (1.8)

$$\begin{aligned} \mathbf{u}_{k+1} + \eta\beta^{-1} D\beta D\mathbf{u}_{k+1} + \tau\nu\alpha^{-1} D\alpha D\mathbf{u}_{k+1} &= \mathbf{f}_k\tau + \mathbf{u}_k + \eta\beta^{-1} D\beta D\mathbf{u}_k, \\ & \quad (k = 0, 1, 2...) \end{aligned} \qquad (3.1)$$

where

$$\mathbf{f}_k(x) := \frac{1}{\tau} \int_{k\tau}^{(k+1)\tau} \mathbf{f}(x,t)dt , \ \beta = \beta(\frac{\eta}{\delta} \cdot x), \ \alpha = \alpha(\frac{\tau}{\nu}\delta \cdot x)$$

and $\delta := \eta + \tau\nu$. For simplifying equation (3.1) we search for a real function $\gamma : \Omega \to \mathbb{R}^1, \gamma(x) \neq 0$ such that γ fulfils the relation

$$\eta\beta^{-1} D\beta\mathbf{v} + \tau\nu\alpha^{-1} D\alpha\mathbf{v} = \delta\gamma^{-1} D\gamma\mathbf{v} .$$

It follows

$$[\eta + \tau\nu - \delta] D\mathbf{v} + \left(\eta\frac{(D\beta)}{\beta} + \tau\nu\frac{(D\alpha)}{\alpha} \right) \mathbf{v} = \delta\frac{(D\gamma)}{\gamma}\mathbf{v}$$

As for $\eta + \tau\nu = \delta$ we have only to set $\gamma = \alpha\beta$. Indeed, it holds

$$\frac{D(\alpha\beta)}{\gamma} = \frac{(D\alpha)\frac{\tau\nu}{\delta}\,\beta + (D\beta)\frac{\eta}{\delta}\,\alpha}{\gamma} = \beta\frac{(D\alpha)}{\alpha}\frac{\tau\nu}{\delta} + \alpha\frac{(D\beta)}{\beta}\frac{\eta}{\delta}.$$

It remains to consider the following problem:

Let $\Omega \subset \mathbb{R}^3$ be a bounded domain with piecewise smooth boundary Γ. There are given $\mathbb{H}$-valued functions

$$\mathbf{f} \in C^1((0,T], L_2(\Omega)),\ \mathbf{g} \in C^1((0,T],\ W_2^{3/2}(\Gamma),\ \mathbf{h} \in W_2^2(\Omega)$$

and real-valued functions $\alpha, \beta, \gamma \in C^\infty(\overline{\Omega})$, $\alpha, \beta, \gamma \neq 0$. We look for a solution of the problem

$$\mathbf{u}_{k+1} + \delta\gamma^{-1}D\gamma D\mathbf{u}_{k+1} = \tau\,\mathbf{f}_k + \mathbf{u}_k + \eta\,\beta^{-1}D\beta D\mathbf{u}_k \qquad (k = 0,1,2,...) \tag{3.2}$$

$$\begin{aligned} \mathbf{u}_k(x) &= \mathbf{g}(x,\tau k) \quad \text{on } \Gamma = \partial\Omega\,,\ k = 0,1,...N-1 \\ \mathbf{u}_0(x) &= \mathbf{h}(x) \quad \text{on } \overline{\Omega}\,. \end{aligned}$$

It follows with $\delta^{-1} =: \mu^2$

$$(\gamma^{-1}D\gamma D + \mu^2)\mathbf{u}_{k+1} = \mu^2(\tau\mathbf{f}_k + \mathbf{u}_k + \beta^{-1}D\beta D\mathbf{u}_k) \qquad (k = 0,1,...,N-1).$$

This equation belongs to the class of **Yukawa type equations**. Such equations are considered in the hypercomplex setting in [14]. Main idea is to factorize the differential operator on the left hand side. We obtain

$$\begin{aligned} \gamma^{-1}(D - i\mu)\gamma(D + i\mu)\mathbf{u}_{k+1} &= \mu^2(\tau\mathbf{f}_k + \mathbf{u}_k + \beta^{-1}D\beta D\mathbf{u}_k) + i\mu\frac{D\gamma}{\gamma}\mathbf{u}_{k+1} \\ &\quad (k = 0,1,...,N{-}1). \end{aligned} \tag{3.3}$$

This follows from the generalized Leibniz rule

$$\gamma^{-1}D(\gamma i\mu\mathbf{u}) = \mu i\, D\mathbf{u} + (\gamma^{-1}i\mu D\gamma)\mathbf{u}\,.$$

The factorization of the left hand side allows it to reduce our considerations to first order problems. In practical calculations, problems can be caused by the zero-order term at time level $(k+1)\tau$ because we have to invert the operator

$$\gamma^{-1}(D - i\mu)\gamma(D + i\mu) - (\gamma^{-1}i\mu D\gamma).$$

Nevertheless for existence results we will study this equation but for simplier calculations it is also of some interest to substitute $(\gamma^{-1}i\mu D\gamma)\mathbf{u_{k+1}}$ by $(\gamma^{-1}i\mu D\gamma)\mathbf{u_k}$.

4. Quaternionic integral operators

Let Ω be a bounded domain in $\mathbb{R}^3$ with the Liapunov boundary $\Gamma := \partial\Omega$ and let $\mathbf{n} = (n_1, n_2, n_3)$ be the unit vector of the outward pointing normal at the point $y \in \Gamma$. The kernel (2.1) generates two important integrals. The Cauchy-type operator is introduced by

$$F_\alpha f(x) := \int_\Gamma e_\alpha(y - x)\mathbf{n}(y)f(y)\,d\Gamma_y \quad x \notin \Gamma.$$

A Teodorescu-type operator is given by

$$T_\alpha f(x) := -\int_\Gamma e_\alpha(x - y)f(y)\,d\Omega_y, \quad x \in \mathbb{R}^3.$$

We have to find operators which correspond to the factorization of the differential operator $(\gamma^{-1}D\gamma D+\mu^2)$ and satisfy the following Borel–Pompeiu formula

$$\mathbf{u}=(D\pm i\mu)T_{\pm i\mu}\mathbf{u}=T_{\pm i\mu}(D\pm i\mu)\mathbf{u}+F_{\pm i\mu}\mathbf{u} \qquad \text{in } \Omega. \tag{4.1}$$

Similar problems were studied in several papers, see e.g. [10], [4], [6] for the three-dimensional case and [1] for the n-dimensional case, $n>3$.
In [6] the following Borel–Pompeiu formula is proved:

$$f(x) \;=\; F_\alpha f(x)+T_\alpha D_\alpha f(x) \quad x\in\Omega,\ f\in C^1(\Omega;\mathbb{H}(\mathbb{C}))\cap C(\overline{\Omega};\mathbb{H}(\mathbb{C})).$$

This is the desired result in the three-dimensional case if we set $\alpha=i\mu$.
In the case of dimensions $n>3$ these operators are given by (see [1])

$$\begin{aligned}(T_{\pm i\mu}\mathbf{u})(x) \;:=\; & -(\frac{i\mu}{2\pi})^{n/2}\int_\Omega |x-y|^{1-n/2}[K_{n/2}(i\mu|x-y|)\frac{x-y}{|x-y|}\\ & -K_{n/2-1}(i\mu|x-y|)]\mathbf{u}(y)dy\end{aligned}$$

and

$$\begin{aligned}(F_{\pm i\mu}\mathbf{u})(x): \;=\; & -(\frac{i\mu}{2\pi})^{n/2}\int_\Gamma |x-y|^{1-n/2}[K_{n/2}(i\mu|x-y|)\frac{x-y}{|x-y|}\\ & -K_{n/2-1}(i\mu|x-y|)]\mathbf{n}(y)\mathbf{u}(y)d\Gamma_y\end{aligned}$$

where $K_s(z)$ denotes McDonald's function. This type of operators was studied also in [2]. The operators $F_{\pm i\mu}$ are called **Cauchy–Bizadse operators** and the operators $T_{\pm i\mu}$ are called **modified Teodorescu transforms.**
Formula (4.1) can be generalized to $\mathbf{u}\in W_2^1(\Omega)$. On the boundary Γ we have then $\mathrm{tr}_\Gamma\mathbf{u}\in W_2^{1/2}(\Gamma)$. In the following we will not distinguish the operators for different dimension because of the same algebraical and mapping properties.

5. A representation formula

Applying the modified Teodorescu transforms on (3.2) we get

$$\begin{aligned}\mathbf{u}_{k+1} \;=\; & T_{-i\mu}\gamma^{-1}T_{i\mu}\mu^2\tau\gamma f_k+\mu^2T_{-i\mu}\gamma^{-1}T_{i\mu}\gamma\mathbf{u}_k+\mu^2T_{-i\mu}\gamma^{-1}T_{i\mu}D\beta\, D\mathbf{u}_k\\ & +T_{-i\mu}\gamma^{-1}\phi_+ +\phi_- - i\mu T_{-i\mu}\gamma^{-1}T_{i\mu}(D\gamma)\mathbf{u}_{k+1},\end{aligned} \tag{5.1}$$

where $\phi_\pm\in\ker(D\pm i\mu)$.
The functions $\phi_\pm$ have to be determined now. Setting $\mathbf{g}_k(x)=\mathbf{g}(x,k\tau)$ $(k=0,1,...N-1)$ and use that $(D\pm i\mu)(F_{\pm i\mu}\mathbf{u})(x)=0$ (cf. [14]) we get $F_{\pm i\mu}T_{\pm i\mu}\mathbf{u}=0$. Furthermore, it follows $F_{-i\mu}\mathbf{u}_{k+1}\mid_\Gamma=F_{-i\mu}\mathbf{g}_{k+1}=\phi_-$. The determination of the second unknown function ϕ_+ is more complicated.
From (5.1) we obtain by using $F_{i\mu}\phi_+=\phi_+$ and $\mathrm{tr}\mathbf{u}_{k+1}=\mathbf{g}_{k+1}$

$$\begin{aligned}Q_{\Gamma,-i\mu}\mathbf{g}_{k+1} \;=\; & \mu^2\mathrm{tr}_\Gamma T_{-i\mu}\gamma^{-1}T_{i\mu}[\tau f_k+(I+\beta^{-1}D\beta D)\mathbf{u}_k]\\ & -i\mu T_{-i\mu}\gamma^{-1}T_{i\mu}(D\gamma)\mathbf{u}_{k+1}+\mathrm{tr}_\Gamma T_{-i\mu}\gamma^{-1}F_{i\mu}\phi_+\end{aligned}$$

Similarly to considerations in [3] one can prove that the operator

$$\mathrm{tr}_\Gamma T_{-i\mu}\gamma^{-1}F_{i\mu} : \mathrm{im}P_{\Gamma,i\mu} \cap W_2^{k-1/2}(\Gamma) \to \mathrm{im}Q_{\Gamma,-i\mu} \cap W^{k+1/2}(\Gamma) \quad (k \geq 1)$$

is an isomorphism, where $P_{\Gamma,i\mu}$ and $Q_{\Gamma,-i\mu}$ is a pair of generalized Plemelj projections which act in corresponding Hardy spaces. It follows

$$\begin{aligned}\phi_+ \;&=\; -F_{i\mu}(\mathrm{tr}_\Gamma T_{-i\mu}\gamma^{-1}F_{i\mu})^{-1}[\mu^2\mathrm{tr}_\Gamma T_{-i\mu}\gamma^{-1}T_{i\mu}\gamma(\tau\mathbf{f}_k + (I+\beta^{-1}D\beta D)\mathbf{u}_k)]\\ &\quad +F_{i\mu}(\mathrm{tr}_\Gamma T_{-i\mu}\gamma^{-1}F_{i\mu})^{-1}[Q_{\Gamma,-i\mu}g_{k+1} + i\mu T_{-i\mu}\gamma^{-1}T_{i\mu}(D\gamma)\mathbf{u}_{k+1}]\end{aligned}$$

Replacing ϕ_+ in (5.1) we get

$$\begin{aligned}\mathbf{u}_{k+1} \;&=\; \mu^2 T_{-i\mu}\gamma^{-1}T_{i\mu}\left[\tau\gamma\mathbf{f}_k + (I+\beta^{-1}D\beta D)\mathbf{u}_k\right] + i\mu T_{-i\mu}\gamma^{-1}T_{i\mu}(D\gamma)\mathbf{u}_{k+1}\\ &\quad -\mu^2 T_{-i\mu}\gamma^{-1}F_{i\mu}(\mathrm{tr}_\Gamma T_{-i\mu}\gamma^{-1}F_{i\mu})^{-1}[\mathrm{tr}_\Gamma T_{-i\mu}\gamma^{-1}T_{i\mu}\gamma(\tau\mathbf{f}_k\\ &\quad +(I+\beta^{-1}D\beta D)\mathbf{u}_k)] + T_{-i\mu}\gamma^{-1}F_{i\mu}(\mathrm{tr}_\Gamma T_{-i\mu}\gamma^{-1}F_{i\mu})^{-1}[Q_{\Gamma,-i\mu}\mathbf{g}_{k+1}\\ &\quad +i\mu T_{-i\mu}\gamma^{-1}T_{i\mu}(D\gamma)\mathbf{u}_{k+1}] + F_{-i\mu}\mathbf{g}_{k+1}\end{aligned}$$

We introduce the projections

$$\mathbb{P}_{i\mu} = F_{i\mu}(\mathrm{tr}_\Gamma T_{-i\mu}\gamma^{-1}F_{i\mu})^{-1}T_{-i\mu}\gamma^{-1} \text{ and } \mathcal{Q}_{i\mu} := I - \mathbb{P}_{i\mu}.$$

$\mathbb{P}_{i\mu}$ is a modified Bergman projection onto the subspace $\ker D + i\mu$ and $\mathcal{Q}_{i\mu}$ is the projection onto the subspace $\gamma(D - i\mu)\overset{\circ}{W}{}_2^1$.
Using these projections the last expression transforms into

$$\begin{aligned}\mathbf{u}_{k+1} \;&=\; \mu^2 T_{-i\mu}\gamma^{-1}T_{i\mu}\gamma[\tau\mathrm{f}_k + (I+\beta^{-1}D\beta D)\mathbf{u}_k]\\ &\quad -\mu^2 T_{-i\mu}\gamma^{-1}\mathbb{P}_{i\mu}T_{i\mu}\gamma[\tau f_k + (I+\beta^{-1}D\beta D)\mathbf{u}_k]\\ &\quad +T_{-i\mu}\gamma^{-1}F_{i\mu}(\mathrm{tr}_\Gamma T_{-i\mu}\gamma^{-1}F_{i\mu})^{-1}\mathrm{tr}T_{-i\mu}D_{-i\mu}\tilde{\mathbf{g}}_{k+1} + F_{-i\mu}\mathbf{g}_{k+1}\\ &\quad -i\mu T_{-i\mu}\gamma^{-1}T_{i\mu}(D\gamma)\mathbf{u}_{k+1}\\ &=\; \mu^2 T_{-i\mu}\gamma^{-1}\mathcal{Q}_{i\mu}T_{i\mu}\gamma[\tau\mathbf{f}_k + (I+\beta^{-1}D\beta D)\mathbf{u}_k]\\ &\quad -i\mu T_{-i\mu}\gamma^{-1}\mathcal{Q}_{i\mu}T_{i\mu}(D\gamma)\mathbf{u}_{k+1}\\ &\quad +F_{-i\mu}\mathbf{g}_{k+1} + T_{-i\mu}\mathbb{P}_{i\mu}\gamma D_{-i\mu}\tilde{\mathbf{g}}_{k+1},\end{aligned}$$

where $\tilde{\mathbf{g}}_{k+1}$ a smooth extension of $\mathbf{g}_{k+1}$ into the domain Ω. We remark that this last expression decomposes our boundary value problem into two problems. The part that contains g_{k+1} solves the homogeneous differential equation with the original (inhomogeneous) boundary values and the other terms are solution of an inhomogeneous differential equation with vanishing boundary data. This is because of the fact that im $\mathcal{Q}_{i\mu}$ consists of all $\mathbb{H}$-valued functions $\mathbf{v}$ with

$$\mathrm{tr}\, T_{-i\mu}\gamma^{-1}\mathbf{v} = 0.$$

This can be proved by applying the generalized Plemelj–Sokkotzkij formulae (cf.[4]). Now we have to consider the operator $R_\mu := -i\mu T_{-i\mu}\gamma^{-1}T_{i\mu}(D\gamma)$. We obtain by using the estimation (cf. [1])

$$\|T_{\pm i\mu}\|_{[L_2,L_2]} \leq \frac{const}{\mu}.$$

The constant depends on the domain and on the dimension of the space.

$$\begin{aligned}\|R_\mu \mathbf{u}\|_{L_2} &\le \frac{C}{\mu} \max_\Omega |\gamma^{-1}| \max_\Omega |D\gamma| \, \|\mathbf{u}\|_{L_2} \\ &= C\sqrt{\eta + \tau\delta} \max_\Omega |\gamma^{-1}| \max_\Omega |D\gamma| \, \|\mathbf{u}\|_{L_2} .\end{aligned}$$

For simplicity we estimated the norms of the coefficients by their maximum norms. Under the condition that

$$\sqrt{\eta + \tau\delta} \max_\Omega |D\gamma| < 1$$

$(I - R_\mu)^{-1}$ exists and we get for u_{k+1} the following representation formula:

$$\begin{aligned}\mathbf{u}_{k+1} &= (I - R_\mu)^{-1}\mu^2 T_{-i\mu}\gamma^{-1} \mathcal{Q}_{i\mu} T_{i\mu}\gamma[\tau \mathbf{f}_k + (I + \beta^{-1} D\beta D)\mathbf{u}_k] \\ &\quad + T_{-i\mu} \mathbb{P}_{i\mu}\gamma D_{-i\mu}\tilde{\mathbf{g}}_{k+1} + F_{-i\mu}\mathbf{g}_{k+1} \\ &= \mu^2 (I - R_\mu)^{-1} T_{-i\mu}\gamma^{-1}\mathcal{Q}_{i\mu} T_{i\mu}\gamma(I + \beta^{-1} D\beta D)\mathbf{u}_k + \mathbf{G}_{k+1}\end{aligned} \tag{5.2}$$

with

$$\mathbf{G}_{k+1} = (I - R_\mu)^{-1}\mu^2 T_{-i\mu}\gamma^{-1}\mathcal{Q}_{i\mu} T_{i\mu}\gamma\tau\mathbf{f}_k + T_{-i\mu}\mathbb{P}_{i\mu}\gamma D_{-i\mu}\tilde{\mathbf{g}}_{k+1} + F_{-i\mu}\mathbf{g}_{k+1}.$$

Equation (5.2) can be used to estimate stability and truncation error. For the truncation error we have to study $|L_\tau \mathbf{u_k} - L\mathbf{u}_k|$ with

$$L\mathbf{u} = \frac{\partial}{\partial t}(\mathbf{u}(x,t) + \eta\beta^{-1} D\beta D\mathbf{u}(x,t)) + \nu\alpha^{-1} D\alpha D\mathbf{u}(x,t)$$

and

$$L_\tau \mathbf{u}_k = \frac{\mathbf{u}_{k+1} - \mathbf{u}_k}{\tau} + \eta\beta^{-1} D\beta D(\frac{\mathbf{u}_{k+1} - \mathbf{u}_k}{\tau}) + \nu\alpha^{-1} D\alpha D\mathbf{u}_{k+1}.$$

Obviously we obtain from

$$\begin{aligned}|L_\tau \mathbf{u_k} - L\mathbf{u}_k| &= |\frac{\mathbf{u}_{k+1} - \mathbf{u}_k}{\tau} - \frac{\partial \mathbf{u}_k}{\partial t} + \eta\beta^{-1} D\beta D(\frac{\mathbf{u}_{k+1} - \mathbf{u}_k}{\tau} - \frac{\partial \mathbf{u}_k}{\partial t}) \\ &\quad + \nu\alpha^{-1} D\alpha D(\mathbf{u}_{k+1} - \mathbf{u}_k)| \le const. \cdot \tau.\end{aligned}$$

The stability can be proved analogously to the method that is demonstrated in [5]. Main problem is to prove that the operator

$$\mu^2 (I - R_\mu)^{-1} T_{-i\mu}\gamma^{-1}\mathcal{Q}_{i\mu} T_{i\mu}\gamma(I + \beta^{-1} D\beta D)$$

is a continuous operator in $L_2(\Omega)$. Then, from the approximation property and the stability we conclude the convergence of the finite difference scheme. The remaining stationary problems at each time level can be treated by using ideas of discrete function theory (see [4]) or discrete potential theory [11] for the general idea and [7] for details.

References

[1] H. Bahmann, K. Gürlebeck, M. Shapiro and W. Sprössig, *On a modified Teodorescu transform* , Integral Transforms and Special Functions **12** Nr. 3 (2001), 213–226.

[2] S. Bernstein, *Fundamental solutions of Dirac type operators.* Banach Center Publications **37**, Banach Center Symposium: Generalizations of Complex Analysis, May 30 - July 1 1994, Warsaw 1996, 159–172.

[3] K. Gürlebeck and W. Sprössig, *Quaternionic Analysis and Boundary Value Problems,* Birkhäuser, Basel, 1990.

[4] K. Gürlebeck and W. Sprössig, *Quaternionic and Clifford calculus for physicists and engineers*, John Wiley, Chichester, 1997.

[5] K. Gürlebeck and W. Sprössig, *Representation theory for classes of initial value problems with quaternionic analysis*, Math. Meth. Appl. Sci. **25** (2002), 1371–1382.

[6] K. Gürlebeck, M. Shapiro and W. Sprössig, *On a Teodorescu transform for a class of metaharmonic functions*, J. Nat. Geom. **21** No.1-2 (2002), 17–38.

[7] A. Hommel, *Fundamentallösungen partieller Differenzenoperatoren und die Lösung diskreter Randwertprobleme mit Hilfe von Differenzenpotentialen.* Dissertation, Bauhaus–Universität Weimar, 1998

[8] G. Kaiser, *Complex-distance potential theory and hyperbolic equations.* In: J. Ryan and W. Sprössig (eds.), *Clifford Analysis*, Progress in Physics **19**, Birkhäuser, Boston–New York, 2000, 135–169.

[9] G. Karch, *Asymptotic Behaviour of Solutions to some Pseudoparabolic Equations*, Math. Meth. Appl. Sci, **20** (1997), 271–289.

[10] V. Kravchenko and M. Shapiro, *Integral representations for spatial models of mathematical physics.* Pitman Research Notes in Math. Series **351**, 1996.

[11] V.S. Ryabenkij, The Method of difference potentials for some problems of continuum mechanics (Russian). Moscow, Nauka, 1987.

[12] A.A. Samarskij, *The Theory of Difference Methods* (in Russian). Moscow, Nauka, 1977.

[13] R.E. Showalter R.E., *Partial differential equations of Sobolev-Galpern type*, Pac. J. Math. **31** (1969), 787–793.

[14] W. Sprössig, *On decomposition of the Clifford Valued Hilbert Space and their Applications to Boundary Value Problems*, Advances in Applied Clifford Algebras **5** No. 2 (1995), 167–185.

[15] B. Wang, *Attractors and Approximate Inertial Manifolds for the generalized BBME*, MMAS **20**, 3 (1997).

K. Gürlebeck
Bauhaus-University Weimar, Coudraystrasse 13b, D 99423 Weimar, Germany
e-mail: `guerlebe@fossi.uni-weimar.de`

W. Sprößig
TU Bergakademie Freiberg, Fakultät für Mathematik und Informatik, Bernhard-von-Cotta-Str. 2, D-09596 Freiberg, Germany
e-mail: `sproessig@math.tu-freiberg.de`

Trends in Mathematics: *Advances in Analysis and Geometry*, 311–343

A Theory of Modular Forms in Clifford Analysis, their Applications and Perspectives

Rolf Sören Kraußhar

Abstract. This chapter is devoted to monogenic, harmonic and polymonogenic Clifford-valued automorphic forms that are related to discrete subgroups of Vahlen groups acting on half-spaces of real and complex Minkowski type spaces. In particular, Eisenstein and Poincaré type series are constructed within this framework.

We discuss their basic properties and provide a short overview about their range of applications to several areas from pure and applied mathematics, as for example, to number theory, to functional analysis and to order theory and pde's on manifolds.

Mathematics Subject Classification (2000). 11F03, 11F55, 30G35, 11G15, 32A25.

Keywords. Clifford analysis, automorphic forms, Vahlen groups, Minkowski type-spaces, function spaces, partial differential equations, order theory, argument principles on conformal manifolds.

1. Introduction

Examples of holomorphic automorphic forms and functions of one complex variable appeared first systematically in the paper [14] by G. Eisenstein (1847) and in lectures from K. Weierstraß in 1863. The general theory of holomorphic automorphic forms of one complex variable has been developed mainly by H. Poincaré, F. Klein and R. Fricke. This theory is not only interesting from the function theoretic point of view; it provides many important applications to other field of mathematics, to physics and to the applied sciences. Areas of applications include number theory, the representation of Riemann surfaces, partial differential equations, function spaces, Yang-Mills theory, quantum gravity [23, 4] and many more.

Already in the first half of the twentieth century higher dimensional (but complex-valued) generalizations have been developed within the framework of several complex variables theory, starting with O. Blumenthal [3] in 1904, continuing with the school of C. L. Siegel and H. Maaß in the thirties and early forties [56, 43].

One of the first authors who considered automorphic forms in a hypercomplex variable was R. Fueter. In 1927 R. Fueter constructed in his paper [18] automorphic forms and functions related to Picard's group in a three-dimensional hypercomplex variable. However, the functions treated in his paper, are not in kernels of Dirac type operators.

In 1949 H. Maaß proceeded to introduce in [44] a further different kind of automorphic forms in a hypercomplex variable. The class of automorphic forms introduced by him consists of complex-valued non-analytic eigenfunctions of the higher dimensional hyperbolic Laplace-Beltrami operator.

Many authors extended these fundamental higher dimensional theories from a large variety of viewpoints. Just to give a few examples, see e.g. works of J. Elstrodt, F. Grunewald and J. Mennicke, [15, 16], E. Freitag and C. F. Hermann [17], A. Krieg [37, 38, 39], O. Richter and H. Skogman [49, 50] which are some of many authors who provided important contributions in this direction.

If one looks for analogous extensions within the framework Clifford analysis and its regularity concepts, then, as far as we know, one finds only very few contributions to higher dimensional versions of the doubly periodic holomorphic Weierstraß functions (as e.g. [13, 19, 21, 23, 51]) before we started in the period of 1998–2002 to fill this gap step by step, and to develop systematically a theory of monogenic and polymonogenic hypercomplex-valued modular forms in the setting of real and complexified Clifford analysis. In [27, 28, 30, 31, 32, 33]) the basis of this theory has been developed. Applications to function spaces, spin geometry and partial differential equations have been developed in joint works with J. Ryan and D. Constales (see [35, 36, 8, 7, 10]). In this chapter we give an overview about the development of the theory of monogenic and polymonogenic automorphic forms within the more general context of Minkowski-type spaces. We furthermore show how an order theory, an argument principle and versions of Rouché's theorems can be extended to the framework of monogenic functions on more general conformally flat manifolds as for example on cylinders and tori. These results in turn extend the theory developed in the recent joint paper with T. Hempfling [26] for the Euclidean space $\mathbb{R}^n$.

2. Groups and Dirac operators on real and complex Minkowski-type spaces

2.1. Clifford Algebras over Minkowski type spaces

Let $\mathbb{K}$ stand for the field of real numbers $\mathbb{R}$ or complex numbers $\mathbb{C}$. Let n be a positive integer, and let p, q be non-negative integers with $p+q = n$. In what follows we consider an n-dimensional vector space over $\mathbb{K}$ that is endowed with a non-degenerate bilinear form of signature (p,q). By $Cl_{p,q}(\mathbb{K})$ we denote its associated Clifford algebra in which one has the following multiplication rules:

$$e_i^2 = 1 \quad \text{for} \quad i = 1, \ldots, p$$
$$e_i^2 = -1 \quad \text{for} \quad i = p+1, \ldots, n$$

and

$$e_i e_j + e_j e_i = 2\delta_{ij} e_i^2.$$

A $\mathbb{K}$-basis for $Cl_{p,q}(\mathbb{K})$ is given by the set $\{e_A : A \subseteq \{1, \ldots, n\}\}$ with $e_A = e_{l_1} e_{l_2} \ldots e_{l_r}$, where $1 \leq l_1 < \ldots, < l_r \leq n, e_\emptyset = e_0 = 1$. The scalar and the vector part of an arbitrary element $a = \sum\limits_{A \in \subseteq \{1,\ldots,n\}} a_A e_A$ where $a_A \in \mathbb{K}$ will be denoted by $Sc(a)$ and by $Vec(a)$. Each element $a \in Cl_{p,q}(\mathbb{C})$ can be written in the form $a = a_0 + i a_1$ with real Clifford numbers a_0, a_1 from $Cl_{p,q}(\mathbb{R})$ where i denotes the complex imaginary unit. We write further $Re(a) = a_0$ and $Im(a) = a_1$. In this notation a complex vector $Z \in \mathbb{C}^{p,q}$ shall be written in the form $Z = X + iY$ where $X, Y \in \mathbb{R}^{p,q}$. If we do not want to specify if the variable is concretely from $\mathbb{R}^{p,q}$ or $\mathbb{C}^{p,q}$ we just use small letters, say for example z.

The "Clifford" conjugation in $Cl_{p,q}(\mathbb{K})$ is defined by $\overline{a} = \sum_A e_A a_A$ where $\overline{e_A} = \overline{e_{l_r}}\, \overline{e_{l-1}} \ldots \overline{e_{l_1}}$, $\overline{e_j} = -e_j$ for $j = 1, \ldots, n$ and $\overline{e_0} = e_0 = 1$. The reversion is defined by $a^* := \sum_A (-1)^{|A|(|A|-1)/2} a_A e_A$, where $|A|$ stands for the cardinality of the set A. Both (anti) automorphisms act on the basis elements e_A; they leave the complex unit i invariant. The "complex" conjugation mapping and element $a = a_0 + i a_1$ from $Cl_{p,q}(\mathbb{C})$ onto $a_0 - i a_1$ shall be denoted by $a^\sharp$ in what follows. The Euclidean (Hermitian) scalar product in $\mathbb{K}^{p,q}$ defined by $\langle z, w \rangle = -\sum\limits_{j=1}^{p} z_j w_j^\sharp + \sum\limits_{j=p+1}^{n} z_j w_j^\sharp$ extends to $Cl_{p,q}(\mathbb{K})$ by regarding $Cl_{p,q}(\mathbb{K})$ as a 2^n-dimensional vector space with the basis elements e_A. In $Cl_{p,q}(\mathbb{K})$ a vector $z \in \mathbb{K}^{p,q}$ is invertible if and only if the expression $\mathcal{N}(z) := zz^* = \sum\limits_{i=1}^{p} z_i^2 - \sum\limits_{i=p+1}^{n} z_i^2$ does not vanish. In this case $z^{-1} = z/\mathcal{N}(z)$. The set $S_0 := \{z \in \mathbb{K}^{p,q}, \mathcal{N}(z) = 0\}$ has in general the geometrical structure of a cone. Only in the case $\mathbb{K} = \mathbb{R}$ with additionally $p = 0$ or $q = 0$ it reduces to a single isolated point. Let us further use the notation $S_{z_0} := \{z \in \mathbb{K}^{p,q}, \mathcal{N}(z - z_0) = 0\}$.

2.2. Half-spaces and related modular groups

Modular forms are usually defined on half-space type domains. The structure of the null cones motivates the consideration of the following half-spaces in $\mathbb{K}^{p,q}$. In the case $\mathbb{K} = \mathbb{R}$ consider

$$H^{p+}(\mathbb{R}^{p,q}) = \{X \in \mathbb{R}^{p,q} \mid \sum_{j=1}^{p} X_j^2 > \sum_{j=p+1}^{n} X_j^2,\ X_1 > 0\}$$

and

$$H^{q+}(\mathbb{R}^{p,q}) = \{X \in \mathbb{R}^{p,q} \mid \sum_{j=1}^{p} X_j^2 < \sum_{j=p+1}^{n} X_j^2,\ X_n > 0\}.$$

Notice that in the case $p = 0$ one obtains $H^{p+}(\mathbb{R}^{0,n}) = \emptyset$ and $H^{q+}(\mathbb{R}^{0,n}) = \{X \in \mathbb{R}^{0,n} \mid X_n > 0\}$. In the case $q = 0$ one gets similarly that $H^{q+}(\mathbb{R}^{n,0}) = \emptyset$ and $H^{p+}(\mathbb{R}^{n,0}) = \{X \in \mathbb{R}^{n,0} \mid X_1 > 0\}$.

In what follows we assume without loss of generality that $p < q$ and use for simplicity the notation $H^+(\mathbb{R}^{p,q})$ for $H^{q+}(\mathbb{R}^{p,q})$. In the complexified case consider in the case $p = 0$ the following half-cone (compare with [1]) as its complexification:

$$H^+(\mathbb{C}^{0,n}) = \{Z = X + iY \in \mathbb{C}^{0,n} \mid X_n^2 > \sum_{j=1}^{n-1} Y_j^2,\ X_n > 0\}.$$

In the case $p = 1, q = n-1$ consider the classical Siegel half-space

$$H^+(\mathbb{C}^{1,n-1}) = \{Z = X + iY \in \mathbb{C}^{1,n-1} \mid Y_n^2 > \sum_{j=1}^{n-1} Y_j^2,\ Y_n > 0\}.$$

In the remaining complexified cases with $p, q \geq 2$, $p < q$ consider the following domains

$$H^+(\mathbb{C}^{p,q}) = \{Z = X + iY \in \mathbb{C}^{p,q} \mid Y_n^2 > \sum_{j=1}^{p} Y_j^2 + \sum_{j=p+1}^{n-1} X_j^2,\ Y_n > 0\}$$

which of course are Siegel type domains as well. Similar domains are considered when $p > q$. One can verify by a direct calculation that these half-spaces have no intersection with the nullcone.
Möbius transformations in $\mathbb{K}^{p,q}$ can be described in terms of the Vahlen groups $SV_{\mathbb{K}}(p,q)$ (see for instance [15, 53]). For convenience, let us recall its definition.

Definition 1. Let $M = \begin{pmatrix} a & b \\ c & d \end{pmatrix}$ be a matrix with coefficients $a, b, c, d \in Cl_{p,q}(\mathbb{K})$ that can all be written as products of vectors from $\mathbb{R}^{p,q}$. Then M belongs to the *Vahlen group* $SV_{\mathbb{K}}(p,q)$ if and only if a, b, c, d satisfy

$$ac^*, cd^*, db^*, bc^* \in \mathbb{K}^{p,q} \quad \text{and} \quad ad^* - bc^* = 1.$$

The Vahlen group acts on $\mathbb{K}^{p,q}$ by its associated Möbius transformation $M\langle z\rangle = (az+b)(cz+d)^{-1}$ where $M = \begin{pmatrix} a & b \\ c & d \end{pmatrix} \in SV_{\mathbb{K}}(p,q)$. The group $SV_{\mathbb{K}}(p,q)$ is generated by the inversion matrix $J := \begin{pmatrix} 0 & -1 \\ 1 & 0 \end{pmatrix}$ and by matrices inducing the translations $z \mapsto z + a$ where $a \in \mathbb{K}^{p,q}$. The group $SV_{\mathbb{C}}(p,q)$ contains the real Vahlen group $SV_{\mathbb{R}}(r,s)$ where $r \leq p$, $s \leq q$, as a subgroup. Each arbitrary discrete subgroup of $SV_{\mathbb{C}}(p,q)$ is contained in a group that is isomorphic to $\Gamma_{\mathbb{C}}^n(\Omega)$ where

$$\Gamma_{\mathbb{C}}^n = \left\langle J := \begin{pmatrix} 0 & -1 \\ 1 & 0 \end{pmatrix}, T_{\omega_1} := \begin{pmatrix} 1 & \omega_1 \\ 0 & 1 \end{pmatrix}, \ldots, T_{\omega_{2n}} := \begin{pmatrix} 1 & \omega_{2n} \\ 0 & 1 \end{pmatrix} \right\rangle$$

where $\omega_1, \ldots, \omega_{2n}$ are $\mathbb{R}$-linear independent vectors from $\mathbb{C}^n$. Of course, the whole group $\Gamma^n_{\mathbb{C}}(\Omega)$ does not act discontinuously for any arbitrary choice of lattice. However, in this paper we restrict ourselves to the particular case where $\omega_1 = e_1, \ldots, \omega_n = e_n, \omega_{n+1} = ie_1, \ldots, \omega_{2n} = ie_n$ which involves discontinous actions. In relation to the half-spaces introduced above the following groups are of special interest.

The group $\Gamma^{0,k}(\mathbb{R}) = \langle J, T_{e_{n-k}}, \ldots, T_{e_{n-1}} \rangle$ leaves $H^{q+}(\mathbb{R}^{p,q})$ invariant when $k \leq q-1$. One observes that the case where $p=0$, $q=n$ (or similarly $p=n$, $q=0$) is special. Under this condition, $H^+(\mathbb{R}^{p,q})$ is invariant under the modular group with the maximal number of translative generators. The situation is different in the complexified case where all the cases turn out to be equivalent. In the complexified case we always have an invariance under the inversion and a translation group consisting of n linear independent translation matrices, independently from the choice of p and q. The space $H^{q+}(\mathbb{C}^{0,n})$ is left invariant under $\Gamma^{0,n}_{\mathbb{C}} = \langle J, T_{e_1}, \ldots, T_{e_{n-1}}, T_{ie_n} \rangle$. Similarly, $H^{q+}(\mathbb{C}^{1,n-1})$ is left invariant under $\Gamma^{1,n-1}_{\mathbb{C}} = \langle J, T_{e_1}, \ldots, T_{e_n} \rangle$. Analogously, for $p, q \geq 2$, $H^+(\mathbb{C}^{p,q})$ is invariant under $\Gamma^{p,q}_{\mathbb{C}} = \langle J, T_{e_1}, \ldots, T_{e_p}, T_{ie_{p+1}}, \ldots, T_{ie_{n-1}}, T_{e_n} \rangle$. All these groups act discontinuously on the respective half-space. These half-spaces serve as definition domains for modular forms related to the associated hypercomplex modular groups.

2.3. Dirac operators on Minkowski type spaces

Some of the first contributions to Dirac operators on Minkowski type spaces have been developed by W. Nef and R. Fueter in the second half of the 1940s. See for instance [47] and [21] pp. 264. Unfortunately, most of their results remained unknown for a rather long time. The theory of Dirac operators on arbitrary Minkowski type spaces had a remarkable renaissance in the 1980s and 1990s. See for example [52, 53], or [5]. The main reference for our purposes is [53]. The Dirac operator in $\mathbb{K}^{p,q}$ is given by

$$D_{\mathbf{K}^{p,q}} := \sum_{j=1}^{p} \frac{\partial}{\partial z_j} e_j - \sum_{j=p+1}^{n} \frac{\partial}{\partial z_j} e_j$$

and functions annihilated by it from the left are called left $\mathbb{K}$-(p,q)-monogenic. When it is clear to which space we refer, then we simply say left monogenic. Sufficiently smooth functions, satisfying $D^h_{\mathbf{K}^{p,q}} f = 0$ shall simply be called left h-monogenic. Notice that

$$D^2_{\mathbf{K}^{p,q}} = \sum_{j=1}^{p} \frac{\partial^2}{z_j^2} - \sum_{j=p+1}^{n} \frac{\partial^2}{z_j^2}$$

which coincides in the cases $\mathbb{R}^{0,n}$ or $\mathbb{R}^{n,0}$ with the real Euclidean Laplacian or, in the cases $\mathbb{C}^{0,n}$ or $\mathbb{C}^{n,0}$ with the complexified Laplacian. The left and right fundamental solution to $D^h_{\mathbb{R}^{p,q}}$ is given by

$$q_{\mathbf{0}}^{(h)}(X) := \begin{cases} \dfrac{X}{\|\mathcal{N}(X)\|^{\frac{n+1-h}{2}}} & h \text{ odd integer with } h < n, \\ \dfrac{1}{\|\mathcal{N}(X)\|^{\frac{n-h}{2}}} & h \text{ even integer with } h < n, \end{cases} \tag{1}$$

and the left and right fundamental solution to $D^h_{\mathbb{C}^{p,q}}$ by

$$q_{\mathbf{0}}^{(h)}(Z) := \begin{cases} \dfrac{Z}{\mathcal{N}(Z)^{\frac{n+1-h}{2}}} & h \text{ odd integer with } h < n, \\ \dfrac{1}{\mathcal{N}(Z)^{\frac{n-h}{2}}} & h \text{ even integer with } h < n. \end{cases} \tag{2}$$

These functions in turn, together with their partial derivatives $q_{\mathbf{m}}^{(h)}(z) := \frac{\partial^{|\mathbf{m}|}}{\partial z^{\mathbf{m}}} q_{\mathbf{0}}^{(h)}(z)$ ($z \in \mathbb{K}^{p,q}$), generalize the classical complex negative power functions to the Clifford analysis setting within the framework of $D^h_{\mathbb{K}^{p,q}}$. Here $\mathbf{m} = (m_1, \ldots, m_n) \in \mathbb{N}_0{}^n$ is a multi-index and $|\mathbf{m}| = m_1 + \ldots + m_n$ is its length.

3. Clifford-analytic Eisenstein series related to translation groups and their applications

The classical Eisenstein series to translation groups in complex analysis are given as Mittag-Leffler series of the negative power functions $1/(z+\omega)^m$ summed up over a one or two dimensional lattice. In the higher dimensional Clifford analysis setting where we consider nullsolutions to $D_{\mathbb{K}^{p,q}}$ in $\mathbb{K}^{p,q}$, the negative power functions are generalized by the functions $q_{\mathbf{m}}^{(h)}(z+\omega)$, as explained in the previous section. Generalizations of the classical Eisenstein series to the context considered here are, roughly speaking, thus given by summations of the expressions $q_{\mathbf{m}}^{(h)}(T\langle z\rangle)$ over the corresponding translation group T.

Indeed, in the simplest case $\mathbb{K}^{p,q} = \mathbb{R}^{0,n}$ or $\mathbb{K}^{p,q} = \mathbb{R}^{n,0}$ one can introduce higher dimensional generalizations in the h-monogenic setting as follows:

Definition 2. Let $k \in \mathbb{N}$ with $1 \leq k \leq n$, $h \in \mathbb{N}$ and $1 \leq h \leq n-1$, and let $\omega_1, \ldots, \omega_k$ be $\mathbb{R}$-linear independent paravectors in $\mathbb{R}^n$. The h-monogenic translative *Eisenstein series associated with the lattice* Ω_k are then defined for all multi-indices $\mathbf{m} \in \mathbb{N}_0^n$ with $|\mathbf{m}| \geq \max\{0, k-n+1+h\}$ by

$$\epsilon_{\mathbf{m},h}^{(k)}(X;\Omega_k) = \sum_{\omega\in\Omega_k} q_{\mathbf{m}}^{(h)}(X+\omega). \tag{3}$$

In the cases $k-n+h \geq 0$ one sets additionally for $\mathbf{m} \in \mathbb{N}_0^n$ with $|\mathbf{m}| = k-n+h$:

$$\epsilon_{\mathbf{m},h}^{(k)}(X;\Omega_k) = q_{\mathbf{0}}^{(h)}(X) + \sum_{\omega\in\Omega_k\setminus\{0\}} \left(q_{\mathbf{0}}^{(h)}(X+\omega) - q_{\mathbf{0}}^{(h)}(\omega)\right). \tag{4}$$

Additionally one can define in the cases with $k-n-1+h \geq 0$ for multi-indices $\mathbf{m} \in \mathbb{N}_0^n$ with $|\mathbf{m}| = k-n-1+h$:

$$\begin{aligned} \epsilon^{(k)}_{\mathbf{m},h;a,b}(X;\Omega_k) &:= q^{(h)}_{\mathbf{m}}(X-a) - q^{(h)}_{\mathbf{m}}(X-b) \\ &+ \sum_{\omega \in \Omega_k \setminus \{0\}} \Big[q^{(h)}_{\mathbf{m}}(X-a+\omega) - q^{(h)}_{\mathbf{m}}(X-b-\omega) \\ &\qquad - q^{(h)}_{\mathbf{m}}(\omega - a) + q^{(h)}_{\mathbf{m}}(\omega - b)\Big], \end{aligned} \tag{5}$$

where a, b are vectors from $\mathbb{R}^n \setminus \Omega_k$ with $a \not\equiv b \bmod \Omega_k$.

Notice that all these series have only point singularities.

In the other cases $\mathbb{R}^{p,q}$ where p and q are both different from zero, we do not obtain a convergent series when considering a summation of the associated expressions $q^{(s)}_{\mathbf{m}}(X+\omega)$ over any arbitrary lattice in $\mathbb{R}^{p,q}$, independently how large $|\mathbf{m}|$ is chosen. This behavior is due to the fact that these expressions do not only have singularities in isolated points but in cones. Therefore, restrictions on the periodicity group need to be made. We get a convergent series (under the same conditions for $|\mathbf{m}|$ as mentioned above), when the lattice Ω is additionally completely contained in $\mathbb{R}^p$ or in $\mathbb{R}^q$. This shows that the configuration $(p,q) = (0,n)$ or $(p,q) = (n,0)$ is special. In these particular cases one obtains even h-monogenic Eisenstein series to n-dimensional periodicity lattices. In the general case working in $\mathbb{R}^{p,q}$ one gets in the best case Eisenstein series for a translation group with $\max\{p,q\}$ translation matrices. In the complexified case however, one has again a more equilibrated relation between the corresponding spaces. In $\mathbb{C}^{p,q}$ one can always define convergent Eisenstein series for a group that has n translation generators. In $\mathbb{C}^{p,q}$ one needs the period lattice to be contained in the linear manifold $e^{i\phi}(\mathbb{R}^p + i\mathbb{R}^q)$ where $\phi \in [0, 2\pi)$ is an arbitrary real parameter. Under this condition we get the same convergence conditions for $|\mathbf{m}|$ as mentioned above. Thus, in the complexified case we can always introduce (independently from the signature) k-fold periodic h-monogenic Eisenstein series, for $1 \leq k \leq n$. We get a divergent series whenever $k > n$.

In the cases $k < n$ these function series provide us with k-fold periodic generalizations of classical trigonometric functions to the Clifford analysis setting (cf. [27, 28]). In the case $k = n$ one obtains generalizations of the elliptic functions to the Clifford analysis setting. Notice that one cannot define n-fold periodic Eisenstein series within the setting of $\mathbb{R}^{p,q}$ whenever p, q are both different from zero.

Some very first contributions to the study of some particular types of functions that belong to that function class of Eisenstein series have been provided by A. C. Dixon [13], R. Fueter [19, 20, 21, 22] and J. Ryan [51]. In these papers the authors introduced the particular series associated with the case $k = n$ (elliptic functions) in the Euclidean spaces $\mathbb{R}^{0,3}, \mathbb{H}, \mathbb{R}^{0,n}$, and studied some basic properties. In [48] T. Qian introduced a general function class defined on particular sector domains and studied Lipschitz pertubations of the $(n-1)$-torus within a cylinder in order to obtain special Fourier multipliers for these particular domains. The special function $\epsilon_{\mathbf{0}}^{(n-1)}$ associated with the orthonormal lattice makes part of that special function class.

A more extensive and systematic study of 1-monogenic Eisenstein series in the Euclidean space $\mathbb{R}^{0,n}$ has been started in [27, 28] within a more general framework. In [32] we studied their complexification to $\mathbb{C}^{0,n}$. h-monogenic Eisenstein series within the framework of $\mathbb{R}^{0,n}$ have been introduced as a special case of Poincaré series in [31] and their complexification to $\mathbb{C}^{0,n}$ in [32]. They in turn contain the holomorphic Cliffordian Weierstrassian functions from [23, 41] which are solutions of the quaternionic Fueter-Sce equation as special cases.

3.1. Function theoretical properties

The Eisenstein series $\epsilon_{\mathbf{m},h}^{(k)}(z,\Omega_k)$ provide a powerful tool to construct a large family of h-monogenic k-fold periodic functions in $\mathbb{K}^{p,q}$ that have the property that the centers of their singularity cones all lie in $S(\mathbb{K}^{p,q})$. Here, and in all that follows, $S(\mathbb{K}^{p,q})$ stands for $\mathbb{R}^p$ or $\mathbb{R}^q$ when $\mathbb{K}=\mathbb{R}$ and for the linear domain manifolds of the form $e^{i\phi}(\mathbb{R}^p+i\mathbb{R}^q)$ with $\phi\in[0,2\pi)$, when $\mathbb{K}=\mathbb{C}$. In particular, the series $\epsilon_{\mathbf{m},h}^{(k)}(z,\Omega_k)$ admit the construction of higher dimensional vector valued variants of many elementary trigonometric within the framework of $D_{\mathbb{K}^{p,q}}^{h}f=0$, as for example (cf. [27, 28]):

$$
\begin{aligned}
\cot_{k,h}(z) &:= \epsilon_{\mathbf{0},h}^{(k)}(z),\\
\tan_{k,h}(z) &:= -\sum_{v\in\mathcal{V}_k(2)\setminus\{0\}} \cot_{k,h}(z+v/2),\\
\csc_{k,h}(z) &:= \frac{1}{2^{n-h-1}}\cot_{k,h}(z/2)-\cot_{k,h}(z),\\
\sec_{k,h}(z) &:= \sum_{v\in\mathcal{V}_k(2)\setminus\{0\}} \csc_{k,h}(z+v/2).
\end{aligned}
$$

Here $k<n$ and $\mathcal{V}_k(2):=\{0,\frac{\omega_1}{2},\ldots,\frac{\omega_k}{2},\ldots,\frac{\omega_1+\ldots+\omega_k}{2}\}$ stands for the canonical system of representatives of the quotient module $\Omega_k/2\Omega_k$. By means of the series $\epsilon_{\mathbf{m},h}^{(k)}$ with $|\mathbf{m}|=1$ one can furthermore construct h-monogenic k-fold periodic generalizations of the cosecant-squared and the secant-squared function, etc.. See also [9] in which we provided a detailed discussion on their relation to their planar counterparts.

The functions

$$
\wp_{\mathbf{r}+\tau(i),h}(z) := \epsilon_{\mathbf{r}+\tau(i),h}^{(n)}(z) \quad i=1,\ldots,n \tag{6}
$$

where $\mathbf{r}\in\mathbb{N}_0^n$ with $|\mathbf{r}|=h-1$ provide natural h-monogenic generalizations of the Weierstraß $\wp$-function to $\mathbb{R}^{0,n}$ or $\mathbb{C}^{p,q}$ $(p+q=n)$. Their partial derivatives generalize the derivatives of Weierstraß $\wp$-functions. Analogues of them in the spaces $\mathbb{R}^{p,q}$ where p,q both differ from zero cannot be defined. All the functions from (6) are generated by partial differentiation from one single h-monogenic primitive $\zeta_{\mathbf{r},h}$

that then satisfies $\frac{\partial}{\partial z_i}\zeta_{\mathbf{r},h}(z) = \wp_{\mathbf{r}+\tau(i),h}(z)$. The function $\zeta_{\mathbf{0},1}$ is the monogenic generalization of the Weierstraß ζ-function which has the form

$$\zeta(z) = q_{\mathbf{0}}(z) + \sum_{\omega\in\Omega_n\backslash\{0\}} \left[q_{\mathbf{0}}(z+\omega) - q_{\mathbf{0}}(\omega) + \sum_{j=1}^{n-1} q_{\tau(j)}(\omega)(z_j + z_n e_n e_j)\right] \quad (7)$$

with $z \in \mathbb{R}^{0,n}$ or $z \in \mathbb{C}^{p,q}$. Here $\tau(j)$ stands for the multi-index having the entry 1 at the j-th position and the entry zero at the remaining ones. The function (7) was introduced in the setting of real quaternions by R. Fueter in [21],[22] and in the setting of real Euclidean space $\mathbb{R}^{0,n}$ by J. Ryan in [51]. The function $\zeta_{\mathbf{r},n-1}(z)$ when $X \in \mathbb{R}^{0,n}$ (n odd) or $z \in \mathbb{R}^{p,q}$ with $p+q=n$ is the holomorphic Cliffordian generalization of the Weierstraß ζ-function that has been introduced in the setting of real quaternions by G. Laville and I. Ramadanoff in the recent work [41]. The functions are in general not n-fold periodic $\zeta_{\mathbf{r},h}$. However, there are vectors $\eta_1^h,\ldots,\eta_n^h$ such that

$$\zeta_{\mathbf{r},h}(z+\omega_j) - \zeta_{\mathbf{r},h}(z) = \eta_j^{(h)}$$

where the ω_j denote the primitive periods of the underlying lattice.

By applying Mittag-Leffler's theorem for real and complexified Minkowski type spaces we can deduce by similar arguments as presented for the monogenic case in the Euclidean space $\mathbb{R}^{0,n}$ in [27, 29] the following structural result. The following theorem points out the central role of this type of Eisenstein series in the study of meromorphic p-fold periodic functions in higher dimensional Minkowski type spaces $\mathbb{K}^{p,q}$.

Theorem 1 (Representation theorem). *Let Ω_k be a k-dimensional lattice in $\mathbb{K}^{p,q}$ that lies completely in $S(\mathbb{K}^{p,q})$. Let $\alpha \in \mathbb{N}_0$ with $\alpha \geq \max\{0, k-n-1+h\}$. Let $f : \mathbb{K}^{p,q}\backslash(\bigcup_{i=1}^{l} S_{a_i+\Omega_k}) \to Cl_{p,q}(\mathbb{K})$ be k-fold periodic. Suppose that all the centers of the singularity cones lie all in $S(\mathbb{K}^{p,q})$. Suppose further that the center points of the singularity cones within an arbitrary period cell modulo Ω_k are all isolated and that its cardinality per period cell is finite. Denote them by $a_1,\ldots,a_l$. Suppose that f is left monogenic in $\mathbb{K}^{p,q}\backslash(\bigcup_{i=1}^{l} S_{a_i+\Omega_k})$. Suppose that the singularity order of a_i with $(1 \leq i \leq l)$ is $N(a_i)$, that $N(a_i)-n+h \geq \alpha$ for all $i \in \{1,\ldots,l\}$ and that the principal parts are given by $\sum_{|\mathbf{m}|=\alpha}^{N(a_i)-n+h} q_{\mathbf{m}}^{(h)}(z-a_i)b_{\mathbf{m}}^{(i)}$. Then there is a left s-entire function $g : \mathbb{K}^{p,q} \to Cl_{p,q}(\mathbb{K})$ with*

$$f(z) = \sum_{i=1}^{l} \sum_{|\mathbf{m}|=\alpha}^{N(a_i)-n+h} \epsilon_{\mathbf{m},h}^{(k)}(z-a_i)b_{\mathbf{m}}^{(i)} + g(z). \quad (8)$$

In the cases $|\mathbf{m}| = \alpha$ with $\alpha = k-n-1+h$ the function $\epsilon_{\mathbf{m},h;a_i,b}^{(k)}$ with an arbitrarily chosen $b \not\equiv a_i \bmod \Omega_k$ and $b \notin \Omega_k$.

The proof is based on applying Mittag-Leffler type theorems for Minkowski type spaces.
We shall see briefly that we get an even stronger result for the class of h-monogenic n-fold periodic functions in the Euclidean space $\mathbb{R}^{0,n}$. The speciality of this function class within the particular framework of $\mathbb{R}^{0,n}$ or $\mathbb{R}^{n,0}$ becomes clear by means of the following theorems providing generalizations of the classical Liouville theorems.

Theorem 2 (1. Liouville theorem). *Every n-fold periodic function that is h-monogenic in a whole period cell is a constant function.*

For 1-monogenic functions this theorem has already been proved by R. Fueter for the quaternionic case (cf. [19]) and by J. Ryan for $\mathbb{R}^{0,n}$ (cf. [51]). Within the framework of the quaternionic Fueter-Sce equation a proof has been provided by G. Laville and I. Ramadanoff in [41]. It can easily be extended to the class of h-monogenic functions since every arbitrary h-monogenic function that is bounded in $\mathbb{R}^n$ also reduces to a constant function as proved for example in [2]. From the compactness of the period cell of an n-fold periodic function the statement thus follows immediately.

In the particular case $k = n$ working in the Euclidean space $\mathbb{R}^n$ the function $g(z)$ from the previous theorem thus reduces to a constant. For the other cases, a similar statement cannot be provided in such a generality.

From this theorem one infers furthermore that in this special case a non-constant n-fold periodic monogenic functions must have singularities in each of its period cells. Moreover, we have:

Theorem 3 (2. Liouville theorem). *The sum of the residues of an n-fold periodic monogenic function within an arbitrary period cell vanishes.*

This statement has been proved for the quaternionic case by R. Fueter (see [19]) and for arbitrary dimensions in [27, 26]. In the case of dealing with non-isolated singularities the notion of a residue has to be understood in the sense of Leray-Norguet residues as introduced by W. Nef in [45, 46] and, more generally in [11]. In particular, each non-constant n-fold periodic monogenic function needs to have at least two point singularities of order $n-1$ or one point singularity of higher order.

Furthermore, we proved recently in [26] the following balance relation between c-points and poles for a particular case:

Theorem 4 (3. Liouville theorem). *Let f be an n-fold periodic monogenic function that has only isolated singularities and that takes all its values in $\mathbb{R}^n$. Moreover, let us suppose that f has only isolated c-points ($c \in \mathbb{R}^n$) and only isolated singularities. Let P be any period cell with no poles and c-points on its surfaces. Let $b_1, \ldots, b_k$ be the poles within P. Let $\varepsilon_1, \ldots, \varepsilon_k > 0$ be sufficiently small such that $B(b_i, \varepsilon_i) \backslash \{b_i\}$ contains no c-points and poles.*

Then

$$\sum_{a \in P} \mathrm{ord}(f - c; a) = - \sum_{i=1}^{k} p(f - a; b_i), \tag{9}$$

where

$$p(f - a; b_i) := \frac{1}{A_n} \int_{\partial B(b_i, \varepsilon_i)} q_0(f(z) - a)(Jf)^*(z) d\sigma(z),$$

A_n standing for the surface measure of the unit ball from R^n and $(Jf)^*$ for the adjunct of the Jacobi matrix.

For the exact definition of the order, see [25].

For the proofs of the second and third Liouville theorem, the use of Cauchy's integral formula for monogenic functions is central. It is thus not clear whether these theorems have similar extensions to the more general function class of h-monogenic functions where $h \geq 2$. Under additional conditions both theorems can be extended to $\mathbb{R}^n$-like linear domain manifolds within the framework of a complexification of the n-fold periodic functions in $\mathbb{C}^n$ satisfying the complexified Dirac equation. Whenever, the cones of the singularities and the manifolds of the c-points have only point intersections with the linear domain manifold, generalizations of Theorem 3 and Theorem 4 can thus be obtained.

Similarly to the classical complex case the functions $\epsilon_{\mathbf{m},h}^{(k)}$ can be characterized in terms of special functional equations that generalize the well-known duplication formula of the classical cotangent. For details, see [34]. By direct rearrangement arguments one can readily deduce:

Theorem 5. *Let Ω_k be a k-dimensional lattice lying completely in $S(\mathbb{K}^{p,q})$. Suppose that $\mathbf{m} \in \mathbb{N}_0^n$ with $|\mathbf{m}| \geq \max\{0, k - n + 1 + h\}$. In these cases*

$$r^{n+|\mathbf{m}|-h} \epsilon_{\mathbf{m},h}^{(k)}(rz) = \sum_{v \in \mathcal{V}_k(r)} \epsilon_{\mathbf{m},h}^{(k)}(z + v/r) \tag{10}$$

where $\mathcal{V}_k(r)$ is the canonical system of representatives of $\Omega_k / r\Omega_k$.

By integration arguments one infers further that (10) is satisfied at least up to a vector constant C also for the cases with $|\mathbf{m}| = k - n + h$, $k \leq n$, $h \leq n - 1$ such that the points of the associated lattice all lie in $S(\mathbb{K}^{p,q})$. For the particular cases mentioned in the previous line where additionally $|\mathbf{m}| + h$ is odd and for all cases dealing with n-dimensional lattices, we managed to show that $C = 0$. See [34] where we provided a detailed proof within the framework of working in the real Euclidean space $\mathbb{R}^{0,n}$. Under the condition that $\Omega_k \subset S(\mathbb{K}^{p,q})$ the proof can easily be adapted to the more general setting working in $\mathbb{K}^{p,q}$, therefore we omit it here. We can say more:

Theorem 6. *Let $h < n$ and $\mathbf{m} \in \mathbb{N}_0^n$ be a multi-index. Let $\Omega_k = \mathbf{Z}\omega_1 + \ldots + \mathbf{Z}\omega_k$ be a k-dimensional lattice lying completely in $S(\mathbb{K}^{p,q})$. Let $\mathbf{m} \in \mathbb{N}_0^n$ with $|\mathbf{m}| \geq \max\{0, k-n+h\}$. Let further $g : \mathbb{K}^{p,q} \to Cl_{p,q}(\mathbb{K})$ be a left or right h-monogenic function with principal parts $q_{\mathbf{m}}^{(h)}(z-\omega)$ at each $\omega \in \Omega_k$. If g satisfies*

$$r^{n+|\mathbf{m}|-h} g(rz) = \sum_{v \in \mathcal{V}_k(r)} g(z + v/r), \tag{11}$$

then there is a $C \in Cl_{p,q}(\mathbb{K})$ such that

$$g(z) = \epsilon_{\mathbf{m},h}^{(k)}(z) + C \tag{12}$$

for all $z \in \mathbb{K}^{p,q} \backslash S_{\Omega_k}$.

Proof. Define the function $s(z) = g(z) - \epsilon_{\mathbf{m},h}^{(k)}(z)$ which is left h-monogenic in the whole space $\mathbb{K}^{p,q}$. Let us first treat the case where $|\mathbf{m}| + h$ is odd. In this case we have the certaincy that

$$r^{n+|\mathbf{m}|-h} s(rz) = \sum_{v \in \mathcal{V}_k(r)} s(z + v/r), \quad s(0) = s_0 \tag{13}$$

with an $s_0 \in \mathbf{Cl}_{p,q}(\mathbb{K})$. Let us assume that $s \not\equiv s_0$. Let $\beta := \|\omega_1 + \ldots + \omega_k\|$. Notice that the maximum principle holds for the class of h-monogenic functions. Therefore, there is an element $c \in \partial B(0, r\beta)$ such that

$$\|s(z)\| < \|s(c)\|$$

for all $z \in \overset{\circ}{B}(0, r\beta)$. Here $B(0, r\beta) = \{z \in \mathbb{K}^{p,q} | \; \|z\| \leq r\beta\}$ stands for a "ball" in $\mathbb{K}^{p,q}$ with radius $n\beta$, $\overset{\circ}{B}(0, r\beta)$ for its open kernel and $\partial B(0, r\beta)$ for its boundary. Notice that

$$\Big\|(c + \sum_{i=1}^{k} \alpha_i \omega_i)/r\Big\| \leq \beta + \frac{r-1}{r}\beta < 2\beta$$

for all $0 \leq \alpha_i < r$, so that we can infer from (13) that

$$\begin{aligned} r^{n+|\mathbf{m}|-h}\|s(c)\| &= \| \sum_{v \in \mathcal{V}_k(r)} s(\frac{c+v}{r})\| \leq \sum_{v \in \mathcal{V}_k(r)} \|s(\frac{c+v}{r})\| < r^k \|s(c)\| \\ &\leq r^{n-h+|\mathbf{m}|}\|s(c)\|, \end{aligned}$$

which provides us with a contradiction, and the assertion is hereby proven for $|\mathbf{m}| + h$ being odd. In the case where $|\mathbf{m}| + h$ is even, consider the first partial derivatives $s_i(z) := \frac{\partial s}{\partial z_i}$ for $i = 1, \ldots, n$. The s_i satisfy definetely the integer multiplication formula

$$r^{n+1+|\mathbf{m}|-h} s_i(rz) = \sum_{v \in \mathcal{V}_p(k)} s_i(z + v/r), \quad s_i(0) = s_{i0}.$$

Applying a similar argument yields finally that $s_i \equiv s_{i0}$ for all $i = 1, \ldots, n$. From $\sharp \mathcal{V}_k(r) = r^k = r^{n+|\mathbf{m}|-h}$ follows $s_{i0} \equiv 0$ and the assertion is hereby proven. □

By similar arguments one can further show that there are no non-constant entire h-monogenic functions satisfying these functional equations. Theorem 6 provides a generalization of Theorem 6 from [34] to the setting of arbitrary Minkowski type spaces $\mathbb{K}^{p,q}$.

3.2. Number theoretical properties

The classical complex-analytic Eisenstein series have many highly interesting number theoretical properties.

Those complex-analytic Weierstraß $\wp$-functions that are associated with lattices that have complex multiplication play a central role in class field theory. Putting $g_2 = 60\sum_{\omega\in\Omega\setminus\{0\}}\omega^{-4}$ and $g_3 = 140\sum_{\omega\in\Omega\setminus\{0\}}\omega^{-6}$, then all division points of the normalized Weierstraß function $\mathcal{P}(z) = \frac{g_2 g_3}{g_2^3-27g_3^2}\wp(z)$ lie all in an abelian Galois field extension of an imaginary quadratic number field, and the trace itself is an element from that proper imaginary number field. The $\wp$ function plays thus a basic role for the construction of algebraic number fields.

The Fourier expansion of the Eisenstein series is explicitly related to representation numbers of sums of divisors and to the Riemann zeta function involving further interesting number theoretical relations.

In [34, 30, 6] we analyzed whether generalizations can be obtained or expected within the framework of the higher dimensional h-monogenic Eisenstein series.

Let us concentrate in this section on the monogenic Eisenstein series in the Euclidean space $\mathbb{R}^{0,n}$. For more general results we refer to the above cited papers.

3.2.1. Lattices with hypercomplex multiplication and hypercomplex division points of the generalized $\wp$-functions in $\mathbb{R}^n$. A two dimensional lattice Ω in the complex plane is said to have complex multiplication whenever there is a $\lambda \in \mathbb{C}\setminus\mathbf{Z}$ so that $\lambda\Omega \subseteq \Omega$. In turn each element $\eta \in \Omega$ where Ω is such a lattice satisfies $\eta\Omega \subseteq \Omega$. Assume without loss of generality that $\Omega = \mathbf{Z} + \mathbf{Z}\tau$ ($\tau \in H^+(\mathbb{C})$). Then Ω has complex multiplication if and only if $\tau \in \mathbb{Q}[\sqrt{-D}]$ where D is a positive square-free integer.

In [34] we were concerned with analyzing which lattices from $\mathbb{R}^n$ generalize those of the complex plane that have complex multiplication within the framework of Clifford analysis. The question of generalizing the complex multiplication of lattices to the Clifford analysis setting is far from being trivial. Notice that if $\lambda \in \mathbb{R}^n$ and if $\omega \in \Omega$ where Ω now denotes a p-dimensional lattice in $\mathbb{R}^n$, then $\lambda\omega$ is not a vector from $\mathbb{R}^n$ anymore. Because of the non-commutativity it is suggestive to focus on lattices that satisfy $\lambda\Omega\mu \subseteq \Omega$ for some vectors $\lambda, \mu \in \mathbb{R}^n$. If $\Omega \subset \mathbb{R}^n$ is an arbitrary lattice, then the relation $\lambda\Omega\mu \subseteq \Omega$ is in general only satisfied when λ, μ are integers. However, there are indeed some special lattices where this relation is also satisfied for some special vectors $\lambda, \mu \in \mathbb{R}^n$. We call these kinds of lattices then lattices with vector multiplication (or with paravector multiplication when working in the paravector formalism) in the general sense. The concept of vector multiplication can be regarded as an adaptation of the general theory of $\mathbf{Z}$-orders in rational Clifford algebras (cf. [16]) to the framework of working in the *vector space* $\mathbb{R}^n$ which is not closed under usual multiplication for general n. The simplest examples for lattices with vector multiplication in the general sense are for instance rectangular lattices. Notice that

$$e_1(\mathbf{Z}\alpha_1 e_1 + \ldots + \mathbf{Z}\alpha_p e_p)\overline{e_1} = \mathbf{Z}\alpha_1 e_1 + \ldots + \mathbf{Z}\alpha_p e_p.$$

Here $\alpha_j \in \mathbb{R}$. Among the lattices that have with vector multiplication in the general sense, a special role is played by those that also satisfy

$$\alpha\eta\Omega\eta \subseteq \Omega \tag{14}$$

for every $\alpha \in \mathbf{Z}$ and every arbitrary $\eta \in \Omega$. Actually, these lattices in turn shall be regarded as the canonical ones generalizing complex lattices with complex multiplication to the Clifford analysis setting. This viewpoint is justified by the following argument.

Suppose that we have the more general situation that there is a vector $\lambda \in \mathbb{R}^n$ and another element $\mu \in Cl_{0,n}(\mathbb{R})$ so that $\lambda\Omega\mu \subseteq \Omega$. Assuming now that f satisfies $f(X+\omega) = f(X)$ for all $\omega \in \Omega_p$, then the function $f_{\lambda,\mu}(X) := \mu f(\lambda X \mu)\lambda$ also satisfies $f_{\lambda,\mu}(X+\omega) = f_{\lambda,\mu}(X)$ for all ω from the same lattice.

However, if $f : \mathbb{R}^n \to Cl_{0n}(\mathbb{R})$ is an arbitrary left and right monogenic function and if λ is an arbitrary vector from $\mathbb{R}^n$, then in general $f_{\lambda,\mu}(X) = \mu f(\lambda X \mu)\lambda$ remains only left and right monogenic, if $\mu = \alpha\lambda$ with a real $\alpha \in \mathbb{R}$. This motivates us to regard lattices with the property (14) as very canonical ones within the theory of hypercomplex multiplication of Clifford-valued periodic functions.

Next one needs to investigate which lattices satisfy (14). In [34] we proceeded one step in this direction. We proved:

Theorem 7. *If the primitive periods* $\omega_1, \ldots, \omega_p$ *from a lattice* $\Omega_p = \mathbf{Z}\omega_1 + \ldots + \mathbf{Z}\omega_n \subset \mathbb{R}^n$ *satisfy the number theoretical conditions*

$$\mathcal{N}(\omega_i) \in \mathbf{Z}, \qquad \langle \omega_i, \omega_j \rangle \in \mathbf{Z} \quad \forall i,j \in \{1,\ldots,p\}, \tag{15}$$

then every integer α *and every vector* η *from that lattice, satisfies*

$$\alpha\eta\Omega\eta \subseteq \Omega. \tag{16}$$

Conversely, if $\Omega_p \subset \mathbb{R}^n$ *is a* p*-dimensional lattice with* (16) *for any* $\alpha \in \mathbf{Z}$ *and any* $\eta \in \Omega_p$*, then the primitive periods satisfy all* (15)*.*

Examples of lattices with (15) are lattices where the primitive periods $\omega_1, \ldots,$ ω_p have the form

$$\omega_i = \sum_{j=1}^{k} \alpha_{ij}\sqrt{m_j}e_j \tag{17}$$

where $\alpha_{ij} \in \mathbf{Z}$ and where the elements m_i are positive integers, as proved in [34].

Remarks. When working in the paravector formalism, one needs to claim that furthermore $2Sc(\omega_i) \in \mathbf{Z}$ and that $\Omega = \overline{\Omega}$ is satisfied in order to have (14), as shown in [34].

Treating the paravector space $\mathbb{R} \oplus \mathbb{R}^n$ in the cases $n = 1, 3, 7$ with numbers from $\mathbb{C}, \mathbb{H}$ or $\mathbb{O}$ respectively, then we have one important speciality. Considering in these cases lattices of the form where the primitive periods satisfy

$$\omega_i = \alpha_{i0} + \alpha_{i_1} \sqrt{m_1} e_1 \tag{18}$$

or

$$\omega_i = \alpha_{i0} + \alpha_{i1}\sqrt{m_1}e_1 + \alpha_{i2}\sqrt{m_2}e_2 + \alpha_{i3}\sqrt{m_1}\sqrt{m_2}e_3 \tag{19}$$

or

$$\begin{aligned}\omega_i = {} & \alpha_{i0} + \alpha_{i1}\sqrt{m_1}e_1 + \alpha_{i2}\sqrt{m_2}e_2 + \alpha_{i3}\sqrt{m_3}e_3 \\ & +\alpha_{i4}\sqrt{m_1}\sqrt{m_2}e_4 + \alpha_{i5}\sqrt{m_1}\sqrt{m_3}e_5 + \alpha_{i6}\sqrt{m_2}\sqrt{m_3}e_6 + \alpha_{i7}\sqrt{m_1}\sqrt{m_2}\sqrt{m_3}e_7\end{aligned} \tag{20}$$

where $\alpha_{ij} \in \mathbf{Z}$, m_1, m_2, m_3 are distinct square-free positive integers and where the e_j shall be identified with the complex or with the corresponding quaternionic or octonionic units, respectively, then we have that each element η from such a lattice satisfies $\eta\Omega \subset \Omega$. In these special cases the lattice is closed under multiplication. Notice that the octonions do not form a Clifford algebra, as they are non-associative.

In the classical complex case one knows furthermore, that (18) is the most general form of a complex lattice that has complex multiplication.

Some first contributions to the quaternionic case were provided by R. Fueter in [20, 21]. The more general theory has been developed recently in [34].

The quaternionic and the octonionic case have also a function theoretic speciality. Whenever Ω is a lattice satisfying (15) and $2Sc(\Omega) \in \mathbf{Z}$, then $\mu f(\lambda X \mu)\lambda$ is left and right monogenic and p-fold periodic with respect to Ω for all arbitrary elements λ, μ from Ω, whenever f is quaternionic (octonionic) left and right monogenic and periodic with respect to Ω.

Let us now turn again to the more general case working in $\mathbb{R}^n$ for arbitrary n.

The fact that both $f_{\alpha\lambda,\lambda}(X)$ and $f(X)$ are left and right monogenic p-fold periodic functions associated to the same period lattice involves explicit relations between both functions. In particular one obtains for the monogenic Weierstraß $\wp_{\tau(i)}$ functions $(i = 1, \ldots, n)$ that

$$\begin{aligned} & \alpha\lambda[\frac{\partial}{\partial X_i}\zeta(\alpha\lambda X\lambda, \Omega_n)]\lambda \\ & \qquad - \frac{1}{[\alpha\mathcal{N}(\lambda)]^{n-2}} \sum_{\alpha^{-1}\lambda^{-1}\omega\lambda^{-1} \in P\backslash\{0\}} \epsilon^{(n)}_{\tau(i),1}(X + \alpha^{-1}\lambda^{-1}\omega\lambda^{-1}) \\ & = \frac{\sum_{h=1}^{n} \left[\Theta_{hi}\alpha\lambda\Big(\sum_{j=1}^{n} n_{h_j}\eta_j^{[1]}\Big)\lambda\right] - \mathcal{N}(\sqrt{\alpha}\lambda)^2 \sum_{h=1}^{n} \Theta_{hi}\eta_h^{[1]}}{\det W}\end{aligned} \tag{21}$$

as proved in [34].

Here, we used the notations $\omega_h = \sum_{j=1}^{n} \omega_{hj} e_j$, $\det\ W = \det \begin{pmatrix} \omega_{11} & \dots & \omega_{1n} \\ \omega_{21} & \dots & \omega_{2n} \\ \dots & & \dots \\ \omega_{n1} & \dots & \omega_{nn} \end{pmatrix}$ and Θ_{hj} stands for the adjoint determinant associated with the element ω_{hj}. Also, P denotes the fundamental period cell.

The relation (21) in turn involves an explicit formula for the trace of the hypercomplex division points of the generalized $\wp$-functions $\wp_{\tau(i)}$. Applying the limit to zero on (21) namely leads to

$$\sum_{\alpha^{-1}\lambda^{-1}\omega\lambda^{-1} \in P\setminus\{0\}} \epsilon^{(n)}_{\tau(i),1}(\alpha^{-1}\lambda^{-1}\omega\lambda^{-1})$$
$$= \frac{[\alpha\mathcal{N}(\lambda)]^{n-2} \sum_{h=1}^{n} \Big[\Theta_{hi}\alpha\lambda\Big(\sum_{j=1}^{n} n_{h_j}\eta_j^{[1]}\Big)\lambda\Big] - \mathcal{N}(\sqrt{\alpha}\lambda)^2 \sum_{h=1}^{n} \Theta_{hi}\eta_h^{[1]}}{\det W} \tag{22}$$

for $i = 1, \dots, n$. If we denote by $\mathcal{R}$ the ring to which the real components of the primitive periods ω_h belong and by $\mathcal{K}$ its associated quotient field, then we can say that the real vector components of the trace of the vector division values of the generalized monogenic $\wp$-functions lie thus all in the field generated by $\mathcal{K}$ and by the real components of the quasi-periodicity constants $\eta_h^{[1]}$ of the monogenic Weierstraß $\wp$-function. In particular, dealing with a lattice of the form (17) where the elements $m_1, \dots, m_n$ are all pairwise distinct positive square-free integers, $\mathcal{K}$ is the multi-quadratic number field $\mathbf{Q}[\sqrt{m_1}, \dots, \sqrt{m_n}]$. Implicitly, we have treated the quaternionic and the octonionic case, too: the series $\epsilon^{(n)}_{\tau(i),1}(X_1e_1 + \dots + X_ne_n)$ can be introduced similarly as in the beginning of Section 3 into the quaternionic and octonionic setting. Simply re-express their formulas for $n = 2, 4, 8$ in the paravector formalism, and then substitute the unit paravectors by the canonical quaternionic and octonionic units. The components of the associated quaternionic or octonionic periods of the special lattice (19) or (20) generate a biquadratic or triquadratic number field, respectively.

For a further progress, a number theoretical analysis of the quasi-periodicity constants (and the relation between them and the periods) needs to be provided. From the generalized Legendre relation for the generalized Weierstraß ζ-function proved in [21, 51] it follows that the periods and the quasi-periodicity constants η_h cannot all be algebraically independent. Could it be that the monogenic Weierstraß $\wp$-function plays a similar role for the construction of field extensions of multi-quadratic number fields as the complex $\wp$-function does for imaginary quadratic number fields?

Such a result would be really of interest for algebraic number theory. Remember that E. Hecke [24] tried managed to construct some higher algebraic number fields by using four-fold periodic Abelian functions in two complex variables. The Abelian functions can be imbedded as special cases within the more general class of four-fold periodic monogenic functions. However, in contrast to Clifford analysis, the periods of the Abelian functions additionally always need to satisfy a special quadratic relation, the so-called Riemann condition. The Riemann condition in turn leads to significant restrictions for the construction of the class fields. E. Hecke was only able to construct subfields. See [24, 21, 22] for more details. In the Clifford analysis setting, the periods of the monogenic elliptic functions do not have to satisfy the Riemann condition. This in turn might be a significant advantage for the construction of class fields.

3.2.2. Relations to divisor sums and Dirichlet series. In classical complex analysis one can use the Weierstraß $\wp$-function as the generating function for the following Eisenstein series

$$G_m(z) = \sum_{(c,d)\in\mathbf{Z}^2\setminus\{(0,0)\}} (cz+d)^{-m} \quad m \in 2\mathbb{N},\ m \geq 4, \quad Im(z) > 0 \tag{23}$$

which have a very interesting Fourier expansion explicitly involving representation numbers of sums of divisors and the Riemann zeta function. In [27] and [30] we showed that similar results can be obtained in the higher dimensional case within the framework of monogenic Eisenstein series. Notice first that (similarly to the complex case) one may restrict oneself to the consideration of lattices of the form $\Omega_{n-1}+\mathbf{Z}\tau$ where $\tau \in H^+(\mathbb{R}^n)$ and where Ω_{n-1} is an $(n-1)$-dimensional lattice lying completely in $span_{\mathbf{R}}\{e_1,\ldots,e_{n-1}\}$. Every arbitrary n-dimensional lattice can be transformed by a simple rotation into a lattice of this special form. Restricting oneself to lattices of this special form, and regarding τ as a hypercomplex variable of the upper half-space $H^+(\mathbb{R}^n)$, then the monogenic Weierstraß $\wp$-function generates the following natural generalization of the series (2) to the framework of Clifford analysis.

Definition 3. (cf. [27, 30]) For a multi-index $\mathbf{m} \in \mathbb{N}_0^n$ with $|\mathbf{m}| \equiv 1(2)$ the monogenic Eisenstein series associated to the generalized monogenic Weierstraß $\wp$-function is defined by

$$G_{\mathbf{m}}(X) := \sum_{(\alpha,\omega)\in\mathbf{Z}\times\Omega_{n-1}\setminus\{(0,0)\}} q_{\mathbf{m}}(\alpha X+\omega) \qquad X \in H^+(\mathbb{R}^n), \quad |\mathbf{m}| \geq 3 \tag{24}$$

where Ω_{n-1} denotes a non-degenerate lattice in $span_{\mathbf{R}}\{e_1,\ldots,e_{n-1}\}$.

More generally, we can introduce h-monogenic generalizations in $H^+(\mathbb{K}^{p,q})$ with respect to a lattice $\Omega_k \subset S(\mathbb{K}^{p,q})$ with $\Omega_k \cap H^+(\mathbb{K}^{p,q}) = \emptyset$ as follows:

$$\begin{aligned} G^{(k)}_{\mathbf{m},h}(z) &:= \sum_{(\alpha,\omega)\in\mathbf{Z}\times\Omega_k\setminus\{(0,0)\}} q^{(h)}_{\mathbf{m}}(\alpha z+\omega), \\ z &\in H^+(\mathbb{K}^{p,q}), \quad |\mathbf{m}| \geq 2+h, \quad |\mathbf{m}|+h \equiv 0 \bmod 2. \end{aligned} \tag{25}$$

Notice that (25) would vanish identically if $|\mathbf{m}|+h \equiv 1 \bmod 2$. The series $G^{(k)}_{\mathbf{m},h}(z)$ is absolutely convergent on $H^+(\mathbb{K}^{p,q})$ whenever $|\mathbf{m}| \geq 2+h$.

Let us first restrict ourselves again to the case of $H^+(\mathbb{R}^n)$ with $k = n-1$. Later, in Section 4, we also need their generalizations. From the monogenicity of $G_{\mathbf{m}}$ and its $(n-1)$-fold periodicity with respect to Ω_{n-1} it follows that $G_{\mathbf{m}}$ can be represented by a normally convergent Fourier series on $H^+(\mathbb{R}^n)$. In what follows let us write a vector $X \in H^+(\mathbb{R}^n)$ in the form $X = \mathbf{X} + X_n e_n$ where $\mathbf{X} \in span_{\mathbb{R}}\{e_1, \ldots, e_{n-1}\}$ and $X_n > 0$. Without loss of generality we may restrict ourselves to the case where $\Omega_{n-1} = \mathbf{Z}e_1 + \ldots + \mathbf{Z}e_{n-1}$ is the orthonormal lattice within the following analysis. As a consequence of the Dirac equation we can furthermore restrict to multi-indices of the form $\mathbf{m} = (m_1, \ldots, m_{n-1}, 0)$.

Theorem 8 (Fourier expansion, cf. [30]**).** *The Fourier series representation of the Eisenstein series $G_{\mathbf{m}}$, $\mathbf{m} = (m_1, \ldots, m_{n-1}, 0)$ with $|\mathbf{m}| \equiv 1(2)$, $|\mathbf{m}| \geq 3$, associated with the orthonormal lattice on the upper half-space reads*

$$G_{\mathbf{m}}(X) = 2\zeta_M^{\Omega_{n-1}}(\mathbf{m}) + A_n(2\pi i)^{|\mathbf{m}|} \sum_{\mathbf{s} \in \mathbf{Z}^{n-1}\setminus\{0\}} \sigma_{\mathbf{m}}(\mathbf{s})(ie_n + \frac{\mathbf{s}}{\|\mathbf{s}\|})e^{2\pi i\langle \mathbf{s}, \mathbf{X}\rangle} e^{-2\pi \|\mathbf{s}\| X_n}. \tag{26}$$

Here, A_n stands for the surface of the unit ball in $\mathbb{R}^n$. Furthermore, $\zeta_M^{\Omega_{n-1}}$ stands for the generalized Riemann zeta function introduced in [27] *which equals in the case $|\mathbf{m}| \equiv 1(2)$*

$$\zeta_M^{\Omega_{n-1}}(\mathbf{m}) = \frac{1}{2} \sum_{\omega \in \Omega_{n-1}\setminus\{0\}} q_{\mathbf{m}}(\omega) \tag{27}$$

and

$$\sigma_{\mathbf{m}}(\mathbf{s}) = \sum_{\mathbf{r}|\mathbf{s}} \mathbf{r}^{\mathbf{m}} \tag{28}$$

where $\mathbf{r}|\mathbf{s}$ means that there is an $\alpha \in \mathbb{N}$ such that $\alpha\mathbf{r} = \mathbf{s}$.

One way to obtain this result is first to expand the subseries $\epsilon_{\mathbf{m}}^{(n-1)}(X)$ ($|\mathbf{m}| \geq 2$) on $H^+(\mathbb{R}^n)$ into a Fourier series of the form

$$\sum_{\mathbf{r} \in L_{n-1}} \alpha_f(\mathbf{r}, x_n) e^{2\pi i \langle \mathbf{r}, \mathbf{X}\rangle}.$$

A direct computation yields

$$\alpha_f(\mathbf{r}, X_n) = \int_{[0,1]^{n-1}} \Big(\sum_{\mathbf{m} \in \mathbf{Z}^{n-1}} q_{\mathbf{m}}(X + \mathbf{m})\Big)\, dX_1 \ldots dX_{n-1} = 0.$$

For $\mathbf{r} \neq \mathbf{0}$ one applies the partial integration method successively (integrating $q_{\mathbf{m}}$ until getting $q_{\mathbf{0}}$ and differentiating the exponential terms) to obtain

$$\alpha_f(\mathbf{r}, X_n) = (2\pi i)^{|\mathbf{m}|}\mathbf{r}^{\mathbf{m}} \int_{\mathbf{R}^{n-1}} q_{\mathbf{0}}(X) e^{-2\pi i \langle \mathbf{r}, \mathbf{X}\rangle} dX_1 \ldots dX_{n-1}.$$

The value of the remaining integral is well-known. See for example [57, 42, 6]. It can be evaluated by applying the residue theorem,

$$\int_{\mathbf{R}^{n-1}} q_0(X)e^{-2\pi i\langle \mathbf{r},\mathbf{X}\rangle}dX_1\ldots dX_{n-1} = \frac{A_n}{2}\Big(ie_n + \frac{\mathbf{r}}{\|\mathbf{r}\|}\Big)e^{-2\pi\|\mathbf{r}\|X_n}$$

so that we finally obtain for $X_n > 0$ that

$$\epsilon_{\mathbf{m}}^{(n-1)}(X) = \frac{A_n}{2}\sum_{\mathbf{r}\in\mathbf{Z}^{n-1}\setminus\{0\}}(2\pi i)^{|\mathbf{m}|}\mathbf{r}^{\mathbf{m}}\Big(ie_n + \frac{\mathbf{r}}{\|\mathbf{r}\|}\Big)e^{-2\pi\|\mathbf{r}\|X_n}e^{2\pi i\langle\mathbf{r},\mathbf{X}\rangle}. \tag{29}$$

By rearrangement arguments we can now rewrite $G_{\mathbf{m}}(X)$ in the form

$$G_{\mathbf{m}}(X) = 2\zeta^{\Omega_{n-1}}(\mathbf{m}) + 2\sum_{\alpha=1}^{\infty}\sum_{\mathbf{m}\in\mathbf{Z}^{n-1}}\epsilon_{\mathbf{m}}(\alpha X) \tag{30}$$

where we used the notations introduced above. Applying (29) to (30) leads then to

$$G_{\mathbf{m}}(X)$$
$$= 2\zeta^{\Omega_{n-1}}(\mathbf{m}) + 2(2\pi i)^{|\mathbf{m}|}\Big(\frac{A_n}{2}\Big)\sum_{\alpha=1}^{\infty}\sum_{\mathbf{r}\in\mathbf{Z}^{n-1}}\mathbf{r}^{\mathbf{m}}\Big(ie_n + \frac{\alpha\mathbf{r}}{\|\alpha\mathbf{r}\|}\Big)e^{2\pi i\langle\alpha\mathbf{r},\mathbf{X}\rangle}e^{-2\pi\|\alpha\mathbf{r}\|X_n}$$

which in turn can be further rearranged into the form (26).

Here we have a nice similarity between the form of the Fourier expansion of the classical Eisenstein series (23) and the structure of the Fourier expansion of the higher dimensional variant defined in (24). The ordinary Riemann zeta function is replaced in the higher dimensional monogenic setting by the expression (27). This expression in turn coincides with the ordinary zeta function, when rewriting it in the special case $n = 2$ in the paravector formalism.

Furthermore, as illustrated in [27, 30], it is closely related to the Epstein zeta function. In a recent paper [6] we illustrated that each vector component of (27) can be written in terms of a finite sum of scalar-valued Dirichlet series of the form

$$\delta(P(\cdot), s) = \sum_{\mathbf{g}\in\mathbf{Z}^{n-1}\setminus\{0\}} P(\mathbf{g})(g_1^2 + \ldots + g_{n-1}^2)^{-s} \tag{31}$$

where P is a real-valued polynomial in $g_1, \ldots, g_{n-1}$ and where s is a complex number satisfying $Re(s) - deg(P) > (n-1)/2$.
The Fourier coefficients $\alpha_f(r)$ for $r \neq 0$ of the complex Eisenstein series

$$\sigma_m(s) = \sum_{r|s} r^{m-1} \tag{32}$$

where $r|s$ means that there is an $\alpha \in \mathbb{N}$ such that $\alpha r = l$ are thus generalized by the expression (28) which in turn can be expressed in terms of (32):

$$\sigma_{\mathbf{m}}(\mathbf{s}) = \mathbf{s}^{\mathbf{m}}\sigma_{-|\mathbf{m}|}(gcd(s_1, \ldots, s_{n-1})). \tag{33}$$

As a consequence of the monogenicity, the monogenic plane wave function appears as a natural generalization of the classical exponential function in the sense of Cauchy-Kowalewski extension. Notice further that infinitely many Fourier coefficients do not vanish, so that the series $G_{\mathbf{m}}(X)$ are indeed non-trivial functions whenever $|\mathbf{m}|$ is odd. In Section 4 of this chapter we will see furthermore, that they provide elementary building blocks for the generation of families of modular forms to larger discrete groups, in particular also to the full modular group which, in turn provides us with an analogy to the classical complex case.

However, the series $G_{\mathbf{m}}(X)$ themselves are not yet modular forms to the full modular group. Only their set of singularities $\mathbb{Q}e_1 + \ldots + \mathbb{Q}e_{n-1}$ is totally invariant under the full hypercomplex modular group Γ_{n-1}. This in fact provides a difference to the complex case.

3.3. Applications to function spaces, pde's and order theory of monogenic functions on some conformal manifolds

3.3.1. Explicit formulas for Bergman and Szegö kernels of strip and rectangular domains. The results outlined in this subsection have been obtained in collaboration with Denis Constales (Ghent University). For details, see our joint papers [8, 7].

Let $\mathcal{D} \subset \mathbb{R}^n$ be a strip or rectangular domain in $\mathbb{R}^n$, and assume without loss of generality that its sides in the first k_1 dimensions $(1 \leq k_1 \leq n)$ $X_1, \ldots, X_{k_1}$ are of positive finite lengths $d_1, \ldots, d_{k_1}$, that its sides in the next $k_2 - k_1$ dimensions $(k_1 + 1 \leq k_2 \leq n)$ are semi-infinite and that its remaining $(n - k_2)$ dimensions extend infinitely in both directions. Without loss of generality we may assume that $\mathcal{D}$ has the form

$$\begin{array}{rl} 0 < X_j < d_j & j = 1, \ldots, k_1 \\ 0 < X_j < \infty & j = k_1 + 1, \ldots, k_2 \\ -\infty < X_j < \infty & j = k_2 + 1, \ldots, n \end{array}$$

Write further $K := \{1, \ldots, k_2\}$ and denote by $A \subset K$ an arbitrary subset.

For any $w \in \mathbb{R}^n$ put $(w^A)_j := (-1)^{j \in A} w_j$ where

$$(-1)^{j \in A} := \begin{cases} 1 & j \notin A \\ -1 & j \in A. \end{cases}$$

The translative Eisenstein series allow us a closed and explicit representation of the Bergman kernel of these types of domains. In terms of the notation introduced above one obtains (cf. [8]):

Theorem 9. *The reproducing Bergman kernel of $\mathcal{D}$ has the following explicit representation*

$$B_{\mathcal{D}}(X, W) = \frac{1}{A_n} \sum_{A \in K, A \neq \emptyset} \sum_{j=1}^{n} (-1)^{|A|+1} (-1)^{j \in A} \epsilon_{\tau(j)}^{(k_1)} (w^A - X) e_j.$$

In the case dealing with a strip domain that is unbounded in at least one dimension, the Bergman kernel is thus given in terms of the first partial derivatives of the monogenic generalized cotangent type functions. In the case of a rectangular domain, bounded in all directions, the Bergman kernel is given in terms of generalizations of the Weierstraß $\wp$-function. This provides a nice analogy to the situation in the classical complex case. We can say more. Remember that the Szegö kernel of a strip domain in the complex plane is given in terms of the complex-analytic cosecant function. In [7] we showed that a similar result can be obtained in the higher dimensional case in terms of the one-fold periodic monogenic generalizations of the cosecant functions $\csc^{(1)}$ introduced in Section 3.1. Precisely speaking, the Szegö kernel of a strip domain bounded in one direction, being without loss of generality of the form

$$D_1 := \{X \in \mathbb{R}^n | 0 < X_1 < d\},$$

is given by

$$S_{D_1}(X, W) = \frac{1}{A_n} \csc^{(1)}(X - W^{\{n\}}, 2d\mathbf{Z}e_1)e_1.$$

The functional equation of the series $\epsilon_{\mathbf{0}}^{(1)}$ (Theorem 5) provides us furthermore with an explicit relationship between the Bergman kernel B_{D_1} and the Szegö kernel S_{D_1}(!) of this domain:

$$B_{D_1}(\frac{X}{2}, \frac{W}{2}) - B_{D_1}(\frac{X+de_1}{2}, \frac{W+de_1}{2}) = -2^{n+1}\frac{\partial}{\partial W_1} S_{D_1}(X, W)$$

The detailed proof can be found in our paper [7]. Extensions to the more general framework working in $\mathbb{K}^{p,q}$ shall be discussed elsewhere.

3.3.2. Boundary value problems on conformally flat cylinders and tori. The results outlined in this subsection have been obtained in collaboration with John Ryan (University of Arkansas). For details, see our joint paper [35].

The main idea is that conformally flat cylinders of infinite extension as well as conformally flat tori can be constructed by the factorization $\mathbb{R}^n/\mathcal{T}_k$, as suggested in a more general but different framework in [55]. In the cases where $k < n$ we are dealing with flat cylinders which shall be denoted by C_k. In the remaining case $k = n$, we get a conformally flat n-torus T_n. Since the universal covering of these manifolds is the Euclidean space $\mathbb{R}^n$, there is a well-defined projection map $p_k : \mathbb{R}^n \to C_k$ $(k < n)$ and $p_n : \mathbb{R}^n \to T_n$. If $U \subset \mathbb{R}^n$ is k-fold periodic with respect to a lattice $\Omega_k \subset \mathbb{R}^n$, then $U' = p_k(U)$ is a subset of C_k, so that each $X \in U$ induces an element $X' = p_k(X) \in C_k$. This map induces further a Dirac operator on C_k, simply viz $D' = p_k(D)$. Functions in the kernel of D'^h are simply called h-cylindrical (toroidal) monogenic. Provided $f : U \to Cl_{0n}(\mathbb{R})$ is k-fold periodic with respect to Ω_k, then p_k induces a well-defined function $f' : U' \to Cl_{0n}(\mathbb{R})$ defined by $f(p_k^{-1}(X'))$ for each $X' \in U'$.

In the cases $k \leq n - h$ the translation invariant Eisenstein series $\epsilon_{\mathbf{0},n}^{(k)}(X - Y)$ induce global Cauchy (Green) kernel functions viz $\epsilon'^{(k)}_{\mathbf{0},h}(X', Y')$ on the associated flat cylinders $C_k = \mathbb{R}^{0,n}/\mathcal{T}_k$ since they are the k-fold periodic periodizations of the Euclidean Cauchy (Green) kernel functions.

Theorem 10 (cf. [35]). *Let $h \leq n-1, k \leq n-h$. Let $G' \subset C_k$ be a domain and $U' \subset G'$ be a compact orientable n-dimensional differentiable manifold whose boundary $\partial U'$ is an n-chain. Let $g' : G' \to Cl_{0,n}(\mathbb{R})$ be cylindrical h-monogenic, that means that ${D'}^h g'(X') = 0$ for all $x' \in G'$. Then for each $Y' \in \overset{\circ}{U'}$ one has*

$$w_{\partial U'}(Y')g'(Y') = \frac{1}{A_n} \int_{\partial U'} \sum_{j=1}^{h} {\epsilon'}^{(k)}_{\mathbf{0},h}(X',Y')d\sigma(X)'D'^{j-1}g'(X') \tag{34}$$

where $w_{\partial U'}(Y')$ denotes the wrapping number of $\partial U'$ at Y' and $d\sigma'(X')$ the oriented surface measure on $\partial U'$.

Also for the case $k = n - h + 1$, in particular for 1-monogenic functions on a torus, a local Cauchy (Green) integral formula is provided by means of the 1-monogenic translative Eisenstein series $\epsilon^{(n)}_{\mathbf{0};a,b}$. More precisely, we obtained:

Theorem 11 (cf. [35]). *Let $G' \subset T_n$ be a domain and $U' \subset G'$ be a compact orientable n-dimensional differentiable manifold whose boundary $\partial U'$ is an n-chain. Let $g' : G' \to Cl_{0,n}(\mathbb{R})$ be a toroidal left monogenic function. Furthermore suppose that $b' = p_k(b)$ does not belong to the closure of $U'_a := p_n(U-a)$. Then for each $y' \in V'$*

$$w_{\partial U'_a}({Y_a}')g'(Y') = \frac{1}{A_n} \int_{\partial U'_a} {\epsilon'}^{(n)}_{\mathbf{0};a,b}(X',Y'_a)d\sigma'(X')G'(X') \tag{35}$$

where $Y'_a = p_n(Y-a)$ and $G(X) = g(X+a)$.

These formulas enable us to study Hardy spaces in this context of flat cylinders and tori, as for example Plemelj projection formulas and Kerzman-Stein formulas. For more details, see [35].

One can prove similar results within the more general framework of working in $\mathbb{K}^{p,q}$, when putting additional restrictions on the extension and the location of the period lattice, so that the associated Eisenstein series converge. A detailed discussion shall be provided elsewhere.

3.3.3. Order theory and an argument principle on conformally flat cylinders and tori. By means of the monogenic Eisenstein series $\epsilon^{(k)}_{\mathbf{0},1}$ one obtained in the particular case $h = 1$ Cauchy integral formulas on C_k for all $1 \leq k \leq n$, where $C_n := T_n$. They in turn can be used further to deduce an argument principle and a generalized version of Rouché's theorem for monogenic functions on conformally flat cylinders and tori. To proceed in this direction we first introduce

Definition 4. Let $k \in \{1,\ldots,n\}$. Suppose that $G' \subseteq C_k$ is an open set and that $f' : G' \to C_k$ is cylindrical (toroidal) left monogenic and that $c' \in G'$ is an isolated zero point on C_k so that there is an $\varepsilon > 0$ such that $B'(c',\varepsilon) \subset C_k$ and $f'|_{B'(c',\varepsilon)\setminus\{c'\}} \neq 0$. Then the integer

$$\operatorname{ord}(f';c') := \frac{1}{A_n} \int\limits_{f'(\partial B'(c',\varepsilon))} \epsilon'^{(k)}_{\mathbf{0}}(X')d\sigma'(X') \tag{36}$$

is called the *order* of f in c. In the case $k = n$ take the function $\epsilon^{(n)}_{\mathbf{0};c,b}$ and choose ε so small that $b' \notin \overline{B'(c',\varepsilon)}$.

In order to show that ord $(f';c')$ is an integer, we use the Cauchy integral formulas on cylinders and tori. In the cases where $k < n$, take $g' \equiv 1$ and replace Y' by $f'(Y')$ and further $\partial U'$ by $\partial f'(U')$. This then leads to

$$\frac{1}{A_n} \int\limits_{\partial f'(U')} \epsilon'^{(k)}_{\mathbf{0}}(X', f'(Y'))d\sigma'(X') = w_{\partial f'(U')}(f'(Y')).$$

Taking now $U' = B'(c',\varepsilon)$ and $Y' = c'$, then $\operatorname{ord}(f',c') = w_{\partial f'(B'(c',\varepsilon))}(0)$ is an integer counting how often the image of the ball around the isolated zero wraps around the image of 0.
For the case $k = n$, put also $G' \equiv 1$, replace Y'_c by $f'(Y'_c)$ and $\partial U'_c$ by $\partial f'(U'_c)$. Afterwards apply the Cauchy integral formula for the torus. Putting furthermore $U'_c = B'(c',\varepsilon)$ and $Y'_c = c'$ leads to the analogous result for the toroidal case.

More generally, in Definition 4 the ball could be replaced by a nullhomologeous $(n-1)$-dimensional cycle parametrizing an $(n-1)$-dimensional surface of an n-dimensional simply connected domain inside G like in the simpler case in Euclidean space (cf. [25]).
Since $f' : G' \to C_k$ is continuously differentiable, we can apply the general transformation rule for differential forms (cf. [58])

$$d\sigma'(f'(X')) = (Jf')^*(X')d\sigma'(X'). \tag{37}$$

Here Jf' denotes the Jacobi matrix and $(Jf')^*$ its adjoint. Notice that f' is a local diffeomorphism (near a'), so that (37) may be rewritten in the form

$$d\sigma(f'(X')) = \det Jf'(X')((Jf')^{-1})^{\top}(X')d\sigma'(X'). \tag{38}$$

This allows us to rewrite (36) in the following form, providing us with an argument principle on conformally flat cylinders and tori.

Theorem 12 (Argument principle on Cylinders and Tori). *Let $G' \subseteq C_k$ be a domain, f' $G' \to C_k$ cylindrical (toroidal) monogenic in G' and $c' \in G'$ an isolated point. Then, under the same conditions for $B'(c',\varepsilon) \subset G'$ as in Definition 4,*

$$\operatorname{ord}(f';c') = \frac{1}{A_n} \int_{\partial B'(c',\varepsilon)} \epsilon'^{(k)}_{\mathbf{0}}(f'(X')) \det Jf'(X')((Jf')^{-1})^{\top}(X')d\sigma'(X').$$

For the case $k = n$ take again $\epsilon^{(n)}_{\mathbf{0};c,b}$.

We also obtain a generalization of Rouché's theorem within this context. To proceed in this direction we introduce a norm on C_k by

$$\|X'\|_{C_k} := \min_{\omega \in \Omega_k} \{\|X+\omega\|\}, \quad X' \in C_k.$$

Theorem 13 (Generalized Rouché's Theorem for isolated zeroes). *Let $G \subset C_k$ be a simply connected domain and let Γ be a nullhomologous $(n-1)$-dimensional cycle parametrizing an $(n-1)$-dimensional surface of an n-dimensional simply connected domain $D \subset G$. Suppose $f, g : G \to C_k$ are cylindrical (toroidal) monogenic functions having only a finite number of zeroes in $\overset{\circ}{D}$ and no zeroes on the boundary ∂D. If*

$$\|f(X) - g(X)\| < \|f(X)\| \quad \forall X \in \partial D, \tag{39}$$

then

$$\sum_{a \in D} \operatorname{ord}(f; a) = \sum_{a \in D} \operatorname{ord}(g; a). \tag{40}$$

For the proof we mention that this relation is a consequence of the fact that the definition of the order which is a topological mapping degree, is invariant under homotopy. To be more explicit, for a real $\lambda \in [0,1]$ we define the function $h_\lambda := f + \lambda(g - f)$. Then we consider

$$\begin{aligned} N(\lambda) &:= \sum_{a \in D} \operatorname{ord}(h_\lambda; a) \\ &= \frac{1}{\sigma_{n+1}} \int_\Gamma \epsilon'^{(k)}_{\mathbf{0}}(f(x) + \lambda g(x) - \lambda f(x))(J(f + \lambda g - \lambda f))^* d\sigma(x). \end{aligned}$$

Here again, take $\epsilon^{(n)}_{\mathbf{0};c,b}$ for the case $k = n$. The integrand depends continuously on λ. Hence, $N(\lambda)$ is a continuous function. Since $N(\lambda) \in \mathbf{Z}$ we conclude that $N(\lambda)$ does not depend on λ. Therefore, $N(\lambda) \equiv$ *const*, thus $N(1) = N(0)$. Notice that we used (37) instead of (38) to simplify our notation.

4. Automorphic forms related to more general hyperbolic polyhedron groups and perspectives

The function series treated in the previous section are totally invariant under the action of translation groups in $\mathbb{K}^{p,q}$. They provide useful building blocks for the construction of automorphic forms to larger discrete subgroups of Vahlen type groups. In order to construct for example monogenic and polymonogenic functions that show an invariance behavior under the action of a translation group and additionally also under a finite rotation group it issuggestive at the very first to sum the series $\epsilon^{(k)}_{\mathbf{m},h}$ or $G^{(k)}_{\mathbf{m},h}$ over that rotation group. Suppose that $u_1, \ldots, u_t$ are $\mathbb{R}$-linear independent unit vectors in $\mathbb{K}^{p,q}$ satisfying $u_j^N = 1$ for a positive integer

$N \in \mathbb{N}$ $(j = 1, \ldots, t)$. We write

$$\mathcal{R}^t = \left\langle \begin{pmatrix} \overline{u_1} & 0 \\ 0 & u_1^{-1} \end{pmatrix}, \ldots, \begin{pmatrix} \overline{u_t} & 0 \\ 0 & u_t^{-1} \end{pmatrix} \right\rangle.$$

Suppose that the parameters k, m, h are chosen so that the series $\epsilon^{(k)}_{\mathbf{m},h}$ or $G^{(k)}_{\mathbf{m},h}$ are normally convergent, as described in Section 3. Under the same conditions, the series

$$R^{(p)}_{\mathbf{m},h}(z) := \sum_{M \in \mathcal{R}^t} \epsilon^{(k)}_{\mathbf{m},h} u(\overline{u}zu)\overline{u}$$

and

$$Q^{(p)}_{\mathbf{m},h}(z) := \sum_{M \in \mathcal{R}^t} G^{(k)}_{\mathbf{m},h} u(\overline{u}zu)\overline{u}$$

are well-defined h-monogenic functions, too. Notice that the centers of the singularity cones of the first series are isolated points in $S(\mathbb{K}^{p,q})$. However, these series do not turn out to be translation invariant for any arbitrary lattice $\Omega \in \mathbb{R}^n$. To have the translation invariance, the primitive periods ω_j and the vectors u_i have to satisfy $u_i \omega_j u_i \in \Omega_p$ $\forall i, j$. Ω_p needs to be a lattice with vector multiplication. The functions $R^{(p)}_{\mathbf{m},h}(z)$ give then examples of h-monogenic modular forms transforming in the way

$$R^{(p)}_{\mathbf{m},h}(z) = uR^{(p)}_{\mathbf{m},h}(\overline{u}zu)\overline{u} \qquad \text{for all} \quad M = \begin{pmatrix} \overline{u} & 0 \\ 0 & u^{-1} \end{pmatrix} \in \mathcal{R}_p$$

and

$$R^{(p)}_{\mathbf{m},h}(z) = R^{(p)}_{\mathbf{m},h}(z + \omega) \qquad \text{for all} \quad \omega \in \Omega_k,$$

as a direct rearrangement argument shows.

Similarly, the series of the form $Q^{(p)}_{\mathbf{m},h}(z)$ transform under translations and rotations. It is more difficult to include also an invariance behavior under inversions. In [31, 32, 33] we proved two construction theorems that meet this end, in which the multiperiodic functions from the previous section can again be used as generating building blocks.

To proceed in this direction let us recall that from the conformal invariance formula for the iterated Dirac operators on Minkowski type spaces (see e.g. [53]) it follows that if $\tilde{f} : H^+(\mathbb{K}^{p,q}) \to \mathbf{Cl}_{pq}(\mathbb{K})$ is a left d-monogenic function $(1 \leq d \leq h)$, then $D_z{}^l{}_{\mathbf{K}^{p,q}}[\overline{q_{\mathbf{0}}^{(h)}(cz+d)^* f(M\langle z \rangle)}] = 0$ for every positive integer l with $l \geq h$ and for all $M \in \Gamma_k$. Here, and in all that follows, Γ_k shall always stand for $\Gamma^{0,k}_{\mathbf{R}}$ in the case $\mathbb{K} = \mathbb{R}$ (where we always assume that additionally $k \leq q-1$) and for $\Gamma^{p,q}_{\mathbb{C}}$ in the complexified case, where $1 \leq k \leq n$. For simplicity we use the abbreviation D^l_z for $D_z{}^l{}_{\mathbf{K}^{p,q}}$, etc..

Starting now with an appropriate monogenic generating function $\tilde{f}$ from Clifford analysis that is totally invariant under the action of the maximal translation group $\mathcal{T}_k$ that is contained in Γ_k, and summing the expressions of the form

$$(\tilde{f}|_h M)(z) := \overline{q_{\mathbf{0}}^{(h)}(cz+d)^*} \tilde{f}(M\langle z \rangle)$$

over a complete set of representatives of right cosets in Γ_k modulo $\mathcal{T}_k$, will lead to several classes of non-trivial Clifford-valued modular forms with respect to Γ_k within the function classes Ker D^l for all positive integers l satisfying $l \geq h$ whenever h is an even positive integer, as we will explain now. In the sequel the notation $M : \mathcal{T}_k \backslash \Gamma_k$ means that M runs through a system of representatives $\mathcal{B}_k$ of the right cosets of Γ_k with respect to $\mathcal{T}_k$, i.e. $\bigcup_{M \in \mathcal{B}_k} \mathcal{T}_k M = \Gamma_k$ and $\mathcal{T}_k M \neq \mathcal{T}_k N$ for $M, N \in \mathcal{B}_k$ with $M \neq N$. More generally as in [31] we may thus formulate:

Theorem 14 (1. Construction theorem). *Let $n \in \mathbb{N}, h \in 2\,\mathbb{N}, h < n$, $k < n-h-1$, and in the particular case $\mathbb{K} = \mathbb{R}$ assume additionally that $k \leq q-1$. The groups Γ_k and $\mathcal{T}_k$ are given as previously mentioned above. Let $\tilde{f} : H^+(\mathbb{K}^{p,q}) \to Cl_{p,q}(\mathbb{K})$ be a bounded and left monogenic function on $H^+(\mathbb{K}^{p,q})$ that is totally invariant under the translation group $\mathcal{T}_k$. Then*

$$f(z) = \sum_{M:\mathcal{T}_k \backslash \Gamma_k} (\tilde{f}|_h M)(z) \qquad z \in H^+(\mathbb{K}^{p,q}) \tag{41}$$

is a Clifford-valued function which is bounded in any compact subset of $H^+(\mathbb{K}^{p,q})$. Moreover, f satisfies $D_z^l f(z) = 0$ for all $l \geq h$ and $f(z) = (f|_h M)(z)$ for all $M \in \Gamma_k$ on the whole half-space.

For the proof in the arbitrary Minkowski type space $\mathbb{K}^{p,q}$ one can proceed in a similar way as for the case in $\mathbb{K}^{0,n}$ which we presented in a detailed form for $\mathbb{K} = \mathbb{R}$ in [31] and for $\mathbb{K} = \mathbb{C}$ in [32].

This construction theorem provides us with non-vanishing hypercomplex modular forms for the whole modular group Γ_p in the classes Ker D^l with $l \geq h$ where h is even, as for example in the class of the Fueter-Sce solutions. To construct concrete examples of vector-valued non-cusp forms within this setting, the series $G_{\mathbf{m},h}^{(k)}$ can be used, as shown in the above cited works for the special case $(p,q) = (0,n)$.

Notice that whenever h is an odd integer, then the construction (41) generates only the zero function. See [31] for details. More generally, whenever we have a group that includes the negative identity matrix, then all functions satisfying $f(z) = (f|_h M)(z)$ for all M from that group vanish identically whenever h is odd.

In order to construct non-trivial h-monogenic examples satisfying $f(z) = (f|_h M)(z)$ for h odd we thus need to restrict ourselves to smaller groups, as for example congruence groups. In the space $\mathbb{K}^{0,n}$ let us for instance consider congruence groups of the form

$$\Gamma_k[s] := \{M \in \Gamma_k |\; M \equiv I \; mod \; s\mathcal{O}\} \tag{42}$$

where s is a positive integer with $s > 2$.
Here I stands for the identity matrix and $\mathcal{O} := \sum_A \mathbf{Z} e_A$. Consider for instance $s = 3$. Taking for example $\tilde{f} \equiv 1$ or $\tilde{f}(z) = G_{\mathbf{m}}(z + e_n)$, then

$$f(z) := \sum_{M:\mathcal{T}_k[3] \backslash \Gamma_k[3]} \overline{q_{\mathbf{0}}^{(h)}(cz+d)^*} \tilde{f}(M < z >) \quad k < n-h-1$$

provides us with non-trivial examples of Clifford-valued modular forms (also for h odd) satisfying

$$f(z) = \overline{q_{\mathbf{0}}^{(h)}(cz+d)^*}\tilde{f}(M<z>) \qquad \forall M \in \Gamma_k[3],$$

as one can show by applying limit arguments. To be precise, in the case where $\tilde{f} \equiv 1$ we obtain

$$\begin{aligned}
\lim_{X_n \to \infty} f(X_n e_n) &= \sum_{M:\mathcal{T}_k[3]\backslash\Gamma_k[3]} \overline{q_{\mathbf{0}}^{(h)}(cX_n e_n + d)^*} \\
&= \sum_{M:\mathcal{T}_k[3]\backslash\Gamma_k[3], c_M \neq 0} \underbrace{\lim_{X_n \to \infty} \overline{q_{\mathbf{0}}^{(h)}(ce_n X_n + d)^*}}_{=0} \\
&+ \sum_{M:\mathcal{T}_k[3]\backslash\Gamma_k[3], c_M = 0} d^*\overline{d^*} \\
&= 1.
\end{aligned}$$

In the case $\tilde{f}(z) = G_{\mathbf{m}}(z+e_n)$, a similar calculation leads to $\lim_{X_n\to\infty} f(X_n e_n) = \lim_{X_n\to\infty} G_{\mathbf{m}}(X_n e_n) = \zeta^{\Omega_{n-1}}(\mathbf{m})$ where $\zeta^{\Omega_{n-1}}(\mathbf{m})$ is the vector-valued generalized Riemann zeta function from Section 3.2.2. Since the expressions $\zeta^{\Omega_{n-1}}(\mathbf{m})$ in turn are the Laurent coefficients of the series $\epsilon_{\mathbf{r},1}^{(n-1)}(z)$ there must be indices $\mathbf{m}$ for which $\zeta^{\Omega_{n-1}}(\mathbf{m}) \neq 0$. Otherwise we would obtain $\epsilon_{\mathbf{r},1}^{(n-1)}(z) \equiv q_{\mathbf{r}}(z)$ within a whole ball, which would be a contradiction.

One possibility to construct also non-trivial modular forms in Ker $D^l{}_{\mathbb{K}^{p,q}}$ ($l \geq h$) where h is odd (in particular for $h = 1$), with respect to the whole modular group Γ_p and even with respect to the *full* modular group Γ_{n-1}, is to consider *two* weight factors (one from the left, another from the right) and to introduce a second auxiliary variable. We exploited this idea recently in [33] in order to construct monogenic and polymonogenic Hilbert modular forms on Cartesian products of the upper half-space.

In what follows we focus on the construction of non-trivial monogenic and polymonogenic modular forms on $H_2^+(\mathbb{K}^{p,q}) = H^+(\mathbb{K}^{p,q}) \oplus H^+(\mathbb{K}^{p,q})$ that transform in the way

$$f(z,w) = (f||_{h,l}M)(z,w) \quad \text{for all } M \in \Gamma_k \tag{43}$$

and whose restriction to the diagonal $z = w$ is at least a non-constant C^∞-function satisfying $f(z,z) = (f||_{h,l}M)(z,z)$ for all $M \in \Gamma_k$.

Here

$$(f||_{h,l}M)(z,w) = \overline{q_{\mathbf{0}}^{(h)}(cz+d)^*} f(M<z>, M<w>) q_{\mathbf{0}}^{(l)}(wc^*+d^*).$$

The following construction theorem provides many non-trivial examples.

Theorem 15 (2. Construction theorem). *Let $k < 2n - (h+l) - 1$. Additionally, assume that $k \leq q-1$ in the case $\mathbb{K} = \mathbb{R}$, or $k \leq n$ in the complexified case. Let $h + l \equiv 0(2)$. Suppose $\tilde{f} : H_2^+(\mathbb{K}^{p,q}) \to Cl_{p,q}(\mathbb{K})$ is a bounded function that satisfies the relation $D_z^h[\tilde{f}(z,w)] = [\tilde{f}(z,w)]D_w^l = 0$ and $\tilde{f}(T\langle z\rangle, T\langle w\rangle) = \tilde{f}(z,w)$ for all $T \in \mathcal{T}_k$ for all $(z,w) \in H_2^+(\mathbb{K}^{p,q})$. Then*

$$f(z,w) := \sum_{M:\mathcal{T}_k\backslash\Gamma_k} (\tilde{f}||_{h,l}M)(z,w) \tag{44}$$

is left h-monogenic in the variable z and right l-monogenic in w and satisfies

$$f(z,w) = (f||_{h,l}M)(z,w) \quad \text{for all } M \in \Gamma_k. \tag{45}$$

for all $(z,w) \in H_2^+(\mathbb{K}^{p,q})$.

The (h,l) biregularity is a consequence of the conformal invariance formulas for Dirac operators which were proved in [53] within the framework of working in general Minkowski-type spaces. In the case $\mathbb{K} = \mathbb{R}$ we get even convergence for $k = q - 1$ whenever $h + l \leq q - 1$. Even for the monogenic case, i.e. $h = l = 1$, we get non-trivial Clifford-valued modular forms for the full modular group provided we are in the half space of the Euclidean space $\mathbb{R}^{0,n}$ of dimension $n \geq 3$. The simplest non-trivial examples for monogenic modular forms for the whole modular group are the biregular Eisenstein series

$$\begin{aligned} \mathcal{E}(z,w) &= \sum_{M:\mathcal{T}_k\backslash\Gamma_k} (1||_{1,1}M)(z,w) \\ &= \sum_{M:\mathcal{T}_k\backslash\Gamma_k} \overline{q_{\mathbf{0}}^{(h)}(cz+d)^* q_{\mathbf{0}}^{(l)}(wc^*+d^*)} \quad (z,w) \in H_2^+(\mathbb{K}^{p,q}). \end{aligned} \tag{46}$$

To verify that these series are non-trivial transcendental functions, consider for instance in the case $\mathbb{K} = \mathbb{R}$ the expression,

$$\begin{aligned} &\lim_{X_n,Y_n\to\infty} \mathcal{E}(e_nX_n, e_nY_n) \\ &= \sum_{M:\mathcal{T}_k\backslash\Gamma_k, c_M\neq 0} \underbrace{\lim_{X_n,Y_n\to\infty} \overline{q_{\mathbf{0}}^{(h)}(ce_nX_n+d)^* q_{\mathbf{0}}^{(l)}(e_nY_nc^*+d^*)}}_{=0} + \sum_{M:\mathcal{T}_p\backslash\Gamma_p, c_M=0} d^*\overline{d^*} \\ &= 2\sum_A e_A\,\overline{e_A} = 2^{k+1} \end{aligned}$$

where we put $z = \sum_{j=1}^n X_je_j$ and $w = \sum_{j=1}^n Y_je_j$. In particular $\mathcal{E}(z,z) \not\equiv 0, \infty$, so that we do indeed have non-trivial modular forms. In the complexified cases a limit argument can be applied, simply choosing the adequate coordinate direction for which we have the positivity condition.

Once again, the series $G_{\mathbf{m}}$ can be used to generate classes of non-trivial examples for hypercomplex modular forms transforming in the way (43) — in this setting, even for obtaining families of non-trivial monogenic modular forms for the full modular group Γ_{n-1} in Euclidean space $\mathbb{R}^{0,n}$ — namely by applying (44) to

$$\tilde{f}(z,w) = G_{\mathbf{m}}(z+e_n)G_{\mathbf{m}}(w+e_n) \tag{47}$$

which is bounded on $H_2^+(\mathbb{K}^{p,q})$. To show the non-triviality we consider again

$$\begin{aligned}\lim_{X_n,Y_n\to\infty} \sum_{M:T_k\backslash\Gamma_k} & \overline{q_{\mathbf{0}}^{(h)}(cX_ne_n+d)^* G_{\mathbf{m}}(M<e_nX_n>+e_n)} \\ & \times\ G_{\mathbf{m}}(M<e_nY_n>+e_n)q_{\mathbf{0}}^{(l)}(Y_ne_nc^*+d^*) \\ = \quad & (-8)\cdot\sum_A e_A^*\|\zeta^{\Omega_{n-1}}(\mathbf{m})\|^2\overline{e_A}^*.\end{aligned}$$

As mentioned above, there must be at least some indices $\mathbf{m}\in\mathbb{N}_0$ of odd length, for which $\zeta(\mathbf{m})$ and hence $\|\zeta(\mathbf{m})\|^2$ is different from zero. For these indices we have then that the limit considered is different from zero proving the non-triviality of the associated series $\sum_{M:T_k\backslash\Gamma_k}(\tilde{f}||_{h,l}M)(z,w)$ and in particular its restriction to the diagonal (z,z) is a non-constant C^∞ automorphic form.

Final remarks

Special variants of the monogenic and polymonogenic automorphic forms discussed in this section admit the treatment of similar problems as outlined in Section 3.3 but within a much more general geometrical context.

Partial results in these directions have been developed in [10] where closed and explicit representation formulas for the Bergman kernel for wedge shaped domains are obtained in terms of more general Eisenstein series, and in [36] where boundary value problems of partial differential equations on more general conformally flat spin manifolds are treated.

Further to this, these families of automorphic forms open the door to deriving explicit argument principles for monogenic functions with isolated a-points on more general conformally flat spin manifolds that can be constructed by factoring $\mathbb{R}^n$ by a discrete subgroup of Vahlen's group, extending our discussion from Section 3.3.3 in a natural and systematic way to a much more general geometrical context; *roughly speaking* simply by replacing the translative Eisenstein series of lowest singularity order by those automorphic forms of lowest singularity order that are associated to the group which parametrizes the spin manifold.

The author is very thankful to Prof. Freitag and the members of his research group from Heidelberg University (Germany) for the very fruitful and helpful discussions on this topic.

References

[1] V. Avanssian. Sur les fonctions harmoniques d' ordre quelconque et leur prolongement analytique dans $\mathbb{C}^n$, in *Séminaire Pierre Lelong – Henri Skoda (Analyse) Années 1980/81 et Colloque de Wimereux, Mai 1981* ed. A. Dold and B. Eckmann, Springer Lecture Notes **919**, Springer Berlin-Heidelberg-New York, 1982, pp. 192–281

[2] F. Brackx. *Teorie van de (k)-monogene funkties.* PhD Thesis University of Ghent, (1973–74).

[3] O. Blumenthal. Über Modulfunktionen von mehreren Veränderlichen, *Math. Ann.* **56** (1903), 509–548 and **58** (1904), 497–527.

[4] S. Carlip. Lectures on (2+1)-Dimensional Gravity. Preprint March 1995, arXiv:gr-qc/9503024 v2.

[5] J. Cnops. *Hurwitz pairs and applications of Möbius transformations*, Habilitation Thesis, Rijksuniversiteit Gent, 1993–1994.

[6] D. Constales and R.S. Kraußhar. Representation formulas for the general derivatives of the fundamental solution to the Cauchy-Riemann operator in Clifford Analysis and Applications, *Z. Anal. Anw.* **21** No.3 (2002), 579–597.

[7] D. Constales and R.S. Kraußhar. Szegö and Polymonogenic Bergman Kernels for half-space and strip domains, and Single-periodic Functions in Clifford Analysis, *Complex Variables* **47** No. 4 (2002), 349–360.

[8] D. Constales and R.S. Kraußhar. Bergman Kernels for rectangular domains and Multiperiodic Functions in Clifford Analysis, *Math. Meth. Appl. Sci.* **25** No. 16–18 (2002), 1509–1526.

[9] D. Constales and R.S. Kraußhar. Closed formulas for the singly-periodic monogenic cotangent, cosecant and cosecant-squared functions in Clifford Analysis, *Journal of the London Mathematical Society*, **67** No. 2 (2003), 401–416.

[10] D. Constales and R.S. Kraußhar. The Bergman Kernels for the half-ball and for fractional wedge-shaped domains in Clifford Analysis, submitted for publication.

[11] R. Delanghe, F. Sommen and V. Souček. Residues in Clifford analysis, in: *Begehr H. and Jeffrey A.(eds.), Partial differential equations with complex analysis*, Pitman Res. Notes in Math. Ser. 262 (1992), 61–92.

[12] R. Delanghe, F. Sommen and V. Souček. *Clifford Algebra and Spinor Valued Functions*, Kluwer, Dordrecht-Boston-London, 1992.

[13] A. C. Dixon. On the Newtonian Potential, *Quarterly Journal of Mathematics* **35** (1904), 283–296.

[14] G. Eisenstein. Genaue Untersuchung der unendlichen Doppelproducte, aus welchen die elliptischen Functionen als Quotienten zusammengesetzt sind, und der mit ihnen zusammenhängenden Doppelreihen (als eine neue Begründung der Theorie der elliptischen Functionen, mit besonderer Berücksichtigung ihrer Analogie zu den Kreisfunctionen), Crelle's Journal, 35, 1847, 153–274.

[15] J. Elstrodt, F. Grunewald and J. Mennicke. Eisenstein Series on three-dimensional hyperbolic space and imaginary quadratic number fields. *J. Reine Angew. Math.* **360** (1985), 160–213.

[16] J. Elstrodt, F. Grunewald and J. Mennicke. Kloosterman sums for Clifford algebras and a lower bound for the positive eigenvalues of the Laplacian for congruence subgroups acting on hyperbolic spaces, *Invent. Math.* **101** No.3 (1990), 641–668.

[17] E. Freitag and C. F. Hermann. Some modular varieties of low dimension, *Adv. Math.* **152** No.2 (2000), 203–287.

[18] R. Fueter. Über automorphe Funktionen in Bezug auf Gruppen, die in der Ebene uneigentlich diskontinuierlich sind, *Crelle's Journal* **137** (1927), 66–77.

[19] R. Fueter. Über vierfachperiodische Funktionen, *Monatshefte Math. Phys.* **48** (1939), 161–169.

[20] R. Fueter. Über die Quaternionenmultiplikation der vierfachperiodischen regulären Funktionen, *Experientia*, **1** (1945), 57.

[21] R. Fueter. *Functions of a Hyper Complex Variable*, Lecture notes written and supplemented by E. Bareiss, Math. Inst. Univ. Zürich, Fall Semester 1948/49.

[22] R. Fueter. Ueber Abelsche Funktionen von zwei Komplexen Variablen. *Ann. Mat. Pura Appl.*, IV Ser. 28 (1949), 211–215.

[23] F. Gürsey and H. Tze. Complex and Quaternionic Analyticity in Chiral and Gauge Theories I, *Annals of Physics* **128** (1980), 29–130.

[24] E. Hecke. Über die Konstruktion relativ Abel'scher Zahlkörper durch Modulfunktionen von zwei Variablen. *Math. Ann.* **74** pp.467.

[25] T. Hempfling. Some remarks on zeroes of monogenic functions, Advances in Appl. Clifford Algebra **11(S2)**, 2001.

[26] T. Hempfling and R.S. Kraußhar. Order Theory for Isolated Points of Monogenic Functions, *Archiv der Mathematik*, **80** (2003), 406–423.

[27] R.S. Kraußhar. *Eisenstein Series in Clifford Analysis*, PhD Thesis RWTH Aachen, Aachener Beiträge zur Mathematik **28**, Wissenschaftsverlag Mainz, Aachen, 2000.

[28] R.S. Kraußhar. Monogenic multiperiodic functions in Clifford analysis, *Complex Variables* **46** No. 4 (2001), 337–368.

[29] R. S. Kraußhar. Generalizations of the complex analytic trigonometric functions to Clifford analysis by Eisenstein series. In: *Functional-Analytic and complex methods, their interactions, and Applications to Partial Differential Equations (Proceedings of the International Graz Workshop, Graz, Austria, 12–16 February 2001)* (eds. H. Florian, N. Ortner, F. Schnitzer, W. Tutschke), World Scientific, River Edge NJ, pp. 438–456.

[30] R.S. Kraußhar. On a new type of Eisenstein series in Clifford analysis, *Z. Anal. Anw.* **20** No. 4 (2001), 1007–1029.

[31] R.S. Kraußhar. Automorphic forms in Clifford Analysis, *Complex Variables* **47** No. 5 (2002), 417–440.

[32] R.S. Kraußhar. Eisenstein Series in Complexified Clifford Analysis, *Computational Methods and Function Theory*, **2** No. 1 (2002), 29–65.

[33] R.S. Kraußhar. Monogenic Modular Forms in Two and Several Real and Complex Vector Variables, to appear in *Computational Methods and Function Theory* **2** No. 2 (2002), 299–318.

[34] R. S. Kraußhar. The Multiplication of the Clifford-analytic Eisenstein Series, *J. Number Theory* **102** (2003), 353–382.

[35] R. S. Kraußhar and J. Ryan. Clifford and Harmonic Analysis on Cylinders and Tori, to appear in *Rev. Mat. Iberoamericana.*

[36] R. S. Kraußhar and J. Ryan. Some Conformally Flat Spin Manifolds, Dirac Operators and Automorphic Forms, submitted for publication. Preprint available on http://arXiv.org/abs/math.AP/0212086

[37] A. Krieg. *Modular Forms on Half-Spaces of Quaternions.* Springer Verlag, Berlin-Heidelberg, 1985.

[38] A. Krieg. Eisenstein series on real, complex and quaternionic half-spaces. *Pac. J. Math.* **133** No.2 (1988), 315–354.

[39] A. Krieg. Eisenstein-Series on the Four-Dimensional Hyperbolic Space, *Journal of Number Theory* **30** (1988), 177–197.

[40] G. Laville and I. Ramadanoff. Holomorphic Cliffordian functions, *Adv. Appl. Clifford Algebr.* **8** No.2 (1998), 323–340.

[41] G. Laville and I. Ramadanoff. Elliptic Cliffordian Functions, *Complex Variables* **45** No.4 (2001), 297–318.

[42] Li, C., McIntosh, A. and T. Qian. Clifford Algebras, Fourier transforms and singular convolution operators on Lipschitz surfaces, *Rev. Mat. Iberoamericana*, **10** (1994), 665–721.

[43] H. Maaß. Zur Theorie der automorphen Funktionen von n Veränderlichen, *Math. Ann.* **117** (1940), 538–578.

[44] H. Maaß. Automorphe Funktionen von mehreren Veränderlichen und Dirichletsche Reihen, *Abh. Math. Sem. Univ. Hamb.* **16** (1949), 53–104.

[45] W. Nef. Über die singulären Gebilde der regulären Funktionen einer Quaternionenvariablen, *Comment. Math. Helv.* **15** (1942–43), 144–174.

[46] W. Nef. Die unwesentlichen Singularitäten der regulären Funktionen einer Quaternionenvariablen, *Comment. Math. Helv.* **16** (1943–44), 284–304.

[47] W. Nef. Funktionentheorie einer Klasse von hyperbolischen und ultrahyperbolischen Differentialgleichungen 2. Ordnung, *Comm. Math. Helv.* **17** (1944), 83–107.

[48] T. Qian. Singular integrals with monogenic kernels on the m-torus and their Lipschitz perturbations. In: *Clifford Algebras in Analysis and Related Topics*, J. Ryan (eds.), CRC Press, Boca Raton, 1996, 157–171.

[49] O. Richter. Theta functions with harmonic coefficients over number fields, *Journal of Number Theory* **95** (2002), 101–121.

[50] O. Richter and H. Skogman. Jacobi theta functions over number fields, preprint, Feb. 2002.

[51] J. Ryan. Clifford Analysis with Generalized Elliptic and Quasi Elliptic Functions, *Appl. Anal.* **13** (1982), 151–171.

[52] J. Ryan. Conformal Clifford manifolds arising in Clifford analysis, *Proc. R. Ir. Acad* **85A** No.1 (1985), 1–23.

[53] J. Ryan. Intertwining operators for iterated Dirac operators over Minkowski-type spaces, *J. Math. Anal. Appl.* **177** No.1, (1993) 1–23.

[54] M. Sce. Osservazione sulle serie di potenzi nei moduli quadratici, *Lincei Rend. Sci. Fis. Mat. et Nat.* **23** (1957), 220–225.

[55] R. Schoen and S.-T. Yau. Conformally flat manifolds, Kleinian groups and scalar curvature, *Inventiones Mathematica* **92** (1988) 47–71.

[56] C.L. Siegel. *Topics in Complex Function Theory Vol. III*, Wiley, New York-Chichester, 1988.

[57] F. Sommen. Some connections between Clifford and Complex Analysis, *Complex Variables* **1** (1982), 97–118.

[58] G. Zöll. Ein Residuenkalkül in der Clifford-Analysis und die Möbiustransformationen für euklidische Räume, PhD Thesis, Lehrstuhl II für Mathematik RWTH Aachen, 1987.

Rolf Sören Kraußhar
Vakgroep Wiskundige Analyse
Universiteit Gent
Galglaan 2
B-9000 Gent
Belgium
e-mail: `krauss@cage.ugent.be`

Trends in Mathematics: *Advances in Analysis and Geometry*, 345–363

Automated Geometric Theorem Proving, Clifford Bracket Algebra and Clifford Expansions

Hongbo Li

Abstract. We report some of our recent progress on using Clifford bracket algebra to simplify automated geometric theorem-proving via the homogeneous model, and on systematic expansions of Clifford expressions with the purpose of finding the shortest expansions. With such simplification and expansion techniques, we are able to finish some pure geometric computation tasks that are too difficult for other algebraic methods to finish even with the aid of a good PC, and we get better results through a readable procedure.

Mathematics Subject Classification (2000). 51K05, 14L24, 15A66, 15A72.

Keywords. Automated theorem-proving, Clifford algebra, bracket algebra, invariant theory, geometric computation.

1. Introduction

Automated theorem proving started in the 1950s as a branch of artificial intelligence (AI). At that time the prevailing approach was database searching and logic programming. In late 1970s, an algebraic approach appeared. It was Wu's method of characteristic sets [18]. The theoretical foundation is no longer logic deduction based on a set of axioms. Geometric theorems are first algebraized, i.e., changed into polynomial equalities and inequalities, then the proving of the conclusion, now algebraized as well, becomes verifying if the zeroes of the conclusion polynomial are included in the zeroes of the hypotheses polynomials. In the 1980s, another algebraic approach – the Gröbner bases method occurred [9]. In the 1990s, several invariant algebraic methods were proposed, including the biquadratic final polynomials [6], Cayley algebra [15], bracket algebra [14], the area method [5], Clifford algebra [10], Grassmann algebra [16], etc.

What is the reason for people to switch from a logic approach to an algebraic one? Because a pure logical approach is very inefficient, even for easy but nontrivial tasks of geometric theorem proving. Why do people continue to develop invariant algebraic methods? In history, invariant algebras were developed after coordinate algebra for the purpose of more advanced and abstract treatment of geometric problems. Since Descartes' analytic geometry, coordinates have been playing a crucial role in the algebraization of geometry. Still, people want more advanced algebraic representations so that the algebraic manipulations become more "geometric", or invariant. Leibniz had dreamed of a geometric calculus dealing directly with geometric objects rather than with number sequences, or coordinates. Gauss realized Leibniz' dream in 2D geometry by his complex numbers method, and Hamilton extended Gauss' 2D numbers to 3D by his quaternions. Then Grassmann, Cayley and Clifford continued the extension to nD numbers by the algebras now bearing their names. In the 19th century, the Cambridge formalist school, led by de Morgan, Peacock, et al., formulated the symbolic algebra used in geometry as follows [19]:

"Symbolic algebra is the science of symbols and their combinations, constructed according to their own rules, which may be applied to arithmetic and other sciences by means of an interpretation."

The algebraic approach to automated geometric theorem proving is an integration of classical symbolic algebra with modern AI techniques, which makes this subject a branch of mathematics. The invariant algebraic approach is a trend for producing short and readable proofs by simplifying algebraic representation and computation. Many difficult geometric computing tasks can only be finished by means of algebraic invariants.

While Cayley algebra and bracket algebra are for invariant computation in projective and affine geometries, Clifford algebra is for invariant computation in metric geometry [8]. Clifford algebra has three major forms: hypercomplex numbers, matrices and abstract symbols. In automated geometric theorem proving, it is the symbolic form that is used.

Invariant algebras can provide succinct representations for geometric entities and relations. Such representations are potentially very helpful in simplifying geometric computation. In applications, however, some difficulties will inevitably occur. They further stimulate the development of invariant algebras.

Difficulty one: Multiple algebraic representations for the same geometric entity or relation.

For example, in the homogeneous model of Euclidean plane [11], let there be two circles or lines passing through points $\mathbf{0},\mathbf{1},\mathbf{2}$ and $\mathbf{0},\mathbf{1}',\mathbf{2}'$ respectively, then their second point of intersection (the first is point $\mathbf{0}$), can be represented uniquely up to scale by

$$\begin{aligned}&\mathbf{012}\cap\mathbf{01'2'}\\ &=\mathbf{1}\cdot\mathbf{2}[\mathbf{011'2'}][\mathbf{021'2'}]\mathbf{0}+[\mathbf{12};\mathbf{1'2'}]_{\mathbf{0}}([\mathbf{021'2'}]\mathbf{1}-[\mathbf{011'2'}]\mathbf{2})\\ &=\mathbf{1'}\cdot\mathbf{2'}[\mathbf{0121'}][\mathbf{0122'}]\mathbf{0}-[\mathbf{12};\mathbf{1'2'}]_{\mathbf{0}}([\mathbf{0122'}]\mathbf{1'}-[\mathbf{0121'}]\mathbf{2'}),\end{aligned}\tag{1.1}$$

which are two equivalent but different representations. Here

$$[\mathbf{12}; \mathbf{1'2'}]_0 = \mathbf{0} \cdot \mathbf{1}\,[\mathbf{021'2'}] - \mathbf{0} \cdot \mathbf{2}\,[\mathbf{011'2'}] = \mathbf{0} \cdot \mathbf{2'}\,[\mathbf{0121'}] - \mathbf{0} \cdot \mathbf{1'}\,[\mathbf{0122'}]. \quad (1.2)$$

The efficiency of a good prover should be independent of different representations of the conclusion. On the other hand, an efficient prover should be able to choose different representations for the same geometric entity in different places of an algebraic expression.

Difficulty two: Multiple expansion results.

(1.2) is the simplest example of *expansion*, i.e., transformation from an abstract form to a simpler one. In invariant algebras, the expansion results are generally not unique. For example, Let $\mathbf{A}, \mathbf{B}, \mathbf{C}$ be three points in the projective plane. If $\mathbf{A} = \mathbf{12} \cap \mathbf{34}$, $\mathbf{B} = \mathbf{1'2'} \cap \mathbf{3'4'}$, $\mathbf{C} = \mathbf{1''2''} \cap \mathbf{3''4''}$, then in the Cayley algebra we have

$$\mathbf{A} = \mathbf{12} \wedge \mathbf{34}, \quad \mathbf{B} = \mathbf{1'2'} \wedge \mathbf{3'4'}, \quad \mathbf{C} = \mathbf{1''2''} \wedge \mathbf{3''4''}. \quad (1.3)$$

Substituting them into the bracket $[\mathbf{ABC}]$, we get

$$[(\mathbf{12} \wedge \mathbf{34})(\mathbf{1'2'} \wedge \mathbf{3'4'})(\mathbf{1''2''} \wedge \mathbf{3''4''})], \quad (1.4)$$

which has 16,847 different expansions into bracket polynomials.

Difficulty three: Simplifications in invariant algebras are difficult.

Simplifications generally include combination, i.e., reduction of the number of terms, and factorization, i.e., decomposition of an expression into the product of smaller ones. For example, the nD bracket algebra over a field $\mathcal{K}$ of characteristic $\neq 2$ is the algebra of $n \times n$ determinants. Since any $2n$ vectors $A_1, \ldots, A_{n-1}$ and $B_1, \ldots, B_{n+1}$ in an nD vector space satisfy the Grassmann-Plücker (**GP**) syzygy

$$\sum_{i=1}^{n+1} (-1)^{i+1} [A_1 \cdots A_{n-1} B_i][B_1 \cdots \check{B}_i \ldots B_{n+1}] = 0, \quad (1.5)$$

combination and factorization of a bracket polynomial are much more difficult than those of a polynomial in coordinate algebra. It is an open problem how to use the **GP** syzygies to reduce the size (degree and number of terms) of a bracket polynomial.

When applying Clifford algebra to automated theorem proving in Euclidean geometry, how should we handle the above difficulties? In this paper, we report some of our most recent work on using Clifford bracket algebra [12] in automated geometric theorem proving via the homogeneous model [11], on how the three problems are tackled, and on the systematic extension of the Clifford expansion formulas of [4], as stimulated by the need of finding the shortest expansions in the Clifford bracket algebra for the homogeneous model. With the powerful expansion and simplification techniques, we are able to finish some pure geometric computing tasks that are too difficult for other algebraic methods to finish even with the aid of a good PC, and we get better results through a *readable* procedure.

2. A theorem proving example

In [12], there are examples in automated geometric theorem proving which are difficult for other algebraic methods to finish in reasonable time but which can be finished with the Clifford bracket algebraic method within a fraction of a second. That version of Clifford bracket algebra should be called *inner-product bracket algebra,* as no Clifford multiplication is involved. In this paper, we are going to develop the inner-product bracket algebra into a sincere Clifford bracket algebra, or the *second version* of Clifford bracket algebra, which has completely different algebraic elements with the first version. The same geometric theorems can be tested once again, and it turns out that the algebraic computations can be finished by hand now – no computer aid is needed.

We use an example in [12] to show how the basic algebraic elements in the second version of Clifford bracket algebra pop up naturally in geometric problems.

Example 1. (Miquel's Theorem) In the plane there are four circles as shown in Figure 1. If points $1, 2, 3, 4$ are co-circular, so are points a, b, c, d.

In the homogeneous model, bold-faced characters are used to denote the null vectors representing the corresponding points in italic font. The hypotheses of the theorem are

$a, b, 1, 2, 3, 4$ are free points in the plane,
$c = 12a \cap 23b, \;\; d = 14a \cap 34b,$
$[\mathbf{1234}] = 0.$

The conclusion is $[\mathbf{abcd}] = 0$.

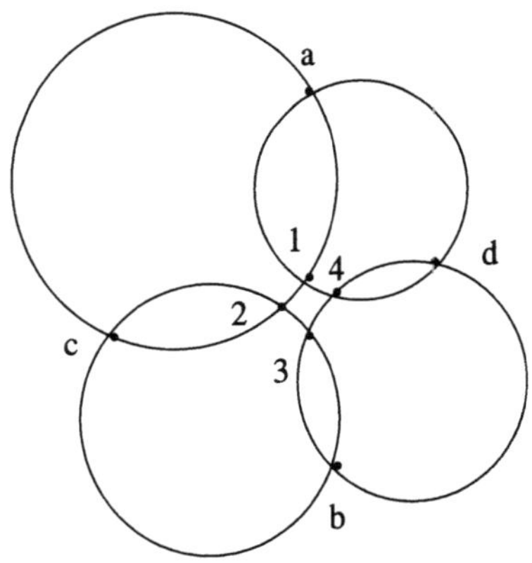

FIGURE 1. Miquel's Theorem.

Below we prove the theorem. Substituting the intersection formula (1.1), i.e.,

$$\begin{aligned} \mathbf{c} &= \mathbf{1} \cdot \mathbf{a}\,[\mathbf{23b1}][\mathbf{23ba}]\,\mathbf{2} + [\mathbf{1a};\mathbf{3b}]_{\mathbf{2}}([\mathbf{23ba}]\,\mathbf{1} - [\mathbf{23b1}]\,\mathbf{a}) \\ \mathbf{d} &= \mathbf{1} \cdot \mathbf{a}\,[\mathbf{43b1}][\mathbf{43ba}]\,\mathbf{4} + [\mathbf{1a};\mathbf{3b}]_{\mathbf{4}}([\mathbf{43ba}]\,\mathbf{1} - [\mathbf{43b1}]\,\mathbf{a}), \end{aligned} \tag{2.6}$$

into [**abcd**], we get

$$
\begin{aligned}
[\mathbf{abcd}] &\overset{\mathbf{c,d}}{=} \underline{\mathbf{1}\cdot\mathbf{a}\,[\mathbf{23ba}][\mathbf{43ba}]}\{\mathbf{1}\cdot\mathbf{a}\,[\mathbf{ab24}][\mathbf{23b1}][\mathbf{43b1}]\\
&\qquad [\mathbf{ab41}][\mathbf{43b1}][\mathbf{1a};\mathbf{3b}]_{\mathbf{2}} - [\mathbf{ab21}][\mathbf{23b1}][\mathbf{1a};\mathbf{3b}]_{\mathbf{4}}\}\\
&\overset{expand}{=} -\mathbf{1}\cdot\mathbf{a}\,[\mathbf{123b}][\mathbf{134b}][\mathbf{24ab}]\\
&\qquad +\mathbf{1}\cdot\mathbf{4}\,[\mathbf{12ab}][\mathbf{123b}][\mathbf{34ab}] - \mathbf{4}\cdot\mathbf{a}\,[\mathbf{12ab}][\mathbf{123b}][\mathbf{134b}]\\
&\qquad -\mathbf{1}\cdot\mathbf{2}\,[\mathbf{134b}][\mathbf{14ab}][\mathbf{23ab}] + \mathbf{2}\cdot\mathbf{a}\,[\mathbf{123b}][\mathbf{134b}][\mathbf{14ab}]\\
&\overset{contract}{=} \mathbf{a}\cdot\mathbf{b}\,[\mathbf{124a}][\mathbf{123b}][\mathbf{134b}]\\
&\qquad +\mathbf{1}\cdot\mathbf{4}\,[\mathbf{123b}][\mathbf{12ab}][\mathbf{34ab}] - \mathbf{1}\cdot\mathbf{2}\,[\mathbf{134b}][\mathbf{14ab}][\mathbf{23ab}]\\
&\overset{coordinates}{=} 0.
\end{aligned}
$$

Explanation of the proof:

The criteria for a good representation are that it should lead to (1) a factored expression, when this is impossible then (2) an expression with minimal number of terms, and when neither is possible then (3) a contractible expression.

By (1.1), **c**, **d** each have two representations. The representations in (2.6) lead to an expression with three monomial factors $\mathbf{1}\cdot\mathbf{a}\,[\mathbf{23ba}][\mathbf{43ba}]$, and are optimal. The first step is eliminating **c**, **d** by choosing suitable representations. In the end of the step, the underlined factors are discarded.

The second step is expanding $[\mathbf{1a};\mathbf{3b}]_{\mathbf{2}}$ and $[\mathbf{1a};\mathbf{3b}]_{\mathbf{4}}$. Each expression has two different expansions, and the criteria for a good expansion are the same as those for a good representation. In this example, no expansions can lead to either a factored expressions or an expression less than five terms. The expansions used in the proof lead to a contractible polynomial, and are optimal.

By *contraction* we mean reducing the number of terms by syzygies. In the third step, the following **GP** syzygy equality

$$-\mathbf{1}\cdot\mathbf{a}\,[\mathbf{24ab}] + \mathbf{2}\cdot\mathbf{a}\,[\mathbf{14ab}] - \mathbf{4}\cdot\mathbf{a}\,[\mathbf{12ab}] = \mathbf{a}\cdot\mathbf{b}\,[\mathbf{124a}] \tag{2.7}$$

is used to reduce the number of terms by two.

The last step is introducing a global coordinate system, changing the expression into a polynomial of coordinates, and computing its remainder with respect to the polynomial corresponding to [**1234**]. The remainder is zero, showing that [**1234**] is a factor of [**abcd**].

So [**abcd**] must be a homogeneous function of [**1234**]. However, this function is difficult to find out using coordinates, nor using inner and outer products only. It took us some efforts to find the integrated invariant hidden in the 3-termed polynomial

$$
\begin{aligned}
p \;=\; & \mathbf{a}\cdot\mathbf{b}\,[\mathbf{124a}][\mathbf{123b}][\mathbf{134b}] + \mathbf{1}\cdot\mathbf{4}\,[\mathbf{123b}][\mathbf{12ab}][\mathbf{34ab}]\\
& -\mathbf{1}\cdot\mathbf{2}\,[\mathbf{134b}][\mathbf{14ab}][\mathbf{23ab}].
\end{aligned} \tag{2.8}
$$

The invariant is $[\mathbf{12ab34}]$, which changes its sign by a left shift, a right shift or a reversion, and which has the following *rational binomial expansion*:

$$[\mathbf{12ab34}] = 2\,\frac{\mathbf{a}\cdot\mathbf{b}\,[\mathbf{124a}][\mathbf{134b}] + \mathbf{1}\cdot\mathbf{4}\,[\mathbf{12ab}][\mathbf{34ab}]}{[\mathbf{14ab}]}. \tag{2.9}$$

With this invariant, factoring (2.8) becomes very easy: the first two terms of p are in the following form:

$$[\mathbf{123b}](\mathbf{a}\cdot\mathbf{b}\,[\mathbf{124a}][\mathbf{134b}] + \mathbf{1}\cdot\mathbf{4}\,[\mathbf{12ab}][\mathbf{34ab}]). \tag{2.10}$$

The first and third terms are in a similar form:

$$[\mathbf{134b}](\mathbf{a}\cdot\mathbf{b}\,[\mathbf{124a}][\mathbf{123b}] - \mathbf{1}\cdot\mathbf{2}\,[\mathbf{14ab}][\mathbf{23ab}]). \tag{2.11}$$

Applying (2.9) to (2.10), we get $[\mathbf{123b}][\mathbf{14ab}][\mathbf{12ab34}]/2$, so

$$p = [\mathbf{14ab}](\frac{1}{2}[\mathbf{123b}][\mathbf{12ab34}] - \mathbf{1}\cdot\mathbf{2}\,[\mathbf{134b}][\mathbf{23ab}]). \tag{2.12}$$

This is the first factorization. Now apply (2.9) once again to (2.12), by the unique correspondence

$$\mathbf{1}\leftrightarrow\mathbf{b},\quad \mathbf{2}\leftrightarrow\mathbf{a},\quad \mathbf{3}\leftrightarrow\mathbf{4}, \tag{2.13}$$

we get the second factorization:

$$p = -[\mathbf{14ab}]\mathbf{3}\cdot\mathbf{b}\,[\mathbf{a21b}][\mathbf{2143}] = \mathbf{3}\cdot\mathbf{b}\,[\mathbf{1234}][\mathbf{12ab}][\mathbf{14ab}]. \tag{2.14}$$

If we apply (2.9) to (2.11) instead of (2.10), we have a similar procedure of factorization. Collecting all the common factors we have discarded, we obtain a nice theorem as follows.

Theorem 2.1. *For* 6 *generalized points* $a, b, 1, 2, 3, 4$ *in the plane, let* $c = 12a \cap 23b$, $d = 14a \cap 34b$, *then*

$$\frac{[\mathbf{abcd}]}{\mathbf{a}\cdot\mathbf{c}\,\mathbf{b}\cdot\mathbf{d}} = \frac{[\mathbf{1234}]}{\mathbf{1}\cdot\mathbf{2}\,\mathbf{3}\cdot\mathbf{4}}\,\frac{[\mathbf{12ab}][\mathbf{34ab}]}{[\mathbf{23ab}][\mathbf{41ab}]}. \tag{2.15}$$

If the juxtaposition of elements in $[\mathbf{12ab34}]$ represents a product, then we get an invariant algebra suitable for geometric computing involving lines and circles. Can we take the juxtaposition as the multiplication in the Clifford algebra $Cl(\mathcal{R}^{3,1})$, and take the square bracket as the dual of the pseudoscalar part of the multiplication? Since $\mathbf{1}, \mathbf{2}, \mathbf{3}, \mathbf{4}, \mathbf{a}, \mathbf{b}$ are in the same 4D space, they satisfy *Cramer's rule*

$$\mathbf{3} = \frac{[\mathbf{34ab}]\mathbf{1} + [\mathbf{13ab}]\mathbf{4} - [\mathbf{134b}]\mathbf{a} + [\mathbf{134a}]\mathbf{b}}{[\mathbf{14ab}]}. \tag{2.16}$$

Substituting (2.16) into the left side of (2.9), and using the formula that for any null vector $\mathbf{1}$ and any sequence of vectors $\mathbf{2}, \mathbf{3}, \ldots, \mathbf{r}$ in a Clifford algebra [13],

$$\mathbf{12}\cdots\mathbf{r1} = 2\sum_{i=2}^{r}(-1)^{r-i}(\mathbf{1}\cdot\mathbf{i})\,\mathbf{12}\cdots\check{\mathbf{i}}\cdots\mathbf{r}, \tag{2.17}$$

we get

$$\begin{aligned}[\mathbf{12ab34}] &= \frac{[\mathbf{34ab}][\mathbf{12ab14}] - [\mathbf{134b}][\mathbf{12aba4}]}{[\mathbf{14ab}]} \\ &= 2\,\frac{\mathbf{1}\cdot\mathbf{4}\,[\mathbf{12ab}][\mathbf{34ab}] + \mathbf{a}\cdot\mathbf{b}\,[\mathbf{124a}][\mathbf{134b}]}{[\mathbf{14ab}]}.\end{aligned} \tag{2.18}$$

So we can indeed take the juxtaposition as the Clifford multiplication! Furthermore, the peculiar symbol $[\mathbf{12};\mathbf{1'2'}]_0$ defined in (1.2) can be taken as $-[\mathbf{01201'2'}]/2$, because by (2.17), they have the same expansion.

The importance of this identification lies in that the Clifford algebraic structure turns out to be unique in constructing more advanced algebraic invariants in Euclidean geometry. Before it shows up, the only basic invariants are inner products and brackets which are determinants of homogeneous coordinates. They prove to be inefficient in doing factorization in the first version of Clifford bracket algebra. The Clifford multiplication introduces two kinds of basic invariants: the scalar part and the dual of the pseudoscalar part of the Clifford multiplication of a sequence of vectors. On one hand, they are complicated polynomials of inner products and brackets when expanded. On the other hand, they occur and function naturally in geometric problems and as a whole represent certain geometric invariants. In the next section we shall establish some properties of the new invariant algebra and its connection with Euclidean geometry.

3. Clifford bracket algebra – second version

The extension of bracket algebra from a vector space to an inner product space is called *Clifford bracket algebra* in [12]. The basic algebraic elements are inner products and brackets of n vectors, where n is the dimension of the space.

Let $\mathcal{K}$ be a field of characteristic $\neq 2$. Let $n \leq m$ be two positive integers, and let ι be a fixed element in $\mathcal{K}-\{0\}$. The n-dimensional *Clifford bracket algebra* generated by m symbols $\mathbf{a}_1, \ldots, \mathbf{a}_m$, is the quotient of the polynomial ring over $\mathcal{K}$ whose indeterminates are elements of $\{\mathbf{a}_i\cdot\mathbf{a}_j, [\mathbf{a}_{i_1}\ldots\mathbf{a}_{i_n}] | 1 \leq i, j, i_1, \ldots, i_n \leq m\}$, modulo the two-sided ideal generated by the following five types of elements:

B1: $[\mathbf{a}_{i_1}\ldots\mathbf{a}_{i_n}]$ if $i_j = i_k$ for some $j \neq k$.

B2: $[\mathbf{a}_{i_1}\ldots\mathbf{a}_{i_n}] - \mathrm{sign}(\sigma)[\mathbf{a}_{i_{\sigma 1}}\ldots\mathbf{a}_{i_{\sigma n}}]$ for any permutation σ of $1, \ldots, n$.

B3: $\mathbf{a}_i\cdot\mathbf{a}_j - \mathbf{a}_j\cdot\mathbf{a}_i$ for $i \neq j$.

GP1: $\displaystyle\sum_{k=1}^{n+1}(-1)^{k+1}\mathbf{a}_j\cdot\mathbf{a}_{i_k}\,[\mathbf{a}_{i_1}\ldots\check{\mathbf{a}}_{i_k}\ldots\mathbf{a}_{i_{n+1}}]$.

GP2: $[\mathbf{a}_{i_1}\ldots\mathbf{a}_{i_n}][\mathbf{a}_{j_1}\ldots\mathbf{a}_{j_n}] - \iota\det(\mathbf{a}_{i_k}\cdot\mathbf{a}_{j_l})_{k,l=1,\ldots,n}$.

To further simplify algebraic manipulations, more advanced invariants are needed. In the second version of Clifford bracket algebra, the basic algebraic elements are the scalar parts and the duals of the pseudoscalar parts of any sequences of vectors, as long as they exist.

Let $\mathcal{K}$ be a field of characteristic $\neq 2$. Let $n \leq m$ be two positive integers, and let ι be a fixed element in $\mathcal{K} - \{0\}$. The n-dimensional *Clifford bracket algebra* generated by symbols $\mathbf{a}_1, \ldots, \mathbf{a}_m$, is the quotient of the polynomial ring over $\mathcal{K}$ whose indeterminates are elements of $\{\langle \mathbf{a}_{i_1} \ldots \mathbf{a}_{i_{2p}} \rangle, [\mathbf{a}_{j_1} \ldots \mathbf{a}_{j_{n+2q-2}}] \mid p, q \geq 1;\ 1 \leq i_1, \ldots, i_{2p}, j_1, \ldots, j_{n+2q-2} \leq m\}$, modulo the two-sided ideal generated by the following seven types of elements: **B1, B2, B3** in which $\mathbf{a}_i \cdot \mathbf{a}_j$ is replaced by $\langle \mathbf{a}_i \mathbf{a}_j \rangle$, **GP1, GP2**, and

DF1: $\langle \mathbf{a}_{i_1} \cdots \mathbf{a}_{i_{2l}} \rangle - \sum_{j=2}^{2l} (-1)^j \langle \mathbf{a}_{i_1} \mathbf{a}_{i_j} \rangle \langle \mathbf{a}_{i_2} \cdots \check{\mathbf{a}}_{i_j} \cdots \mathbf{a}_{i_{2l}} \rangle$.

DF2: $[\mathbf{a}_{i_1} \cdots \mathbf{a}_{i_{n+2l-2}}] - \sum_{1 \leq \sigma_{2l} \leq n+2l-2} \text{sign}(\sigma_{2l}, \check{\sigma}_{2l}) \langle \sigma_{2l}(\mathbf{a}_i) \rangle\ [\check{\sigma}_{2l}(\mathbf{a}_i)]$. Here σ_{2l}, $\check{\sigma}_{2l}$ is a partition of $1, 2, \ldots, n+2l-2$ into two subsequences of length $2l$ and $n-2$ respectively.

It can be proved that all the properties on the scalar part and the dual of the pseudoscalar part of the Clifford multiplication of a sequence of vectors in a Clifford algebra, can be derived from the above seven kinds of generators. This justifies the term "Clifford bracket algebra".

In the homogeneous model of Euclidean geometry [11], the square of any vector representing a geometric point is zero, and the corresponding Clifford bracket algebra is called *null bracket algebra*. We use the homogeneous model of the Euclidean plane to show the geometric interpretations of the basic algebraic elements in the null bracket algebra.

First of all, for any vectors $\mathbf{a}_1, \mathbf{a}_2$ representing two points,

$$\mathbf{a}_1 \cdot \mathbf{a}_2 = -\frac{1}{2}|a_1 a_2|^2. \tag{3.19}$$

For any three points a_1, a_2, a_3,

$$\begin{array}{rclcl} \langle \mathbf{e}\mathbf{a}_1\mathbf{a}_2\mathbf{a}_3 \rangle & = & (a_3 - a_2) \cdot (a_1 - a_2) & = & |a_1 a_2||a_2 a_3| \cos \angle(a_2 a_3, a_2 a_1), \\ {[\mathbf{e}\mathbf{a}_1\mathbf{a}_2\mathbf{a}_3]} & = & 2S_{a_1 a_2 a_3} & = & |a_1 a_2||a_2 a_3| \sin \angle(a_2 a_3, a_2 a_1). \end{array} \tag{3.20}$$

Here $\mathbf{e}$ represents the Gaussian point at infinity of the Euclidean plane, $\angle(a_2 a_3, a_2 a_1)$ is the oriented angle from vector $a_2 a_3$ to vector $a_2 a_1$, and $S_{a_1 a_2 a_3}$ is the signed area of the oriented triangle $a_1 a_2 a_3$. For any four points a_1, a_2, a_3, a_4,

$$\begin{array}{rcl} {[\mathbf{e}\mathbf{a}_1\mathbf{a}_2\mathbf{e}\mathbf{a}_3\mathbf{a}_4]} & = & -4S_{a_1 a_3 a_2 a_4} \\ & = & 2|a_1 a_2||a_3 a_4| \sin \angle(a_3 a_4, a_1 a_2), \\ \langle \mathbf{e}\mathbf{a}_1\mathbf{a}_2\mathbf{e}\mathbf{a}_3\mathbf{a}_4 \rangle & = & 2(a_2 - a_1) \cdot (a_4 - a_3) \\ & = & 2|a_1 a_2||a_3 a_4| \cos \angle(a_3 a_4, a_1 a_2). \end{array} \tag{3.21}$$

Here $S_{a_1 a_3 a_2 a_4}$ is the signed area of oriented quadrilateral $a_1 a_3 a_2 a_4$.

By means of the relations

$$\begin{aligned}
\mathbf{a}_1 \cdot \mathbf{a}_3 \langle \mathbf{a}_1\mathbf{a}_2\mathbf{a}_3\mathbf{a}_4 \rangle &= \langle \mathbf{a}_1\mathbf{a}_2\mathbf{a}_3 \rangle_1 \cdot \langle \mathbf{a}_1\mathbf{a}_3\mathbf{a}_4 \rangle_1 \\
\mathbf{a}_1 \cdot \mathbf{a}_3 [\mathbf{a}_1\mathbf{a}_2\mathbf{a}_3\mathbf{a}_4] &= \langle \mathbf{a}_1\mathbf{a}_2\mathbf{a}_3 \rangle_1 \cdot \langle \mathbf{a}_1\mathbf{a}_3\mathbf{a}_4 \rangle_{\tilde{3}} \\
[\mathbf{a}_1\mathbf{a}_2\mathbf{a}_3\mathbf{a}_1\mathbf{a}_4\mathbf{a}_5] &= 2 \langle \mathbf{a}_1\mathbf{a}_2\mathbf{a}_3 \rangle_1 \cdot \langle \mathbf{a}_1\mathbf{a}_4\mathbf{a}_5 \rangle_{\tilde{3}} \\
\langle \mathbf{a}_1\mathbf{a}_2\mathbf{a}_3\mathbf{a}_1\mathbf{a}_4\mathbf{a}_5 \rangle &= 2 \langle \mathbf{a}_1\mathbf{a}_2\mathbf{a}_3 \rangle_1 \cdot \langle \mathbf{a}_1\mathbf{a}_4\mathbf{a}_5 \rangle_1
\end{aligned} \tag{3.22}$$

and

$$\begin{aligned}
\langle \mathbf{a}_1 \cdots \mathbf{a}_{2l+1} \rangle_{\tilde{3}}^2 &= \langle \mathbf{a}_1 \cdots \mathbf{a}_{2l+1} \rangle_{\tilde{1}}^2 \\
&= -2^{2l-1}(\mathbf{a}_1 \cdot \mathbf{a}_2)(\mathbf{a}_2 \cdot \mathbf{a}_3) \cdots (\mathbf{a}_{2l} \cdot \mathbf{a}_{2l+1})(\mathbf{a}_{2l+1} \cdot \mathbf{a}_1),
\end{aligned} \tag{3.23}$$

$$\langle \mathbf{a}_1 \cdots \mathbf{a}_{2l} \rangle^2 + [\mathbf{a}_1 \cdots \mathbf{a}_{2l}]^2 = 2^{2l-2}(\mathbf{a}_1 \cdot \mathbf{a}_2)(\mathbf{a}_2 \cdot \mathbf{a}_3) \cdots (\mathbf{a}_{2l-1} \cdot \mathbf{a}_{2l})(\mathbf{a}_{2l} \cdot \mathbf{a}_1), \tag{3.24}$$

it can be proved that

$$\begin{aligned}
\langle \mathbf{a}_1\mathbf{a}_2\mathbf{a}_3\mathbf{a}_4 \rangle &= -8\frac{\rho_{a_1a_2a_3}\rho_{a_1a_3a_4}S_{a_1a_2a_3}S_{a_1a_3a_4}}{|a_1a_3|^2} \cos\angle(N^{a_1}_{a_1a_2a_3}, N^{a_1}_{a_1a_3a_4}) \\
[\mathbf{a}_1\mathbf{a}_3\mathbf{a}_2\mathbf{a}_4] &= -8\frac{\rho_{a_1a_2a_3}\rho_{a_1a_3a_4}S_{a_1a_2a_3}S_{a_1a_3a_4}}{|a_1a_3|^2} \sin\angle(N^{a_1}_{a_1a_2a_3}, N^{a_1}_{a_1a_3a_4}) \\
\langle \mathbf{a}_1\mathbf{a}_2\mathbf{a}_3\mathbf{a}_1\mathbf{a}_4\mathbf{a}_5 \rangle &= 8\rho_{a_1a_2a_3}\rho_{a_1a_4a_5}S_{a_1a_2a_3}S_{a_1a_4a_5} \cos\angle(N^{a_1}_{a_1a_2a_3}, N^{a_1}_{a_1a_4a_5}) \\
[\mathbf{a}_1\mathbf{a}_2\mathbf{a}_3\mathbf{a}_1\mathbf{a}_4\mathbf{a}_5] &= 8\rho_{a_1a_2a_3}\rho_{a_1a_4a_5}S_{a_1a_2a_3}S_{a_1a_4a_5} \sin\angle(N^{a_1}_{a_1a_2a_3}, N^{a_1}_{a_1a_4a_5}).
\end{aligned} \tag{3.25}$$

Here $\rho_{a_1a_2a_3}$ is the radius of circle $a_1a_2a_3$, and $N^{a_1}_{a_1a_2a_3}$ is the outward normal direction of the circle at point a_1.

Using $\mathbf{a}_1\mathbf{a}_2\mathbf{a}_1 = 2(\mathbf{a}_1 \cdot \mathbf{a}_2)\mathbf{a}_1$ and the relations

$$\begin{aligned}
[\mathbf{a}_1\mathbf{a}_2\mathbf{a}_3\mathbf{a}_4\mathbf{a}_1\mathbf{a}_5 \cdots \mathbf{a}_{2l+5}] &= 2(\langle \mathbf{a}_1\mathbf{a}_2\mathbf{a}_3\mathbf{a}_4 \rangle [\mathbf{a}_1\mathbf{a}_5 \cdots \mathbf{a}_{2l+5}] \\
&\quad +[\mathbf{a}_1\mathbf{a}_2\mathbf{a}_3\mathbf{a}_4] \langle \mathbf{a}_1\mathbf{a}_5 \cdots \mathbf{a}_{2l+5} \rangle), \\
\langle \mathbf{a}_1\mathbf{a}_2\mathbf{a}_3\mathbf{a}_4\mathbf{a}_1\mathbf{a}_5 \cdots \mathbf{a}_{2l+5} \rangle &= 2(\langle \mathbf{a}_1\mathbf{a}_2\mathbf{a}_3\mathbf{a}_4 \rangle \langle \mathbf{a}_1\mathbf{a}_5 \cdots \mathbf{a}_{2l+5} \rangle \\
&\quad -[\mathbf{a}_1\mathbf{a}_2\mathbf{a}_3\mathbf{a}_4] [\mathbf{a}_1\mathbf{a}_5 \cdots \mathbf{a}_{2l+5}]), \\
[\mathbf{a}_1\mathbf{a}_2\mathbf{a}_3\mathbf{a}_1\mathbf{a}_4\mathbf{a}_5\mathbf{a}_1\mathbf{a}_6 \cdots \mathbf{a}_{2l+6}] &= 2([\mathbf{a}_1\mathbf{a}_2\mathbf{a}_3\mathbf{a}_1\mathbf{a}_4\mathbf{a}_5]\langle \mathbf{a}_1\mathbf{a}_6 \cdots \mathbf{a}_{2l+6} \rangle \\
&\quad +\langle \mathbf{a}_1\mathbf{a}_2\mathbf{a}_3\mathbf{a}_1\mathbf{a}_4\mathbf{a}_5 \rangle[\mathbf{a}_1\mathbf{a}_6 \cdots \mathbf{a}_{2l+6}]), \\
\langle \mathbf{a}_1\mathbf{a}_2\mathbf{a}_3\mathbf{a}_1\mathbf{a}_4\mathbf{a}_5\mathbf{a}_1\mathbf{a}_6 \cdots \mathbf{a}_{2l+6} \rangle &= 2(\langle \mathbf{a}_1\mathbf{a}_2\mathbf{a}_3\mathbf{a}_1\mathbf{a}_4\mathbf{a}_5 \rangle\langle \mathbf{a}_1\mathbf{a}_6 \cdots \mathbf{a}_{2l+6} \rangle \\
&\quad -[\mathbf{a}_1\mathbf{a}_2\mathbf{a}_3\mathbf{a}_1\mathbf{a}_4\mathbf{a}_5][\mathbf{a}_1\mathbf{a}_6 \cdots \mathbf{a}_{2l+6}]),
\end{aligned} \tag{3.26}$$

we find that all the basic algebraic elements in the null bracket algebra are the sines and cosines of the sums of some oriented angles, multiplied by rational monomials of distances, radii and signed areas.

4. Clifford expansion

Changing a Clifford expression into an equivalent expression of inner and outer products is called *Clifford expansion*. In Clifford bracket algebra, like in other invariant algebras, factored and shortest expansions are extremely important. The proof of Miquel's Theorem is a typical example.

The earliest work on Clifford expansion is [4], where two fundamental dimension-free expansion formulas are established. The first formula is *Caianiello's first identity* [3]:

Theorem 4.1. *Let $\mathbf{a}_1, \ldots, \mathbf{a}_k$ be vectors. Let $\langle \mathbf{a}_1 \cdots \mathbf{a}_k \rangle_{k-2l}$ be the $(k-2l)$-graded part of the Clifford multiplication of the k vectors. Then*

$$\langle \mathbf{a}_1 \cdots \mathbf{a}_k \rangle_{k-2l} = \sum_{1 \leq \sigma_{2l} \leq k} \mathrm{sign}(\sigma_{2l}, \breve{\sigma}_{2l}) \langle \sigma_{2l}(\mathbf{a}) \rangle \, \Lambda \breve{\sigma}_{2l}(\mathbf{a}). \tag{4.27}$$

Here the summation is over all partitions of the sequence $1, 2, \ldots, k$ into two subsequences $\sigma_{2l}, \breve{\sigma}_{2l}$ of length $2l, k-2l$ respectively, $\sigma_{2l}(\mathbf{a})$ is the Clifford multiplication of the $\mathbf{a}$'s with σ_{2l} as indices, and $\Lambda \breve{\sigma}_{2l}(\mathbf{a})$ is the outer product of the $\mathbf{a}$'s with $\breve{\sigma}_{2l}$ as indices.

The scalar $\langle \mathbf{a}_1 \cdots \mathbf{a}_{2l} \rangle$ is traditionally called the *pseudo-Pfaffian* of the vectors. It has the following recursive expansion:

$$\langle \mathbf{a}_1 \cdots \mathbf{a}_{2l} \rangle = \sum_{i=2}^{2l} (-1)^i \langle \mathbf{a}_1 \mathbf{a}_i \rangle \, \langle \mathbf{a}_2 \cdots \breve{\mathbf{a}}_i \cdots \mathbf{a}_{2l} \rangle. \tag{4.28}$$

In our application, the grade is either 0 or the dimension n of the base vector space. The expansion given by (4.27) turns out to be too long. For example, in 4D null bracket algebra, the bracket $[\mathbf{a}_1\mathbf{a}_2\mathbf{a}_3\mathbf{a}_4\mathbf{a}_5\mathbf{a}_6]$ is expanded into $C_6^2 = 15$ terms by (4.27). On the other hand, the expansion we use in proving in Miquel's Theorem is binomial.

Even if the expansion is restricted to be polynomial, Caianiello's first identity still gives too many terms. When the grade is n, as long as the characteristic of the field is not 2, we have the following much shorter expansions:

Theorem 4.2.

$$\begin{aligned}
&\langle \mathbf{a}_1 \cdots \mathbf{a}_{n+2l} \rangle_n \\
= &\ \frac{1}{C_l^k} \sum_{1 \leq \sigma_{2k} \leq m} \operatorname{sign}(\sigma_{2k}, \check{\sigma}_{2k}) \langle \sigma_{2k}(\mathbf{a}) \rangle \, \langle \check{\sigma}_{2k}(\mathbf{a}) \mathbf{a}_{m+1} \cdots \mathbf{a}_{n+2l} \rangle_n \\
&+ \sum_{i=1}^{k} (-1)^{i+1} \frac{C_k^i}{C_l^i} \left(\sum_{m+1 \leq \tau_{2i} \leq n+2l} \operatorname{sign}(\tau_{2i}, \check{\tau}_{2i}) \langle \tau_{2i}(\mathbf{a}) \rangle \, \langle \mathbf{a}_1 \cdots \mathbf{a}_m \check{\tau}_{2i}(\mathbf{a}) \rangle_n \right) \\
= &\ \frac{1}{C_l^k} \sum_{m+1 \leq \tau_{2k} \leq n+2l} \operatorname{sign}(\tau_{2k}, \check{\tau}_{2k}) \langle \tau_{2k}(\mathbf{a}) \rangle \, \langle \mathbf{a}_1 \cdots \mathbf{a}_m \check{\tau}_{2k}(\mathbf{a}) \rangle_n \\
&+ \sum_{i=1}^{k} (-1)^{i+1} \frac{C_k^i}{C_l^i} \left(\sum_{1 \leq \sigma_{2i} \leq m} \operatorname{sign}(\sigma_{2i}, \check{\sigma}_{2i}) \langle \sigma_{2i}(\mathbf{a}) \rangle \, \langle \check{\sigma}_{2i}(\mathbf{a}) \mathbf{a}_{m+1} \cdots \mathbf{a}_{n+2l} \rangle_n \right).
\end{aligned} \tag{4.29}$$

It can also be written as a change of $(0, n)$*-typed summation:*

$$\begin{aligned}
&\sum_{1 \leq \sigma_{2k} \leq m} \operatorname{sign}(\sigma_{2k}, \check{\sigma}_{2k}) \langle \sigma_{2k}(\mathbf{a}) \rangle \, \langle \check{\sigma}_{2k}(\mathbf{a}) \mathbf{a}_{m+1} \cdots \mathbf{a}_{n+2l} \rangle_n \\
= &\sum_{i=0}^{k} \sum_{m+1 \leq \tau_{2i} \leq n+2l} (-1)^i C_{l-i}^{k-i} \operatorname{sign}(\tau_{2i}, \check{\tau}_{2i}) \langle \tau_{2i}(\mathbf{a}) \rangle \, \langle \mathbf{a}_1 \cdots \mathbf{a}_m \check{\tau}_{2i}(\mathbf{a}) \rangle_n.
\end{aligned} \tag{4.30}$$

To see that the number of terms is greatly reduced, let us check the simplest case $m = n + 2l - 1$ and $k = l$. Then

$$\langle \mathbf{a}_1 \cdots \mathbf{a}_{n+2l} \rangle_n = \sum_{1 \leq \sigma_{2l} \leq n+2l-1} \operatorname{sign}(\sigma_{2l}, \check{\sigma}_{2l}) \langle \sigma_{2l}(\mathbf{a}) \rangle \, \langle \check{\sigma}_{2l}(\mathbf{a}) \mathbf{a}_{n+2l} \rangle_n. \tag{4.31}$$

So it is the sequence $1, 2, \ldots, n + 2l - 1$ that is partitioned instead of the whole sequence $1, 2, \ldots, n + 2l$ in (4.27). The number of terms is reduced by C_{n+2l-1}^n. When m takes other values, the number of terms can be further reduced.

Below we prove Theorem 4.2. By symmetry we only need to prove the first equality in (4.29). The theorem is true in the following three special cases, whose proof can be found in [13].

Case 1. $k = 1$ and any $0 \leq m \leq n + 2l$.

Case 2. $m = n + 2l$ and any $1 \leq k \leq l$.

Case 3. $m = n + 2l - 1$ and any $1 \leq k \leq l$.

Now assume that the first equality in (4.29) is true for $k = r - 1$ and any $0 \leq m \leq n + 2l$, where $r \leq l$. We prove that it is true for $k = r$ and any $0 \leq m \leq n + 2l - 2$. There are three cases to consider:

Case 4. $2r \leq m$.

Let σ_{2r-2} be a sequence of $2r-2$ elements in $1,\ldots,m$, and let its remainder in $1,\ldots,m$ be $\chi=\chi_{m-2r+2}$. Let τ_2 be a subsequence of 2 elements in $m+1,\ldots,n+2l$, and let its remainder in $m+1,\ldots,n+2l$ be $\pi_{n+2l-m-2}$. Let ϕ_{2i} be a subsequence of $2i$ elements in $\pi=\pi_{n+2l-m-2}$, and let its remainder in π be $\psi=\psi_{n+2l-m-2-2i}$. By the hypothesis of the induction,

$$\begin{aligned}
&\langle \mathbf{a}_1\cdots\mathbf{a}_{n+2l}\rangle_n\\
=\ &\frac{1}{C_l^{r-1}}\sum_{1\le\sigma_{2r-2}\le m}\mathrm{sign}(\sigma_{2r-2},\chi)\langle\sigma_{2r-2}(\mathbf{a})\rangle\,\langle\chi(\mathbf{a})\mathbf{a}_{m+1}\cdots\mathbf{a}_{n+2l}\rangle_n\\
&+\sum_{i=1}^{r-1}(-1)^{i+1}\frac{C_{r-1}^i}{C_l^i}\left(\sum_{m+1\le\tau_{2i}\le n+2l}\mathrm{sign}(\tau_{2i},\pi)\langle\tau_{2i}(\mathbf{a})\rangle\,\langle\mathbf{a}_1\cdots\mathbf{a}_m\pi(\mathbf{a})\rangle_n\right)\\
=\ &\frac{1}{C_l^{r-1}}\sum_{1\le\sigma_{2r-2}\le m}\mathrm{sign}(\sigma_{2r-2},\chi)\langle\sigma_{2r-2}(\mathbf{a})\rangle\,\langle\chi(\mathbf{a})\mathbf{a}_{m+1}\cdots\mathbf{a}_{n+2l}\rangle_n\\
&+\sum_{i=2}^{r-1}(-1)^{i+1}\frac{C_{r-1}^i}{C_l^i}\sum_{m+1\le\tau_{2i}\le n+2l}\mathrm{sign}(\tau_{2i},\pi)\langle\tau_{2i}(\mathbf{a})\rangle\,\langle\mathbf{a}_1\cdots\mathbf{a}_m\pi(\mathbf{a})\rangle_n\\
&+\frac{r}{l}\sum_{m+1\le\tau_{2i}\le n+2l}\mathrm{sign}(\tau_{2i},\pi)\langle\tau_{2i}(\mathbf{a})\rangle\,\langle\mathbf{a}_1\cdots\mathbf{a}_m\pi(\mathbf{a})\rangle_n\\
&-\frac{1}{l}\sum_{m+1\le\tau_{2i}\le n+2l}\mathrm{sign}(\tau_{2i},\pi)\langle\tau_{2i}(\mathbf{a})\rangle\,\langle\mathbf{a}_1\cdots\mathbf{a}_m\pi(\mathbf{a})\rangle_n,
\end{aligned}\tag{4.32}$$

and

$$\begin{aligned}
&\langle \mathbf{a}_1\cdots\mathbf{a}_m\pi(\mathbf{a})\rangle_n\\
=\ &\frac{1}{C_{l-1}^{r-1}}\sum_{1\le\sigma_{2r-2}\le m}\mathrm{sign}(\sigma_{2r-2},\chi)\langle\sigma_{2r-2}(\mathbf{a})\rangle\,\langle\chi(\mathbf{a})\pi(\mathbf{a})\rangle_n\\
&+\sum_{i=1}^{r-1}(-1)^{i+1}\frac{C_{r-1}^i}{C_{l-1}^i}\left(\sum_{\{\phi_{2i}\}\subseteq\{\pi\}}\mathrm{sign}(\phi_{2i},\psi)\langle\phi_{2i}(\mathbf{a})\rangle\,\langle\mathbf{a}_1\cdots\mathbf{a}_m\psi(\mathbf{a})\rangle_n\right).
\end{aligned}\tag{4.33}$$

First, let ω_{2i+2} be the ascending sequence formed by elements in $\{\tau_2, \phi_{2i}\}$. Then its remainder in $m+1, \ldots, n+2l$ is $\check{\omega}_{2i+2} = \pi$. We have

$$\begin{aligned}
&-\frac{1}{l}\sum_{i=1}^{r-1}(-1)^{i+1}\frac{C_{r-1}^i}{C_{l-1}^i}\sum_{m+1\le\tau_2\le n+2l}\ \sum_{\{\phi_{2i}\}\subseteq\{\pi\}}\mathrm{sign}(\tau_2,\pi)\mathrm{sign}(\phi_{2i},\psi)\langle\tau_2(\mathbf{a})\rangle \\
&\qquad\qquad \langle\phi_{2i}(\mathbf{a})\rangle\,\langle\mathbf{a}_1\cdots\mathbf{a}_m\psi(\mathbf{a})\rangle_n \\
= &-\frac{1}{l}\sum_{i=1}^{r-1}(-1)^{i+1}\frac{C_{r-1}^i}{C_{l-1}^i}\sum_{m+1\le\omega_{2i+2}\le n+2l}\ \sum_{\{\tau_2\}\subseteq\{\omega_{2i+2}\}}\mathrm{sign}(\tau_2,\pi)\mathrm{sign}(\phi_{2i},\psi) \\
&\qquad\qquad \langle\tau_2(\mathbf{a})\rangle\,\langle\phi_{2i}(\mathbf{a})\rangle\,\langle\mathbf{a}_1\cdots\mathbf{a}_m\psi(\mathbf{a})\rangle_n \\
= &\sum_{i=1}^{r-1}(-1)^i\frac{(i+1)C_{r-1}^i}{lC_{l-1}^i}\sum_{m+1\le\omega_{2i+2}\le n+2l}\mathrm{sign}(\omega_{2i+2},\check{\omega}_{2i+2})\langle\omega_{2i+2}(\mathbf{a})\rangle \\
&\qquad\qquad \langle\mathbf{a}_1\cdots\mathbf{a}_m\check{\omega}_{2i+2}(\mathbf{a})\rangle_n \\
= &\sum_{i=2}^{r}(-1)^{i+1}\frac{C_{r-1}^{i-1}}{C_l^i}\sum_{m+1\le\tau_{2i}\le n+2l}\mathrm{sign}(\tau_{2i},\check{\tau}_{2i})\langle\tau_{2i}(\mathbf{a})\rangle\langle\mathbf{a}_1\cdots\mathbf{a}_m\check{\tau}_{2i}(\mathbf{a})\rangle_n.
\end{aligned} \tag{4.34}$$

Second, let ξ_2 be a subsequence of 2 elements in χ, and let its remainder in χ be η_{m-2r}. Let ζ_{2r} be the ascending sequence formed by elements in $\{\sigma_{2r-2}, \xi_2\}$. Then the remainder of ζ_{2r} in $1, \ldots, m$ is η_{m-2r}. By the conclusion in Case 1,

$$\begin{aligned}
&\sum_{m+1\le\tau_2\le n+2l}\mathrm{sign}(\tau_2,\pi)\langle\tau_2(\mathbf{a})\rangle\langle\chi(\mathbf{a})\pi(\mathbf{a})\rangle_n \\
= &\ (l-r+1)\langle\chi(\mathbf{a})\mathbf{a}_{m+1}\cdots\mathbf{a}_{n+2l}\rangle_n \qquad\qquad (4.35)\\
&-\sum_{\{\xi_2\}\subseteq\{\chi\}}\mathrm{sign}(\xi_2,\eta_{m-2r})\langle\xi_2(\mathbf{a})\rangle\langle\eta_{m-2r}(\mathbf{a})\mathbf{a}_{m+1}\cdots\mathbf{a}_{n+2l}\rangle_n.
\end{aligned}$$

By the conclusion in Case 2,

$$\begin{aligned}
&\frac{1}{lC_{l-1}^{r-1}}\sum_{1\le\sigma_{2r-2}\le m}\ \sum_{\{\xi_2\}\subseteq\{\chi\}}\mathrm{sign}(\sigma_{2r-2},\chi)\mathrm{sign}(\xi_2,\eta_{m-2r})\langle\sigma_{2r-2}(\mathbf{a})\rangle\,\langle\xi_2(\mathbf{a})\rangle \\
&\qquad\qquad \langle\eta_{m-2r}(\mathbf{a})\mathbf{a}_{m+1}\cdots\mathbf{a}_{n+2l}\rangle_n \\
= &\ \frac{1}{lC_{l-1}^{r-1}}\sum_{1\le\zeta_{2r}\le m}\ \sum_{\{\xi_2\}\subseteq\{\zeta_{2r}\}}\mathrm{sign}(\zeta_{2r},\eta_{m-2r})\mathrm{sign}(\sigma_{2r-2},\xi_2)\langle\sigma_{2r-2}(\mathbf{a})\rangle\,\langle\xi_2(\mathbf{a})\rangle \\
&\qquad\qquad \langle\eta_{m-2r}(\mathbf{a})\mathbf{a}_{m+1}\cdots\mathbf{a}_{n+2l}\rangle_n \\
= &\ \frac{1}{C_l^r}\sum_{1\le\zeta_{2r}\le m}\mathrm{sign}(\zeta_{2r},\check{\zeta}_{2r})\langle\zeta_{2r}(\mathbf{a})\rangle\,\langle\check{\zeta}_{2r}(\mathbf{a})\mathbf{a}_{m+1}\cdots\mathbf{a}_{n+2l}\rangle_n.
\end{aligned} \tag{4.36}$$

Using (4.33), (4.34), (4.35), (4.36), we get

$$
\begin{aligned}
&-\frac{1}{l}\sum_{m+1\le\tau_2\le n+2l}\mathrm{sign}(\tau_2,\pi)\langle\tau_2(\mathbf{a})\rangle\,\langle\mathbf{a}_1\cdots\mathbf{a}_m\pi(\mathbf{a})\rangle_n\\
= &-\frac{1}{lC_{l-1}^{r-1}}\sum_{1\le\sigma_{2r-2}\le m}\ \sum_{m+1\le\tau_2\le n+2l}\Big(\mathrm{sign}(\sigma_{2r-2},\chi)\mathrm{sign}(\tau_2,\pi)\langle\sigma_{2r-2}\rangle\,\langle\tau_2(\mathbf{a})\rangle\\
&\langle\chi(\mathbf{a})\pi(\mathbf{a})\rangle_n\Big)-\frac{1}{l}\sum_{i=1}^{r-1}(-1)^{i+1}\frac{C_{r-1}^{i}}{C_{l-1}^{i}}\sum_{m+1\le\tau_2\le n+2l}\ \sum_{\{\phi_{2i}\}\subseteq\{\pi\}}\mathrm{sign}(\tau_2,\pi)\\
&\qquad\mathrm{sign}(\phi_{2i},\psi)\langle\tau_2(\mathbf{a})\rangle\,\langle\phi_{2i}(\mathbf{a})\rangle\,\langle\mathbf{a}_1\cdots\mathbf{a}_m\psi(\mathbf{a})\rangle_n\\
= &-\frac{l-r+1}{lC_{l-1}^{r-1}}\sum_{1\le\sigma_{2r-2}\le m}\mathrm{sign}(\sigma_{2r-2},\chi)\langle\sigma_{2r-2}\rangle\,\langle\chi(\mathbf{a})\mathbf{a}_{m+1}\cdots\mathbf{a}_{n+2l}\rangle_n\\
&+\frac{1}{lC_{l-1}^{r-1}}\sum_{1\le\sigma_{2r-2}\le m}\ \sum_{\{\xi_2\}\subseteq\{\chi\}}\mathrm{sign}(\sigma_{2r-2},\chi)\mathrm{sign}(\xi_2,\eta_{m-2r})\langle\sigma_{2r-2}(\mathbf{a})\rangle\\
&\qquad\langle\xi_2(\mathbf{a})\rangle\langle\eta_{m-2r}(\mathbf{a})\mathbf{a}_{m+1}\cdots\mathbf{a}_{n+2l}\rangle_n\\
&+\sum_{i=2}^{r}(-1)^{i+1}\frac{C_{r-1}^{i-1}}{C_l^i}\sum_{m+1\le\tau_{2i}\le n+2l}\mathrm{sign}(\tau_{2i},\check{\tau}_{2i})\langle\tau_{2i}(\mathbf{a})\rangle\langle\mathbf{a}_1\cdots\mathbf{a}_m\check{\tau}_{2i}(\mathbf{a})\rangle_n\\
= &-\frac{1}{C_l^{r-1}}\sum_{1\le\sigma_{2r-2}\le m}\mathrm{sign}(\sigma_{2r-2},\chi)\langle\sigma_{2r-2}\rangle\,\langle\chi(\mathbf{a})\mathbf{a}_{m+1}\cdots\mathbf{a}_{n+2l}\rangle_n\\
&+\frac{1}{C_l^r}\sum_{1\le\zeta_{2r}\le m}\mathrm{sign}(\zeta_{2r},\check{\zeta}_{2r})\langle\zeta_{2r}(\mathbf{a})\rangle\,\langle\check{\zeta}_{2r}(\mathbf{a})\mathbf{a}_{m+1}\cdots\mathbf{a}_{n+2l}\rangle_n\\
&+\sum_{i=2}^{r}(-1)^{i+1}\frac{C_{r-1}^{i-1}}{C_l^i}\sum_{m+1\le\tau_{2i}\le n+2l}\mathrm{sign}(\tau_{2i},\check{\tau}_{2i})\langle\tau_{2i}(\mathbf{a})\rangle\langle\mathbf{a}_1\cdots\mathbf{a}_m\check{\tau}_{2i}(\mathbf{a})\rangle_n.
\end{aligned}
\tag{4.37}
$$

Substituting (4.37) into the last term of (4.32), we get the first equality of (4.29) for $k=r$.

Case 5. $2r-2\le m$ but $2r>m$.

In this case, the right side of (4.35) contains only the first term, and (4.36) does not occur. The middle term on the right side of the last equality in (4.37) vanishes. The previous proof is still valid.

Case 6. $2r-2>m$.

In this case the first term on the right side of the last equality in (4.32) vanishes, so does the first term on the right side of (4.33). The proof can be finished by substituting (4.33), (4.34) into the last term of (4.32). This finishes the proof.

Theorem 4.2 in its form (4.2), which is on the change of $(0, n)$-typed summation, can be extended to other types of summations.

Theorem 4.3. (Change of (0,0)-typed summation)

$$\begin{aligned}&\sum_{m+1\leq\sigma_{2l-m}\leq 2k+2l} \operatorname{sign}(\sigma_{2l-m},\check{\sigma}_{2l-m})\langle \mathbf{a}_1\cdots\mathbf{a}_m\sigma_{2l-m}(\mathbf{a})\rangle\,\langle\check{\sigma}_{2l-m}(\mathbf{a})\rangle\\ &=\sum_{i=0}^{[\frac{m}{2}]}\sum_{1\leq\pi_{2i}\leq m} C_{k+l-m}^{k-i}\operatorname{sign}(\pi_{2i},\check{\pi}_{2i})\langle\pi_{2i}(\mathbf{a})\rangle\,\langle\check{\pi}_{2i}(\mathbf{a})\mathbf{a}_{m+1}\cdots\mathbf{a}_{2k+2l}\rangle.\end{aligned} \tag{4.38}$$

where the combinatorial symbol is extended to any integers k, l:

- *For any $0 \leq l \leq k$, $C_k^l = \frac{k!}{l!(k-l)!}$.*
- *For any $l > k$, or any $l < 0 \leq k$, $C_k^l = 0$.*
- *For any $l \geq k > 0$, $C_{-k}^{-l} = (-1)^{k+l}C_{l-1}^{k-1}$.*

Theorem 4.4. (Change of $(0, n)$-typed summation, general case) *Let $i \leq 2k$ and $j \leq n + 2l$. Let $u = n + 2k + 2l - i - j$. Then for any vectors $\mathbf{a}_1, \ldots, \mathbf{a}_i, \mathbf{b}_1, \ldots, \mathbf{b}_j$, $\mathbf{c}_1, \ldots, \mathbf{c}_u$,*

$$\begin{aligned}&\sum_{1\leq\sigma_{2k-i}\leq u}\operatorname{sign}(\sigma_{2k-i},\check{\sigma}_{2k-i})\langle\mathbf{a}_1\cdots\mathbf{a}_i\sigma_{2k-i}(\mathbf{c})\rangle\langle\mathbf{b}_1\cdots\mathbf{b}_j\check{\sigma}_{2k-i}(\mathbf{c})\rangle_n\\ =\;&(-1)^{i(j+1)}\sum_{m=[\frac{i+1}{2}]}^{\min(k,\,[\frac{i+j}{2}])}\sum_{1\leq\tau_{2m-i}\leq j} C_{k+l-m}^{l}\operatorname{sign}(\tau_{2m-i}^{\dagger},\check{\tau}_{2m-i})\\ &\qquad\langle\mathbf{a}_1\cdots\mathbf{a}_i\tau_{2m-i}^{\dagger}(\mathbf{b})\rangle\langle\check{\tau}_{2m-i}(\mathbf{b})\mathbf{c}_1\cdots\mathbf{c}_u\rangle_n.\end{aligned} \tag{4.39}$$

Theorem 4.5. ($(0, n)$-typed summation, special case) *Let $k, l \geq 0$, and let $u = n + 2k + 2l$. Define*

$$\begin{aligned}&T_{k,l}(\mathbf{a}_1\ldots\mathbf{a}_m)\\ &=\sum_{m+1\leq\sigma_{2k-m}\leq u}\operatorname{sign}(\tau_{2k-m},\check{\tau}_{2k-m})\langle\mathbf{a}_1\cdots\mathbf{a}_m\sigma_{2k-m}(\mathbf{a})\rangle\,\langle\check{\sigma}_{2k-m}(\mathbf{a})\rangle_n\end{aligned} \tag{4.40}$$

for any $0 \leq m \leq 2k$. Then

$$\begin{aligned}&T_{k,l}(\mathbf{a}_1\ldots\mathbf{a}_m)\\ &=\begin{cases}0, & \text{if}\quad m=2i+1, 0\leq i\leq k-1;\\ C_{k+l-i}^{l}\langle\mathbf{a}_1\cdots\mathbf{a}_{2i}\rangle\langle\mathbf{a}_{2i+1}\cdots\mathbf{a}_u\rangle_n, & \text{if}\quad m=2i, 0\leq i\leq k.\end{cases}\end{aligned} \tag{4.41}$$

Theorem 4.6. (Change of (n,n)-typed summation, general case) *Let $i \leq n+2k$ and $j \leq n+2l$. Let $u = 2n+2k+2l-i-j$. Then for any vectors $\mathbf{a}_1,\ldots,\mathbf{a}_i,\mathbf{b}_1,\ldots,\mathbf{b}_j$, $\mathbf{c}_1,\ldots,\mathbf{c}_u$,*

$$\begin{aligned}
&\sum_{1\leq\sigma=\sigma_{n+2k-i}\leq u} \mathrm{sign}(\sigma,\check{\sigma})\langle \mathbf{a}_1\cdots\mathbf{a}_i\sigma(\mathbf{c})\rangle_n \,\langle \mathbf{b}_1\cdots\mathbf{b}_j\check{\sigma}(\mathbf{c})\rangle_n \\
&= (-1)^{(n+i)(j+1)} \sum_{m=\max(0,\,[\frac{i+1-n}{2}])}^{\min(k,\,[\frac{i+j-n}{2}])} \sum_{1\leq\tau=\tau_{n+2m-i}\leq j} C^l_{k+l-m}\,\mathrm{sign}(\tau^\dagger,\check{\tau}) \\
&\qquad \langle \mathbf{a}_1\cdots\mathbf{a}_i\tau^\dagger(\mathbf{b})\rangle_n\,\langle\check{\tau}(\mathbf{b})\mathbf{c}_1\cdots\mathbf{c}_u\rangle_n.
\end{aligned} \tag{4.42}$$

Theorem 4.7. ((n,n)-typed summation, special case) *Let $k,l \geq 0$, and let $u = 2n+2k+2l$. Define*

$$U_{k,l}(\mathbf{a}_1\ldots\mathbf{a}_m) = \sum_{m+1\leq\sigma=\sigma_{n+2k-m}\leq u} \mathrm{sign}(\sigma,\check{\sigma})\langle \mathbf{a}_1\cdots\mathbf{a}_m\sigma(\mathbf{a})\rangle_n\,\langle\check{\sigma}(\mathbf{a})\rangle_n \tag{4.43}$$

for any $0 \leq m \leq n+2k$. Then

$$U_{k,l}(\mathbf{a}_1\ldots\mathbf{a}_m) = \begin{cases} 0, & \text{if } m<n; \\ 0, & \text{if } m=n+2i-1,\ 1\leq i\leq k; \\ C^l_{k+l-i}\langle\mathbf{a}_1\cdots\mathbf{a}_{n+2i}\rangle_n\,\langle\mathbf{a}_{n+2i+1}\cdots\mathbf{a}_u\rangle_n, & \text{if } m=n+2i,\ 0\leq i\leq k. \end{cases} \tag{4.44}$$

Caianiello's second identity [3, 4] is on the expansion of the Clifford multiplication of two extensors (outer products of vectors). Recently, this identity was implemented into the Maple package **CLIFFORD** [1] to simplify algebraic computation, called *Rota-Stein formula* [7].

Theorem 4.8. (Caianiello's second identity)

$$\begin{aligned}
&\langle(\mathbf{a}_1\wedge\cdots\wedge\mathbf{a}_r)(\mathbf{b}_1\wedge\cdots\wedge\mathbf{b}_s)\rangle_{r+s-2l} \\
&= \sum_{1\leq\sigma_l\leq r}\sum_{1\leq\tau_l\leq s} \mathrm{sign}(\check{\sigma}_l,\sigma_l)\mathrm{sign}(\tau_l,\check{\tau}_l)\,\langle\Lambda\sigma_l(\mathbf{a})\Lambda\tau_l(\mathbf{b})\rangle\,\Lambda\check{\sigma}_l(\mathbf{a})\wedge\Lambda\check{\tau}(\mathbf{b}).
\end{aligned} \tag{4.45}$$

The above identity should be extended to arbitrarily finitely many extensors, just as Caianiello's first identity. The extension was obtained by us recently [13].

Theorem 4.9. *Let $r_1, \ldots, r_k$ be positive integers. For $1 \leq i \leq k$, let $s_i = \sum_{h=1}^{i} r_h$. Let $\mathbf{a}_1, \ldots, \mathbf{a}_{s_k}$ be vectors, and let $A_{r_i}^{(i)} = \mathbf{a}_{s_{i-1}+1} \wedge \cdots \wedge \mathbf{a}_{s_i}$. Then for any $l \geq 0$,*

$$\begin{aligned}
&\langle A_{r_1}^{(1)} \cdots A_{r_k}^{(k)} \rangle_{s_k - 2l} \\
&= \sum_{\substack{i_1 + \cdots + i_k = 2l, \\ 0 \leq i_j \leq \min(l, r_j),\ 1 \leq j \leq k}} \quad \sum_{\substack{s_{j-1}+1 \leq \sigma_{i_j}^{(j)} \leq s_j, \\ 1 \leq j \leq k}} \operatorname{sign}(\sigma_{i_1}^{(1)}, \ldots, \sigma_{i_k}^{(k)}, \\
&\qquad \check{\sigma}_{i_1}^{(1)}, \ldots, \check{\sigma}_{i_k}^{(k)}) \, \langle \Lambda\sigma_{i_1}^{(1)}(\mathbf{a}) \cdots \Lambda\sigma_{i_k}^{(k)}(\mathbf{a}) \rangle \, \Lambda\check{\sigma}_{i_1}^{(1)}(\mathbf{a}) \wedge \cdots \wedge \Lambda\check{\sigma}_{i_k}^{(k)}(\mathbf{a}).
\end{aligned} \tag{4.46}$$

When s_k is even,

$$\begin{aligned}
&\langle A_{r_1}^{(1)} \cdots A_{r_k}^{(k)} \rangle \\
&= \sum_{\substack{i_2 + \cdots + i_k = r_1, \\ 0 \leq i_j \leq r_j,\ 2 \leq j \leq k}} \quad \sum_{\substack{s_{j-1}+1 \leq \sigma_{i_j}^{(j)} \leq s_j, \\ 2 \leq j \leq k}} \operatorname{sign}(\sigma_{i_2}^{(2)}, \ldots, \sigma_{i_k}^{(k)}, \check{\sigma}_{i_2}^{(2)}, \ldots, \check{\sigma}_{i_k}^{(k)}) \\
&\qquad \langle \Lambda\check{\sigma}_{i_2}^{(2)}(\mathbf{a}) \cdots \Lambda\check{\sigma}_{i_k}^{(k)}(\mathbf{a}) \rangle \, A_{r_1}^{(1)} \cdot (\Lambda\sigma_{i_2}^{(2)}(\mathbf{a}) \wedge \cdots \wedge \Lambda\sigma_{i_k}^{(k)}(\mathbf{a})).
\end{aligned} \tag{4.47}$$

From the above expansion formulas, we can obtain some nice summation formulas. The following are two examples.

Theorem 4.10. *Let $A_r = \mathbf{a}_1 \wedge \cdots \wedge \mathbf{a}_r$, $B_s = \mathbf{b}_1 \wedge \cdots \wedge \mathbf{b}_s$. Then*

$$\sum_{1 \leq \sigma_t \leq r} \operatorname{sign}(\sigma_t, \check{\sigma}_t) \langle \Lambda\sigma_t(\mathbf{a}) \, B_s \, \Lambda\check{\sigma}_t(\mathbf{a}) \rangle_{r+s-2l} = (-1)^{st} b(t, r, l) \langle B_s A_r \rangle_{r+s-2l}, \tag{4.48}$$

where $b(t, r, l)$ is the coefficient of x^t in $(1-x)^l(1+x)^{r-l}$.

Theorem 4.11. *For vectors $\mathbf{a}_1, \ldots, \mathbf{a}_r$,*

$$\sum_{1 \leq \sigma_t \leq r} \operatorname{sign}(\sigma_t, \check{\sigma}_t) \langle \sigma_t(\mathbf{a}) \check{\sigma}_t(\mathbf{a}) \rangle_{r-2l} = c(t, r, l) \langle \mathbf{a}_1 \cdots \mathbf{a}_r \rangle_{r-2l}, \tag{4.49}$$

where $c(t, r, l)$ is the coefficient of x^t in $(1+x^2)^l(1+x)^{r-2l}$.

Coming back from the expansions into polynomials in Clifford bracket algebra, let us consider the expansions into rational polynomials in null bracket algebra. Many rational binomial expansion and summation formulas have been discovered by us in the procedure of automated geometric theorem proving. Still, a complete classification of all such expansions is far from being established. The following are two rational binomial expansions and two binomial summations in 4D null bracket algebra. They are essential in an algebraic proof of Miquel's Five-Star Theorem [12] by hand.

$$
\begin{aligned}
& [\mathbf{a}_1\mathbf{a}_2\mathbf{a}_3\mathbf{a}_4\mathbf{a}_1\mathbf{a}_2\mathbf{a}_5\mathbf{a}_6][\mathbf{a}_1\mathbf{a}_2\mathbf{a}_3\mathbf{a}_5] \\
= \; & 2(\mathbf{a}_2\cdot\mathbf{a}_3[\mathbf{a}_1\mathbf{a}_2\mathbf{a}_5\mathbf{a}_6][\mathbf{a}_1\mathbf{a}_2\mathbf{a}_5\mathbf{a}_1\mathbf{a}_3\mathbf{a}_4] - \mathbf{a}_2\cdot\mathbf{a}_5[\mathbf{a}_1\mathbf{a}_2\mathbf{a}_3\mathbf{a}_4][\mathbf{a}_1\mathbf{a}_2\mathbf{a}_3\mathbf{a}_1\mathbf{a}_5\mathbf{a}_6]), \\
& \langle\mathbf{a}_1\mathbf{a}_2\mathbf{a}_3\mathbf{a}_4\mathbf{a}_1\mathbf{a}_2\mathbf{a}_5\mathbf{a}_6\rangle[\mathbf{a}_1\mathbf{a}_2\mathbf{a}_3\mathbf{a}_5] \\
= \; & 2(\mathbf{a}_2\cdot\mathbf{a}_3\langle\mathbf{a}_1\mathbf{a}_2\mathbf{a}_5\mathbf{a}_6\rangle[\mathbf{a}_1\mathbf{a}_2\mathbf{a}_5\mathbf{a}_1\mathbf{a}_3\mathbf{a}_4] - \mathbf{a}_2\cdot\mathbf{a}_5[\mathbf{a}_1\mathbf{a}_2\mathbf{a}_3\mathbf{a}_4]\langle\mathbf{a}_1\mathbf{a}_2\mathbf{a}_3\mathbf{a}_1\mathbf{a}_5\mathbf{a}_6\rangle), \\
& \langle\mathbf{a}_1\mathbf{a}_2\mathbf{a}_3\mathbf{a}_4\rangle[\mathbf{a}_1\mathbf{a}_2\mathbf{a}_3\mathbf{a}_1\mathbf{a}_5\mathbf{a}_6] - \langle\mathbf{a}_1\mathbf{a}_2\mathbf{a}_3\mathbf{a}_1\mathbf{a}_5\mathbf{a}_6\rangle[\mathbf{a}_1\mathbf{a}_2\mathbf{a}_3\mathbf{a}_4] \\
= \; & -2\mathbf{a}_1\cdot\mathbf{a}_2\,\mathbf{a}_2\cdot\mathbf{a}_3[\mathbf{a}_1\mathbf{a}_3\mathbf{a}_4\mathbf{a}_1\mathbf{a}_5\mathbf{a}_6], \\
& \langle\mathbf{a}_1\mathbf{a}_2\mathbf{a}_3\mathbf{a}_4\rangle\langle\mathbf{a}_1\mathbf{a}_2\mathbf{a}_3\mathbf{a}_1\mathbf{a}_5\mathbf{a}_6\rangle + [\mathbf{a}_1\mathbf{a}_2\mathbf{a}_3\mathbf{a}_1\mathbf{a}_5\mathbf{a}_6][\mathbf{a}_1\mathbf{a}_2\mathbf{a}_3\mathbf{a}_4] \\
= \; & -2\mathbf{a}_1\cdot\mathbf{a}_2\,\mathbf{a}_2\cdot\mathbf{a}_3\langle\mathbf{a}_1\mathbf{a}_3\mathbf{a}_4\mathbf{a}_1\mathbf{a}_5\mathbf{a}_6\rangle.
\end{aligned}
\tag{4.50}
$$

5. Conclusion

To find more efficient algebraic methods in automated geometric theorem proving we need invariant algebras, and to simplify algebraic manipulations we need more advanced invariants. Under this environment Clifford bracket algebra is proposed, whose basic algebraic elements are the scalar and pseudoscalar parts of the Clifford multiplications of vectors. Geometrically, this algebra is equipped with not only the basic invariants such as distances and areas, but also trigonometric functions of sums of oriented angles. The need to find factored and shortest expansions of the Clifford multiplication into inner and outer products leads to the study of Clifford expansions into polynomials and rational polynomials. Some important results are established, including the simplifications and extensions of Caianiello's expansion formulas. Some rational binomial expansion and summation formulas in null bracket algebra are also established. The classification of all such formulas is to be completed in the future.

References

[1] R. Ablamowicz. **CLIFFORD**, Maple package, http://math.tntech.edu/rafal

[2] M. Barnabei, A. Brini, G.-C. Rota. On the Exterior Calculus of Invariant Theory. *J. Algebra* **96**: 120–160, 1985.

[3] P. Bravi and A. Brini. Remarks on Invariant Geometric Calculus, Cayley-Grassmann Algebras and Geometric Clifford Algebras. Manuscript.

[4] E. Caianiello. *Combinatorics and Renormalization in Quantum Field Theory.* Benjamin, Reading, MA, 1973.

[5] S. C. Chou, X. S. Gao and J. Z. Zhang. *Machine Proofs in Geometry.* World Scientific, Singapore, 1994.

[6] H. Crapo and J. Richter-Gebert. Automatic Proving of Geometric Theorems. In: *Invariant Methods in Discrete and Computational Geometry*, N. White (ed.), pp. 107–139, Kluwer Academic Publishers, 1994.

[7] B. Fauser. A Treatise on Quantum Clifford Algebra, Univ. of Konstanz, Germany, 2002. Available at http://clifford.physik.uni-konstanz.de/~fauser

[8] D. Hestenes and G. Sobczyk. *Clifford Algebra to Geometric Calculus*, Kluwer, Dordrecht, 1984.

[9] B. Kutzler and S. Stifter. On the Application of Buchberger's Algorithm to Automated Geometry Theorem Proving. *J. Symbolic Computation* **2**: 389–398, 1986.

[10] H. Li. New Explorations in Automated Theorem Proving in Geometries, Ph. D. dissertation, Peking Univ., Beijing, 1994.

[11] H. Li, D. Hestenes and A. Rockwood. Generalized Homogeneous Coordinates for Computational Geometry. In *Geometric Computing with Clifford Algebras*, G. Sommer (ed.), pp. 27–60, Springer, Heidelberg, 2001.

[12] H. Li. Automated Theorem Proving in the Homogeneous Model with Clifford Bracket Algebra. In: *Applications of Geometric Algebra in Computer Science and Engineering*, L. Dorst et al. (eds), pp. 69–78, Birkhauser Boston, 2002.

[13] H. Li. Clifford Expansions and Summations. *MM Research Report* **21**, 2002. Available at http://www.mmrc.iss.ac.cn/annal/~li1.pdf

[14] B. Sturmfels and N. White. Gröbner Bases and Invariant Theory. *Adv. Math.* **76**: 245–259, 1989.

[15] B. Sturmfels and W. Whiteley. On the Synthetic Factorization of Homogeneous Invariants. *J. Symbolic Computation* **11**: 439–454, 1991.

[16] D. Wang. Geometric reasoning with geometric algebra. In: *Geometric Algebra with Applications in Science and Engineering*, E. Bayro-Corrochano et al (eds), pp. 89–109, Birkhauser, Boston, 2001.

[17] N. White. The Bracket Ring of Combinatorial Geometry I. *Trans. Amer. Math. Soc* **202**: 79–103, 1975.

[18] W. T. Wu. *Basic Principles of Mechanical Theorem Proving in Geometries, Volume I: Part of Elementary Geometries.* Science Press, Beijing 1984; Springer 1994.

[19] I. Yaglom. *Felix Klein and Sophus Lie*, Birkhauser, Boston, Basel, 1988.

Hongbo Li
Academy of Mathematics and System Sciences
Chinese Academy of Sciences
Beijing 100080
China
e-mail: hli@mmrc.iss.ac.cn

Trends in Mathematics: *Advances in Analysis and Geometry*, 365–376

Quaternion-valued Smooth Orthogonal Wavelets with Short Support and Symmetry

Lizhong Peng and Jiman Zhao

Abstract. In this paper, we define the quaternion-valued multiresolution analysis of $L^2(\mathbb{R}, H)$. We give the properties of the scaling functions, wavelet functions and their corresponding low-pass and high-pass filters, and present a sufficient condition for the existence of the quaternion-valued wavelet. By solving the system of equations, we obtain some kinds of low-pass filters and high-pass filters with short support and symmetry of smooth orthogonal wavelets. We also construct quaternion-valued wavelets of a quaternion variable on $L^2(H, H)$.

Keywords. Quaternion-valued multiresolution analysis, scaling function, wavelet function, low-pass filter, high-pass filter.

1. Introduction

Wavelets, multi-wavelets, multi-band wavelets, and matrix (vector) valued wavelets have been extensively studied. Many papers and books on these topics appeared, for example [5], [6], [10], [11], [13], [14], [16], etc.

It is well known that an orthogonal wavelet function with compact support and certain regularity can not have any symmetry (see [5], [6]), but multi-wavelets, multi-band wavelets, and matrix (vector) valued wavelets may have symmetry properties as well as other good properties.

In this paper we construct quaternion-valued wavelets, which, among other good properties, have the symmetry properties. In [8], by identifying the quaternion-valued function with the duplex complex matrix-valued function, J. X. He and B. Yu introduced quaternion-valued multiresolution analysis of $L^2(\mathbb{R}, H)$, and constructed three kinds of quaternion-valued scaling functions and wavelet functions using a method similar to what is given in [15] (in matrix viewpoint). Only the trivial one of them had the symmetry property. In this paper we offer an approach that directly constructs quaternion-valued scaling and wavelet functions rather than by identifying the quaternion with the duplex complex matrix.

Firstly we give the definition of quaternion-valued multiresolution analysis of $L^2(\mathbb{R}, H)$, scaling function, wavelet function and the corresponding low-pass and high-pass filters. Then we study the properties of the scaling function and its corresponding low-pass filter, and present a sufficient condition for the existence of quaternion-valued wavelets. Considering the orthonormal property of the scaling function, and by adding the compacted support, symmetry, proper smoothness to the scaling function, we get a system of equations for the coefficients of the corresponding low-pass filter. By solving these equations, we obtain low-pass filters. The cascade algorithm is still valid if substituting the real-valued coefficients with quaternion-valued coefficients. Finally we get the smooth scaling functions with short support and symmetry. We also get a formula to give the corresponding high-pass filters and orthogonal wavelets with smoothness, short support and symmetry. By using quaternion-valued wavelets on $L^2(\mathbb{R}, H)$, we also construct quaternion-valued wavelets of a quaternion variable on $L^2(H, H)$. In this case, there are one scaling function and 15 wavelet functions.

As a special Clifford algebra, the quaternion algebra H over R is an associative, non-commutative and division algebra. It has the basis: e_0, e_1, e_2, e_3, which satisfies $e_1^2 = e_2^2 = -1, e_3 = e_1e_2$, and $e_1e_2 = -e_2e_1$.

Let $a = \sum_{k=0}^3 a_ke_k$ and $b = \sum_{k=0}^3 b_ke_k$. Then their product ab is written as

$$\begin{aligned} ab &= (a_0b_0 - a_1b_1 - a_2b_2 - a_3b_3)e_0 + (a_1b_0 + a_0b_1 - a_3b_2 + a_2b_3)e_1 \\ &\quad +(a_2b_0 + a_3b_1 + a_0b_2 - a_1b_3)e_2 + (a_3b_0 - a_2b_1 + a_1b_2 + a_0b_3)e_3. \end{aligned} \tag{1.1}$$

The conjugate quaternion to a is defined by $\overline{a} = a_0e_0 - \sum_{j=1}^3 a_je_j$ and the norm by $|a| = (a\overline{a})^{\frac{1}{2}}$.

It is easy to prove that

$$|ab| = |a||b|. \tag{1.2}$$

For more details about quaternions, see ([7]), and for general Clifford algebras, see ([1]).

Similar to the definition of Clifford modules in [1], we have the following definition.

Definition 1.1. A *quaternion module* $L^2(\mathbb{R}, H)$ is denoted by

$$L^2(\mathbb{R}, H) = \{f : \mathbb{R} \to H, f(x) = \sum_{k=0}^{3} f_k(x)e_k \mid f_k \in L^2(\mathbb{R}), k = 0, 1, 2, 3\}.$$

The inner product of $L^2(\mathbb{R}, H)$ is denoted by

$$(f, g) = \int_{\mathbb{R}} f(x)\overline{g}(x)dx.$$

Like in the classical case, we observe that for $f(x), g(x) \in L^2(\mathbb{R}, H)$, the Planchette formula is still valid, i.e.,

$$\int_{\mathbb{R}} f(x)\overline{g}(x)dx = \int_{\mathbb{R}} \hat{f}(\xi)\overline{\hat{g}}(\xi)d\xi, \tag{1.3}$$

where the Fourier transform for each real-valued function in quaternion-valued functions is defined as in [6] with $\frac{1}{\sqrt{2\pi}}$. For continuous wavelet transforms on $L^2(\mathbb{R}, H)$, $L^2(\mathbb{R}^2, H)$, and for Clifford-valued admissible wavelets, see [9], [17], [2], [3], [4].

2. Quaternion-valued Multiresolution Analysis

In [12], Marius Mitrea introduced one kind of Clifford-valued multiresolution analysis. There he constructs Clifford-valued higher dimensional orthogonal wavelets-like bases. Instead of high dimension, here we define quaternion-valued multiresolution analysis (QMRA) of one dimension on $L^2(\mathbb{R}, H)$, which is more similar to the conventional multiresolution analysis (MRA).

Definition 2.1. A *multiresolution analysis* of $L^2(\mathbb{R}, H)$ (QMRA) consists of a sequence $\{V_m\}_{m\in\mathbb{Z}}$ of embedded closed subspaces of $L^2(\mathbb{R}, H)$ satisfying the following conditions:

1. $\cdots \subset V_{-2} \subset V_{-1} \subset V_0 \subset V_1 \subset V_2 \subset \cdots$
2. $\overline{\bigcup_{j\in\mathbb{Z}} V_j} = L^2(\mathbb{R}, H)$, $\bigcap_{j\in\mathbb{Z}} V_j = \{0\}$
3. $f(x) \in V_j$ if and only if $f(2x) \in V_{j+1}, \forall j \in \mathbb{Z}$
4. there exists a function $\varphi \in V_0$ such that $\{\varphi_{0,n}(x) = \varphi(x-n) : n \in \mathbb{Z}\}$ is an orthonormal basis for V_0, that is, for all $f \in V_0$,

$$\|f\|^2 = \int_{-\infty}^{+\infty} |f(x)|^2 dx = \sum_{n\in\mathbb{Z}} |(f, \varphi_{0,n})|^2.$$

The function φ is called the *quaternion-valued scaling function* (or father wavelet) for the QMRA.

If $\{V_j\}_{j\in\mathbb{Z}}$ is a multiresolution analysis of $L^2(\mathbb{R}, H)$ and if V_0 is the closed subspace generated by integer translation of a single function φ, then we say that φ *generates* the multiresolution analysis. Then it follows from 3. that $\{\varphi_{j,n}(x) = 2^{\frac{j}{2}}\varphi(2^j x - n)\}_{n\in\mathbb{Z}}$ is an orthonormal basis for V_j.

The real importance of a QMRA lies in the fact that it enables us to construct an orthonormal basis for $L^2(\mathbb{R}, H)$. In order to prove this statement, first we assume that $\{V_j\}_{j\in\mathbb{Z}}$ is a QMRA of $L^2(\mathbb{R}, H)$.

Since $V_m \subset V_{m+1}$, we define W_m as the orthogonal complement of V_m in V_{m+1}. Like in the conventional MRA, we have $f \in W_m$ if and only if $f(2\cdot) \in W_{m+1}$ and

$$L^2(\mathbb{R}, H) = \oplus_{m\in\mathbb{Z}} W_m.$$

Now we are going to show that there is a function $\psi \in W_0$ such that $\{\psi_{0,n}, n \in \mathbb{Z}\}$ is an orthonormal basis for W_0 (then it follows from the definition of QMRA that $\{\psi_{j,k}, j, k \in \mathbb{Z}\}$ is an orthonormal basis of $L^2(\mathbb{R}, H)$).

Following the conventional approach, by using the non-commutativity product (1.1) and the Plancherel formula (1.3) of quaternions carefully, we obtain the following two propositions (omit the arguments).

Proposition 2.2. *For any function $f \in L^2(\mathbb{R}, H)$, the following conditions are equivalent:*

1. *The system $\{f_{0,n}(x) = f(x-n), n \in \mathbb{Z}\}$ is orthonormal.*
2. $\sum_{k\in\mathbb{Z}} |\hat{f}(\xi + 2k\pi)|^2 = 1$, *a.e.* $\xi \in \mathbb{R}$.

Proposition 2.3. *For any two functions $f, g \in L^2(\mathbb{R}, H)$, the set of functions $\{f_{0,n}(x) = f(x-n), n \in \mathbb{Z}\}$ and $\{g_{0,n}(x) = g(x-n), n \in \mathbb{Z}\}$ are orthogonal, i.e., $(f_{0,n}, g_{0,m}) = \delta_{n,m}, n, m \in \mathbb{Z}$, if and only if $\sum_{k\in\mathbb{Z}} \hat{f}(\xi + 2k\pi)\bar{\hat{g}}(\xi + 2k\pi) = 0$, a.e. $\xi \in \mathbb{R}$.*

From Definition 2.1, we have

$$\varphi(x) = \sqrt{2}\textstyle\sum_{n\in\mathbb{Z}} h_n \varphi(2x-n), \tag{2.1}$$

$$\psi(x) = \sqrt{2}\textstyle\sum_{n\in\mathbb{Z}} g_n \varphi(2x-n), \tag{2.2}$$

where $h_n, g_n, n \in \mathbb{Z}$ are quaternion-valued constants, and $\{h_n\}$ and $\{g_n\}$ are called *low-pass filter* and *high-pass filter*, respectively.

Using the above two propositions, we obtain the following so-called orthogonality conditions (which are similar to [6] and [16]):

$$|H(\xi)|^2 + |H(\xi+\pi)|^2 = 1, \text{ a.e. } \xi \in \mathbb{R}, \tag{2.3}$$

$$G(\xi)\overline{H}(\xi) + G(\xi+\pi)\overline{H}(\xi+\pi) = 0, \text{ a.e. } \xi \in \mathbb{R}, \tag{2.4}$$

$$|G(\xi)|^2 + |G(\xi+\pi)|^2 = 1, \text{ a.e. } \xi \in \mathbb{R}, \tag{2.5}$$

where

$$H(\xi) = \frac{1}{\sqrt{2}}\sum_{n\in\mathbb{Z}} h_n e^{-in\xi}, \; G(\xi) = \frac{1}{\sqrt{2}}\sum_{n\in\mathbb{Z}} g_n e^{-in\xi};$$

the above are also called quaternion-valued low-pass filter and high-pass filter, respectively.

In conventional MRA, the high-pass filter $\{g_n\}$ can be obtained simply from the low-pass filter $\{h_n\}$, the existence of correspondent high-pass filter needs to be proved, and it needs also to be designed after the low-pass filter in our case. Its existence can be proved by a similar argument for the case of a matrix-valued situation in [16]. By the existence of $G(\xi)$, we obtain the existence of a quaternion-valued wavelet function.

Proposition 2.4. *Let $\psi(x)$ be a quaternion-valued function, let its Fourier transform satisfy*

$$\hat{\psi}(\xi) = G(\frac{\xi}{2})\hat{\varphi}(\frac{\xi}{2})$$

where $G(\xi)$ fulfills (2.4)), (2.5). Then $\{\psi_{0,n}(x), n \in \mathbb{Z}\}$ forms an orthonormal basis for W_0.

Using the properties of the scaling function φ and the product of quaternions, we have the following proposition on the coefficients $h_n, n \in \mathbb{Z}$ of the low-pass filter.

Proposition 2.5. (a) $\sum_n h_n = \sqrt{2}$

(b) $\sum_n h_n \overline{h}_{n+2k} = \delta_{0,k},\ k \in \mathbb{Z}$

(c) $\sum_n (-1)^n h_n = 0$

(d) $\sum_n (-1)^n n^l h_n = 0,\ l = 0, 1, 2, \ldots, p-1$

where $p \leq [m/2] - 1$, m is the length of the low-pass filter, and it is not to hard to see that (d) *is equivalent to*

(e) $\frac{d^k H(\xi)}{d\xi^k}|_{\xi=\pi} = 0, k = 0, 1, \ldots, p-1.$

Remark. (a) and (b) imply that

$$(f)\ \sum_n h_{2n} = \sum_n h_{2n+1} = \frac{1}{\sqrt{2}}.$$

Similar to classical wavelet theory, (e) can be explained as the vanishing moments of the scaling function and the wavelet function, which guarantees the smoothness of them.

3. Construction of smooth symmetric orthogonal wavelets with short support

In [8], the authors gave some design of wavelets by using the method given in [15], there the scaling function has no symmetry. In this section, by solving the system of equations, we construct some smooth orthogonal wavelets with short support and symmetry.

To get such scaling functions, first we construct the corresponding low-pass filters. Denote

$$H(\xi) = \frac{1}{\sqrt{2}} \sum_{n=0}^{2N+1} h'_n e^{-in\xi} = \frac{1}{\sqrt{2}} (\sum_{n=0}^{N} h_n e^{-in\xi} + \sum_{k=N+1}^{2N+1} h_{2N+1-k} e^{-ik\xi}), \tag{3.1}$$

where $h'_n = h_n$, $n = 0, \ldots, N$; $h'_n = h_{2N+1-n}$, $n = N+1, \ldots, 2N+1$; while $h_n = h_{n0} + h_{n1}e_1 + h_{n2}e_2 + h_{n3}e_3$ are quaternions, $n = 0, \ldots, N$, $N \in \mathbb{Z}$.

Now using Proposition 2.5 (adding proper vanishing moments) we solve the system of equations on $\{h_n\}_{n=0}^N$.

Design 1. Let $N = 1$, there are four coefficients satisfying the following equations:

$$2(h_0 + h_1) = \sqrt{2}, \quad 2(|h_0|^2 + |h_1|^2) = 1, \quad h_0\overline{h_1} + h_1\overline{h_0} = 0,$$

i.e.,

$$h_0 + h_1 = \frac{\sqrt{2}}{2}, \tag{3.2}$$

$$|h_0|^2 + |h_1|^2 = \frac{1}{2}, \tag{3.3}$$

$$h_{00}h_{10} + h_{01}h_{11} + h_{02}h_{12} + h_{03}h_{13} = 0. \tag{3.4}$$

(3.2), (3.3) and (3.4) yield the following equation:

$$h_{00}^2 + h_{01}^2 + h_{02}^2 + h_{03}^2 = \frac{1}{\sqrt{2}} h_{00}. \tag{3.5}$$

Thus we obtain

$$h_{10} = \tfrac{1}{\sqrt{2}} - h_{00}, \quad h_{11} = -h_{01},$$

$$h_{12} = -h_{12}, \qquad h_{13} = -h_{03},$$

which satisfies

$$h_{01}^2 + h_{02}^2 + h_{03}^2 = \frac{1}{\sqrt{2}} h_{00} - h_{00}^2, \tag{3.6}$$

where $h_{00}, h_{01}, h_{02}, h_{03}$ are real-valued. So there are three free parameters.

From (3.6), we know that $h_{00} \in (0, \frac{1}{\sqrt{2}}]$, and when $h_{00} = \frac{\sqrt{2}}{4}, \frac{\sqrt{2}}{2}$, the right side of (3.6) reaches its maximal value $\frac{1}{8}$ and its minimal value 0, respectively. If $h_{00} = \frac{\sqrt{2}}{2}$, the solutions of the above equations degenerate to the classical Haar low-pass filter.

Remark. In this design, there are no solutions if we add vanishing moments.

Design 2. When $N = 2$, there are 6 coefficients. Using 2 vanishing moments (p=2), we have the following equations:

$$2(h_0 + h_1 + h_2) = \sqrt{2}, \tag{3.7}$$

$$2(|h_0|^2 + |h_1|^2 + |h_2|^2) = 1, \tag{3.8}$$

$$h_0\overline{h_2} + h_1\overline{h_2} + h_2\overline{h_1} + h_2\overline{h_0} = 0, i.e.\ Re(h_0\overline{h_2}) + Re(h_1\overline{h_2}) = 0, \tag{3.9}$$

$$5h_0 - 3h_1 + h_2 = 0. \tag{3.10}$$

(3.7) and (3.10) yield

$$h_1 = \frac{\sqrt{2}}{8} + h_0, \tag{3.11}$$

$$h_2 = \frac{3}{8}\sqrt{2} - 2h_0. \tag{3.12}$$

Substituting (3.11) and (3.12) into (3.8), we get

$$h_{00}^2 + h_{01}^2 + h_{02}^2 + h_{03}^2 - \frac{5}{24}\sqrt{2}h_{00} = \frac{1}{32}, \tag{3.13}$$

Now substituting (3.11) and (3.12) into (3.9), we obtain

$$-h_{00}^2 - h_{01}^2 - h_{02}^2 - h_{03}^2 + \frac{\sqrt{2}}{8} h_{00} + \frac{3}{128} = 0, \tag{3.14}$$

(3.13) and (3.14) yield

$$h_{00} = -\frac{3}{64}\sqrt{2}. \tag{3.15}$$

Substituting (3.15) into (3.13), we obtain

$$h_{01}^2 + h_{02}^2 + h_{03}^2 = \frac{15}{32 \times 64}. \tag{3.16}$$

So the solutions are as follows:

$$\begin{aligned} &h_{00} = -\tfrac{3}{64}\sqrt{2}, && h_{01}^2 + h_{02}^2 + h_{03}^2 = \tfrac{15}{32\times 64},\\ &h_{10} = \tfrac{5}{64}\sqrt{2}, && h_{11} = h_{01},\ h_{12} = h_{02},\ h_{13} = h_{03},\\ &h_{20} = \tfrac{15}{32}\sqrt{2}, && h_{21} = -2h_{01},\ h_{22} = -2h_{02},\ h_{23} = -2h_{03}, \end{aligned}$$

where $h_{01},\ h_{02},\ h_{03}$ are real-valued.

There are two free parameters in these solutions.

If we set $h_{01} = h_{02} = h_{03}$, the we can get one solution:

$$\begin{aligned} h_0 &= -\frac{3}{64}\sqrt{2} - \frac{\sqrt{10}}{64}e_1 - \frac{\sqrt{10}}{64}e_2 - \frac{\sqrt{10}}{64}e_3;\\ h_1 &= \frac{5}{64}\sqrt{2} - \frac{\sqrt{10}}{64}e_1 - \frac{\sqrt{10}}{64}e_2 - \frac{\sqrt{10}}{64}e_3;\\ h_2 &= \frac{15}{32}\sqrt{2} + 2\frac{\sqrt{10}}{64}e_1 + 2\frac{\sqrt{10}}{64}e_2 + 2\frac{\sqrt{10}}{64}e_3. \end{aligned}$$

Design 3. When $N = 3$, there are 8 coefficients. Using 3 vanishing moments (p=3), we solve the following equations:

$$2(h_0 + h_1 + h_2 + h_3) = \sqrt{2}, \tag{3.17}$$

$$2(|h_0|^2 + |h_1|^2 + |h_2|^2 + |h_3|^2) = 1, \tag{3.18}$$

$$h_0\overline{h}_2 + h_1\overline{h_3} + h_2\overline{h}_3 + h_3\overline{h_2} + h_3\overline{h}_1 + h_2\overline{h_0} = 0,$$

i.e.,

$$Re(h_0\overline{h}_2) + Re(h_1\overline{h}_3) + Re(h_2\overline{h}_3) = 0, \tag{3.19}$$

$$-7h_0 + 5h_1 - 3h_2 + h_3 = 0, \tag{3.20}$$

$$-343h_0 + 215h_1 - 117h_2 + 37h_3 = 0. \tag{3.21}$$

(3.17) and (3.20) yield

$$2h_0 - h_1 + h_2 = \frac{\sqrt{2}}{8}, \tag{3.22}$$

(3.20) and (3.21) yield

$$14h_0 - 5h_1 + h_2 = 0. \tag{3.23}$$

By (3.22) and (3.23), we have

$$h_1 = \frac{\sqrt{2}}{32} + 3h_0. \tag{3.24}$$

Substituting (3.24) into (3.22), we have

$$h_2 = \frac{5\sqrt{2}}{32} + h_0. \tag{3.25}$$

Substituting (3.24) and (3.25) into (3.20), we obtain

$$h_3 = \frac{10\sqrt{2}}{32} - 5h_0. \tag{3.26}$$

(3.24), (3.25) , (3.26), (3.18) yield

$$h_{00}^2 + h_{01}^2 + h_{02}^2 + h_{03}^2 - \frac{42\sqrt{2}}{32 \times 18} h_{00} = \frac{130}{32^2 \times 18}. \tag{3.27}$$

(3.24), (3.25) , (3.26), (3.19) yield

$$-h_{00}^2 - h_{01}^2 - h_{02}^2 - h_{03}^2 + \frac{15\sqrt{2}}{32 \times 19} h_{00} = -\frac{120}{32^2 \times 19}. \tag{3.28}$$

By (3.27) and (3.28), we obtain

$$h_{00} = -\frac{155\sqrt{2}}{32 \times 528}. \tag{3.29}$$

Substituting (3.29) into (3.27), we get

$$h_{01}^2 + h_{02}^2 + h_{03}^2 = \frac{1583470}{32^2 \times 528^2}. \tag{3.30}$$

Thus by (3.24), (3.25), (3.26), (3.29) and (3.30), we obtain the solutions of the equations:

$$h_{00} = -\tfrac{155\sqrt{2}}{32\times 528}, \qquad h_{01}^2 + h_{02}^2 + h_{03}^2 == \tfrac{1583470}{32^2\times 528^2},$$

$$h_{10} = \tfrac{\sqrt{2}}{32} + 3h_{00} = \tfrac{63\sqrt{2}}{32\times 528}, \qquad h_{11} = 3h_{01},\ h_{12} = 3h_{02},\ h_{13} = 3h_{03},$$

$$h_{20} = \tfrac{5\sqrt{2}}{32} + h_{00} = \tfrac{2485\sqrt{2}}{32\times 528}, \qquad h_{21} = h_{01},\ h_{22} = h_{02},\ h_{23} = h_{03},$$

$$h_{30} = \tfrac{10\sqrt{2}}{32} - 5h_{00} = \tfrac{6055\sqrt{2}}{32\times 528}, \quad h_{21} = -5h_{01},\ h_{22} = -5h_{02},\ h_{23} = -5h_{03},$$

where, h_{01}, h_{02}, h_{03} are real-valued.

There are two free parameters in these solutions.

If we set $h_{01} = h_{02} = h_{03} > 0$, the we get one solution:

$$h_0 = -0.01297366845 + 0.04299920158e_1 + 0.04299920158e_2 + 0.04299920158e_3,$$

$$h_1 = 0.00527316846 + 0.12899760476e_1 + 0.12899760476e_2 + 0.12899760476e_3,$$

$$h_2 = 0.20799720066 + 0.04299920158e_1 + 0.04299920158e_2 + 0.04299920158e_3,$$

$$h_3 = 0.5068100805 - 0.21499600794e_1 - 0.21499600794e_2 - 0.21499600794e_3.$$

Denote

$$G(\xi) = \frac{1}{\sqrt{2}} \sum_{n=0}^{2N+1} g_n e^{-in\xi}, \tag{3.31}$$

and set

$$G(\xi) = e^{-i(2N+1)\xi}\overline{H}(\xi+\pi), \tag{3.32}$$

where $H(\xi)$ is defined by (3.1). Then it can be written in the form

$$G(\xi) = \frac{1}{\sqrt{2}} \sum_{n=0}^{2N+1} (-1)^{1-n}\overline{h'}_{2N+1-n}e^{-in\xi}.$$

Comparing with (3.31), we get

$$g_n = (-1)^{1-n}\overline{h'}_{2N+1-n},\ n = 0,\cdots,2N+1,$$

Therefore $g_n = (-1)^{1-n}\overline{h}_n$, $n = 0,\cdots,N$; $g_n = (-1)^{1-n}\overline{h}_{2N+1-n}$, $n = N+1,\cdots,2N+1$.

If $H(\xi)$ is one of the above designs, we can prove by direct calculation that $G(\xi)$ given by (3.32) satisfies (2.4) and (2.5). Therefore, by Proposition 2.4, we get three kinds of wavelets $\psi(x) = \sqrt{2}\sum_{n=0}^{2N+1} g_n\varphi(2x-n)$.

We say that there is also a cascade algorithm in our case. In order to get the cascade algorithm, first we prove the following theorem.

Theorem 3.1. *Suppose $\phi(x) \in L^1(\mathbb{R}, H)$ with compact support in $[-M, M]$, and satisfying $\int_{\mathbb{R}} \phi(x)dx = 1$.*

(1) *If f is a continuous quaternion-valued function, then for all $x \in \mathbb{R}$,*

$$\lim_{j\to\infty} 2^j \int_R f(x+y)\overline{\phi}(2^j y)dy = f(x). \tag{3.33}$$

(2) *If f is uniformly continuous, then (3.33) constitutes uniform convergence.*

(3) *If f is Hölder continuous with exponent α :*

$$|f(x) - f(y)| \le C|x-y|^{\alpha},$$

then the convergence is exponentially fast in j :

$$|f(x) - 2^j \int_{\mathbb{R}} f(x+y)\overline{\phi}(2^j y)dy| \le C2^{-j\alpha}.$$

Proof. By the properties of ϕ, we have

$$\begin{aligned} &|f(x) - 2^j \textstyle\int_{\mathbb{R}} f(x+y)\overline{\phi}(2^j y)dy| \\ &= |2^j \textstyle\int_{\mathbb{R}}[f(x) - f(x+y)]\overline{\phi}(2^j y)dy| \\ &= |\textstyle\int_{\mathbb{R}}[f(x) - f(x+2^{-j}z)]\overline{\phi}(z)dz|. \end{aligned}$$

Denote

$$\sum_{r=0}^{3} h_r(z) = [f(x) - f(x+2^{-j}z)]\overline{\phi}(z),$$

then by (1.1), we have

$$|h_r(z)| \le 4|f(x) - f(x+2^{-j}z||\phi(z)|, \tag{3.34}$$

thus

$$\begin{aligned}
&|f(x) - 2^j \textstyle\int_{\mathbb{R}} f(x+y)\overline{\phi}(2^j y)dy| \\
&= |\textstyle\sum_{r=0}^3 \int_{\mathbb{R}} h_r(z)dze_r| = (\sum_{r=0}^3 |\int_{\mathbb{R}} h_r(z)dz|^2)^{\frac{1}{2}} \\
&\leq \textstyle\sum_{r=0}^3 |\int_{\mathbb{R}} h_r(z)dz| \leq \sum_{r=0}^3 \int_{\mathbb{R}} |h_r(z)|dz \\
&\leq 16 \textstyle\int_{\mathbb{R}} |f(x) - f(x+2^{-j}z)||\phi(z)|dz \text{ by } (3.34) \\
&\leq 16\|\phi\|_1 sup_{u \leq 2^{-j}M}|f(x) - f(x+u)|.
\end{aligned}$$

The theorem is proved. □

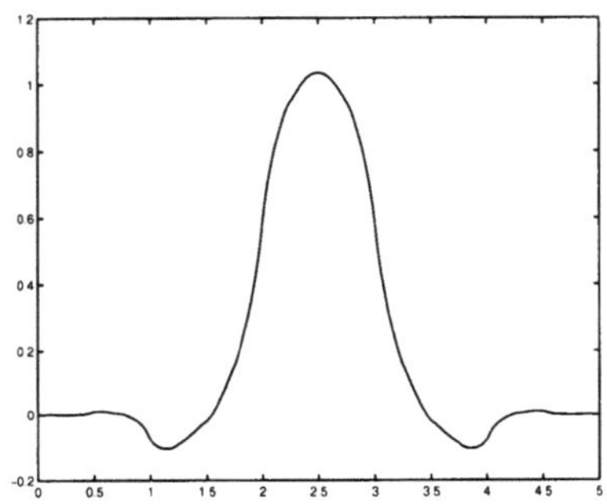

$\varphi_0(6)$

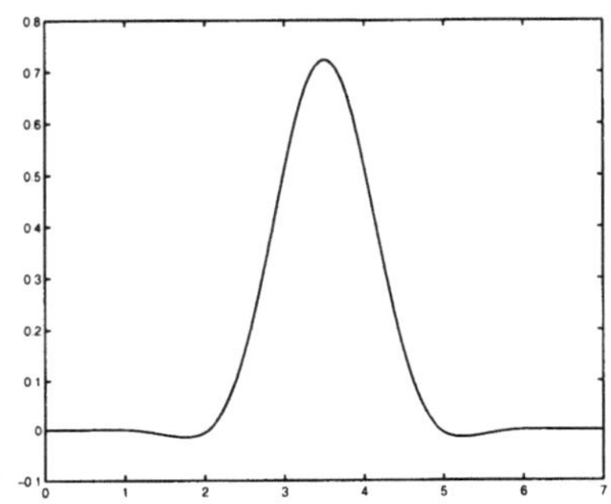

$\varphi_0(8)$

Using this theorem, (1.1) and the same technique as in [6], we can deduce our cascade algorithm, i.e.,

$$\eta_{j+1}(x) = \sqrt{2}\sum_n h_n \eta_j(2x-n),\ \eta_0(x) = \chi_{[0,\frac{1}{2}]}(x),\ \varphi(x) = \lim_{j\to+\infty} \eta_j(x),$$

where h_n are the quaternion-valued coefficients of the low-pass filter. Therefore the corresponding scaling functions can be obtained by this algorithm.

As the quaternion-valued cascade algorithm is more complicated than the conventional one, it is difficult to draw the graphs of these scaling and wavelets functions. The above two graphs are the two scaling functions which are obtained from the conventional cascade algorithm by using the real part of coefficients in the above special low-pass filters ($N = 2$ and $N = 3$). We can get some idea about the smoothness of the scaling functions and wavelet functions we designed.

Finally, by using the above quaternion-valued wavelets on $L^2(\mathbb{R}, H)$, we are going to construct the quaternion-valued wavelets of quaternion variable on $L^2(H, H)$. For the quaternion variable $xe_0 + ye_1 + ue_2 + ve_3$, there are one scaling function and 15 wavelet functions. Let $\psi^0 = \phi$, $\psi^1 = \psi$. For $\epsilon \in \{0,1\}^4$, we define

$$\Psi^\epsilon(x,y,u,v) = \psi^{\epsilon_1}(x)\psi^{\epsilon_2}(y)\psi^{\epsilon_3}(u)\psi^{\epsilon_4}(v).$$

Then there are a total of 16 functions supported on a cube, where the function $\Psi^{0,0,0,0}$ is the scaling function, the other 15 functions Ψ^ϵ, $\epsilon \in \{0,1\}^4 \backslash (0,0,0,0)$

are the wavelet functions. We can also select different scaling function and wavelet function for each component of the quaternion variable, we obtain the functions $\Psi^{\epsilon}(x, y, u, v)$ supported on a rectangle.

References

[1] F. Brackx, R. Delanghe, and F. Sommen, *Clifford Analysis*. Pitman Publ., 1982.

[2] F. Brackx, F. Sommen, Clifford-Hermite Wavelets in Euclidean Space, *The Journal of Fourier Analysis and Applications,* **6** (**3**) (2000), 209–310.

[3] F. Brackx, F. Sommen, *The continuous wavelet transform in Clifford analysis*, Clifford Analysis and Its Applications, NATO ARW Series, F.Brackx, JSR. Chisholm, V.Soucek eds., Kluwer Acad. Publ., Dordrecht, The Netherlands (2001) 9–26.

[4] F. Brackx, F. Sommen, The generalized Clifford-Hermite continuous wavelet transform, Advances in Applied Clifford Algebras **11** (S) (2001), 219–231. (Proceedings of the Cetraro conference in October 1998).

[5] C. K. Chui, *An Introduction to Wavelets,* Academic Press, New York, 1992.

[6] I. Daubechies, *Ten Lectures on Wavelets*, Soc. Indust. & Appl. Math., Philadelphia, 1992.

[7] K. Gürlebeck, W. Sprößig, *Quaternion Analysis and Elliptic Boundary Value Problems,* Birkhäuser Verlag, Basel, Berlin, 1990.

[8] J. X. He, B. Yu, *Wavelet Analysis of Quaternion-valued Functions*, Preprint.

[9] J. X. He, B. Yu, Continuous Wavelet Transforms on the Space $L^2(\mathbb{R}, H; dx)$, to appear in *Appl. Math. Letters.*

[10] Q. T. Jiang, Orthogonal multiwavelets with optimum time-frequency resolution, *IEEE Trans. on Signal Proc., Special Issue on "Filter banks and wavelets: Theory and applications"*, **46** (1998), 830–844.

[11] S. Mallat, *A Wavelet tour of signal processing,* 2nd ed. (English), San Diego, Academic Press. xxiv, p. 637, 1999.

[12] M. Mitrea, *Clifford wavelets, Singular Integrals and Hardy space,* Springer-Verlag, Berlin, Heidelberg, 1994.

[13] L. Z. Peng, Y. G. Wang, *Construction of Compactly Suppotred Orthonormal Wavelets with Beautiful Structure*, Research Report, No. 90, Peking University, 2002.

[14] G. Strang, V. Strela, Short Wavelets and Matrix Dilation Equations, *IEEE Trans. Signal Processing,* **43** (1995), 108–115.

[15] A. T. Walden, A. Z. Serroukh, Wavelet analysis of matrix-valued time-series, *Proc. R. Soc. Lond. A,* **458** (2002), 157–179.

[16] X. G. Xia, W. Suter, Vector-valued wavelets and vector filter banks, *IEEE, Trans. Signal Processing,* **44** (1996), 508–518.

[17] J. M. Zhao, L. Z. Peng, Quaternion-valued Admissible Wavelets Associated with the 2-dimensional Euclidean Group with Dilations, *J. Natural Geometry,* **20** (**2**) (2001), 21–32.

Acknowledgment

The authors are grateful to the referees for the useful suggestions. The second author would like to thank Prof. Hongbo Li for his encouragement and financial support.

Lizhong Peng
LMAM, School of Mathematical Sciences
Peking University
Beijing 100080
P. R. China.
e-mail: lzpeng@pku.edu.cn

Jiman Zhao
MM Key Lab., Academy of Mathematics and System Sciences
Chinese Academy of Sciences
Beijing 100080
P. R. China.
e-mail: jmz70@sina.com

MIX
Papier aus verantwortungsvollen Quellen
Paper from responsible sources
FSC® C105338

If you have any concerns about our products,
you can contact us on
ProductSafety@springernature.com

In case Publisher is established outside the EU,
the EU authorized representative is:
Springer Nature Customer Service Center GmbH
Europaplatz 3, 69115 Heidelberg, Germany

Printed by Libri Plureos GmbH
in Hamburg, Germany